Intelligent Systems in Process Engineering

Paradigms from Design and Operations

This book is a compilation of Advances in Chemical Engineering, Volumes 21 and 22.

Intelligent Systems in Process Engineering

Paradigms from Design and Operations

Edited by

GEORGE STEPHANOPOULOS

CHONGHUN HAN

Laboratory for Intelligent Systems in Process Engineering
Department of Chemical Engineering
Massachusetts Institute of Technology
Cambridge, Massachusetts

ACADEMIC PRESS

San Diego   New York   Boston   London   Sydney   Tokyo   Toronto

Copyright © 1996 by ACADEMIC PRESS, INC.

Academic Press, Inc.
A Division of Harcourt Brace & Company
525 B Street, Suite 1900, San Diego, California 92101-4495

United Kingdom Edition published by
Academic Press Limited
24-28 Oval Road, London NW1 7DX

Library of Congress Cataloging-in-Publication Data

Intelligent systems in process engineering : paradigms from design and
   operations / edited by George Stephanopoulous, Chonghun Han.
          p.   cm.
   Includes bibliographical references and index.
   ISBN 0-12-666240-1 (alk. paper)
   1. Chemical process control. 2. Intelligent control systems.
I. Stephanopoulos, George. II. Han, Chonghun.
TP155. 75. I54   1995
660'.281'0281'028563--dc20   95-43504
                                        CIP

PRINTED IN THE UNITED STATES OF AMERICA
95 96 97 98 99 00 QW 9 8 7 6 5 4 3 2 1

"All virtue is one thing: Knowledge"

PLATO

However,

"The chance of the quantum theoretician is not the ethical freedom of the Augustinian"

NORBERT WIENER

George Stephanopoulos dedicates
this editorial work to
Eleni-Nikos-Elvie
with love and gratitude

Chonghun Han dedicates
this editorial work to
Jisook and *Albert*

CONTENTS

PART I: PRODUCT AND PROCESS DESIGN

Modeling Languages: Declarative and Imperative Descriptions of Chemical Reactions and Processing Systems

Christopher J. Nagel, Chonghun Han, and George Stephanopoulos

Automation in Design: The Conceptual Synthesis of Chemical Processing Schemes

CHONGHUN HAN, GEORGE STEPHANOPOULOS, AND JAMES M. DOUGLAS

Symbolic and Quantitative Reasoning: Design of Reaction Pathways through Recursive Satisfaction of Constraints

MICHAEL L. MAVROVOUNIOTIS

PART II: PROCESS OPERATIONS

Nonmonotonic Reasoning: The Synthesis of Operating Procedures in Chemical Plants

CHONGHUN HAN, RAMACHANDRAN LAKSHMANAN, BHAVIK BAKSHI, AND GEORGE STEPHANOPOULOS

Inductive and Analogic Learning: Data-Driven Improvement of Process Operations

PEDRO M. SARAIVA

Empirical Learning through Neural Networks: The Wave-Net Solution

ALEXANDROS KOULOURIS, BHAVIK R. BAKSHI, AND GEORGE

STEPHANOPOULOS

Reasoning in Time: Modeling, Analysis, and Pattern Recognition of Temporal Process Trends

BHAVIK R. BAKSHI AND GEORGE STEPHANOPOULOS

Intelligence in Numerical Computing: Improving Batch Scheduling Algorithms through Explanation-Based Learning

Matthew J. Realff

CONTRIBUTORS TO PART I

Numbers in parentheses indicate the pages on which the authors' contributions begin.

JAMES M. DOUGLAS, *Department of Chemical Engineering, University of Massachusetts, Amherst, Massachusetts 01003* (43)

CHONGHUN HAN, *Laboratory for Intelligent Systems in Process Engineering, Department of Chemical Engineering, Massachusetts Institute of Technology, Cambridge, Massachusetts 02139* (1, 43)

KEVIN G. JOBACK, *Molecular Knowledge Systems, Inc., Nashua, New Hampshire 03063* (257)

MICHAEL L. MAVROVOUNIOTIS, *Department of Chemical Engineering, Northwestern University, Evanston, Illinois 60208* (147)

CHRISTOPHER J. NAGEL, *Molten Metal Technology, Inc., Waltham, Massachusetts 02154* (1, 187)

GEORGE STEPHANOPOULOS, *Laboratory for Intelligent Systems in Process Engineering, Department of Chemical Engineering, Massachusetts Institute of Technology, Cambridge, Massachusetts 02139* (1, 43, 187, 257)

CONTRIBUTORS TO PART II

Numbers in parentheses indicate the pages on which the authors' contributions begin.

BHAVIK R. BAKSHI, *Department of Chemical Engineering, Ohio State University, Columbus, Ohio 43210* (313, 347, 485)

CHONGHUN HAN, *Laboratory for Intelligent Systems in Process Engineering, Department of Chemical Engineering, Massachusetts Institute of Technology, Cambridge, Massachusetts 02139* (313)

ALEXANDROS KOULOURIS, *Laboratory for Intelligent Systems in Process Engineering, Department of Chemical Engineering, Massachusetts Institute of Technology, Cambridge, Massachusetts 02139* (437)

RAMACHANDRAN LAKSHAMANAN, *Department of Chemical Engineering, University of Edinburgh, Edinburgh, Scotland, United Kingdom* (313)

MATTHEW J. REALFF, *School of Chemical Engineering, Georgia Institute of Technology, Atlanta, Georgia 30332* (549)

PEDRO M. SARAIVA, *Department of Chemical Engineering, University of Coimbra, 3000 Coimbra, Portugal* (377)

GEORGE STEPHANOPOULOS, *Laboratory for Intelligent Systems in Process Engineering, Department of Chemical Engineering, Massachusetts Institute of Technology, Cambridge, Massachusetts 02139* (437, 485)

PROLOGUE

The adjective "intelligent" in the term "intelligent systems" is a misnomer. No one has ever claimed that an intelligent system in an engineering application possesses the kind of intelligence that allows it to *induce* new knowledge (1), or "to contemplate its creator, or how it evolved to be the system that it is" (2). Åström and McAvoy (3) have suggested terms such as "knowledgeable" and "informed" to accentuate the fact that these software systems depend on large amounts of (possibly) fragmented and unstructured knowledge. For the purposes of this book, the term "intelligent system" always implies a computer program, and although the quotation marks around the adjective intelligent may be dropped occasionally, no one should perceive it as a computer program with attributes of human-like intelligence. Instead, the reader should interpret the adjective as characterizing a software artifact that possesses a computational procedure, an algorithm, which attempts to "model and emulate," and thus automate, an engineering task that used to be carried out *informally* by a human. Whether or not this models the actual cognitive process in a human is beyond the scope of this book.

In the wide spectrum of engineering activities, collectively known as *process engineering* and encompassing tasks from product and process development through process design and optimization to process operations and control, so-called intelligent systems have played an important role. Ten years ago the broad introduction of knowledge-based expert systems created a pop culture that started affecting many facets of process engineering work. Expert systems were followed by their cousins, fuzzy systems, and the explosion in the use of neural networks. During the same period, the object-oriented programming (OOP) paradigm, one of the most successful "products" of artificial intelligence, has led to a revolutionary rethinking of programming practices, so that today OOP is the paradigm of choice in software engineering. After 10 years of work, 15 books/monographs/ edited volumes, over 700 identified papers in archival research and professional journals, 65 reviews/tutorial/industrial survey papers, about 150 Ph.D. theses, and several thousand industrial applications worldwide (4), the area of what is known as "intelligent systems" has turned from fringe to mainstream in a large number of process engineering activities. These include monitoring and analysis of process operations, fault diagnosis, supervisory control, feedback control, scheduling and planning of process operations, simulation, and process and product design. The early emphasis on tools and methodologies, originated by research in artificial intelligence, has given place to more integrative approaches, which focus more on the engineering problem and its characteristics. So, today, one does not encounter as frequently as 10 years ago conference sessions with titles including terms such as "expert systems," "knowledge-based systems," or "artificial intelligence." Instead one sees many more mature contributions, from both the academic and industrial worlds, in mainstream engineering sessions, with significant components of what one would have earlier termed "intelligent systems." The evolving complementarity in the use of approaches from artificial intelligence, systems and control theory, mathematical programming, and statistics is a strong indication of the maturity that the area of intelligent systems is reaching.

A. THE CURRENT SETTING

The explosive growth of academic research and industrial practice in the synthesis, analysis, development, and deployment of intelligent systems is a natural phase in the saga of the Second Industrial Revolution. If the First Industrial Revolution in 18th century England ushered the world into an era characterized by machines that extended, multiplied, and leveraged human *physical capabilities*, the Second, currently in progress, is based on machines that extend, multiply, and leverage human *mental abilities* (5). The thinking man, *Homo sapiens*, has returned to its Platonic roots where "all virtue is one thing, knowledge." Using the power and versatility of modern computer science and technology, software systems are continuously developed to preserve knowledge for it is perishable, clone it for it is scarce, make it precise for it is often vague, centralize it for it is dispersed, and make it portable for it is difficult to distribute. The implications are staggering and have already manifested themselves, reaching the most remote corners of the earth and the inner sancta of our private lives. In this expanding pervasiveness of computers, intelligent systems can affect and are affecting the way we educate, entertain, and govern ourselves, communicate with each other, overcome physical and mental disabilities, and produce material wealth. Computer-based deployment of "knowledge" has been thrust by modern sociologists into the center of our culture as the force most effective in resolving inequities in the distribution of biological, historical, and material inheritance. But what is the tangible evidence? Software systems have been composed to do the following (5–7): (i) harmonize chorales in the style of Johann Sebastian Bach and automate musical compositions into new territories; (ii) write original stanzas and poems with thematic uniformity, which could pass as human creations for about half of the polled readers; (iii) compose original drawings and "photographs" of nonexistent worlds; (iv) "author" complete books.

Equally impressive are the results in engineering and science. Characterized as "knowledgeable," "informed," "expert," "intelligent," or any other denotation, software systems have expanded tremendously the scope of automation in scientific and engineering activities (4, 8–13).

B. THE THEORETICAL SCOPE AND LIMITATIONS OF INTELLIGENT SYSTEMS

"So what?" a skeptic may ask. Are the above examples manifestations of the computer's long-awaited, human-like intelligence? No one familiar with Gödel's theorem of incompleteness would ask such a question (14,15), for this theorem states that it is not possible to create a formal system that is both consistent and complete. As such, you cannot create a software system based on some sort of a formal system, i.e., a consistent set of axioms, which can reflect upon itself and discover (not invent) a new dimension of knowledge (1).

Indeed, whenever you focus your attention on any of the so-called intelligent systems, and you take the time to learn the mechanisms they use to generate their marvelous and wondrous behavior, you come up with the anti-climactic realization that everything is quite ordinary and perfectly expectable with no surprises or mystical insights. Such reaction reminds us of how Sherlock Holmes reacted when a man questioned the brilliance of his deductive reasoning in solving one of his cases:

> Mr. Jabez Wilson laughed heavily. "Well, I never!" said he. "I thought at first that you had done something clever, but I see that there was nothing in it, after all." "Begin to think, Watson," said Holmes, "that I made a mistake in explaining. 'Omne ignotum promagnifico,' you know, and my poor little reputation, such as it is, will suffer shipwreck if I am so candid."

Similarly, Alan Turing, the father of the digital computer and creator of the Turing Test for checking the "intelligence" of a machine, put it this way:

> The extent to which we regard something as behaving in an intelligent manner is determined as much by our own state of mind and training as by the properties of the object under consideration. If we are able to explain and predict its behavior or if there seems to be little underlying plan, we have little temptation to imagine intelligence. With the same object, therefore, it is possible that one man would consider it as intelligent and another would not; the second man would have found out the rules of its behavior.

C. The Character of the Ten Paradigms

All the paradigms of intelligent systems in this volume have plans and assume extensive amounts of knowledge. As such they are ordinary computer programs and they emulate a precise computational procedure, which uses a predefined set of data. In the Aristotelian form, "all instruction given or received (by the intelligent systems) by way of argument, proceeds from preexistent knowledge." Consequently, one should see all cases put forward by the individual chapters as nothing more than paradigms for new uses of the computer. Every one of them carries out deduction from a predefined set of knowledge, using explicit reasoning strategies. The reader should not search for inductive generation of new knowledge, even when the terms "induction" and "inductive reasoning" have been loosely employed in some chapters. Instead, the reader should see each chapter as a computer-based paradigm in *capturing, articulating,* and *utilizing* various forms of knowledge. As a result, the reader will notice that the ten chapters of this volume serve as a paradigm of an integrative attitude to the modeling and processing of knowledge. Nowhere in this volume will the reader find artificial debates on the superiority of a numerical over a symbolic approach or vice versa. On the contrary, the engineering methodologies advanced by the individual chapters indicate that *all available knowledge should be acquired, modeled, and used* within a framework that requires interaction and/or integration of processing methodologies from artificial intelligence, systems and control theory, operations research, statistics, and others.

It is this integrative attitude that today characterizes most of the work in the area of "intelligent systems for process engineering," as the editors of this volume have indicated in a recent review article (4). It is this need for integrative approaches that has moved the applications of artificial intelligence into the mainstream of engineering activities. This is certainly the pivotal feature that characterizes the ten paradigms discussed in the subsequent chapters.

The ten chapters of this volume advance ten distinct paradigms for the use of ideas and methodologies from artificial intelligence in conjunction with techniques from various other areas. They represent the culmination of research efforts which started in 1986 at the *Laboratory for Intelligent Systems in Process Engineering* (LISPE) of the Chemical Engineering Department at MIT and currently are spread over a half a dozen academic institutions. Each chapter, as the corresponding title indicates, is centered around two themes. The first theme (represented by the first part of a title) is drawn from the artificial intelligence techniques discussed in the specific chapter, while the second theme (represented by the second part of the title) focuses on a process engineering problem. It should be noted, though, that it is the process engineering problem, its formulation and characteristics, that sets the tone for every chapter. The various components of the corresponding intelligent systems serve specific needs. Nowhere will the reader find the "a technique in search for a problem to solve" attitude, which has led to the distortion of several engineering problems and the malignant proliferation of techniques. As a result, even if future developments suggest a change in the techniques used, the formulation of the engineering problems may retain the bulk of the essential features proposed by each of the ten chapters.

D. THEMES COVERED BY THE TEN CHAPTERS

Let us now give a brief synopsis of the themes advanced by each of the ten chapters. The five chapters of Part I advance paradigms which are related to product and process design, while the five chapters of Part II focus on aspects of process operations.

Part I: Product and Process Design

Chapter 1. MODELING LANGUAGES:
        Declarative and Imperative Descriptions of Chemical Reactions and Processing Systems

To model is to represent reality, and modeling as an essential task of any engineering activity is always *contextual.* Within the scope of differing engineering contexts, the same physical entity, e.g., molecule, chemical reaction, or process flowsheet, is represented with a broad variety of models. An enormous amount of effort is expended in the development and maintenance of a *Babel of models*, sporting different languages and being at cross purposes with each other, although like their biblical counterpart they share a common progenitor—in this case, the fundamentals of chemistry/physics and the principles of chemical engineering science. Creating a language that supports the expeditious generation of consistent models has become the key to unlocking the power of computer-aided tools and unleashing the explosive synergism between human and computer. However, a modeling language is of little use if it only creates representations of physical entities as "things unto themselves" without meaningful semantic designation to what it purports to represent. Furthermore, the model of an entity should contain all knowledge that has some bearing on the representation of that entity, be that *declarative* or *imperative* (procedural) in character. Chapter 1 describes two modeling languages; LCR (*L*anguage for *C*hemical *R*easoning) to represent molecules and chemically reactive systems, and MODEL.LA. (*MODEL*ing *LA*nguage) for the representation of processing systems. Both are based on the same principles and have, to a large extent, a common structure. Both have been based on ideas and techniques which originated in artificial intelligence, and both have been implemented in a similar object-oriented programming environment.

Chapter 2. *AUTOMATION IN DESIGN:*
The Conceptual Synthesis of Chemical Processing Schemes

If you really know how to carry out an engineering task, then you can instruct a computer to do it automatically. This self-evident truism can be used as the litmus test of whether a human "really" knows how to, say, design an engineering artifact. Experience has shown that engineers have been able to automate the process of design in very few instances, thus demonstrating the presence of serious flaws in (a) their understanding of how to do design and/or (b) their ability to clearly articulate the design methodology, both of which can be traced to the inherent difficulty of making the "best" design decisions. The pivotal element in automating the design process is *modeling the design process itself*, which includes the following modeling tasks: (1) modeling the *structure of design tasks* that can take you from the initial design specifications to the final engineering artifact; (2) representing the *design decisions* involved in each task, along with the assumptions, simplifications, and methodologies needed to frame and make the design decisions; (3) modeling the *state of the evolving design*, along with the underlying rationale. Chapter 2 shows how one can use ideas and techniques from artificial intelligence, e.g., symbolic modeling, knowledge-based systems, and logic, to construct a computer-implemented model of the design process itself. Using Douglas' hierarchical approach as the conceptual model of the design process, this chapter shows how to generate models of the design tasks' structure, design decisions, and the state of design, thus leading to automation of large segments of the synthesis of chemical processing schemes. The result is a *human-aided, machine-based* design paradigm, with the computer "knowing" how the design is done, what the scope of design is, and how to provide explanations and the rationale for the design decisions and the resulting final design. Such a paradigm is in sharp contrast with the traditional *computer-aided, human-based* prototype, where the computer carries out numerical calculations and data fetching from files and databases, but has no notion of how the design is done, knowledge resting exclusively in the province of the individual human designer.

Chapter 3. *SYMBOLIC AND QUANTITATIVE REASONING:*
Design of Reaction Pathways through Recursive Satisfaction
of Constraints

Given a fixed, predetermined set of elementary reactions, to compose reaction pathways (mechanisms) which satisfy given specifications in the transformation of available raw materials to desired products is a problem encountered quite frequently during research and development of chemical and biochemical processes. As in the assembly of a puzzle, the pieces (available reaction steps) must fit with each other (i.e., satisfy a set of constraints imposed by the precursor and successor reactions) and conform with the size and shape of the board (i.e., the specifications on the overall transformation of raw materials to products). Chapter 3 draws from *symbolic and quantitative reasoning* ideas of AI which allow the systematic synthesis of artifacts through a *recursive satisfaction of constraints* imposed on the artifact as a whole and on its components. The artifacts in this chapter are mechanisms of catalytic reactions and pathways of biochemical transformations. The former require the construction of *direct* mechanisms, without cycles or redundancies, to determine the basic legitimate chemical transformations in a reacting system. The latter are the chemical engines of living cells, and they represent legitimate routes for the biochemical conversion of substrates to products either desired from a bioprocess or essential for cell survival. The algorithms discussed in this chapter could be used in one of the following two settings: (a) Synthesize alternative pathways of chemical/biochemical reactions as a means to interpret overall transformations which are experimentally observed. (b) Synthesize reaction pathways in the course of exploring a new, alternative production route. This chapter discusses examples in both directions. Although it is concerned only with constraints on the directionality and stoichiometry of elementary reactions, the ideas can be extended to include other types of constraints arising, for example, from kinetics or thermodynamics.

Chapter 4. *INDUCTIVE AND DEDUCTIVE REASONING:*
The Case of Identifying Potential Hazards in Chemical Processes

All reasoning carried out by computers is *deductive*; i.e., any software has all the necessary data, stored in various forms in a database, and possesses all the necessary algorithms to operate on the set of data and *deduce* some results. Many researchers in the area of cognitive psychology make similar claims on the reasoning mechanisms of human beings. The fact remains, though, that both humans and machines can use very simple "algorithms" on small sets of data and produce results which could not have been visible to the "naked eye" of direct reasoning. In such cases, we tend to talk about the *inductive* capabilities of either of the two. These ideas are nowhere more prominent than in the area of *hazards identification and analysis*. One often hears, "if I knew that the conversion of A to B could be catalyzed by the presence of C then I would have foreseen the last disaster, and have done something about it," with the speaker converting a problem of *inductive* identification (i.e., induce the possibility of a hazard from the list of chemicals) into an issue of deductive reasoning. Chapter 4 demonstrates that the identification of hazards is essentially an interplay between inductive and deductive reasoning. Through inductive reasoning one attempts to generate all potential hazardous top-level events which can be justified by the presence of a set of chemicals. The reasoning is called inductive because it has the potential to generate specific knowledge that was not "visible" ahead of time. Once the potentially harmful top-level events have been identified, deductive reasoning attempts to "walk" through the processing scheme, its unit operations, and their design or operating characteristics (assumptions, or decisions) and generate the preconditions which would enable the occurrence of a specific top-level event. The inductive reasoning procedures operate on a set of chemicals and create in an *exhaustive, bottom-up* manner many alternative reaction pathways, some of which could lead to a hazard, e.g., release of large amounts of energy over a short period of time. On the other hand, the deductive reasoning procedures are *goal-directed* and operate in a *top-down* manner. Chapter 4 develops the detailed framework for the implementation of these ideas which, among other benefits, offers the following advantages: (a) formalizing the hazards identification problem and unifying the methodological approaches at any stage of the design activities and (b) systematizing the generation and evaluation of mechanisms for the prevention of hazards, or containment of their effects.

Chapter 5. *SEARCHING SPACES OF DISCRETE SOLUTIONS:*
The Design of Molecules Possessing Desired Physical Properties

Strings of letters make words. From words to verses and stanzas, a poet composes a work with its own dynamic behavior, e.g., emotional impact on the reader, which transcends the character of its components. In an analogous manner, atoms form functional groups and these in turn yield molecules with distinct behavior, e.g., physical properties. It takes a Homeric or Shakespearean genius to convert letters to an epic with a predefined desired impact. It suffices to efficiently search a space of combinatorial alternatives in order to identify the molecules which satisfy the desired constraints on a set of physical properties. Often the requisite scientific knowledge is fragmented, dispersed, and nonformalized, making the deductive search for the desired molecules inefficient or impossible. The inductive "genius" of a scientist or engineer is needed to break the impasse in such cases. By evolution or revolution one needs to respond to tighter and shifting product specifications and identify new solvents, pharmaceuticals, imaging chemicals, herbicides and pesticides, refrigerants, polymeric materials, and many others. Chapter 5 sketches the characteristics of an intelligent, computer-aided tool to support the synthetic search for the desired molecules. With functional groups as the "letters" of an alphabet, automatic and interactive procedures compose and screen classes of potential molecules. The automatic synthesis algorithm defines and searches the space of discrete solutions (molecules) through a hierarchical sequence of the space's representations. However, one should never overestimate the effectiveness of search algorithms in locating the desired solutions. Quite frequently one needs to resort to human-driven, abductive jumps. Chapter 5 also describes how automatic search can become interwoven with effective man–machine interaction. Thus, the resulting computer-aided tool, the *Molecule Designer*, constitutes a paradigm of an intelligent system with two distinct but integrated and complementary capabilities. Examples of the synthesis of refrigerants, solvents, polymers, and pharmaceuticals illustrate the logic and features of the design procedures in the *Molecule Designer*.

Part II: Process Operations

Chapter 6: *NONMONOTONIC REASONING:*
The Synthesis of Operating Procedures in Chemical Plants

The inherent difficulty of planning a sequence of actions to take you from one point to another usually increases as more obstacles are placed in your way. The number of these obstacles (constraints) that you must circumvent determines the complexity of the task, because any time you run into one of them you must *backtrack* and try an alternative step or path of steps. Such *serial* (or *linear* or *monotonic*) construction of a plan is fraught with pitfalls and repeated backtracking. The more the constraints, the more inefficient the monotonic planning. If, on the other hand, an action-step (a Clobberer) leads to the violation of a constraint, then *do not backtrack*. Take another action-step (a White Knight) which, when it precedes a Clobberer, negates the impact of the Clobberer, and you never need to backtrack. So the more constraints the more efficient your planning process. Such *nonserial* (or *nonlinear*, or *nonmonotonic*) reasoning has become the essence of all modern and efficient planners, whether they are *logic-based* and *explicit*, or *implicit* enumerators of alternative plans. The purpose of Chapter 6 is twofold: (i) To introduce the ideas of *nonmonotonic reasoning* in the planning of process operations. (ii) To demonstrate how nonmonotonic planning can be used to synthesize operating procedures for chemical processes, either off-line for standard tasks (e.g., routine start-up or shut-down), or on-line for real-time response to large departures from desired conditions. It is shown that hierarchical modeling of process operations and operators is essential for the efficient deployment of nonmonotonic planning, and that the tractability of the resulting algorithms is strictly dependent on the form of the operators. In this regard, the modeling needs in this chapter draw heavily from the material of Chapter 1. Nonmonotonic planners handle with superb efficiency constraints on (a) the temporal ordering of operations, (b) avoidable mixtures of chemical species, and (c) bounding quantitative conditions on the state of a process. Consequently, they could be used to generate explicitly all feasible operating procedures, leaving a far smaller search space for the selection of the optimum procedure by a numerical optimizer.

Chapter 7. INDUCTIVE AND ANALOGICAL LEARNING:
Data-Driven Improvement of Process Operations

Informed and systematic observation of naturally generated data can lead to the formulation of interesting and effective generalizations. While some statisticians believe that experimentation is the only safe and reliable way to "learn" and achieve operational improvements in a manufacturing system, other statisticians and all the empirical machine learning researchers contend that by looking at past historical records and sets of examples, it is possible to extract and generate important new knowledge. Chapter 7 draws from *inductive and analogical learning* ideas in an effort to develop systematic methodologies for the extraction of structured new knowledge from operational data of manufacturing systems. These methodologies do not require any a priori decisions/assumptions either on the character of the operating data (e.g., probability density distributions) or on the behavior of the manufacturing operations (e.g., linear or nonlinear structured quantitative models), and they make use of *instance-based learning* and *inductive symbolic learning* techniques developed in artificial intelligence. They are aimed to be complementary to the usual set of statistical tools that have been employed to solve analogous problems. Thus, one can see the material of Chapter 7 as an attempt to fuse statistics and machine learning in solving specific engineering problems. The framework developed in this chapter is quite generic and can be used to generate operational improvement opportunities for manufacturing systems (a) which are simple or complex (with internal structure), (b) whose performance is characterized by one or multiple objectives, and (c) whose performance metrics are categorical (qualitative) or continuous (real numbers). A series of industrial case studies illustrates the learning ideas and methodologies.

Chapter 8. EMPIRICAL LEARNING THROUGH NEURAL NETWORKS:
The Wave-Net Solution

Empirical learning is an ever-lasting and ever-improving procedure. Although *neural networks* (NN) captured the imagination of many researchers as an outgrowth of activities in artificial intelligence, most of the progress was accomplished when empirical learning through NNs was cast within the rigorous analytical framework of the *functional estimation problem*, or *regression*, or *model realization*. Independently of the name, it has been long recognized that, due to the inductive nature of the learning problem, to achieve the desired accuracy and generalization (with respect to the available data) in a dynamic sense (as more data become available) one needs to seek the unknown approximating function(s) in functional spaces of varying structure. Consequently, a recursive construction of the approximating functions at multiple resolutions emerges as a central requirement and leads to the utilization of wavelets as the basis functions for the recursively expanding functional spaces. Chapter 8 fuses the most attractive features of a NN, i.e., representational simplicity, capacity for universal approximation, and ease in dynamic adaptation, with the theoretical soundness of a recursive functional estimation problem, using wavelets as basis functions. The result is the *Wave-Net* (*Wave*let *Net*work), a multiresolution hierarchical NN with localized learning. Within the framework of a Wave-Net where adaptation of the approximating function is allowed, we have explored the use of the L^∞ error measure as the design criterion. One may cast any form of data-driven empirical learning within the framework of a Wave-Net to address a variety of modeling situations encountered in engineering problems, such as design of process controllers, diagnosis of process faults, and planning and scheduling of process operations. Chapter 8 discusses the properties of a Wave-Net and illustrates its use on a series of examples.

Chapter 9. REASONING IN TIME:
Modeling, Analysis, and Pattern Recognition of Temporal
Process Trends

The plain record of a variable's numerical values over time does not invoke appreciable levels of cognitive activity in a human. Although it can cause a fervor of numerical computations by a computer, the levels of cognitive appreciation of the variable's temporal behavior remain low. On the other hand, if one presents the human with a graphical depiction of the variable's temporal behavior, the level of cognition increases and a wave of reasoning activities is unleashed. Nevertheless, when the human is presented with scores of graphs depicting the temporal behavior of interacting variables, his/her reasoning abilities are severely tested. In such a case, the computer will happily continue crunching numbers without ever rising above the fray and thus developing a "mental" model, interpreting correctly the temporal interactions among the many variables. Reasoning in time is very demanding, because time introduces a new dimension with significant levels of additional freedom and complexity. While the real-valued representation of variables in time is completely satisfactory for many engineering tasks (e.g., control, dynamic simulation, planning and scheduling of operations), it is very unsatisfactory for all those tasks which require decision-making via logical reasoning (e.g., diagnosis of process faults, recovery of operations from large unsolicited deviations, "supervised" execution of start-up or shut-down operating procedures). To improve the computer's ability to reason efficiently in time, we must first establish new forms for the representation of temporal behavior. It is the purpose of Chapter 9 to examine the engineering needs for temporal decision-making and to propose specific models which encapsulate the requisite temporal characteristics of individual variables and composite processes. Through a combination of analytical techniques, such as *scale-space filtering, wavelet-based, multiresolution decomposition of functions,* and modeling paradigms from artificial intelligence, Chapter 9 develops a concise framework that can be used to model, analyze, and synthesize the temporal trends of process operations. Within this framework, the modeling needs for logical reasoning in time can be fully satisfied, while maintaining consistency with the numerical tasks carried out at the same time. Thus, through the modeling paradigms of this chapter, one may put together intelligent systems which use consistent representations for their logical-reasoning and numerical tasks.

Chapter 10. *INTELLIGENCE IN NUMERICAL COMPUTING: Improving Batch Scheduling Algorithms through Explanation-Based Learning*

Learning comes from reflection upon accumulated experience and the identification of patterns found among the elements of past experience. All numerical algorithms used in scientific and engineering computing are based on the same paradigm: *execute a predetermined sequence of calculation tasks and produce a numerical answer.* The implementation of the specific numerical algorithm is oblivious to the experience gained during the solution of a specific problem and, in the next encounter, a different, or even the same problem is solved through the execution of exactly the same sequence of calculation steps. The numerical algorithm makes no attempt to reflect upon the structure and patterns of the results it produced, or to reason about the structure of the calculations it has performed. Chapter 10 shows that this need not be the case. By allowing an algorithm to reflect upon and reason with aspects of the problems it solves and its *own structure of computational tasks, the algorithm can learn* how to carry out its tasks more efficiently. Such *intelligent numerical computing* represents a new paradigm, which will dominate the future of scientific and engineering computing. But, in order to unlock the computer's potential for the implementation of truly intelligent numerical algorithms, the *procedural* depiction of a numerical algorithm must be replaced by a *declarative* representation of the algorithmic logic. Such a requirement upsets an established tradition and imposes new educational challenges, which most educators and educational curricula have not, as yet, even recognized. This chapter shows how one can take a branch and bound algorithm, used to identify optimal schedules of batch operations, and endow it with the ability to learn to improve its own effectiveness in locating the optimal scheduling policies for flowshop problems. Given that most batch scheduling problems are NP-hard, it becomes clear how important it is to improve the effectiveness of algorithms for their solution. Using the Ibaraki framework, a branch and bound algorithm is declaratively modeled as a *discrete decision process*. Then explanation-based machine learning strategies can be employed to uncover patterns of generic value in the experience gained by the branch and bound algorithm from solving specific instances of scheduling problems. The logic of the uncovered patterns (i.e., new knowledge) can be incorporated into the control strategy of the branch and bound algorithm when the next problem is to be solved.

GEORGE STEPHANOPOULOS AND CHONGHUN HAN

REFERENCES

1. Stephanopoulos, G., Computers, Systems, Languages and Other Fragments. *CAST Newsletter*, Spring (1994).
2. Antsaklis, P. J. and Passino, K. M., eds., "An Introduction to Intelligent and Autonomous Control." Kluwer, Norwell, MA, 1993.
3. Åström, K. J. and McAvoy, T. J., Intelligent Control, *J. Proc. Cont.* **2**(3), 115 (1992).
4. Stephanopoulos, G. and Han, C., "Intelligent Systems in Process Engineering: A Review," *Proc. PSE,* Kyongju, Korea, 1994.
5. Kurtzweil, R., "The Age of Intelligent Machines." MIT Press, Cambridge, MA, 1990.
6. Mandelbrot, B. B., "The Fractal Geometry of Nature." W. H. Freeman, New York, NY, 1983.
7. Davis, P. J. and Hersh, R., "Descartes' Dream: The World According to Mathematics." Harcourt Brace Jovanovich, San Diego, CA, 1986.
8. Stephanopoulos, G. and Mavrovouniotis, M.L., eds., Artificial Intelligence in Chemical Engineering—Research and Development, *Comp. Chem. Eng.* **12**(9/10), (1988).
9. Stephanopoulos, G., Artificial Intelligence and Symbolic Computing in Process Engineering Design. *In* "Foundations of Computer-Aided Process Design" (J. J. Siirola, I. E. Grossmann, and G. Stephanopoulos, eds.), p. 21, Elsevier, New York, 1989.
10. Stephanopoulos, G., Artificial Intelligence: What Will Its Contributions Be to Process Control? *In* "The Second Shell Process Control Workshop" (D. M. Prett, C. E. Garcîa, and B. L. Ramaker, eds.), p. 591, Butterworths, Stoneham, MA, 1990.
11. Stephanopoulos, G., Brief Overview of AI and Its Role in Process Systems Engineering. *In* "Process Systems Engineering, Vol. I," CACHE, 1992.
12. Mavrovouniotis, M., ed., "Artificial Intelligence in Process Engineering." Academic Press, San Diego, CA, 1990.
13. Quantrille, T. E. and Liu, Y. A., "Artificial Intelligence in Chemical Engineering." Academic Press, San Diego, CA, 1991.
14. Hofstadter, D. R., "Gödel, Escher, Bach: An Eternal Golden Braid." Vintage Books, New York, 1980.
15. Penrose, R., "The Emperor's New Mind." Oxford University Press, Oxford, UK, 1989.

MODELING LANGUAGES: DECLARATIVE AND IMPERATIVE DESCRIPTIONS OF CHEMICAL REACTIONS AND PROCESSING SYSTEMS

Christopher J. Nagel, Chonghun Han,
and George Stephanopoulos

Laboratory for Intelligent Systems in Process Engineering
Department of Chemical Engineering
Massachusetts Institute of Technology
Cambridge, Massachusetts 02139

1

To model is to represent reality, and modeling as an essential task of any engineering activity is always *contextual*. Within the scope of differing engineering contexts, the same physical entity, such as a molecule, chemical reaction, or process flowsheet, is represented with a broad variety of models. An enormous amount of effort is expended in the development and maintenance of a *Babel of models*, sporting different languages and being at cross-purposes with each other, although like their biblical counterparts share a common progenitor: in this case, the fundamentals of chemistry and physics and the principles of chemical engineering science. Creating a language that supports the expeditious generation of consistent models has become the key to unlocking the power of many computer-aided tools, and unleashing the explosive synergism between human and computer. However, a modeling language is of little use if it only creates representations of physical entities as "things onto themselves" without meaningful semantic designation to what it purports to represent. Furthermore, the model of an entity *should contain all knowledge* that has some bearing to the representation of that entity, whether that is *declarative* or *imperative* (procedural) in character. In this chapter we will describe two modeling languages: LCR (Language for Chemical Reasoning) to represent molecules and chemically reactive systems, and MODEL.LA. (MODELing LAnguage) for the representation of processing systems. Both are based on the same principles and have, to a large extent, a common structure. Both have been based on ideas and techniques, which originated in artificial intelligence, and both have been implemented in a similar object-oriented programming environment.

I. Introduction

Since the early efforts, computer-aided modeling of physical systems has been generally organized around unilateral computations that perform predetermined operations on fixed inputs to produce the desired outputs. This is still a common practice since it helps engineers and scientists to manage tedious calculations and coordinate and control diverse numerical tasks, by producing models that use very efficiently computer resources. However, a series of inherent weaknesses have pointed out the disadvantages of the traditional approach: (1) the time and cost associated with computer model development are high; (2) the resulting models are difficult to document and maintain adequately; (3) the reuse of computer-aided models is minimal, as these models tend to be task-specific and are often intrinsically linked to solution procedures; (4) the models cannot be synthesized automatically by the computer in the course of automatic execution of an engineering task; and (5) for interactive modeling the modeler is required to be highly skilled in programming. As a result of these weaknesses the duplication of modeling efforts has been enormous. Accumulated modeling knowledge is almost impossible to use, since the underlying modeling context (purpose, assumptions, simplifications) has never been documented and rationalized. So, why must every new modeling effort start from scratch (Meyer, 1987). Furthermore, automatic generation of models at higher abstractions, cannot be done. Finally, the fragmentation of modeling efforts and the ad hoc character of their computer implementation have led to internal inconsistencies among the various models used in different process engineering tasks. A typical example of this phenomenon is found in the diversity of models used to support process-control-related engineering tasks (Stephanopoulos, 1990) such as design of process controllers, controller adaptation mechanisms, optimization of process operations, and diagnosis of process faults.

A. The Five Premises of a Modeling System

To overcome the weaknesses discussed above, any modeling system should be built on the following premises.

1. Articulate All Declarative Knowledge

The model of an entity, whether a processing unit, a molecule, or a chemical reaction, should contain explicitly all relevant information, including the following (Stephanopoulos *et al.*, 1990a):

(a) *Underlying assumptions.* The form of the relationships that describe the behavior of a given entity depends on a series of assumptions, such as the operational mode; the assumed mechanisms for mass, energy, or momentum transfer; the mechanistic pathway for a chemical reaction; and the constitutive models for the estimation of physical properties.

(b) *Simplifications* made by the modeler (human, or another computer program) in order to limit the model's validity over a given range of conditions, or to underscore the relative importance of various physicochemical phenomena.

(c) *Scope of engineering task*, i.e., what the model is intended for. Typical examples are (1) the variety of models required by the different levels of the hierarchical synthesis of process flowsheets (Stephanopoulos, 1990a, 1990b); (2) the variety of models needed for feedback and adaptive control, diagnosis, or planning of process operations (see the fifth chapter in this volume).

(d) *Relationships among the various variables and parameters.* These relationships express the fundamental principles of physics and chemistry as well as the assumed mechanisms for rate and equilibrium phenomena, and the correlations for the estimation of physical properties. These relationships could be quantitative, qualitative, or semiquantitative (e.g., order-of-magnitude, or ordinal), depending on the level of the available knowledge, and could express static or dynamic behavior.

2. Separate Declarative from Procedural Knowledge

The declarative knowledge that is articulated in a given model represents the "what is" knowledge about the modeled entity. Since the same declarative model could be used for a variety of engineering tasks, it is essential that it be separated from the procedural knowledge, i.e., the "how to" knowledge of a particular engineering methodology. Previous modeling approaches relied on a tight integration of declarative and procedural knowledge with the sequential modular simulators (e.g., ASPEN; see Evans *et al.*, 1979) exemplifying the tightest integration, and the equation-oriented simulators (e.g., SPEEDUP; Perkins and Sargent, 1982; Pantelides, 1988) offering a weaker but still dominant integration.

3. Hierarchical and Multiview Representation of Entities

A modeling system should allow the representation of the various entities at any level of detail, using multiple, coexisting abstractions, which can communicate with each other. This is a critical requirement as it sets the desired modeling system apart from many other modeling environments. For example, a processing system could be represented in any of the following abstractions (Fig. 1): (a) an overall plant, (b) a network with generalized reaction and separation sections, (c) a network of reactors and abstract separation sections, or (d) a network of processing units. All models, depicting the four abstraction views in Fig. 1, should be consistent with each other and should allow the transfer of information from more abstract to more detailed representations and vice versa. In addition, two distinct and different versions of the same abstraction (e.g., Fig. 2) have many modeling components in common. The modeling system should allow common handling of the common modeling elements. For example, the modeling relations for the generalized reaction and separation sections

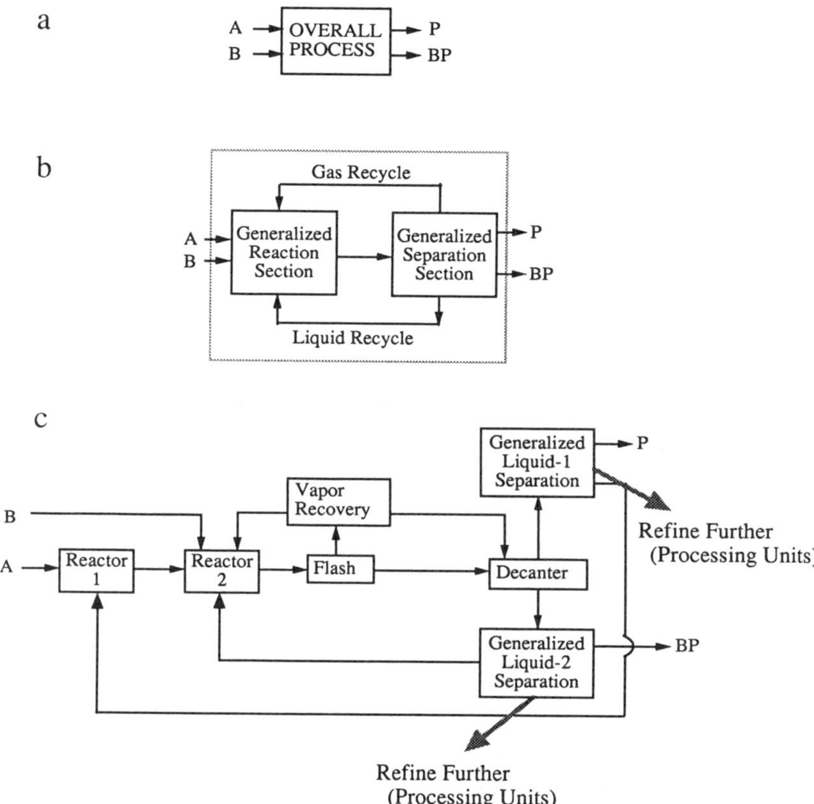

FIG. 1. Three abstraction levels for the representation of a processing scheme.

of the two versions in Fig. 2 are the same. If we change the models in one of the versions, the representation of the other version should be updated automatically.

4. Automatic Generation and Modification of Models

Once the modeler has described the elements of the modeling premise 1—i.e., made the underlying assumptions and simplifications and defined the scope of the engineering task—the modeling system should automatically generate the modeling relationships and their elements. In other words, the modeling language should possess declarative representations of relationships, of the terms that made up such relationships, of the variables that make up the terms, and of the semantic connections that

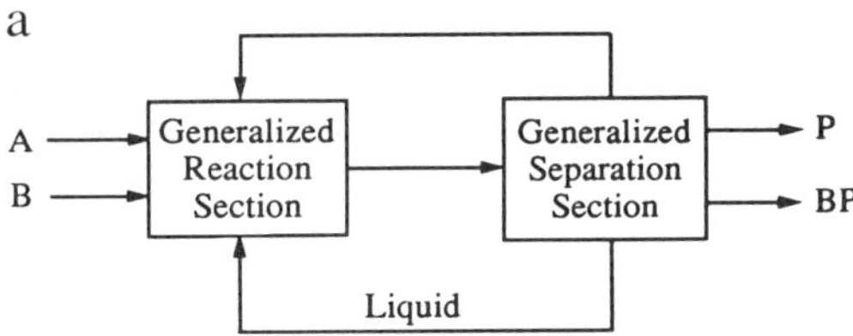

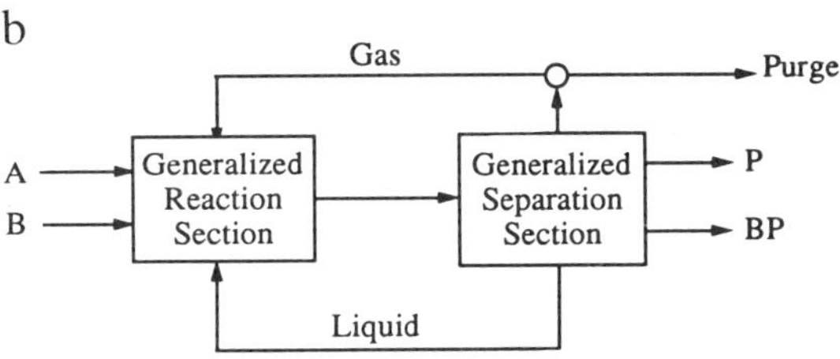

FIG. 2. Two distinct versions of a plant at the same abstraction.

assign physical meaning to the variables, terms, and relationships. Furthermore, once the modeling system has developed the model of an entity, it should be capable of updating the model automatically, if the modeling assumptions and simplifications have changed.

5. Simulation and Reasoning

The modeling relationships of the various entities should possess enough granularity to allow structural reorganization for efficient numerical simulation and logical reasoning. For example, given the static quantitative model of a process, the modeling system should possess the necessary modeling information to allow the construction of the corresponding cause and effect, directed graph, or automatically analyze the degrees of freedom and generate the simplest numerical procedures for simulation.

The preceding five premises imply requirements that go beyond the scope of traditional computer-aided modeling systems. The modeling language should allow interaction with the modeler at a high level of abstraction, while the generated models should be explicit, transparent, fully documented, and usable without the need to consult manuals for assumptions or input/ output language conventions.

B. Review of Modeling Systems for Process Simulation

In recent years several modeling systems have appeared in the literature, all of them aimed at eliminating the perceived modeling bottleneck in generating static or/ and dynamic simulations of processing systems. Consequently, all the modeling systems to be discussed in the subsequent paragraphs do not conform to all of the five premises, discussed in the previous section, and consequently they cannot be used outside the simulation tasks that they were designed to serve.

1. ASCEND (*A*dvanced *S*ystem for *C*omputations in *EN*gineering *D*esign)

As its name implies, ASCEND is a general computer-aided environment with a modeling language that was developed to support the expeditious generation of mathematical models that are needed for the simulation and design of processing systems (Piela, 1989; Piela *et al.*, 1991). It follows an object-oriented paradigm, but it is strongly typed. Its main features can be summarized as follows:

a. Building Blocks. ASCEND uses three types of generic modeling elements: `model`, `atom`, and `type`. Models are structured entities that are built hierarchically from (1) instances of other models, atoms, or/ and types; and (2) relationships between these instances. Atoms are primitive variables, which are used to represent physical quantities, design variables, etc. Types are elementary declarations of real, integer, string, Boolean, etc. They are predefined in ASCEND.

b. Relationships. Several operators define relationships among the various elements of ASCEND. Three of them are used to define inheritance relationships: (1) the ***refines*** operator is used to link a subclass to a class in an inheritance hierarchy; (2) the ***is-a*** operator is used to declare an object as the instance of a class (or, a type); and (3) the ***is-refined-to*** operator changes the type associated with a previously declared instance. Two relationships are used to group instances together: (1) the ***are-alike*** operator coerces all members of a group of instances to take the type associated with the most refined instance; (2) the ***are-the-same*** operators is similar to the previous one, but merges the operands into a single structure.

c. General Features. ASCEND uses the inherited mechanism of the object-oriented paradigm in order to provide expeditious extension of its "vocabulary." It contains both instances and classes, but instances are not allowed to have local attributes. ASCEND generates mathematical relationships that are expressed as implicit equations or solved assignments. It uses a fairly concise syntax to express a fairly broad set of modeling statements.

2. OMOLA (<u>O</u>bject-oriented <u>MO</u>deling <u>LA</u>nguage)

OMOLA (Andersson, 1989; Nilsson, 1993) was designed to support the model generation and simulation needs during the computer-aided synthesis and analysis of control systems. It is oriented toward the creation of structured dynamic systems. Its main features can be summarized as follows:

a. Modeling Elements. All models are created as subclasses of the mother class, `Model`, thus inheriting all the attributes of the mother class. It is possible to create specialized versions of an existing model, M1, by creating a subclass from the class, M1. The description of each model is organized around four modeling elements; `Terminal`, `Parameter`, `Variable`, and `Realization`. The first is used as a point of communication between the models of the various entities. `Parameter` and `Variable` are used to describe model attributes that are time-independent and time-varying, respectively. `Realization` is the element that is used to define the mathematical relationships describing the behavior of an entity. Each of the four modeling elements is described by a structured class with a series of attributes. Specialized subclasses emanate from `Terminal` and `Realization`, thus allowing an expansion in the vocabulary of the language itself.

b. Functionality. OMOLA allows the representation of complex structured systems through the aggregation of individual models. The language offers no semantic relationships to link the generated models or their attributes to any physical concepts. Consequently, it cannot be used to automate the generation of models from a set of physical modeling assumptions. Also, it does not allow for the articulation of qualitative or semiquantitative knowledge.

3. MODASS (<u>MOD</u>eling <u>AS</u>Sistant)

MODASS (Sørlie, 1990) is more of a toolbox for the construction of models. Although it allows declarative description of models and their elements, it also permits the incorporation of subroutines as model components.

a. Modeling Elements. The modeling elements of MODASS are assembled in a hierarchy of classes. At the top is the class, `model`, containing very few common attributes. At the next level we encounter the subclass, `Model-Element`, which contains the basic building blocks of the actual model. Further specializations (i.e., subclasses) provide narrow specifications of the sets of equations used for modeling specific entities. To represent the definitional boundary of an entity and the gateways for the information exchange between different entities, MODASS uses the classes, `Process-Terminal` and `Boundary-Process`, respectively. Specializations of the latter class allow the tailoring of the modeling element to mass, energy, and information transfer.

b. Functionality. MODASS has very limited network of semantic relationships among the different modeling elements, but it offers a satisfactory set of tools to the user for (1) the navigation through the tree of library models, (2) the symbolic manipulation of the modeling equations, and (3) the examination of models for inconsistencies and errors. Its lack of meaningful semantic relationships among the various modeling elements and the physical entities they represent makes MODASS a modeling system with limited scope, and largely outside the framework of the five premises, discussed in Section I.A.

4. Hybrid Phenomena Theory

Hybrid phenomena theory (HPT) (Woods, 1993) is not a modeling language in strict terms, but a theoretical framework for the generation of models that capture the topological, phenomenological, and behavioral aspects of the entity being modeled. The *topological* model of an entity includes a set of objects, (e.g., pipes, vessels, valves) and a set of logical relationships among the objects. The *phenomenological* model of the entity contains a set of objects, describing instances of physical phenomena that occur in the system being modeled. These phenomena are active processes such as heat transfer, material accumulation, and convective flow. The *behavioral* model is a state-space model providing a quantitative description of the entity's dynamic (or, static) behavior. The topological, phenomenological, and behavioral modeling components of an entity are tightly integrated and information flows among them as needed.

a. Modeling Elements. The object-oriented character of HPT places the `quantity` as the mother-class of the subclasses `constant`, `parameter`, and `variable`, whose further specialization has produced the subclass `state-variable`. A second hierarchy of classes, emanating from the mother-class `object`, captures the information available for various physical objects, such as materials (represented by class `stuff` and its specializations), processing units (from class `process-equipment`), devices (from class `device`). A hierarchy of `conditions` provides the logical mechanism for monitoring the values of data and taking action when specific conditions are met.

b. Functionality. The modeling elements, described above, are used as the building blocks for the construction of higher-level modeling objects, such as `views` and `phenomena`, which describe physical interactions. The topological model of HPT provides a Boolean representation of the model's components, whereas the phenomenological model supplies a qualitative description of the entity's behavior. It is the behavioral, state–space model that carries the full quantitative description.

5. Critique of Modeling Systems for Simulation

All the systems, described in the previous paragraphs, were created with a very specific purpose in mind: to expedite the generation and modification of quantitative models to support the needs of static or dynamic simulation or design. As a result, they are very difficult (if possible at all) to be used for other engineering tasks, which may have a different use for the generated models (e.g., diagnosis). Their common weakness results from two essential deficiencies: first, they cannot capture all forms of knowledge (e.g., qualitative or semiquantitative relationships); second, they do not possess a rich set of semantic relations, which allow intelligent response to queries about (a) the assumptions that produced a given model, (b) the interrelationship of physical phenomena captured in the modeling relations, (c) the propagation of information flow among different views of the same entity, etc. In Section IV we will see how the modeling language MODEL.LA. can overcome these deficiencies.

C. MODELING SYSTEMS IN CHEMISTRY

With the exception of computational chemistry, computer-aided chemical reasoning has not enjoyed the success experienced in other scientific disciplines. This has often been attributed to the conceptual nature of chemistry and the inexactness of known relationships (Dugundji and Ugi,

1973). However, the synthesis of new chemical reaction paths and processing alternatives that support their production remains an attractive objective within the general field of process synthesis. The availability of raw materials and energy, changing ecological and health considerations, and shifting requirements of the market, reinforce each other: to create the need to identify new routes or new processes to obtain existing chemicals or develop new chemicals to meet perceived requirements.

Chemists were the first to look into computer-aided organic synthesis (CAOS) in an attempt to answer Woodward's question (Woodward, 1956, 1963) and furthered by Corey (1967): "How does a chemist choose a pathway for the synthesis of a large organic molecule, given either the great diversity of organic structures and reactions, or, in contrast, the critical importance of each step to ultimate success?" The works of Corey, Wipke, Ugi, Hendrickson, Jorgensen, Moreau, Gelernter, and others support the CAOS effort (see Table I). Excellent reviews can be found in the works of Vernin and Chanon (1986) and Wipke *et al.* (1974).

For any problem of computer-aided organic synthesis, the quality and quantity of required data will vary, the most appropriate strategies may be different, but the questions central to the advancement of computer-aided chemical reasoning remain the same:

(a) How should the molecules and reactions be represented?
(b) What is the necessary degree of descriptional completeness, and at what cost can it be achieved?

Just as Lavoisier stated 200 years ago, "It is time to rid chemistry of obstacles of every kind...this reform must be brought about by perfecting the language," we postulate that a high-level (computer) language must be developed for meaningful advancement to occur in computer-aided chemical reasoning.

Domain-specific modeling languages are "very high level" and of "special purpose." They have a *theme*, that is, the class of ideas that is optimized to communicate. They allow the user to employ terms and constructs that lie closer to the informal terminology and modes of speech customary in the discussion of domain-specific problems; however, in constructing such a language, one should strike for domain-specific generality of the language. Thus, the same modeling language should satisfy all the needs: chemical synthesis, reaction analysis, pathway optimization, and innovative design of reaction or processing networks. The implication of this requirement is clear: *A modeling language in chemical reasoning should be fully declarative, and in no way should its generality be compromised by the specificity of the methodologies of the chemical tasks themselves.*

TABLE I

COMPUTER AIDS IN CHEMISTRY

Number	Authors	Program name	First published
1	Corey-Wipke	OCSS	1969
2	Corey	LHASA	1971
3	Ugi-Gasteiger	—	1971
4	Hendrickson	—	1971
5	Bersohn	—	1971
6	Weise	AHMOS	1973
7	Gelernter	SYNCHEM	1973
8	Barone-Chanon	SOS	1973
9	Powers	DINASYN	1973
10	Brownscombe	EXTRUS	1973
11	Brownscombe	HEXARR	1973
12	Wipke	SECS	1974
13	Ugi-Gasteiger	CICLOPS	1974
14	Benedek	SIMUL	1974
15	Dubois	SYNOPSYS	1975
16	Whitlock	—	1976
17	Donova	HEDOS	1976
18	Powers	REACT	1977
19	Govind	REPAS	1977
20	Djerassi	REACT	1977
21	Pensak	LHASA (Du Pont)	1977
22	Kaufmann	PASCOP	1978
23	Moreau	MASSO	1978
24	Gelernter	SYNCHEM2	1978
25	Ugi-Gasteiger	EROS	1978
26	Yoneda	GRACE	1978
27	Gasteiger	PSYCHE	1979
28	Weise	GSS	1979
29	Barone-Chanon	SAS	1979
30	Stolow-Joncas	LHASA (Educ)	1980
31	Agnihotri	CHIRP	1980
32	Jorgensen	CAMEO	1980
33	Gund	SECS (Merck)	1980
34	Hendrickson	SYNGEN	1981
35	Hippe	SCANSYNTH	1981
36	Hippe	SCANPHARM	
37	Hippe	SCANMAT	
38	Zefirov	FLAMINGOES	1981
39	Ghose	—	1981
40	Schubert	ASSOR	1981
41	Zin	—	1982
42	Erdos	ASR	1983
43	Kaufmann	PSYCHO	1984
44	Seidel	—	1984
45	Hara	PFP	1984
46	Wipke	SST	1984
47	Cense	MICROSYNTHESE	1985
48	Barone-Chanon	TAMREAC	1985

In the absence of any formal modeling language for the representation of chemicals and chemical reactions, it is important to state a concise set of requirements that the desired language should satisfy. These requirements, consistent with the five premises of Section I.A are

1. Represent molecules, reactions, and reaction pathways at various levels of detail.
2. Allow easy extension to new classes of molecules and reactions.
3. Allow the contextual representation of chemicals' reactivity, i.e., reactivity influenced by the relative position of atoms, the surrounding conditions and presence or absence of specific molecules.
4. Support all scientific or empirical knowledge brought together during the synthesis of reaction pathways.

In Section III we will see how LCR (Language for Chemical Reasoning) meets these requirements.

II. LCR: A Language for Chemical Reasoning

In this section we will describe the components of a modeling language, called LCR, that was developed to represent the knowledge about chemically reacting systems. First, we will describe the *basic modeling elements*, which constitute the building blocks for the representation of the declarative knowledge about molecules, reactions, and pathways. Second, we will discuss the semantic relationships among the basic modeling elements. These semantic relationships establish the "meaning" behind the linguistic expressions, defining knowledge about molecules and reactions. Third, we will present the syntax used by the language for the description of chemically reacting systems. LCR was implemented on a Symbolics 3650. It consists of approximately 50,000 lines of LISP (excluding code for the interfaces). Extensive discussion on the use of LCR for pathway generation can be found in Nagel (1991). The fourth chapter in this volume also discusses how LCR has been used to generate the reaction-based potential hazardous events in a chemical plant.

A. MODELING ELEMENTS OF LCR

A single chemical structure may exhibit multiple chemical behaviors, depending on the conditions in the surrounding environment. As a result, LCR uses two groups of modeling elements; the first provides those elements that are needed to describe the structure of the molecule (e.g., atoms, bonds, connectivity), and the second class contains those elements that characterize its chemical behavior and are related to the atom's electronic configuration. .

1. Modeling Elements Defining Chemical Structures

LCR uses a graph theoretic representation of molecular structures, wherein nodes are atoms and edges are bonds. Three modeling elements are employed to create and describe chemical structures. Each modeling element has been implemented as a *class*, in an object-oriented programming environment, and is described by a set of attributes and a set of procedures.

a. Modeling Element 1: `atom`. An atom is the building block of chemical structures? Each `atom` encapsulates its own structural information, modeling relationships, and modeling assumptions. The attributes describing the object class, `atom`, are shown in Table II. Additional attributes may be associated with the `atom` class to refine the class description. For example, the attributes, formal-charge, oxidation-state, steric-hindrance, chirality, may be included into an atom's description to refine the class description. Several of the attributes given in Table II warrant further explanation: Attributes "identifier" and "old-identifier" are pointers used to trace an atom's lineage; attribute "database-atom" contains an object `db-atom` that manages invariant information indigenous to a particular atomic species (e.g., atomic-symbol, atomic-number, atomic-weight). The attribute "parent-abc" associates an atom to a given chemical structure; its value provides the vehicle for moving between the atom representation and the "parent-abc" representation. "Parent-group" is used similarly. An `atom` may have an association with a functional group or set of groups, as well as a parent-abc. As in the attribute case of "parent-abc," the value of "parent-group" provides a conduit for accessing information residing at the `group` representational level. As will be shown later, all representational levels can be accessed from any individual level. The freedom of information flow between levels enhances representational expressiveness and facilitates efficient reasoning.

A partial listing of the methods operating on the class `atom` is also shown in Table II. Methods with the "compute" prefix (e.g., *compute-parent-groups*, *compute-hybridization*, and *compute-formal-charge*) are used to evaluate attribute values. Predicate methods (i.e., methods returning a Boolean value) are represented by the "-p" suffix. These methods are

TABLE II

SELECT ATTRIBUTES AND METHODS FOR `atom`, `bond`, AND `atom-bond-configuration` CLASSES

atom	bond	atom-bond-configuration
atom attributes	**bond attributes**	**atom-bond-configuration attributes**
identifier	identifier	identifier
old-identifier	character	atoms
name	strength	bonds
type	length	empirical-formula
chiral-p	type	molecular-weight
formal-charge	atoms	charge
electronegativity	parent-abc	equivalent-atoms
electron-withdrawing-substituent-pacidity	progenitors	equivalent-bonds
	db-bond	weakest-bond
connectivity	**bond-methods**	heat-of-formation
hybridization	methods	entropy-of-formation
conjugated-p	find-bond-chains	free-energy-of-formation
p-orbitals	cleave-bond	homo
open-approach-p	compute-bond-strength	lumo
radical-orbital	remove-bond	progenitors
bond-angle	add-bond	environment
parent-abc	modify-bond	...
progenitors	create-bond	**atom-bond-configuration-methods**
parent-groups	**bond-selectors**	setup-atom-bond-graph-descriptor
neighbor-atoms	same-bond-type-p	atom-bond-graph-descriptor
neighbor-groups	bond-equivalence-p	setup-atom-bond-graph-from-old-graph
bonds	alpha-bonds	setup-atom-bond-configuration
database-atom	beta-bonds	make-graph-from-symmetric-connectivity-matrix
atom methods	gamma-bonds	
compute-connectivity-number	equivalent-bonds-p	make-bonds-from-connectivity-list
compute-parent-groups	terminal-bond-p	make-atom-bond-configuration
compute-neighbor-groups	internal-neighbor bonds	equivalent-atom-bond-configuration-p
compute-hybridization	terminal-bond-for-additions-p	make-old-arc-to-new-arc
compute-p-orbitals		identify-bond-printed-representation
compute-formal-charge		correct-bond-number-p
compute-radical-orbital		atom-symmetry-identification
identify-electron-withdrawing-substituent		bond-symmetry-identification
find-atom-chains		enumerate-all-atom-chains
atom-backbone		identify-atom-bond-configuration-environment
atom-degree		identify-atom-bond-configuration
higher-degree-atom-p		compute-equivalent-bonds
compute-open-approach-p		compute-weakest-bond
identify-conjugated-p		compute-equivalent-atoms
atom-specific-selections		...
create-atom		
abstract-grouping		
1,2-mobile-atom-p		
1,5-mobile-atom-p		
mobile-univalent-atom-p		

helpful when a characteristic of the atom is used by several different procedures, whether they are internal or external to the object class that contains them. For example, various operations may require knowledge about an atom's mobility as in the case of radical rearrangements; rearrangement alternatives are accessed using methods illustrated by *1,2-mobile-atom-p* and *1,5-mobile-atom-p*.

The remaining methods evaluate properties of an atom or properties of the parent structure. For example, *atom-backbone* identifies the skeleton of interest; *find-atom-chains* enumerates all the paths emanating from the specified atom; *abstract-atom-grouping* produces metagroups around reaction centers. Partitioning of the parent structure into active and inactive sites enables efficient manipulation during pathway construction, as we will see in Section III. New atoms are constructed using the method, *create-atom*. Selector functions, implemented as methods, are also built into the class `atom`. These methods provide an efficient means of collecting atom, which contain some specified desired property. Illustrative examples of selector methods include *collect-sp2-atoms*, *collect-oxygen-atoms*, *collect-beta-neighbors*, and *collect-terminal-atoms*.

b. Modeling Element 2: bond. Atoms are connected by bonds. Like atoms, each bond has associated with it structural information and models that manipulate this information to deduce new features that describe it. Bonds know the atoms that describe it. It is through these elements (bonds) that information is transferred from entity to entity, and new connectivity information is deduced.

A partial listing of the describing attributes and methods operating on object class `bond` is presented in Table II. Notice that the `atom`'s attribute "value" allows information flow from the `bond` representation to an `atom` representation, whereas the "parent-abc" attribute value allows movement between various levels of abstraction within the representation (e.g., groups and structures). In addition to evaluating attribute values, `bond` methods are used to facilitate the construction of chemical structures. To achieve this purpose, *cleave-bond*, *add-bond*, *remove-bond*, *modify-bond*, and *create-bond* are provided. Like atom selectors, bond selectors are used to collect bonds exhibiting a designated property to facilitate processing.

c. Modeling Element 3: atom-bond-configuration. Atoms and bonds containing connectivity information and spatial relations represent an instance of the `atom-bond-configuration` class. All chemical structures can be defined by an instance of the `atom-bond-`

`configuration` class. This structure, although in the strictest sense is not a primitive, has been elevated to a primitive to reduce complexity in subsequent reasoning. Associated with the `atom-bond-config-uration` class are all methods and properties indigenous to an abstract chemical structure, such as physical and spatial properties. Thus, each specialized chemical structure is constructed from an atom-bond-config-uration. An atom-bond-configuration in conjunction with the constituent atoms and bonds allows us to capture and isolate the structural features of a configuration from the feature that characterizes its behavior.

Generic chemical structures are represented by the object class atom-bond-configuration. These attributes describing this class include "name," "identifier," "atoms," "bonds," "empirical formula," and "frontier molecu-lar orbital" (FMO). Since atoms and bonds are objects, the values of the attributes describing these entities are the set of objects constituting the atom list and the bond list, respectively. As a consequence, the values of the attributes describing these objects are easily accessible to proced-ures invoked by the atom-bond-configuration class. For example, the evaluation of an atom-bond-configuration's highest occupied molecular orbital (HOMO) requires evaluation of the atomic orbitals (AOs) that compose it. This information, which is resident in the atoms compos-ing the atom-bond-configuration, is accessed through selector functions that are applied by the *compute-HOMO* procedure of the class atom-bond-configuration.

Attributes common to an `atom-bond-configuration` as well as methods operating on the instances of this class, are shown in Table II. Methods of particular interest include general setup methods, such as *map-old-abc-to-new-abc*, *equivalent-atom-bond-configuration-p*, and *iden-tify-atom-bond-configuration-environment*. Setup methods, as expected, provide a means for instantiation. *Map-old-abc-to-new-abc* is a utility method that maintains the system pointers, which is an important feature when competing pathways are simultaneously analyzed. *Equivalent-atom-bond-configuration-p* determines when two instances of the atom-bond-con-figuration are equivalent, whereas *identify-atom-bond-configuration-en-vironment* accesses information on the reaction environment. Although the reaction environment is specified for most pathways of synthetic interest, in the broader domain of computer-aided chemical reasoning many sys-tems exist in which the environment is not known a priori.

2. Modeling Elements Defining Reactive Behavior of Chemicals

The reactive behavior of a chemical structure is determined by the interaction between low-level transformations of an instance of atom-

`bond-configuration` (`abc`), such as bond cleavage, bond formation, electron distribution, and the conditions in the surrounding environment. To capture all the necessary information, LCR used the following classes of modeling elements.

a. Modeling Element 4: `chemical-behavior`. The electronic state that characterizes chemical reactivity is captured in chemical behavior. This may result from internal or external influences or both. Assessment of these electronic states characterize radical, nucleophilic, and electrophilic behavior or combinations thereof.

The set of behaviors that defines the spectrum of a species' reactivity is determined by its electronic state, which is dictated by internal and external influences. These influences may be established by the ground state, the excited state, or the effect of an external environment on the state. The chemical-behavior class uses a set of operations that define the chemical behavior of a chemical structure and proceed as follows:

Step 1. A set of operations, S, is used to deduce and assess the structural character, s, of an `abc` instance: $abc \overset{S}{\to} s$.

Step 2. Given s, a set of operations, I evaluates the electronic character i of the molecular structure: $s \overset{I}{\to} i$.

Step 3. Given a set of k species with structure $s_1, s_2, \ldots, s_k$, and electronic characters, $i_1, i_2, \ldots$, a set of operations, E, assess the electronic character, e, of the surrounding environment,

$$\{s_1, s_2, \ldots, s_k; i_1, i_2, \ldots, i_k\} \overset{E}{\to} e.$$

Step 4. Given s, i, and e of a specific chemical structure, a set of operations B establishes the set of potential instances of `chemical-behavior` (or `cb`) for the specific chemical structure:

$$\{s, i, e\} \overset{B}{\to} cb.$$

The operations S, I, E, and B are methods invoked by the instances of the class `chemical-behavior` and form the basis for the assignment of specific reactivity of a given `abc`. During their execution call on other methods, they are encapsulated by the structural classes, `atom`, `bond`, and `abc`. An important feature of `cb` is that detail can be managed and persued on demand to assist in an evaluation. The structure permits the association of a set of potential behaviors with a species. These tasks can be performed at run time and the assignment made *dynamically*. For example, whether an alcohol acts as a bulk solvent, weak acid, weak nucleophile, or form an alcoholate anion, is dynamically assigned by `chemical-behavior` (cb) once the reaction environment is known.

b. Modeling Element 5: `reaction-environment`. The set of properties characterizing the environment in which a reaction is occurring is contained in reaction-environment. Attributes typical of this element include; "temperature," "pressure," "wavelength," "surface type," "species concentration," "pH," "species present," etc. Methods built into the modeling element evaluate attribute values and access information in other modeling elements.

c. Modeling Element 6: `ab-initio-operator`. Low-lying transformations that are characterized by bond cleavage, bond formation, and electron distribution are captured in the modeling element `ab-initio-operator`. These transformations may be grouped as mass transfer operations and energy transfer operations for abstraction purposes. Axioms, physical laws, rules, or constraints, obtained from the physical sciences are encoded to prevent the generation of an atom-bond-configuration obtained by applying infeasible transformations. Such knowledge may include conservation of mass, conservation of energy, charge support, Pauli exclusion principle, or valence constraints.

A specific chemical transformation or rearrangement is accomplished by calling on a set of base operations, called `ab-initio-operators` (K_{ai}'s), to perform the necessary elementary structural changes and electron redistribution. The base operations selected are evaluated on the reaction sites of the specified reactant(s). These low-lying operations have embedded models or constraints, such as conservation of mass and energy, that prevent the generation of infeasible structures as specified by the scope of knowledge that defines them. For example, if we have not specified operators capable of constructing delocalized bonds, then we should not expect structures that emanate from this knowledge. We have found that the knowledge embedded in these models is nearly always structural in nature reflecting for example, knowledge about orbital symmetry, atom coordination, or charge. As will be discussed later, K_{ai} operators can be used directly to generate the upper bound of theoretically feasible transformations.

d. Modeling Element 7: `composite-operator`. The class `composite-operator` contains knowledge about user specifications and mechanistic operations. Generic rate information, such as relative magnitude of rate constants (e.g., rate constants of photochemical reactions lie between

10^{10} and 10^{12} s^{-1}) and electronic state information (e.g., excited, radical, charged) are also contained in `composite-operator`. This information is used to limit an operator's range of applicability.

The class `composite-operator` contains procedures that transform a set of chemical species of predisposed behaviors into a set of products, provided that an optional set of prespecified conditions is satisfied. These conditions can be specified by the user or imposed by the system. They may encompass virtually any symbolically encodable concept: structural character, chemical behavior, spatial orientation, reaction conditions, free-energy requirements, enthalpy considerations, toxicity, etc. An instance of the `composite-operator` (e.g., K) calls on the following procedures to carry out the corresponding tasks:

$K_{get\text{-}sites}$, to identify the potential sites for reaction, by assessing the chemical-behavior of each `atom-bond-configuration`.

K_t, to transform a set of potential reaction sites, utilizing instances (e.g., K_{ai}) of `ab-initio-operator`. Successful applications of K_{ai} operators generate new instances of `atom-bond-configuration`.

K_f, to return a Boolean value of "true," when the encoded set of prespecified conditions are achieved, and "false," otherwise.

LCR uses a set of predefined instances of K to represent a set of known transformations utilizing specific ab-initio-operators. New instances K of composite operators can be easily built by the user through the aggregation of different sets of ab-initio-operators, using the facilities of LCR.

3. Modeling Elements for Reactions and Pathways

The successful application of a composite operator on a chemical structure, or a set of structures, constitutes a reaction:

$$\text{R: } \{abc_1\} \xrightarrow{K} \{abc_2\}.$$

a. Modeling Element 8: `reaction`. This modeling class captures the information that is specific to a given transformation. An instance of `reaction` contains information about the species determining which are the reactants and which are the products. It also contains a reference to the instance of the composite operator K, which is responsible for the transformation, and the instance of `reaction-environment`, which supplies the reaction conditions.

Instances of `reaction` constitute the building blocks for the construction of a reaction pathway P, which is modeled as an ordered list of

```
reactions:
```

$$\mathrm{P} \triangleq \{R_1, R_2, \ldots, R_n\}, \text{ where } R_1: \{\mathtt{abc}_{i-a}\} \xrightarrow{K_i} \{\mathtt{abc}_i\}.$$

The transitivity of semantic relations in LCR (see Section I.C) allows the system to automatically construct an instance of `reaction` as the abstract representation of a pathway. For example, if R_1: A $\rightarrow$ C and R_2: B $\rightarrow$ C, then LCR constructs automatically R_3: A $\rightarrow$ C and includes it in the library of reactions for future pathway contraction.

b. Modeling Element 9: `context`. A context is a consistent set of assumptions that characterize a species behavior or structural character. If one wants to change the description of some modeling elements (i.e., associate a new set of assumptions with them), but also preserve the former version, then one assigns a new `context` to the new version, keeping all other information the same. The use of context is particularly helpful when a reactant can exist in several forms, such as resonance structures, or keto enol forms. If a context is associated with a pathway, assumptions leading to the selection of a particular behavior can be modified and the pathway adjusted accordingly without reinitializing the entire system.

4. Modeling Elements to Describe Quantitative Relationships

In all the previous modeling elements, the need arises for establishing equations, inequalities, Boolean, ordinal or order-of-magnitude relationships among the attributes (variables and parameters) of the modeling elements. To capture these quantitative relationships, LCR uses the modeling classes `constraint` and `generic-variable` as well as their various subclasses. Since these modeling elements have been borrowed from the modeling language MODEL.LA., we will defer their description to Section IV. For the time being it suffices to say that `generic-variable` provides a structured representation of any variable or parameter, and `constraint` does the same for a set of different types of relationships.

5. Subclasses of Modeling Elements

Each of the nine modeling elements described above possesses a basic structure of attributes and methods and are inherited by any of their instances. Nevertheless, we have found that the construction of specialized subclasses, emanating from a specific class, enriches the vocabulary of

LCR and brings it closer to the linguistic constructs used by chemists and engineers. For example, a functional group can be described by an instance of the class `atom-bond-configuration`, but having a distinct modeling element, called `group`, with its own specialized attributes and methods, it facilitates the identification of such groups in molecules and streamlines the reasoning around such specific structures. Nevertheless, in order to preserve the fact that a group is also in atom-bond-configuration, we create the class `group`, as a subclass of the class `atom-bond-configuration`. Inheritance mechanisms of object-oriented programming allow the modeling class, `group`, to inherit all attributes and methods of the mother-class, `atom-bond-configuration`. Using this inheritance mechanism, we have developed expanded trees of modeling subclasses for four of the basic modeling elements: `atom-bond-configuration`, `chemical-behavior`, `ab-initio-operator`, and `composite-operator`.

a. The `atom-bond-configuration` *Classes.* Four main subclasses emanate from the `abc` class: (1) `group`, representing a substructure of the atom-bond-configuration—Additional groups may be added to or built from the list of conventional functional groups (e.g., acid from oxo and alcohol); (2) `ion`, representing any atom-bond-configuration that has a net electrical charge; (3) `radical`, representing any atom-bond-configuration that has an open shell; and (4) `molecule`, representing any atom-bond-configuration that is not an ion or a radical. Further specialization of these primary subclasses, either through the mixing of the primary subclasses (e.g., the class radical-ion may be formed by mixing radical and ion) or by specializing the class description (e.g., creating a class organic under molecule), allows extension of the generic abc-class.

Consider, for example, specialization of the class `molecule` (Fig. 3) into `organic-molecule` and `inorganic-molecule` and suppose inorganic-molecule contains a method for electron counting. This method is inherited to `organometallic-molecule`; how the method is used and combined with other methods is controlled by `organometallic-molecule`. Likewise, if methods and attributes of `organic-molecule` facilitate reasoning about the organometallic system, they can be included as well.

b. The `chemical behavior` *Classes.* Chemical behavior is classified into two main subclasses: `external-influences` and `internal-influences` (Fig. 4). Unlike the abc-class, the cb-class structure provides a means for communicating attribute values between classes, independent

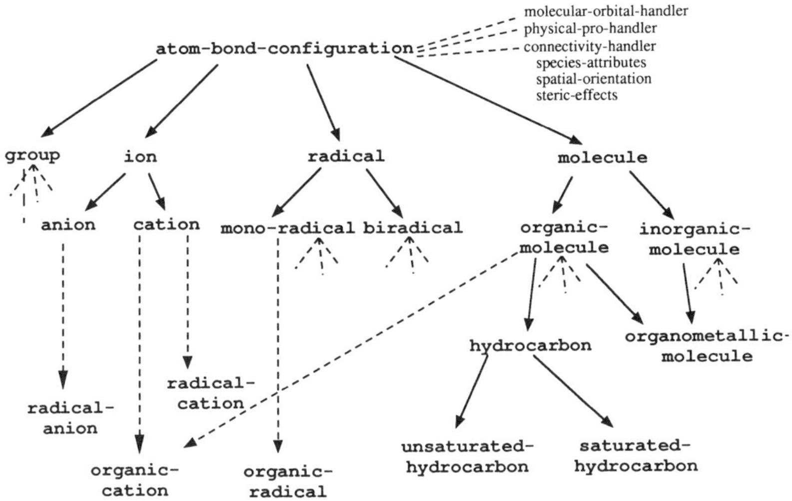

FIG. 3. Hierarchy of atom-bond-configuration subclasses.

of their position within the class structure. This is necessary because chemical behavior is often a combination of effects (e.g., the ground state of a species is influenced by its inherent electronic structure and its external environment). LCR provides a means of combining these effects. These operations constitute the basis for reactivity assignment. They are distributed throughout the hierarchy and are used to classify electronic states. The classes composing this hierarchy categorize electronic state and not the derived properties associated with an electronic state. Nucleophilicity, for example, is derived from the class pertinent-occupied-molecular-orbital (POMO). Radical behavior is derived from the class singly-occupied-molecular-orbital (SOMO).

As in the abc hierarchy, each class composing the chemical behavior hierarchy contains only those methods and attributes that pertain to it. Concepts derived at a particular class are used in classes of higher specialization so that more sophisticated concepts can be deduced and discriminating properties elucidated at the proper level. For example, the class POMO contains methods for evaluating the POMO, whether it is the HOMO, the n-HOMO, or any other occupied molecular orbital. These attributes may then be used at more specialized level to expand on the features of an electronic state that gives rise to a particular conceptual behavior. Nucleophilicity illustrates this point, because it may arise from either a negatively charged ion or a neutral atom (i.e., nucleophilicity can

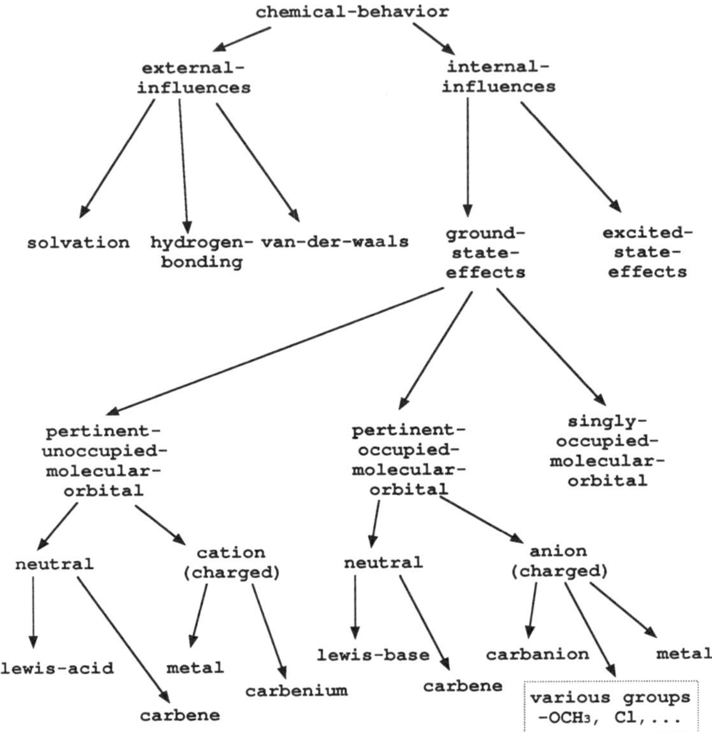

Fig. 4. Hierarchy of chemical-behavior subclasses.

result from lone pairs that are bonded, as in the case of the lone pairs on nitrogen; or nonbonded, as in the case of the electrons comprising a π bond).

An important distinction between the abc and cb hierarchy is that the cb hierarchy of classes is not treated as task specialization by operators calling for the assignment of chemical behavior. Since cb is a set of potential behaviors, {b}, it is necessary to associate behaviors to a species that lies on the same class level [e.g., pertinent unoccupied molecular orbital (PUMO) and POMO] as well as on different, more specialized, levels of the same class (e.g., Lewis-base and POMO). This is necessary because the species dynamically assumes the requisite behavior—(weak) nucleophile or Lewis base—when it is required by the procedure $K_{get\text{-}sites}$.

c. The ab-initio-operator *Classes*. Five primary subclasses were developed to model elementary transformations (Fig. 5) (1) bond-

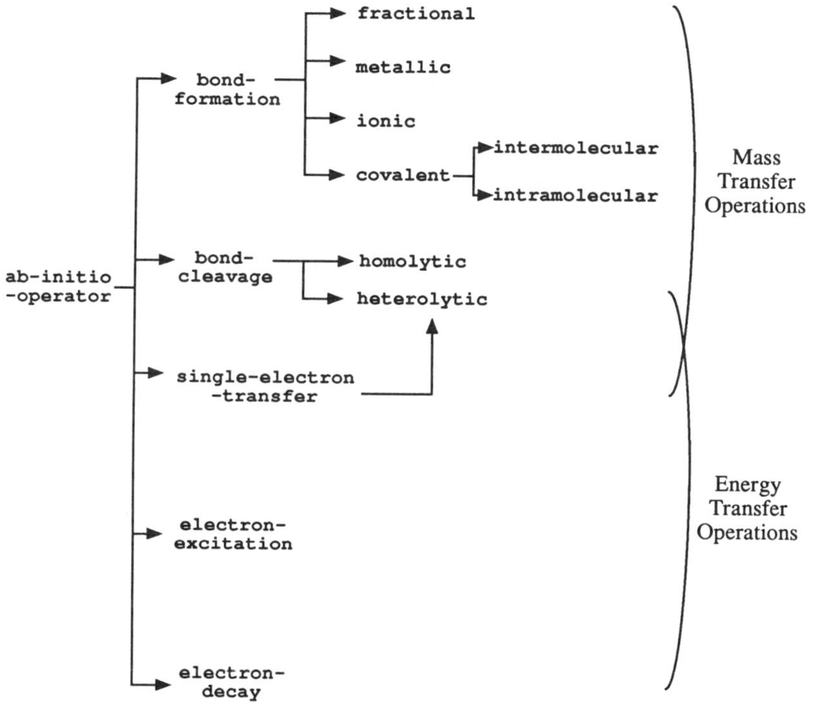

FIG. 5. Hierarchy of ab-initio-operator subclasses.

formation, (2) bond-cleavage, (3) single-electron-transfer, (4) electron-excitation, and (5) electron-decay. The operations stem from two types of operations: mass transfer and energy transfer. Each subclass contains methods that prevent the generation of theoretically infeasible structures. As in the abc-class, the subclasses of ab-initio-operator can be mixed, together with their methods, to form various new classes. For example, the subclass heterolytic-bond-cleavage is formed by combining the parent classes of single-electron-transfer and bond-cleavage. Together, these operations afford the generation of any elementary reaction mechanism.

d. The composite-operator *Classes.* The hierarchy of composite-operator classes is shown in Fig. 6. As described earlier, composite operators are built from sets of ab-initio-operators. They utilize spatial

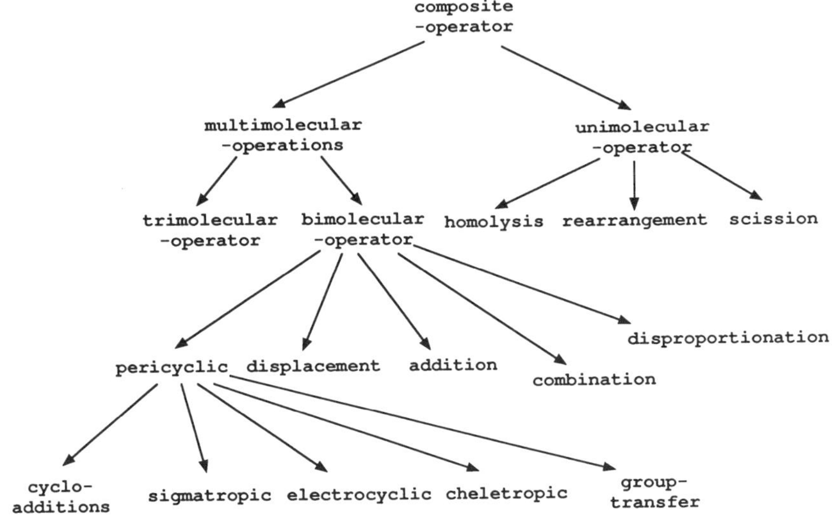

Fig. 6. Hierarchy of composite-operator subclasses.

operators, which modify the orientation of an abc. Although many hierarchies can be developed utilizing various classification schemes, for efficiency reasons, we have chosen a two-staged classification built on particle reaction requirements and specialized by structural reaction requirements. Because of this formulation, there are only two primary subclasses extending from composite-operator: multimolecular-operator and unimolecular-operator. Each subclass is then further specialized according to the structural requirements of the composite operator. For example, the bimolecular operation of radical coupling requires two radical substrates.

This classification limits the combinatorial explosion encountered in conventional computer-aided synthesis approaches, because composite operators know a priori when they apply. For example, the application of bimolecular operators to a list of chemical species results in the selective application of radical-operator to radicals contained in the list and molecule-operator to species that are molecules.

B. Semantic Relations among Modeling Elements in LCR

If the modeling elements of LCR represent the modes of a network, the semantic relationships constitute the links (edges) of the network. There

exist different types of links, with each one establishing a different interpretation on the relationship of the connected nodes. Thus, whereas the
modeling objects carry by themselves no assertional importance, the semantic relationships establish assertions and thus impose interpretation,
i.e., meaning, to the representational objects.

1. Entity-Attribute Semantics

The first two semantic relationships establish the structure of a modeling element; they allow the declaration of the attributes and methods of a
modeling element. Although all object-oriented systems possess, by definition, the first two semantic relationships and no special computer-aided
provisions are needed, they have been included here for completeness.

a. Semantic Relationship 1: "is-attribute-of." This relationship enables the
association of a modeling object as an attribute of an other object. It is
useful when the present scope of LCR's modeling elements needs to be
extended. Without this relationship a modeling would have remained a
conventional low-level data structure. For example, the attributes of the
object representing hydrocarbons are declared as follows:

```
hydrogen-atoms      is-attribute-of      hydrocarbon
aromatic-p          is-attribute-of      hydrocarbon
```

b. Semantic Relationship 2: "is-method-of." This relationship declares that
a given computational procedure is "owned" by a particular modeling
object, and that the procedure describes part of the object's behavior or
computational ability. The following are typical examples of the use of this
semantic relationship:

```
compute-aliphatic-p              is-method-of      hydrocarbon
collect-longest-carbon-chain     is method-of      hydrocarbon
```

2. Specialization Semantics

The following two semantic relationships establish the links between a
basic modeling element and its derivative subclasses of modeling objects
and instances. Both of them are isomorphic mappings.

a. Semantic Relationship 3: "is-a." This is needed to establish *sub/
superset* links and is used to define various new subclasses of modeling
objects, emanating from a specific class. Thus, starting with a small

number of basic modeling elements, this semantic relationship allows a modeling language to be expanded and remain open-ended. Each new class of modeling objects is a structure isomorphic to the modeling element from which it was derived, but with more specializations. Typical examples of its use are

```
molecule        is-a       abc
hydrocarbon     is-a       organic-molecule
addition        is-a       bimolecular operator
```

b. Semantic Relationship 4: "is-a-member-of." This relationship is needed to relate isomorphic structures, emanating from the same class of modeling objects, with identical specialization. For example, assume that specific assumptions on the modeling of free radicals lead to the class of modeling objects called `radical`. Then if specific free radicals, `radical-1` and `radical-2`, are modeled under the same set of assumptions, the resulting models will have the same internal structure as `radical`, but different assigned names and values to the various variables or attributes describing it. Consequently

```
radical-1      is-a-member-of      radical
radical-2      is-a-member-of      radical
```

3. Specification Semantics

Specification mappings, described by the following six semantic relationships, are used to specify the value of attributes of various modeling elements. They express, (a) binary relations, such as whole/part links, (b) communication lines among modeling objects, or (c) the value of simple describing properties.

a. Semantic Relationship 5: "is-composed-of." It defines the link between a modeling object and other modeling objects in which the former is decomposed. For example, to represent a `free-radical-reaction`, one may want to disaggregate it to the level of `initiation-reaction`, `propagation-reaction`, and `termination-reaction`. In such case

```
free-radical-reaction    is composed-of    {initiation-reaction;
                                            propagation-reaction;
                                            termination reaction}
```

b. Semantic Relationship 6: "is-part-of." This relationship defines the link between a modeling object and the modeling object that is containing it. Continuing with the example presented above, we can specify that

> initiation-reaction   *is-part-of*   free-radical-reaction

In particular, this semantic relation is established when the attribute "parent" of the modeling element `initiation-reaction` receives the value `free-radical-pathway`. The *"is-part-of"* relationship deals with structural containment—the *"is-a"* relationship describes conceptual containment. It is obvious that one of the properties of the relation is *transitivity*, and that *"is-composed-of"* and *"is-part-of"* are symmetrical relationships.

c. Semantic Relationship 7: "is-attached-to." This is the first of two semantic relationships used to elucidate the structural connectivity among specific modeling elements, such as *atoms* to form *molecules*. A typical example is

> `oxygen-1`   *is-attached-to*   `(carbon-1; carbon-2)`

The *is-attached-to* semantic relationship establishes linkages between different modeling elements, allowing flow of information from one to the other(s). Furthermore, since atoms know their membership into various functional groups, auxiliary links are automatically established; thus LCR creates automatically the following semantic connection:

> `oxygen-1`   *is-attached-to*   `oxo-1`

or

> `oxygen-1`   *is-attached-to*   `carboxyl-1`

d. Semantic Relationship 8: "is-connected-by." This is the symmetric relationship of *is-attached-to*. Using the previous examples we have

> `oxygen-1`   *is-connected-by*   `carbon-1`
> `oxygen-1`   *is-connected-by*   `carbon-2`

c. Semantic Relationship 9: "is-described-by." At a basic level, the state of a modeling object is described by the values of a set of variables. Moreover, its function or behavior is described by mathematical relationships (or procedures). This semantic relationship is used to declare the variables and mathematical relationships which describe the state of a modeling object, for example:

$$\begin{array}{lll} \texttt{oxygen-1} & \textit{is-described-by} & \text{``hybridization-oxygen-1''} \\ \texttt{S}_n\texttt{2} & \textit{is-described-by} & \textit{K-get-sites-}S_n\textit{2.} \end{array}$$

The first example indicates that the object oxygen-1 is described by the value of the attribute "hybridization", and the second example shows that the behavior of object, S_n2 (an instance of class `reaction`) is described by the procedure K-get-sites-S_n2.

f. Semantic Relationship 10: "is-describing." This is the inverse of the previous relationship:

$$\begin{array}{lll} \text{``hybridization-oxygen-1''} & \textit{is-describing} & \texttt{oxygen-1} \\ \textit{K-get-sites-}S_n\textit{2} & \textit{is-describing} & \texttt{S}_n\texttt{2} \end{array}$$

LCR has the appropriate mechanisms to keep symmetrical semantic links properly updated, when one of them changes (e.g., is created, deleted, or modified).

g. Semantic Relationship 11: "is-characterized-as." This semantic relationship specializes a class by defining the character (i.e., value) of some of its properties. It represents the semantic relation between a modeling object and an attribute of its description. For example, if the class `pertinent-occupied-molecular-orbital` is described as homolytic, and nucleophilic, semantic links like

$$\begin{array}{lll} \text{POMO} & \textit{is characterized-as} & \texttt{homolytic} \\ \text{POMO} & \textit{is-characterized-as} & \texttt{nucleophilic} \end{array}$$

are established between the modeling objects and the attributes of its description. This semantic relationship is particularly important in defining the context of a particular pathway. For example, after a pathway is constructed, the user may wish to change the characterization of an intermediate (e.g., Lewis acid versus nucleophile) so that the implication of the change can be investigated. This is accomplished easily with the semantic relationship *is-characterized-as*. The axioms of commutativity and merging are supported by this relation.

4. Aggregation / Disaggregation Semantics

Aggregation (or, abstraction) is the process by which details of a modeling object are left unspecified in favor of a less cluttered description of its structure. Disaggregation (or refinement) is the opposite process. Thus, abstraction maps a set of modeling objects into an object described by a simpler modeling element.

a. Semantic Relationship 12: "is-disaggregated-in." This semantic relation exists between modeling elements located in different contexts and responds to the need of breaking down systems into smaller, more tractable components, where additional detail can be added. For example, a theoretically feasible but unsubstantiated pathway can be characterized by a set of assumptions that lead to the structural character defining the constituent intermediates. This same pathway may be characterized by another set of assumptions, at a later stage, that leads to a different set of intermediates. Different contexts allow us to encapsulate the distinct representation. However, if we seek to transfer information from the pathway to the species in which it is broken we have to create a communication route. The semantics links

global-pathway-1	*is-disaggregated-in*	initiation-pathway-1
global-pathway-1	*is-disaggregated-in*	propagation-pathway-1
global-pathway-1	*is-disaggregated-in*	termination-pathway-1

provide the communication vehicle.

Notice that the *is-disaggregated-in* link is designed to support communications between objects of the *same type in different contexts*. In this way it is differentiated from *is-composed-of*. The scope of *is-composed-of* is restricted to objects of any type provided they are in the *same context*. The behavior of *is-disaggregated-in* is similar to that of *is-composed-of* with respect to the properties that it supports. This semantic link obeys the axioms of transitivity, commutativity, and merging.

b. Semantic Relationship 13: "is-abstracted-by." It is the inverse of the previous one, and using the same example, we have

{initiation-pathway-1, propagation-pathway-1, termination-pathway-1}

is-abstracted-by   global-pathway-1

5. Properties of LCR's Semantic Relationships

The semantic relations of LCR establish how different objects relate to one another. They were formally defined to obey the following three axioms:

1. *Transitivity*:

> If (O1 "semantic-relation" O2) and (O2 "semantic-relation" O3)
> then
> (O1 "semantic-relation" O3)

where "semantic-relation" applies to the relations `is-a`, `is-composed-of`, and `is-part-of`, and O1, O2, and O3 are arbitrary modeling objects of this system.

For example,

If (`Methyl-1` *is-part-of* `Ethyl-1`) and (`Ethyl-1` *is-part-of* `Ethane-1`) then
(`Methyl-1` *is-part-of* `Ethane-1`)

2. *Commutativity*:

$$
\begin{array}{l}
(\text{THE } \textit{``attribute-1''} \text{ OF } \text{O1 IS } A_1) \\
\cdots\cdots\cdots\cdots\cdots\cdots\cdots\cdots\cdots \\
(\text{THE } \textit{``attribute-1''} \text{ of } \text{O1 IS } A_{j-1}) \\
(\text{THE } \textit{``attribute-1''} \text{ of } \text{O1 IS } A_j) \\
\cdots\cdots\cdots\cdots\cdots\cdots\cdots\cdots\cdots \\
(\text{THE } \textit{``attribute-1''} \text{ of } \text{O1 IS } A_n)
\end{array}
$$

is the same as

$$
\begin{array}{l}
(\text{THE } \textit{``attribute-1''} \text{ of } \text{O1 IS } A_1) \\
\cdots\cdots\cdots\cdots\cdots\cdots\cdots\cdots\cdots \\
(\text{THE } \textit{``attribute-1''} \text{ of } \text{O1 IS } A_j) \\
(\text{THE } \textit{``attribute-1''} \text{ of } \text{O1 IS } A_{j-1}) \\
\cdots\cdots\cdots\cdots\cdots\cdots\cdots\cdots\cdots \\
(\text{THE } \textit{``attribute-1''} \text{ of } \text{O1 IS } A_n)
\end{array}
$$

where *"attribute-1"* is any specific attribute establishing one of the following semantic relationships: *"is-composed-of," "is-attached-to," "is-connected-by," "is-described-by," "is-describing,"* and *"is-characterized-as."* In other words, the order in which attributes are listed is irrelevant.

3. *Merging*:

$$
\begin{array}{l}
(\text{THE } \textit{``attribute-1''} \text{ OF } \text{O1 IS } A_1) \\
\cdots\cdots\cdots\cdots\cdots\cdots\cdots\cdots\cdots \\
(\text{THE } \textit{``attribute-1''} \text{ of } \text{O1 IS } A_{j-1}) \\
(\text{THE } \textit{``attribute-1''} \text{ of } \text{O1 IS } A_j) \\
\cdots\cdots\cdots\cdots\cdots\cdots\cdots\cdots\cdots \\
(\text{THE } \textit{``attribute-1''} \text{ of } \text{O1 IS } A_n)
\end{array}
$$

is the same as

$$(THE\ \textit{"attribute-1"}\ of\ O1\ IS\ SET.OF\ (A_1 \ldots A_{j-1}\ A_j \ldots A_n))$$

where *"attribute-1"* is any specific attribute establishing one of the following semantic relationships: ***"is-composed-of,"*** ***"is-attached-to,"*** ***"is-connected-by,"*** ***"is-described-by,"*** ***"is-describing,"*** or ***"is-characterized-as."*** Hence, attributes of the same concept can be merged. For example,

```
(THE  Neighbor-atoms of Carbon-1 IS Hydrogen-17)
(THE  Neighbor-atoms of Carbon-1 IS Nitrogen-21)
(THE  Neighbor-atoms of Carbon-1 IS Carbon-2)
```

is the same as

```
(THE  Neighbor-atoms of Carbon-1
      IS THE SET OF
      (Hydrogen-1, Nitrogen-21, Carbon-2))
```

C. Syntax of LCR

Grammars are intended to capture what we may call the *structure* or *syntax* of languages, as opposed to their meaning or semantics. The syntax of a programming language is a strict, precise set of rules that describe the string of symbols that constitute legal statements and specify how a statement breaks down into its constituent parts. The syntax description is done using formal grammar, often referred as *metalanguage* (i.e., special language used to describe other languages). The description language used here is an extended BNF (Backus–Naur form or Backus–normal form), which is more compact than BNF. The BNF (Naur, 1963) is a widely used formal method developed by computer scientists for the precise syntactic description of computer languages. The BNF grammars conventionally have four elements: (1) a *terminal vocabulary* that corresponds directly to the vocabulary of allowed tokens of the language to be defined; (2) a *nonterminal vocabulary*, conventionally enclosed in pointed brackets ($\langle$ and $\rangle$); (3) a *set of production rules* defining way of building up phrases (represented by nonterminal symbols) from terminal and nonterminal vocabulary items; and (4) a *correspondence* indicating which nonterminal symbol is associated with the "master" or "sentence" phrase type of the language. In defining the grammar we will follow the standard practice of writing a production beginning with the "master" phrase type at the top of the grammar.

Two important observations can be made about BNF grammar rules. First, BNF rules can be defined in a recursive manner; that is, the phrase name can appear on both sides of the symbol "::= ." This property can be observed in the rules defining "⟨structural-def⟩," "⟨input-def⟩," "⟨output-def⟩," etc. This recursive property allows an infinite number of statements to be described by a finite number of rules. Second, a BNF description of a grammar is complete when every phrase name has been defined reaching the level of terminal vocabulary.

1. Explanation of Notation

::=　　Is defined as.

[]　　Encloses optional unit.

{}　　Indicates mandatory choice.

⟨ ⟩　　Unit that is described separately.

...　　Indicates repetition of syntactic signs.

|　　Separator for alternatives.

*　　What the braces enclose may appear any number of times (including zero).

+　　What the braces enclose may appear any non-zero number of times (must appear at least once).

[[ ]]　　Double brackets indicate that any number of the alternatives enclosed may be used, and those may occur in any order, but each alternative may be used at most once unless followed by a star.

2. Examples of Syntax

Word in capital letters: reserved word.

Lowercase words: metalinguistic variable names.

⟨modeling-element⟩::= ⟨modeling-element-name⟩[⟨attribute⟩]*.
⟨modeling-element-name⟩::= (⟨modeling-class⟩|⟨atom⟩|
　　　　　　　　⟨bond⟩|⟨reaction⟩|
　　　　　　　　⟨reaction-environment⟩|⟨context⟩)
　　　　　　　　[attribute⟩]*[⟨method⟩]$^+$.
⟨attribute⟩::= ⟨attribute-name⟩ IS-ATTRIBUTE-OF
　　　　　⟨modeling-element⟩.
⟨method⟩::= ⟨method-name⟩ IS-METHOD-OF ⟨modeling-element⟩.
⟨class⟩::= (⟨class-name⟩ IS-A {[⟨modeling-class⟩]$^+$[⟨class⟩]$^+$}).
⟨instance⟩:: +(⟨instance-name⟩ IS-MEMBER-OF {[⟨class⟩]$^+$}.

⟨abc⟩::= ⟨abc-input⟩.
⟨abc-input⟩::= be-matrix|atom-table bond-table.
⟨abc⟩::= ⟨group⟩|⟨ion⟩|⟨radical⟩|⟨molecule⟩.
⟨group⟩::= [atom]⁺[bond]⁺.
⟨ion⟩::= [atom]⁺[bond].
⟨radical⟩::= [atom]⁺[bond].
⟨molecule⟩::= [atom]⁺[bond]⁺.
⟨cb⟩::= {[⟨behavior⟩]⁺}.
⟨behavior⟩::= (IS-CHARACTERIZED-BY {⟨Internal-Influences⟩}
                [⟨External-Influences⟩]).
⟨Internal-Influences⟩::= (IS-CHARACTERIZED-BY
                {⟨Ground-state-effects⟩
                ⟨Excited-state-effects⟩}).
⟨Ground-state-effects⟩::= (IS-CHARACTERIZED-BY
                {⟨PUMO⟩|⟨POMO⟩|⟨SOMO⟩}).
⟨Excited-state-effects⟩::= (IS-CHARACTERIZED-BY {$\sigma^*|\pi^*$}).
⟨External--Influences⟩::= (IS-CHARACTERIZED-BY
                {[⟨solvation⟩][⟨hydrogen-bonding⟩]
                [⟨Van der Waals⟩]}.
⟨composite-operator⟩::= ⟨inputs⟩⟨enabling-conditions⟩
                ⟨abinitio-operator⟩⟨output⟩.
                [⟨composite-operator⟩].
⟨input⟩::= ⟨react-cond⟩⟨abc⟩⟨cb⟩.
⟨output⟩::= ⟨abc⟩.
⟨abc⟩::= {[(Instance-of-abc)]⁺}.
⟨cb⟩::= {[(Instance-of cb)]⁺}.
⟨reaction-cond⟩::= {(Instance-of reaction-conditions)}.
⟨enabling-conditions⟩::= {[⟨conditionals⟩]⁺}.
⟨conditionals⟩::= {[[(Instance-of user preferences)]⁺
                {[(Instance-of physical requirements)]⁺}]}.
⟨abinitio-operator⟩::= {[⟨Kai⟩]⁺}.
⟨K$_{ai}$-operator⟩::= Bond formation|Bond Cleavage|
                Single Electron Transfer|Electron Excitation|
                Electron Decay.
⟨K$_{ai}$⟩::= ⟨K$_{ai}$-operator⟩⟨ − ⟨K$_{ai}$-input⟩⟨K$_{ai}$-enabling-cond⟩⟨output⟩.
⟨K$_{ai}$-input⟩::= {[⟨reaction-center⟩]⁺}.
⟨reaction-center⟩::= {[⟨atom⟩]⁺}.
⟨K$_{ai}$-enabling-cond⟩::= {[⟨K$_{ai}$-condition⟩]⁺}.
⟨K$_{ai}$-condition⟩::= {[physico-chemical-requirements]⁺}.
⟨physical-requirements⟩::= (IS-ABSTRACTING
                ⟨physio-chemical-requirements⟩).

III. Formal Construction of Representations for Chemicals and Reactions

The nine basic modeling elements of LCR, although they offer a fairly limited vocabulary, are generic enough to allow (1) *infinite extensibility* of LCR's vocabulary through a finite set of rules, and (2) *representation and analysis* of any potential reaction pathway.

In this section we will examine the mechanisms that LCR possesses to achieve these objectives.

A. Extension of LCR's Modeling Objects

In Section II.B we discussed the modeling subclasses emanating from the four basic elements: `atom-bond-configuration`, `chemical-behavior`, `ab-initio-operator`, and `composite-operator`. Let us examine what mechanisms LCR possesses for such linguistic extensibility, which is imperative for the representation of the widely diverse knowledge of chemistry.

1. Extension of the `atom-bond-configuration` Hierarchical Tree

Suppose it is advantageous to reason about cyclic aliphatic species directly. For this purpose, we create the modeling class `cyclic-aliphatic` and link it to the parent class `aliphatic` (Fig. 7), thereby establishing a set membership:

cyclic-aliphatic   *is-a*   aliphatic

Since `aliphatic` is a specialization of the class `hydrocarbon`, it is advantageous to create a link between `aliphatic` and `hydrocarbon` (Fig. 7):

aliphatic   *is-a*   hydrocarbon

This allows the attributes describing hydrocarbon to be inherited by aliphatic. Similarly, the link established between `aliphatic` and `cyclic-aliphatic` allows this information to be accessed from `cyclic-aliphatic` directly. Having created these links, any attribute or method residing in the class `aliphatic` or any of its superclasses is inherited by `cyclic-aliphatic`. For example, if methods *cyclic-p*,

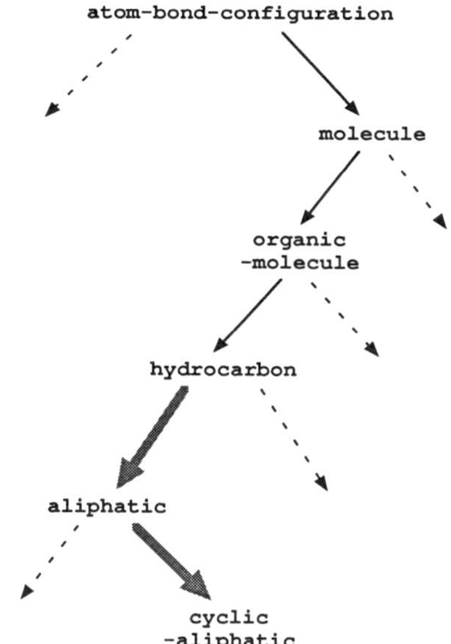

FIG. 7. Extending the atom-bond-configuration tree of subclasses.

compute-nonbonded-strain, and *compute-torsional-strain* reside in atom-bond-configuration and methods *compute-skeleton-chains* and *compute-longest-carbon-chain* reside in organic-molecule, each of these methods as well as any additional descriptive attributes are directly accessible from the modeling class cyclic-aliphatic; no duplication of modeling effort is required. How methods and attributes are combined or mixed into the cyclic-aliphatic class is user-controlled, through LCR's semantic relationships.

It may also be desirable to extend the utility of the cyclic-aliphatic class and its descriptors through the addition of class-specific methods and attributes (Table III). For example, the user may wish to include the following attributes in the class description:

ring-strain	*is-attribute-of*	cyclic-aliphatic
number-of-rings	*is-attribute-of*	cyclic-aliphatic
ring-sizes	*is-attribute-of*	cyclic-aliphatic
ring-number	*is-attribute-of*	cyclic-aliphatic
torsional-strain	*is-attribute-of*	cyclic-aliphatic
nonbonded-strain	*is-attribute-of*	cyclic-aliphatic

. . .

TABLE III

SELECT ATTRIBUTES AND METHODS FOR `cyclic-aliphatic`
AND `excited-state-effects` CLASSES

cyclic-aliphatic	excited-state-effects
cyclic-aliphatic attributes	**excited-state-effects attributes**
identifier (inherited from super class)	identifier (inherited from super class)
old-identifier (inherited from super class)	old-identifier (inherited from super class)
name (inherited from super class)	bonding-orbital (inherited from super class)
parent-abc (inherited from super class)	anti-bonding-orbital (inherited from super class)
progenitors (inherited from super class)	dominate-behavior (inherited from super class)
…(inherited from super class)	competitive-behavior (inherited from super class)
ring-strain	…(inherited from super class)
number-of-rings	S_0
ring-sizes	S_1
ring-number	S_2
torsional-strain	S_3
nonbonded-strain	T_0
bridged-atoms	T_1
…	T_2
cyclic-alphatic methods	…
compute-identifier (inherited from super class)	**excited-state-effects methods**
compute-old-identifier (inherited from super class)	compute-identifier (inherited from super class)
compute-parent-abc (inherited from super class)	compute-old-identifier (inherited from super class)
compute-progenitors (inherited from super class)	compute-bonding-orbital (inherited from super class)
…(inherited from super class)	compute-anti-bonding-orbital (inherited from super class)
compute-name (super class method modified)	compute-dominate-behavior (inherited from super class)
compute-ring-strain	compute-competitive-behavior (inherited from super class)
compute-number-of rings	…(inherited from super class)
compute-ring-sizes	compute-S_0
compute-ring-number	compute-S_1
compute-torsional-strain	compute-S_3
compute-nonbonded-strain	compute-T_0
compute-bridged-atoms	compute-T_1
monocyclic-ring-p	compute-T_2
bicyclic-ring-p	compute-T_3
fused-ring-p	assess-energy-gap
…	allowed-transition-p
	forbidden-transition-p
	fluorescence-p
	phosphorescence-p
	internal-conversion-p
	internal-system-crossing-p
	…

The values of these attributes are then established by the methods that have been associated with `cyclic-aliphatic`. Methods indicative of this class include the following:

compute-ring-strain	*is-method-of*	`cyclic-aliphatic`
compute-number-of-rings	*is-method-of*	`cyclic-aliphatic`
compute-ring-sizes	*is-method-of*	`cyclic-aliphatic`
compute-nonbonded-strain	*is-method-of*	`cyclic-aliphatic`
compute-bridged-atoms	*is-method-of*	`cyclic-aliphatic`
monocyclic-ring-p	*is-method-of*	`cyclic-aliphatic`
bicyclic-ring-p	*is-method-of*	`cyclic-aliphatic`
fused-ring-p	*is-method-of*	`cyclic-aliphatic`

In these methods, those with the "compute-" prefix determine attribute values; those with the "-p" suffix are predicate methods returning Boolean values. The remaining methods are invoked on instances of the class `cyclic-aliphatic` and are designed to raise the abstraction level (i.e., they allow the user to communicate using higher-level concepts). These methods may be called from outside of the class `cyclic-aliphatic` provided they operate on instances of the `cyclic-aliphatic` class. The method *bridged-atoms* illustrates this concept; it calls *bridged-atom-p*, a predicate method of atom, from the class `cyclic-aliphatic`. Likewise, if reactivity assessment requires knowledge about ring structure, we can access that information by applying the semantic relation *is-describing* to an instance of `cyclic-aliphatic` (ICA), as shown below:

bridged-atoms-ICA	*is-describing*	(`atom-1...atom-n`)

Notice also that although the methods, *compute-nonbonded-strain* and *compute-torsional-strain* were inherited by `cyclic-aliphatic`, more sophisticated versions of these methods were supplied to `cyclic-aliphatic`. This would have been unnecessary if the original versions of these methods supported fully *all* cyclic configurations (i.e., a robust method given at any ring configuration). We have found that the ability to incrementally improve methods accelerates model development.

2. Extension of the `chemical behavior` *Hierarchical Tree*

Suppose that we choose to embellish the modeling class `chemical-behavior` by developing the subclass `excited-state-effects`. We begin by developing a subclass architecture consistent with the existing organization: characterization of the electronic state. In this spirit, we have specialized the class `internal-influences` into the modeling

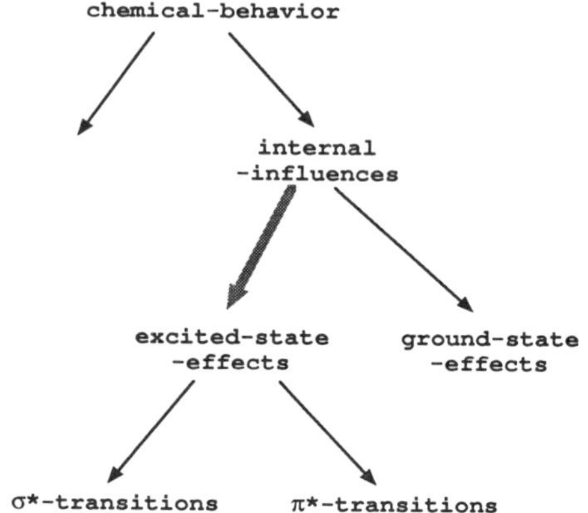

FIG. 8. Extending the chemical-behavior tree of subclasses.

subclasses ground-state-effects and excited-state-effects. We do this using the semantic relation *is-a* (Fig. 8):

excited-state-effects　　　　*is-a*　　　　internal-influence

Further, we specialize excited-state-effects according to the accompanying electronic transitions: transitions to σ^* and transition to π^*; and link these subclasses to excited-state-effects as shown below (see also Fig. 8):

σ^*-transitions　　　　*is-a*　　　　excited-state-effects
π^*-transitions　　　　*is-a*　　　　excited-state-effects

The addition of descriptive attributes and methods to each of these new modeling classes further characterizes each class. To characterize the modeling class excited-state-effects, we add attributes and methods descriptive of singled (S) and triplet (T) states. The states are refined further by energy level. For example ground states (S_0, T_0) are differentiated from their excited states (S_1, T_1) as well as from their higher states $(S_2, T_2,$ and $S_3, T_3)$. As before, we link these attributes using the semantic relationship *is-attribute-of*:

S_0 (or, S_1, S_2, S_3)　　　　*is-attribute-of*　　　　excited-state-effects
T_0 (or, T_1, T_2, T_3)　　　　*is-attribute-of*　　　　excited-state-effects

Additional attributes may also be added. The degree to which it is advantageous to add additional class descriptors (e.g., attributes, methods) is defined by the scope of the modeling efforts.

Methods characterizing excited-state-effects are linked to their associated class using the semantic relation *is-method-of*. These methods can be grouped according to whether the result is used by the attribute's compute method:

compute-S_0 (or, -S_1,-S_2,-S_3) ***is-method-of*** `excited-state-effects`
compute-T_0 (or, -T_1,-T_2,-T_3) ***is-method-of*** `excited-state-effects`

or by methods used to derive additional properties of the attribute space. The latter group may include generic methods such as *assess-energy-gap*, which computes the energy difference between two energy states (e.g., S_1 and S_2 or S_1 and T_1); or predicate functions that access feasibility. Several of these are shown below:

assess-energy-gap	***is-method-of***	`excited-state-effects`
allowed-transition-p	***is-method-of***	`excited-state-effects`
forbidden-transition-p	***is-method-of***	`excited-state-effects`
fluorescence-p	***is-method-of***	`excited-state-effects`
internal-conversion-p	***is-method-of***	`excited-state-effects`
internal-system-crossing-p	***is-method-of***	`excited-state-effects`

Together these elements give the modeling class' utility. Notice that the capability of moving electrons between energy levels requires that `atom` or `bond` have attributes that describe atomic and molecular orbitals. Given these attributes, the movement of an electron from one energy level to another can be orchestrated by methods residing in `atom`, `bond`, and `ab-initio-operator` (e.g., *ionization* or *single-electron-transfer*). In the latter case, a well-defined abstraction barrier is established to help maintain program modularity. Clearly, if the modeling effort does not require analysis of electron movement, the functionality need not be included. If it is required, LCR has the capability to expand and accommodate the modeling effort regardless of scope. A state description of `excited-state-effects`, after attribute and method addition, is shown in Table III.

Similarly, the subclasses of `excited-state-effects`, i.e., σ^*-transitions and π^*-transitions, are described by the transitions which characterize them. These include σ to σ^*, n to σ^*, n to π^*, and π to π^* transitions. The corresponding attributes and methods are linked to their respective classes as described earlier. Very briefly, they are shown below:

$\sigma \rightarrow \sigma^*$	***is-attribute-of***	σ^*-`transitions`
n $\rightarrow \sigma^*$	***is-attribute-of***	σ^*-`transitions`
...		
$\pi \rightarrow \pi^*$	***is-attribute-of***	π^*-`transitions`
n $\rightarrow \pi^*$	***is-attribute-of***	π^*-`transitions`
...		

and

compute-σ → σ-transition*	***is-method-of***	σ*-transitions
compute-n → σ*-*transition*	***is-method-of***	σ*-transitions
. . .		
compute-π → π-transition*	***is-method-of***	π*-transitions
compute-n → π*-*transition*	***is-method-of***	π*-transitions
. . .		

We can also specialize a modeling class by characterizing several of its dominant or critical properties. For selected operations, such as composite or ab initio operations, specialization can further refine the search space and afford greater (search) efficiency.

The semantic relationship ***is-characterized-as*** is used for this purpose. It provides a window for viewing the more interesting behaviors/ structures associated with a class, such as

excited-state-transition	***is-characterized-as***	$\pi \to \pi^*$
excited-state-transition	***is-characterized-as***	$n \to \pi^*$

or

electronic-state	***is-characterized-as***	excimer
electronic-state	***is-characterized-as***	exciplex

Using this semantic relation, the user may investigate alternative pathways by specifying a different characterization. Consistency is maintained by the multiple-context modeling utility of LCR.

3. Extension of the `ab-initio-operator` Hierarchical Tree

Our attention now turns to the operators: methods that perform the actual chemical transformations. Bond formation, bond cleavage, electron excitation, and electron decay are a few such examples. These operators lie at the heart of LCR.

We will illustrate the development of these operators by creating the ab initio modeling element, `covalent-bond`. We will establish in this development: (1) the communication protocol between `covalent-bond` and `ab-initio-operator`, (K_{ai}); (2) the criteria on enabling conditions; (3) how criteria can be relaxed, tightened, augmented, or supple-

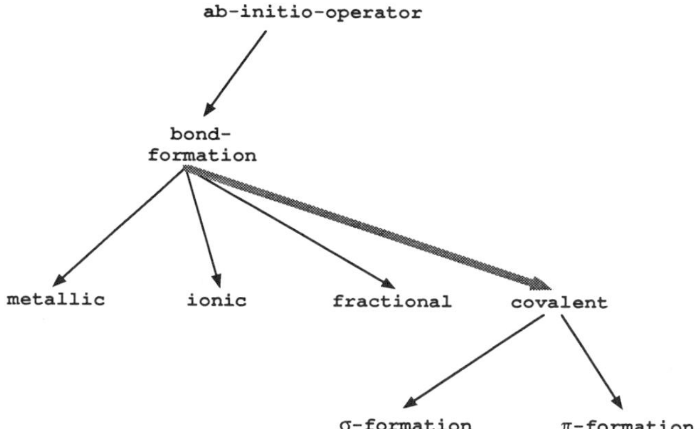

FIG. 9. Extending the ab-initio-operator tree of subclasses.

mented; and (4) how additional functionality can be added to covalent-bond (reflecting advances in our understanding of the bonding mechanism).

We initiate the development of covalent-bond by creating the modeling element and establishing the normal superclass links using the semantic relationship, *is-a* (Fig. 9):

covalent-bond	*is-a*	ab-initio-operator
σ-formation	*is-a*	covalent-bond
π-formation	*is-a*	covalent-bond

However, we will not specify the attributes of covalent-bond, σ-formation, or π-formation as we did previously, where the attributes were used for purposes of bookkeeping. How the attributes are chosen and ultimately used depends on the character and generality of the supporting encoded methods.

To effect a transformation, covalent-bond requires the use of specific methods. In particular, we need to establish generic methods that are responsible for primary transformations. The dominant method of covalent-bond is *make-covalent-bond. Make-covalent-bond* establishes a covalent-bond between two reaction centers when enabling conditions are achieved. It also performs all the necessary bookkeeping to ensure the bond is recognizable by the instance of abc. Expressed in pseudocode convention, these methods take the following form:

```
METHOD: make-covalent-bond
                input
                initialize
                if   feasible-p
                     then enable-transformation
                     return
                end
                return
```

where the method *enable-transformation* performs the transformation and the method, *feasible-p*, represents the enabling conditions, $\langle K_{ai}$-enabling-cond$\rangle$, of the transformation.

The syntax for the definition of the subclass was given in Section II.B and is reproduced by the following seven (7) lines in Backus–Naur form (BNF):

1. $\langle K_{ai} \rangle ::= \langle K_{ai}$-operator$\rangle \langle K_{ai}$-input$\rangle \langle K_{ai}$-enabling-cond$\rangle \langle$output$\rangle$
2. $\langle K_{ai}$-operator$\rangle ::=$ Bond formation|Bond Cleavage|Ionization| Single Electron Transfer|Electron Excitation|Electron Decay
3. $\langle K_{ai}$-input$\rangle ::= \{[\langle$reaction-center$\rangle]^+\}$
4. $\langle$reaction-center$\rangle ::= \{[\langle$atom$\rangle]^+\}$
5. $\langle K_{ai}$-enabling-cond$\rangle ::= \{[\langle K_{ai}$-condition$\rangle]^+\}$
6. $\langle K_{ai}$-condition$\rangle ::= \{[$physicochemical-requirements$]^+\}$
7. $\langle$physical-requirements$\rangle ::=$ (IS-ABSTRACTED-BY $\langle$physicochemical-requirements$\rangle$)

Line 1 indicates that the definition of K_{ai} depends on the following four modeling objects: $\langle K_{ai}$-operator$\rangle$, $\langle K_{ai}$-input$\rangle$, $\langle K_{ai}$-enabling-cond$\rangle$, and $\langle$output$\rangle$. In turn, $\langle K_{ai}$-operator$\rangle$ is defined by a set of attributes (line 2), and $\langle K_{ai}$-input$\rangle$ is defined by a list of $\langle$reaction-centers$\rangle$ (see line 3) with each element of the list defined as an instance of the basic modeling element, `atom`.

Continuing with our task, we recognize that `covalent-bond` requires the specification of $\langle K_{ai}$-enabling-cond$\rangle$, $\langle K_{ai}$-input$\rangle$, and $\langle$output$\rangle$. We will focus on $\langle$output$\rangle$ and $\langle K_{ai}$-enabling-cond$\rangle$ since the specification of $\langle K_{ai}$-input$\rangle$ is transparent (i.e., a list of one or more reaction centers).

The $\langle$output$\rangle$ of `covalent-bond` is an object that represents a covalent bond between two reaction centers. This object (i.e., instance of `covalent-bond`) is not a single (independent) entity. It knows membership and has pointers that connect it to the new `abc`, of which it is a part. The new `abc` recognizes it as a covalent bond. The situation of a nonnil $\langle$output$\rangle$ being returned is dependent on the properties of the reaction centers and level of sophistication encoded into $\langle K_{ai}$-enabling-cond$\rangle$.

The new `abc` recognizes it as a covalent bond. The situation of a nonnil ⟨output⟩ being returned is dependent on the properties of the reaction centers and level of sophistication encoded into ⟨K_{ai}-enabling-cond⟩.

Specific criteria must be satisfied for a covalent bond to be formed between two atoms. These criteria are based on the physical laws of nature and tempered by our knowledge of the bonding system. They are encoded into the procedure ⟨K_{ai}-enabling-cond⟩.

The general algorithm for ⟨K_{ai}-enabling-cond⟩ is

```
⟨K_ai-enabling-cond⟩:
    input
    initialize
    if  feasible-p
        then true
        return
    end
    return
```

where *feasible-p* represents a list of predicates procedures specifying the criteria for feasibility. The pseudocode procedural form of ⟨K_{ai}-enabling-cond⟩ for the modeling class, `covalent-bond`, is given below:

```
⟨covalent-bond-enabling-cond⟩
    input: reaction-center-pair
    initialize
            reaction-center-1  ← first element of reaction-center-pair
            reaction-center-2  ← last element of reaction-center-pair
    if  favorable-interatomic-distance-p and
        potentially-stable-bond-formation-p and
        available-bonding-electron-p and
        proper-orbital-symmetry-p and
        conservation-laws-upheld-p
        then true
        return
    end
    return
```

The feasibility criteria, following the "`if`" construct, are implemented as predicate methods. They perform the following evaluations: the method *favorable-interatomic-distance*-p assesses the interatomic distances between the two reaction centers to determine whether bond formation is likely; the method *potentially-stable-bond-formation*-p verifies that the steric repulsion between the reactions centers is less than the potential energy of the system and that the entropic term for bond formation is less than the enthalpy term for bond formation; *available-bonding-electron*-p checks the availability of bonding electrons; *proper-orbital-symmetry*-p determines whether the bonding orbitals are in phase; and *conservation-law-upheld*-p validates whether mass, energy, and momentum have been conserved. These methods are associated with `covalent-bond` using the semantic relationship ***is-method-of***, as we have shown previously, specifically:

favorable-interatomic-distance-p	***is-method-of***	`covalent-bond`
potentially-stable-bond-formation-p	***is-method-of***	`covalent-bond`
available-bonding-electron-p	***is-method-of***	`covalent-bond`

Notice that the method *conservation-laws-upheld*-p has not been associated with `covalent-bond`. This method has been associated with K_{ai}, allowing it to be accessed by all subclass modeling elements of K_{ai}.

We can also specialize these methods according to physical requirements and chemical requirements. This provides an abstraction barrier between what is true (i.e., the physical laws) and what is believed to be true (i.e., current theory). For example, suppose that the methods *favorable-interatomic-distance*-p and *potentially-stable-bond-formation*-p are physical requirements whereas the methods *available-bonding-electron*-p and *proper-orbital-symmetry*-p are chemical requirements. We represent this to `covalent-bond` by associating the top-level methods (i.e., methods associated with K_{ai}), chemical requirements, and physical requirements, to `covalent-bond` directly (i.e., we override the method inherited by the mother model class). This is accomplished using the semantic relationship **is-method-of**, as was demonstrated earlier:

physical-requirement	***is-method-of***	`covalent-bond`
chemical-requirement	***is-method-of***	`covalent-bond`

We then tailor these methods by associating the methods of `covalent-bond` to them. This is achieved using the semantic relationship ***is-composed-of***:

physical-requirement ***is-composed-of*** (favorable-interatomic-distance-p,
                                       potentially-stable-bond-formation-p)
chemical-requirement ***is-composed-of*** (available-bonding-electron-p,
                                       proper-orbital-symmetry-p)

4. *Extension of the* `composite-operator` *Hierarchical Tree*

Suppose now that we wish to add an additional operator to the model class `composite-operator`, K. For example, consider the addition of radical-disproportionation, a transformation that captures transformations among radicals:

$$\cdot C_2H_5 + \cdot C_2H_5 \longrightarrow C_2H_4 + C_3H_8$$

$$\longrightarrow C_2H_6 + C_3H_8.$$

As before, the addition of the subclass, $K_{radical-disproportionation}$, to the model class hierarchy requires that links be established between `composite-operator` and `bimolecular-operator, bimolecular-operator` and `radical-operator`, and `radical-operator` and `radical-disproportionation`:

bimolecular-operator	*is-a*	composite-operator
radical-operator	*is-a*	bimolecular-operator
radical-disproportionation	*is-a*	radical-operator

Like K_{ai}, the modeling element K performs transformations. This necessitates that procedures be developed and associated with K. Unlike K_{ai}, the procedures of K are based on known transformations. To identify the procedures required of $K_{radical-disproportionation}$, we return to the BNF syntax presented in Section II.D and reproduce the following links:

1. ⟨composite-operator⟩::= ⟨inputs⟩⟨enabling-conditions⟩
    ⟨ab-initio-operator⟩⟨output⟩
    [⟨composite-operator⟩]
2. ⟨input⟩::= ⟨react-cond⟩⟨abc⟩⟨cb⟩.
3. ⟨output⟩::= ⟨reaction⟩,
4. ⟨abc⟩::= {[(instance-of abc)]⁺}.
5. ⟨cb⟩::= {[(instance-of cb)]⁺}.
6. ⟨reaction-cond⟩::= {(instance-of reaction-conditions)}.
7. ⟨enabling-conditions⟩::= {[⟨conditionals⟩]⁺}.
8. ⟨conditionals⟩::= [[[(instance-of user preferences)]⁺
    {[(instance-of physical requirements)]⁺}]].

The syntax of K identifies the elements of any subclass modeling element (e.g., $K_{radical-disproportionation}$). These are elements ⟨inputs⟩ ⟨enabling-conditions⟩ ⟨ab-initio-operator⟩ ⟨output⟩. Each of these is defined by simpler entities, as shown above. (The decomposition of ⟨ab-initio-operator⟩ was given in the previous section.)

In addition, we have seen that a composite operator constructs new reactions through the execution of three methods, K_f, $K_{\text{get-sites}}$, and K_t. In the context of radical disproportionation reactions:

$$R_1 + R_2 \rightarrow R'_1 + R'_2,$$

these three procedures do the following:

1. K_f evaluates the feasibility of radical disproportionation (e.g., R_1 and R_2 are both radicals; energy in the reaction environment is less than that required to cleave the newly formed bond; relative stability of products, etc.)
2. $K_{\text{get-sites}}$ identifies potential reaction sites in R_1 and R_2.
3. K_t supplies a sequence of ab initio operators and applies these operators to the reaction sites identified by $K_{\text{get-sites}}$ to obtain the desired transformation.

The feasibility predicate, K_f, for $K_{\text{radical-disproportionation}}$ is similar in spirit to $\langle K_{\text{ai-enabling-cond}} \rangle$ with the exception that the user can encode personal preferences. For example, suppose for safety reasons that the user's interest is focused on a substituted product. This preference could be encoded into K_f by passing in a predicate test, e.g., *substituted-p*, that would be applied to reaction-centers.

METHOD: K_f
    input: substrate-list
    when substrates are radicals
        collect radical centers into potential-reaction-centers
        evaluate potential-bonds between potential-reaction-centers
            for each bond in potential-bonds evaluate bond-energy
            when bond is stable in reaction-environment
            and user-preferences satisfied
                collect potential-reaction-center-pair
                    into potential-reaction sites
                return potential-reaction-sites
            return
        return
    end
    return

Notice the usage of inputs (i.e., $\langle \text{abc} \rangle$, $\langle \text{cb} \rangle$, and $\langle \text{reaction-condition} \rangle$) in K_f. Substrates and substrate-list are sets of $\langle \text{abc} \rangle$'s; radicals are identified using $\langle \text{cb} \rangle$; bond stability is assessed in the context of $\langle \text{reaction-condition} \rangle$ (i.e., the values of the instance of `reaction-environment`). Additionally note that multiple products may be found since there may be several favorable sites; potential-reaction-sites are collected on the basis of thermodynamic stability relative to the reaction-environment.

The method $K_{\text{get-sites}}$ is responsible for producing products, given a set of potential reaction sites. It applies K_t to potential reaction center combinations and assesses the relative stabilities of the products. The procedure for $K_{\text{get-sites}}$ is given below:

> METHOD: $K_{\text{get-sites}}$
> **input:** potential-reaction-sites
> **for** each reaction-center-pair in potential-reaction-sites
>   apply K_t to reaction-center-pair
>   **collect** potential-structure-pair into potential-structures
>   **return** potential-structures
> **sort** potential-structure-pair by stability
> **when** stability-acceptable-p of potential-structure-pair
>   **collect** potential-structure-pair into product-pairs
>   **return** product-pairs
> **end**
> **return**

The call to K_t enables the desired transformation. This transformation (encoded below by K_t) begins by cleaving the bond connecting a mobile moiety (e.g., H) to a parent atom adjacent to reaction-center-1. Products are then produced by creating new bonds between these centers using generic methods (e.g., *make-covalent-bond* and *cleave-bond*) to achieve bond formation and bond cleavage. K_t uses these generic methods to access knowledge contained in $\mathbf{K}_{\text{ai}}$.

> METHOD: K_t
> **input:** reaction-center-pair
> **initialize**   reaction-center ← first element of reaction-center-pair
>   reaction-center ← second element of reaction-center-pair
> cleave-bond mobile moiety adjacent to reaction-center-1
> biradical ← reaction-center-1 structure
> product-1 ← apply make-covalent-bond to mobile moiety
>   and reaction-center-2
> product-2 ← apply make-covalent-bond to biradical
> **end**
> **return**

By changing the transformation code and by adding selector functions we can enhance K_t to enable additional disproportionations. These may include transformations depicting disproportionation of such radicals as R_3CO_2 and R_2HCO_2.

Finally, we associate the procedures K_f, K_t, and $K_{\text{get-sites}}$ to the modeling class $\mathbf{K}_{\textbf{radical-disproportionation}}$ using the semantic relation **is-method-of**:

K_f	*is-method-of*	radical-disproportionation
$K_{\text{get-sites}}$	*is-method-of*	radical-disproportionation
K_{ft}	*is-method-of*	radical-disproportionation

We also establish a second (redundant) link using the semantic relationship *is-described-by*. In addition, LCR establishes automatically the following three semantic links, which are the symmetrical relationships of the preceding three:

radical-disproportionation	*is-described-by*	K_t
radical-disproportionation	*is-described-by*	$K_{\text{get-sites}}$
radical-disproportionation	*is-described-by*	K_f

Now, the modeling class, $\mathbf{K}_{\textbf{radical-disproportionation}}$, is fully recognized and usable by **K**. An evaluation of $\mathbf{K}_{\textbf{radical-disproportionation}}$ on R_1 and R_2 will return an instance of the class `reaction` or set of instances (reflecting multiple favored products). Each `reaction` constructed will contain values for the attributes (e.g., reactants, products, stoichiometry, reaction-environment, enabling conditions) as specified by the modeling element `reaction` (see Section II.A).

B. The "Model Class Decomposition Digraph" (MCDD)

As the examples of the previous section have indicated, the modeling objects in LCR's hierarchies are built from simpler ones by linking them into composite entities. The mechanism for the linkage is the declaration of the semantic relationship between two modeling entities. Thus, starting from one of the nine basic modeling classes (the mother-class), we can create a modeling subclass and endow it with new specialized attributes and methods. Each new attribute could be modeled as an instance of an existing or a new modeling class. This mechanism can be applied recursively until desired representation has been achieved. The assumptions that guide the subclassing of a new modeling element from an existing one, and the specification of its attributes form the scope of the *modeling context*.

Let a modeling object be represented by a node, and let an *edge* represent its semantic link to another modeling object (i.e., another node). Then, the gradual and modular definition of a model can be represented by an evolving directed graph, which is called *model class decomposition digraph* (MCDD). The MCDD represents the net result of the model class decomposition taking into account all the components created at that time. The basic idea is simple and is schematically depicted in Fig. 10. At state 1, the model root A has only one constituent entity called B. At state 2, the entity C has been incorporated into the substructure of B; but C is not a simple entity, it is a composite object itself (e.g., a covalent bond *is-described-by* a set of mathematical components). Consequently, its whole structure is pasted into B's structure. Stage 3 shows that B's substructure has been further specified by incorporating an additional composite object called D. Following with B's definition, stages 4–6 show that a new constitutive part, named E, is being gradually created. This presents a contrast with the previous incorporation of entities C and D; they had predefined substructures. This process will continue until A's structure has been completely specified.

The modularity of the model class definition process allows the modeler to make use of preexisting classes or to gradually define new ones. This feature combined with the communicative properties of the semantic relations makes possible the modification or upgrading of a model class with the minimum of effort.

MCDDs are made up of nodes and edges. The digraph's *nodes* have the particularity of being classes; each one of them corresponds to a constituent entity of the composite object that the model class is representing at the time. Pairs of nodes in a digraph are linked by *edges*. In a MCDD edges describe the semantic relations that victualed the entities represented by the nodes. Any edge of a MCDD can be associated with any of the following specification semantic relations: *is-composed-of*, *is-connected-by*, and *is-described-by*. For example,

A MCDD is a digraph $G = (X, U_R)$, where

(a) $X = \{x_1, x_2, \ldots, x_n\}$ is the set of nodes representing classes of modeling objects.
(b) $U_R = \{(x_i, x_j) \in X \times X / x_i R x_j\}$ is the set of edges connecting nodes x_i and x_j through the three specification oriented semantic relationships, mentioned above.

A MCDD can be characterized as an acyclic digraph whose starting node, S, has a nil predecessor [i.e., $\Gamma^-(S) = 0$] labeled with the name of the system the model is intended to represent. The successor nodes of S, $\Gamma^+(S)$, are labeled with the names of the first-level components of the model. Each of these first-level constituent entities may itself be decomposed into second-level entities, which will be represented by new successor nodes in the digraph, i.e., $\Gamma^+[\Gamma^+(S)]$. Thus, the model of any node, x_i, is the

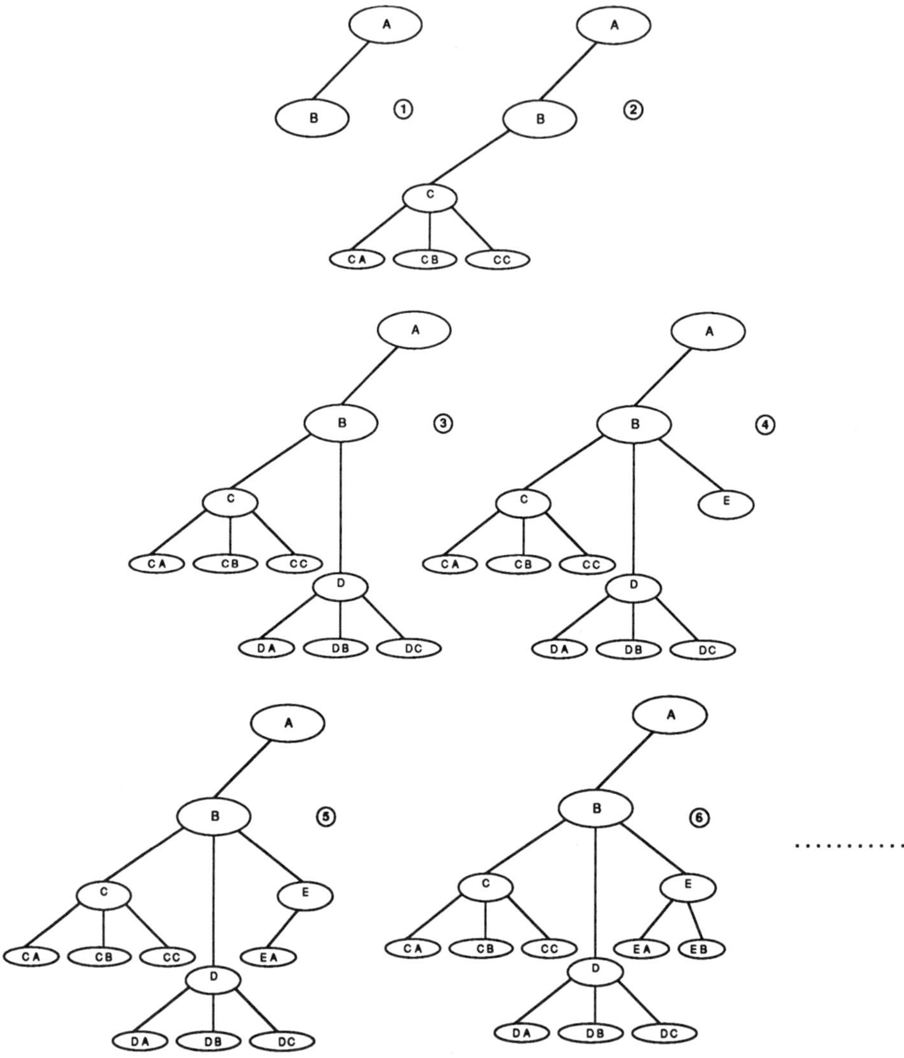

FIG. 10. The evolution of an MCDD's structure. (Reprinted from Stephanopoulos, G., Henning, G., and Leone, H. MODEL.LA A modeling language for process engineering. Part I, *Comp. Chem. Eng.* **14**, Page 813, Copyright 1990, with kind permission from Elsevier Science Ltd, The Boulevard, Langford Lane, Kidlington 0X5 1GB, UK.)

subgraph of the MCDD generated by x_i and its descendants of $(\Gamma^+(x_i) \cup \{x_i\})$;

$$\text{MCDD}(x_i) = \left(\{\hat{\Gamma}^+_{\text{MCDD}}(x_i) \cup \{x_i\}\}, U_R(x_i)\right),$$

where $\hat{\Gamma}^+_{\text{MCDD}}(x_i)$ is the set of descendants of x_i in MCDD and $U_R(x_i)$ is given by

$$U_R(x_i) = \left\{(x_a, x_b) \in \{\hat{\Gamma}^+_{\text{MCDD}}(x_i) \cup \{x_i\}\} X \{\hat{\Gamma}^+_{\text{MCDD}}(x_i) \cup \{x_i\}\} / x_a R x_b\right\}.$$

Formally, a MCDD represents a hierarchical decomposition of a model into constituent entities. Since this decomposition must ultimately terminate, a MCDD should not have directed circuits. Thus, MCDDs are digraphs that have the following properties:

Strict hierarchy. No component node can appear in the path that has as initial end-point one of its constituent entity nodes. Consequently, no component can have a decomposition which eventually contains a constituent entity of the same type.

Uniformity. Any two nodes with the same label have attached to them the same constituent entities. This property implies that any two components of the same type have the same isomorphic decomposition structures.

These properties are particularly important because they guarantee the development of a reasonable model and easy access to its different views, components, etc. An additional property, resulting from the uniformity axiom is that all MCDDs have the capability to specify the digraphs that are associated with the instances of the model class. The instantiation of a MCDD generates a *model decomposition digraph* (MCD). MCDs, like MCDDs, do not have directed circuits. They are finite graphs, obtained as a result of the instantiation of another finite graph.

C. Generation and Representation of Reaction Pathways

In the previous sections we discussed how we can use LCR's mechanisms to add new modeling elements, or to generate representations from existing modeling elements. In this section we will show how to use LCR for the construction of reaction pathways. We will employ examples from the domain of the *free-radical chemistry*, such as oxidation of butane.

LCR provides a method for generating all possible reactions and identifying the resulting pathways, given a set of substrates and a reaction environment. The keyword arguments (i.e., arguments with a colon prefix) accepted by this method are given below:

(find-all-pathways   :substrates   :operators   :override-environment
                     :initiator-p)

The method tests for an override reaction environment allowing the user to investigate the influence of various environments without manually resetting the environment attribute of each substrate. [Recall that an abc (i.e., a substrate) "knows" its environment.] The user may also identify whether an initiator is present or believed to be present. This allows the user to investigate the effect of various reaction trajectories without knowing specifics concerning the initiator mechanism. The keyword *:operations* allows the user to specify the types of transformation to be used; the default value is `composite-operator`. The user may specify other operators as well. This focuses the elucidation of pathways. For example, by supplying $K_{free-radical}$ to the keyword argument the user can investigate pathways of free-radical origin.

Alternatively, the user may wish to investigate all theoretically feasible pathways subject to a set of preferences. This is accomplished by supplying *:operators* with K^*, a modified composite operator, which has encoded the user preferences. Recall that K^* as an instance of the `composite-operator`, generates reactions by executing the methods K_t, K_f, and $K_{get-sites}$. K^* contains no encoded transformations within K_t but rather applies the complete set of instances of ab-initio-operator (i.e., K_{ai}) directly. The K_f procedure reflects user preferences. This allows the user to investigate pathways having prespecified features ($\Delta G < 10$ kcal/mol, stereocenters, etc.). Generation of these pathways is accomplished as follows:

(find-all-pathways   *:substrates* (`butane oxygen`)   *:operators* K^*
                     :initiator-p nil)

Similarly, if there are no preferences, theoretically feasible reactions can be generated by calling find-all-pathways with *:operators* $K_{ab-initio}$ directly:

(find-all-pathways   *:substrates* (`butane oxygen`)   *:operators* $K_{ab-initio}$
                     :initiator-p nil)

This functionality allows the user to control the scope for the identification of potential pathways. Pathways connecting two states are identified using the method *find-all-pathways-connecting*. Its general form is

(find-all-pathways-connecting *:substrates   :desired-products*
                                    :operators   :initiator-p
                                    :strategy)

where substrates may be a class allowing the user to investigate pathways leading from a desired product to a class or classes of substrates. Alternatively, *:substrates* can be left unspecified. Once a control (e.g., synthesis) strategy has been specified, the method, *find-all-connecting-pathways*, chains on K_f and K_t methods of K, using the semantic relation **is-described-by**, and appropriate selector functions. The current implementation of LCR allows chaining only in the forward direction (i.e., synthetic direction) when $K_{ab-initio}$ or K^* is supplied to *:operators*.

Let us return now to the oxidation of butane. We will illustrate the generation of feasible pathways using K^*. In this example, K^* was restricted to thermodynamically viable free-radical pathways; products were restricted to their homologous series. The generation call was

(find-all-pathways   *:substrates* (butane oxygen)   *:operators* K^*
                        :initiator true)

Figure 11a presents several of the pathways identified when *s*-BuOH is formed during the initiation process. The attributes of the initiation reaction, `initiation-1`, for the oxidation of butane are presented in Fig. 12a, where objects are denoted by #⟨object-name⟩. Note that each of these objects can in turn be expanded or disaggregated using the semantic relationships. Figure 12b illustrates one of the many pathways, pathway-1, generated during the oxidation of butane into *s*-BuOH and AcEt. The expansion of the object, #⟨initiation-1⟩, an element in the value of the attribute, "composing-reactions," results in the description shown in Fig. 12a. The user has access to this information at every step of the synthetic process using the semantic relationships provided by LCR (see Section III.E).

By changing the values of the object, `reaction-environment`, one can generate new trajectories (denoted in Fig. 11b by dashed lines), corresponding to a higher-energy environment. Figure 11c presents the reaction pathways resulting from yet another `reaction-environment`. The various pathways stemming from these environments are combined in Fig. 13. The assumption set and conditions specifying each environment are managed by the context generation utility of LCR, discussed in the next section.

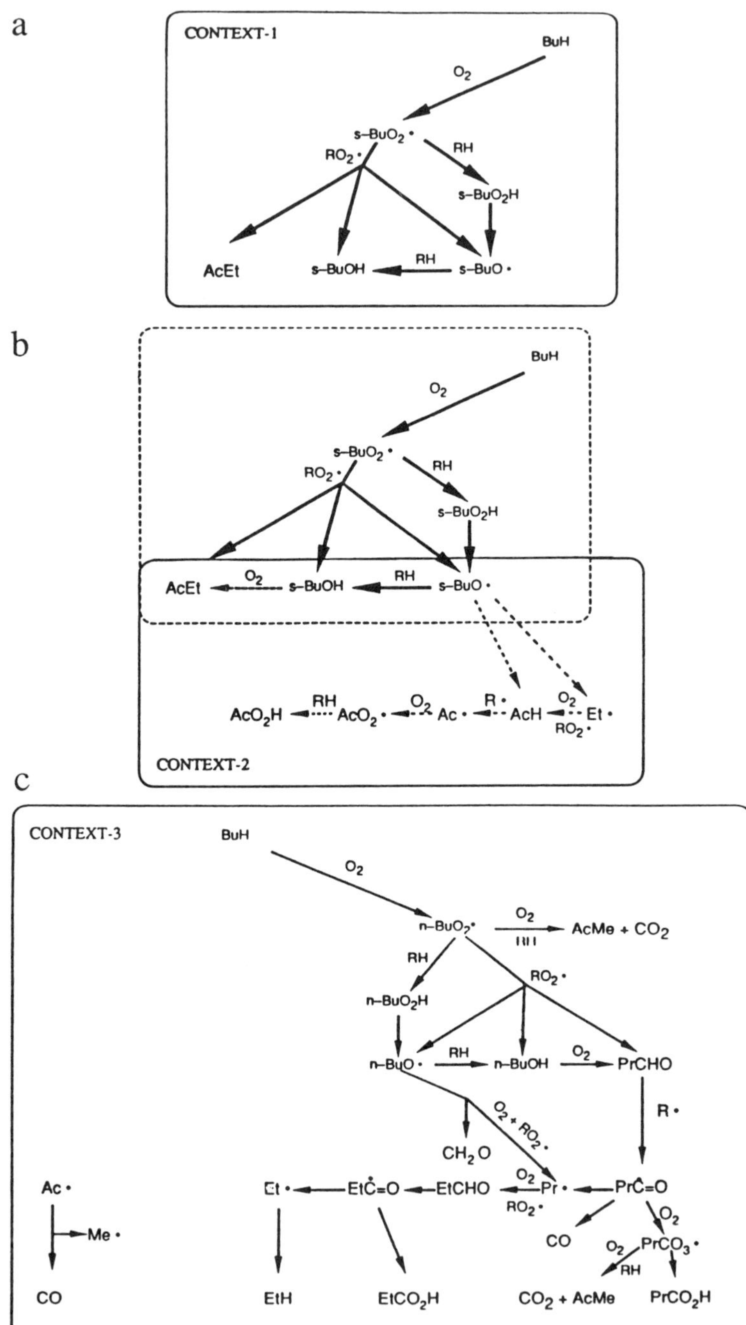

FIG. 11. Three distinct abstractions in describing the oxidation of butane.

a

Reaction Object

identifier:	*unbound*
name:	initiation-1
reactants:	(#<$C_4H_{10}O_2$>)
products:	(#<C_4H_9O> #<HO>)
stoichiometry:	((#<$C_4H_{10}O_2$> . -1)
	(#<C_4H_9O> . +1)
	(#<HO> . +1))
reaction-environment:	#<reaction-environment-1>
enabling-conditions:	K_f
composing-transformations:	K_t
composing-reactions:	*unbound*
rate-expression:	#<rate-expression-1>
equilibrium-constant:	#<equilibrium-constant-1>
context:	#<context-1>

b

Pathway Object

identifier	*unbound*
name	pathway-1
reactants	(# <C_4H_{10}> #<O_2>)
products	(#<$CH_3COCH_2CH_3$>)
stoichiometry	unbound
competing-pathways	(<pathway-2>
	#<pathway-3>)
composing-reactions	(#<initiation-1>
	#<initiation-2>
	#<abstraction-1>
	#<disproportionation-1> ...)
global-rate-expression	#<composite-rate-exp-1>
global-equilibrium-constant	*unbound*

Fig. 12. The description of (a) a reaction object and (b) a pathway object during the oxidation of butane.

By indexing and storing new reactions in their entirety, i.e., by storing the complete objects making up reactions, chemical structures, and chemical conditions, whether they were generated by K, or discovered through the use of K^* or K_{ai}, we have given LCR the ability to learn. Since a context is associated with each reaction, retrieval of specific reactions (from the ever-expanding library of chemical reactions) and incorporation in the evolving generation of pathways can be focused to the desired context, in order to suit specific needs.

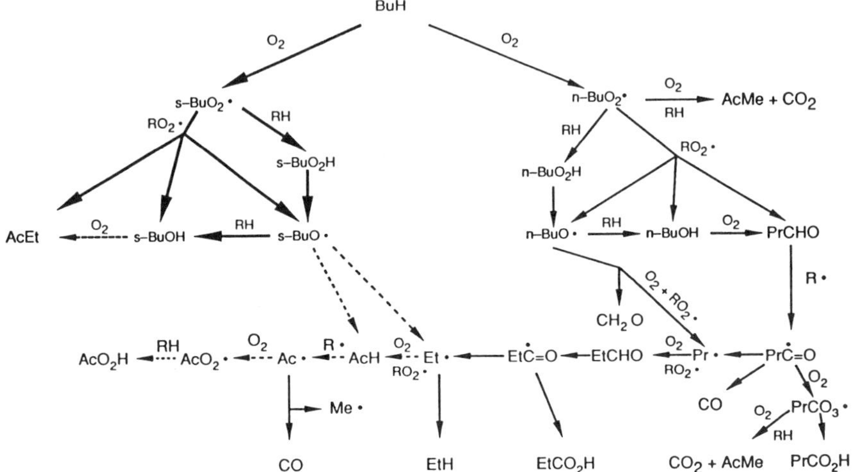

FIG. 13. Linking distinct contexts used for the representation of butane oxidation.

D. Creation of Contextual Reaction Models

The generation and representation of molecular structures, reaction, and pathways is always done within the scope of some "context." In Fig. 11 we see three (3) distinct networks of reactions, occurring during the oxidation of butane, and corresponding to three (3) distinct "contexts" of operating conditions. In general, the representation of any object within LCR depends on a set of assumptions that can be classified into the following three categories:

1. *Assumptions on structural components.* They declare component/ subcomponent relations or level combinations of subcomponents: Chemical A has been conjectured as a `cyclic-aliphatic` molecule; bond between atoms A_1 and A_2 is assumed to be `covalent-bond`; pathways P is assumed to be a serial concatenation of pathways, P_1, P_2, and P_3.
2. *Assumptions on behavior or functionality.* These are relations defining: reactivity of various reaction centers in a molecule, heats of reaction, bond strength between two atoms, etc.
3. *Assumptions on the values of variables*, describing attributes of molecular structures or/ and reactions.

These assumptions are expressed in terms of generalized constraints, which represent Boolean, qualitative, semiquantitative, or quantitative relationships among the modeling elements. The set of generalized constraints forms the *"context"* within which a model is valid, and are viewed as hard constraints taking on a Boolean character; thus, they are either satisfied or violated (never satisfied to a certain degree).

Let us assume that a model X developed within CONTEXT-1, is to be modified. CONTEXT-2 is created as a "child" of CONTEXT-1, that will be referred to as the "parent-context." In principle, CONTEXT-2 "inherits" all the assumptions that are valid in CONTEXT-1. Then, in order to account for the modifications of model X it is possible to change the inherited assumptions with additions and deletions. The changes are covered by the following rule:

All assumptions in CONTEXT-1 are "inherited" by CONTEXT-2 unless overridden in CONTEXT-2.

Generalizing this rule, we can say that:

The assumptions true in a given context are all the assumptions that are true in the parent-context, minus the assumptions that have been specifically deleted in it, plus any assumption that has been specifically added in it.

Following with this hypothetical case, let us assume that a different representation of X is required. Now, two options are given to us: (1) if the new representation is seen as an alternative to the one already introduced in CONTEXT-2, a new context called CONTEXT-3, should be created as a child of CONTEXT-1; (2) If the modification is intended to change model X built in CONTEXT-2, CONTEXT-3 should be created as a child of CONTEXT-2. The first choice is analogous to the situation found in butane oxidation. The assumptions corresponding to the free-radical oxidation of butane are encapsulated in CONTEXT-1 (Fig. 11a). Given that two distinct global pathways have to be analyzed, CONTEXT-1 gives rise to CONTEXT-2 and CONTEXT-3, which encapsulate the assumptions that generate the representations of Fig. 11b and 11c, respectively. CONTEXT-2 and CONTEXT-3 siblings both have information in common with CONTEXT-1, but there is no direct relation between them. The second situation corresponds to many multistep syntheses. During the synthesis of a complex molecule the chemist makes assumptions regarding how the molecule should be constructed, and these assumptions give rise to new assumptions. For example, assumptions regarding the specification of a carbon skeleton with correct key group orientation may be kept in CONTEXT-A. This context may give rise to CONTEXT-B placing restrictions (i.e. assumptions) on the latent functionality of the skeleton. These restrictions, in turn, may give rise to CONTEXT-C, i.e., assumptions regarding the need for certain protective groups.

Generalizing, we can state that contexts are always arranged in hierarchical manner. The relations among them are established in a direct graph, generated by the context relation (CR) on the set of contexts, C. The relation CR is read as *"gives-rise-to,"* and the digraph is called MODEL ABSTRACTION TREE (MAT). It is a tree because each node representing a context has only one predecessor. A MAT is defined by

$$\text{MAT} = (C, U_R),$$

where

$$C = \{\text{CONTEXTS}\},$$

$$U_R = \{(C_i, C_j) \in C \times C / C_i CR C_j\}.$$

This tree of contexts is similar to the "class precedence list" used in hierarchical inheritance mechanisms. An example of such tree is presented in Fig. 14, where CONTEXT-1 gives rise to CONTEXT-2 and CONTEXT-3, and CONTEXT-2 gives rise to CONTEXT-4. All contexts represent different reaction pathways for the same overall reaction.

1. Communication between Contextual Models

Contexts can be related as (a) siblings (e.g., CONTEXT-2 and CONTEXT-3 in Fig. 14) or (b) "parent"–"child" (e.g., CONTEXT-1 and CONTEXT-2 in Fig. 14).

In the first case, the contexts do not need to communicate with each other, but in the second case, information must pass from the parent-context to the child-context. The operations associated with the modification of the parent-context to produce the child-context, presently available in LCR are

1. *MODIFY*: an operation applied to an individual assumption in order to alter its value.
2. *DELETE*: an operation applied to a single assumption in order to alter its value.
3. *ABSTRACT*: an operation that is applicable to the set of assumptions representing an object model. The representation of this object is abstracted to obtain a less detailed description of its structure.
4. *DISAGGREGATE*: is the opposite to ABSTRACT. It is applied to the assumption set, representing a model, but its purpose is to introduce more refinement.

Whereas MODIFY and DELETE are simple operations, ABSTRACT and DISAGGREGATE are quite complex. Whichever of the last two operations is performed during the modification of a new context, it should allow the user to bind together components that are conceptually related in the two contexts. Thus, the ABSTRACT and

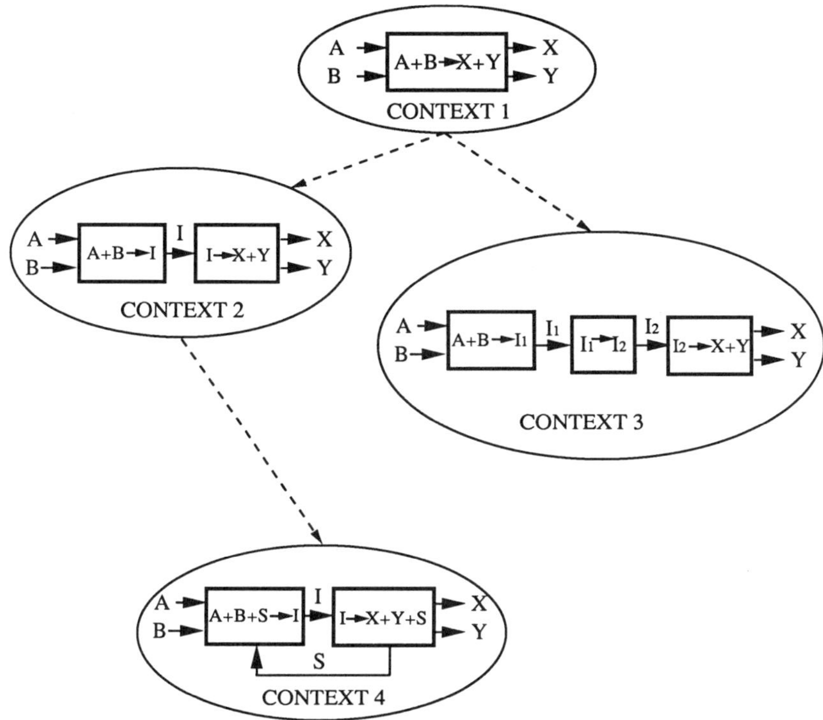

FIG. 14. A tree of contextual inheritances.

DISAGGREGATE operations should provide the framework for the communication across contexts. The symmetric operations of abstraction and disaggregation can be described by an algorithm that considers them in a very general sense, and can be summarized as follows:

```
ABSTRACTION (NEW-CONTEXT NEW-UNIT OLD-UNITS)
     ESTABLISHED-STRUCTURAL-COMPATIBILITY
     (NEW-CONTEXT NEW UNIT OLD-UNITS)
     ESTABLISH-BEHAVIORAL-COMPATIBILITY
     (NEW-CONTEXT NEW-UNIT OLD-UNITS)
END
DISAGGREGATION (NEW-CONTEXT NEW-UNITS OLD-UNIT)
     ESTABLISH-STRUCTURAL-COMPATIBILITY
     (NEW-CONTEXT NEW-UNITS OLD-UNIT)
     ESTABLISH-BEHAVIORAL-COMPATIBILITY
     (NEW-CONTEXT NEW-UNITS OLD-UNIT)
END
```

Since contexts have only one parent there is no need to specify the name of the context where the old-units are located. With the name of a new context, its parent context will be uniquely specified. The details of the above algorithms will be discussed in the following paragraphs, where we will make extensive use of the semantic links, *is-disaggregated-in* and *is-abstracted-by*.

2. *Structural Compatibility among Contextual Models*

Structural compatibility analysis operates on pathways that are located in two different contexts, bound together by a CR relation. During the disaggregation operation, a pathway that exists in the parent context is substituted by a set of pathways in the next context. The structural compatibility operation will establish the following semantic link:

P_{old} *is-disaggregated-in* $(SET.OF(P_1,...,P_n,))$ in NEW-CONTEXT

Since a parent-context may have several children, it is always necessary to specify the name of the child-context when the disaggregation occurs. For example, during the oxidation of butane, the context `butane-oxidation-reaction-environment` gives rise to the contexts `butane-oxidation-termination-1` and `butane-oxidation-termination-2`. So, the relation

```
butane-pathways is-disaggregated-in
     (SET.OF (initiation, propagation, termination))
          in butane-oxidation-termination-1
```

provides an unambiguous link and avoids any confusion with models located in the other contexts (`initiation, propagation,...,`etc.) The abstraction procedure is just the inverse operation:

$(SET.OF (P_1,...,P_n,))$ *is-abstracted-by* P_{new} in NEW-CONTEXT

Now, a set of P values are abstracted in a single pathway. Figure 15a shows an example of the abstraction and disaggregation processes between CONTEXT-i and CONTEXT-j. If CONTEXT-j is created as a child of CONTEXT-i, the structural-compatibility operation will generate the link

P *is-disaggregated-in* $(SET.OF (P_1 \, P_2))$ in CONTEXT-j

On the other hand, if CONTEXT-i is a child of CONTEXT-j, the structural compatibility operation would generate the following link:

$(SET.OF (P_1, P_2))$ *is-abstracted-by*   P   in CONTEXT-i

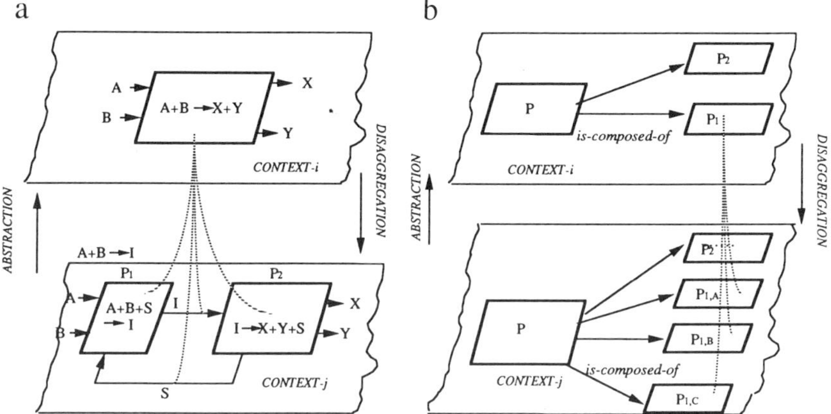

FIG. 15. Aggregation and disaggregation of representations of reacting systems.

However, there are more complex situations that need further analysis. Specifically, the case of pathways that are being disaggregated or abstracted and that themselves are not simple pathways from a structural point of view, but are also "part-of" other pathways in the parent context. What happens with these semantic links in the new context? A typical situation is depicted in Fig. 15b, where pathway P_1, which is *part-of* pathway P in CONTEXT-i, is disaggregated in pathways $P_{1,A}$, $P_{1,B}$ and $P_{1,C}$ in CONTEXT-j. The structural-compatibility procedure, apart from establishing **is-disaggregated-in** links across the contexts, should also establish **is-part-of** links between pathways $P_{1,A}$, $P_{1,B}$ and $P_{1,C}$, and pathway P.

More formally, the structural-compatibility procedure should perform the following checking operation every time a pathway P_{old} is disaggregated in a new context.

IF P_{old} *is-part-of* ?PATHWAY-X in OLD-CONTEXT
THEN ?PATHWAY-Y *is-part-of* ?PATHWAY-X in NEW-CONTEXT

where ?PATHWAY-Y is one of the pathways in which P_{old} is disaggregated, i.e., the following link exists between them:

P_{old} *is-disaggregated-in* ?PATHWAY-Y in NEW-CONTEXT

In order to avoid the generation of redundant links that the checking operation can produce due to the transitivity axiom, ?PATHWAY-X, will be restricted to be the predecessor of P_{old} that is related to P_{old} by means of a *is-composed of-* link. This condition is met by the only member of the following set

$$\{x \mid x \in A \wedge \{\{y \mid x \text{ } \textbf{\textit{is-composed-of}} \text{ } y\} \cap A\} = \phi\}$$

where

$$A = \{x \mid x \text{ } \textbf{\textit{is-composed-of}} \text{ } P_{old}\}$$

Let us now consider the inverse situation, depicted in Fig. 15b, where pathways $P_{1,A}$, $P_{1,B}$, and $P_{1,C}$, that are part of the pathway P in CONTEXT-j are abstracted in pathway P_1 in CONTEXT-i. In this case, the structural-compatibility procedure will establish **_is-abstracted-by_** links across the contexts and a **_is-composed-of_** link between pathways P and P_1 in CONTEXT-i. In order to consider situations like the one presented above, the following checking operation is also carried out every time a set of pathways $\{P_i, i = 1, \ldots, n\}$ **_is-abstracted-by_** the pathway P_{new} in a new context.

IF ?PATHWAY-X **_is-composed-of_** $\{P_i, i = 1, \ldots, n\}$ in OLD-CONTEXT
THEN ?PATHWAY-X **_is-composed-of_** P_{new} in NEW-CONTEXT

where $\{P_i, i = 1, \ldots, n\}$ **_is-abstracted-by_** P_{new} in NEW-CONTEXT.

For the same reasons expressed above, that is, to avoid the creation of redundant links, ?PATHWAY-X will be restricted to be the pathway that is a predecessor of the members of the set $\{P_i, i = 1, \ldots, n\}$, and which is also related to them by a **_is-composed-of_** link. Like before, ?PATHWAY-X is the only member of the set

$$\{x \mid x \in P \wedge \{\{y \mid x \text{ } \textbf{\textit{is-composed-of}} \text{ } y\} \cap P\} = \phi\}$$

where

$$P = \{x \mid x \text{ } \textbf{\textit{is-composed-of}} \text{ } \{P_i, i = 1, \ldots, n\}\text{old}\}$$

E. Case Study: Ethane Pyrolysis

To demonstrate the utility of LCR and the interaction between various modeling elements, semantic relationships, and supporting methods, let us first consider the pyrolysis of ethane forming principally ethylene and hydrogen. Although this example is relatively simple, it highlights the functionality of LCR and underscores various issues that require resolution for computer implementation to be successful.

1. Initialization

We begin by initializing the instance of `atom-bond-config-uration`, which represents `ethane` (Table IV). Each attribute in Table IV whose value is an object can in turn be expanded, using the semantic relationships of LCR. For example, the attribute "empirical-formula" contains a list of `ab-atom` instances, whose expansion makes accessible those properties of an atom, which are independent of its environment (Table V), and include electronegativity, valence electrons, and atom-weight. Similarly, expansion of the instances `atom` in the attribute "atoms" describes substrate-dependent properties of an atom (Table VI), which include "bonds," "hybridization," "neighbor-atoms," "neighbor-groups," etc. Similarly, Fig. 16 shows the attributes of a particular instance of `bond`, describing the characteristics of a specific bond in ethane, while Table VII shows the attributes describing the group "methyl-1".

2. Generation of Pathways

The identification of various free-radical pathways and underlying elementary reactions involved in the pyrolysis of ethane are evaluated by applying the procedure, FIND-ALL-PATHWAYS, to the substrates, and associating with that call the designated instance of `reaction-environment` and the group of composite operators to be used (e.g., K, K^*, $K_{ab\text{-}initio}$). The call to FIND-ALL-PATHWAYS is shown below:

> (FIND-ALL-PATHWAYS
> :substrates (`ethane`) :*operators* $K_{free\text{-}radical}$
> :*override-environment* `reaction-environment`)

The method FIND-ALL-PATHWAYS begins by calling the composite operators that constitute the domain of free-radical operations on the substrates specified by the keyword argument :*substrates*. These operators loop over the substrate list and evaluate properties specified by $K_{free\text{-}radical}$. Homolytic dissociation of `ethane` is initiated when FEASIBLE-P, a compound predicate method of $K_{initiation}$ ($K_{initiation}$ is an operation of $K_{free\text{-}radical}$), is evaluated on the sites identified by $K_{get\text{-}sites}$. These sites are passed to FEASIBLE-P, which searches the `reaction-environment` to establish potential radical forming processes: thermal cleavage, photochemical cleavage, and oxidation–reduction processes. Attributes describing `reaction-environment` establish thermal cleavage as a likely initiation mechanism; photochemical cleavage

TABLE IV

Attribute Values of an Instance of Ethane

identifier:	"C2H6-T0779"
name:	"Ethane"
atoms:	(#⟨ATOM C-T0764⟩ #⟨ATOM H-T0765⟩ #⟨ATOM H-T0766⟩ #⟨ATOM H-T0767⟩ #⟨ATOM C-T0768⟩ #⟨ATOM H-T0769⟩ #⟨ATOM H-T0770⟩ #⟨ATOM H-T0771⟩)
bonds:	(#⟨BOND b-T0772⟩ #⟨BOND b-T0773⟩ #⟨BOND b-T0774⟩ #⟨BOND b-T0775⟩ #⟨BOND b-T0776⟩ #⟨BOND b-T0777⟩ #⟨BOND b-T0778⟩)
empirical-formula:	((#DB-ATOM 414000575⟩.6) (#⟨DB-ATOM 414001073⟩.2))
empirical-formula-string:	"C2H6"
molecular-weight:	30.07
charge:	0.0
terminal-skeleton-atoms:	(#⟨ATOM C-T0764⟩ #⟨ATOM C-T0768⟩)
equivalent-atoms:	((#⟨ATOM C-T0768⟩ #⟨ATOM C-T0764⟩) (#⟨ATOM H-T0771⟩ #⟨ATOM H-T0765⟩ #⟨ATOM H-T0766⟩ #⟨ATOM H-T0767⟩ #⟨ATOM H-T0769⟩ #⟨ATOM H-T0770⟩))
equivalent-bonds:	((#⟨BOND b-T0775⟩) (#⟨BOND b-T0778⟩ #⟨BOND b-T0772⟩ #⟨BOND b-T0773⟩ #⟨BOND b-T0774⟩ #⟨BOND b-T0776⟩ #⟨BOND b-T0777⟩))
weakest-bond:	(#⟨BOND b-T0775⟩)
weakest-bond-strength:	82.6
weakest-bond-strength-ratio:	1.0
ordered-eq-bonds:	((#⟨BOND b-T0775⟩) (#⟨BOND b-T0778⟩ #⟨BOND b-T0772⟩ #⟨BOND b-T0773⟩ #⟨BOND b-T0774⟩ #⟨BOND b-T0776⟩ #⟨BOND b-T0777⟩))
groups:	(#⟨GROUP methyl-1⟩#⟨GROUP methyl-2⟩ #⟨GROUP terminal-sp3-methylene-1⟩ #⟨GROUP terminal-sp3-methylene-2⟩)
group-bonds:	#⟨BOND b-T0775⟩ #⟨BOND b-T0772⟩ #⟨BOND b-T0776⟩)
meta-groups:	#⟨GROUP ethyl-1⟩ #⟨GROUP ethyl-2⟩)
methyl-carbon:	NIL
primary-carbons:	(#⟨ATOM C-T0764⟩ #⟨ATOM C-T0768⟩)
secondary-carbons:	NIL
tertiary-carbons:	NIL
terminal-carbons:	(#⟨ATOM C-T0764⟩ #⟨ATOM C-T0768⟩)
backbones:	(#⟨ATOM C-T0764⟩ #⟨ATOM C-T0768⟩)
backbone-length:	2
progenitor:	NIL
environment:	#⟨reaction-environment-1 11235723⟩
⋮	⋮

TABLE V

DATA FROM AN EXTERNAL DATABASE FOR atom "CARBON"

Name:	"Carbon"
atomic-symbol:	"C"
atomic-number:	6
atom-weight:	12.001
valence:	4
row:	"1"
column	"4a"
orbitals:	*unbound*
valence-electrons:	4
electronegativity:	*unbound*

and initiation by oxidation–reduction processes are eliminated because their attribute values are unbound or nil (i.e., nontrue).

Using this knowledge, $K_{initiation}$ constructs a list of potentially-cleavable-bonds by applying the method IDENTIFY-WEAKEST-BONDS to the bond representation of each substrate contained in the substrate-list (i.e., ethane). IDENTIFY-WEAKEST-BOND returns a list of bonds,

TABLE VI

SELECT ATTRIBUTES OF AN ATOM

identifier:	"C-T0764"
old-identifier:	NIL
free-valence:	0
diradical-p:	NIL
formal-charge:	0.0
electron-withdrawing-substituent-p:	NIL
connectivity-number:	4
hybridization:	"sp3"
conjugated-p:	NIL
p-orbitals:	NIL
open-approach-p:	T
parent-molecule:	#⟨ORGANIC-MOLECULE C2H6-T0779⟩
progenitors:	NIL
parent-groups:	(#⟨GROUP methyl-1⟩ #⟨GROUP terminal-sp3-methylene-1⟩ #⟨GROUP ethyl-1⟩ #⟨GROUP ethyl-2⟩)
neighbor-atoms:	(#⟨ATOM C-T0768⟩ #⟨ATOM H-T0767⟩ #⟨ATOM H-T0766⟩ #⟨ATOM H-T0765⟩)
neighbor-groups:	(#⟨GROUP methyl-2⟩ #⟨GROUP terminal-sp3-methylene-2⟩)
bonds:	(#⟨BOND b-T0775⟩ #⟨BOND b-T0774⟩ #⟨BOND b-T0773⟩ #⟨BOND b-T0772⟩)
DB-atom:	#DB-ATOM 414001073⟩

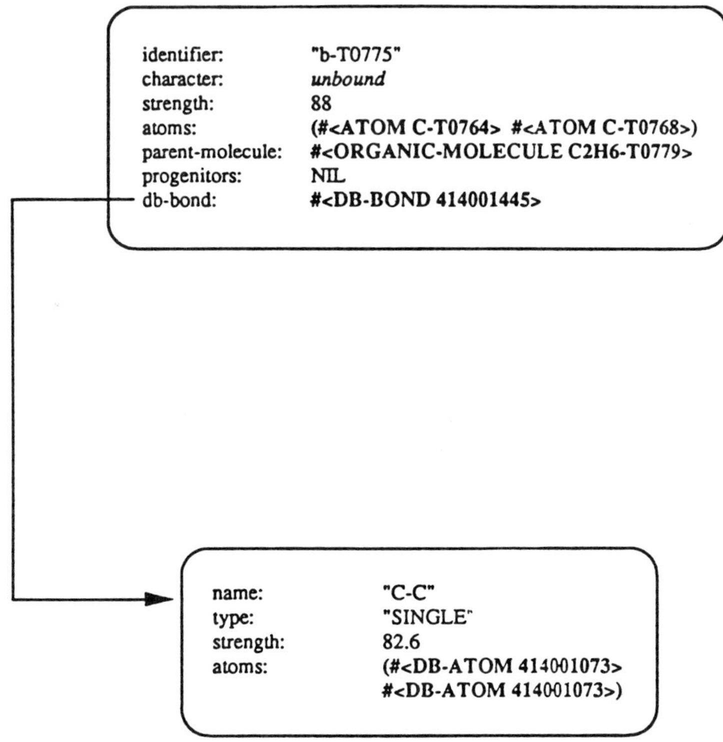

FIG. 16. Description of a bond with reference to data from an external database.

TABLE VII

SELECT ATTRIBUTES OF METHYL GROUP

identifier:	"methyl-1"
string-identifier:	"-CH3"
type:	unbound
neighbor-groups:	(#⟨GROUP methyl-2⟩
	#⟨GROUP terminal-sp3-methylene-2⟩
	#⟨GROUP ethyl-1⟩ #⟨GROUP ethyl-2⟩)
comprising-atoms:	(#⟨ATOM C-T0764⟩ #⟨ATOM H-T0767⟩
	#ATOM H-T0766⟩ #⟨ATOM H-T0765⟩)
connecting-bonds:	#⟨BOND b-T0775⟩
connecting-atoms:	#⟨ATOM C-T0768⟩
group-weight:	15.035001
parent-molecule:	#⟨ORGANIC-MOLECULE C2H6-T0779⟩
DB-group:	#⟨DB-GROUP methyl⟩

arranged in increasing strength. This is shown below:

IDENTIFY-WEAKEST-BOND applied to `bond-abstraction` $\Rightarrow$
  `((bond-1) (bond-2 bond-3 bond-4 bond-5 bond-6 bond-7))`

where

`bond-1` = carbon-carbon single bond connecting `methyl-1` to `methyl-2`
`bond-2` to `bond-4` = carbon-hydrogen single bond on `methyl-1`
`bond-3` to `bond-7` = carbon-hydrogen single bond on `methyl-2`

Bond selection is based on relative strength. A default energy difference of 10 kcal/ mol is used by IDENTIFY-WEAKEST-BOND, but this value can be changed by the user. Bonds of equivalent strength are listed together.

For each system of reactants, IDENTIFY-WEAKEST-BOND identifies the weakest substrate `bond`, and compares it with others in the system to ensure minimal global bond strength. Selected instances of `bond` are then appended to potentially-cleavable-bonds.

The weakest system `bond` is identified by applying the method IDENTIFY-ABSOLUTE-WEAKEST to the bond representation of `ethane`. The value of *weakest-bond* is the carbon–carbon single bond connecting `methyl-1` to `methyl-2`. Specifically

IDENTIFY-ABSOLUTE-WEAKEST-BOND
applied to potentially-cleavable-bonds `bond-1`

Once selected, it is associated with the global queue *weakest-bond*. Specifically

weakest-bond `bond-1`

This queue prevents $K_{initiation}$ from processing the weakest bond of individual molecules when the energy difference between those bonds is greater than the default value. It also ensures that the initiation process is focused on substrates that are most likely to cleave in `reaction-environment`. For example, in a complex molecule or set of species where one particular bond is substantially weaker than others in the system (e.g., a peroxy bond), knowledge of *weakest-bond* focuses the initiation process on that bond. Once potentially-cleavable-bonds and *weakest-bond* become known to $K_{initiation}$, the method, FEASIBLE-P evaluates the energy in `reaction-environment` and determines whether there is sufficient energy for bond cleavage. The method SUFFICIENT-THERMAL-ENERGY-P performs this evaluation. When FEASIBLE-P evaluates true, $K_{initiation}$ calls K_t on the bond selected for cleavage. Application of bond cleavage operations, operators of $K_{ab-initio}$, to the weakest bond of `ethane` produces a `microhomolysis-reaction`. Each `micro-reaction` has associated with it reactants, products, and reaction stoichiometry. In addition, the bond selected for cleavage is explicit, as are various other properties of the `reaction`. Methods of kinetic operator are used to identify these properties; they include COLLECT-PRODUCTS, COMPUTE-STOICHIOMETRY, COMPUTE-BOND-FAVORABILITY, ASSESS-THERMODYNAMIC-FAVORABILITY, and COMPUTE-EQUILIBRIUM-CONVERSION.

Information pertinent to the transformation is associated with `micro-reaction` to make it easily accessible by methods and operations external to `micro-reaction`. Information essential to the external environment includes:

```
micro-homolysis-reaction
      reactants:        (ethane)
      products:         (methyl-radical-1  methyl-radical-2)
      stoichiometry:    ((ethane.-1)
                        (methyl-radical-1.+1)
                        (methyl-radical-2.+1))
      bond-cleaved:     bond-1
      environment:      reaction-environment
```

Since multiple bonds may reside in potentially-cleavable-bonds, each possibly leading to a `micro-reaction`, $K_{initiation}$ creates a `global-reaction` that acts as a place holder for managing high-level information: bond-queue and `micro-reactions`. `Global-reactions` maintain a record of this information and the association of a `micro-reaction` corresponding to a particular bond cleavage:

```
global-homolysis-reaction
    bonds-to-be-cleaved: (bond-1 ... bond-n)
    weakest-bond: (bond-1)
    micro-reaction: (micro-homolysis-1 ... micro-homolysis-n)
```

K_t, the function responsible for performing a particular transformation, calls methods contained in $K_{ab-initio}$. Principal methods invoked by K_t to effect homolytic bond cleavage of `ethane` are GENERAL-SCISSION and CLEAVE-BOND. The methods often utilize helping functions, contained in various modeling elements (e.g., IDENTIFY-BOND-TYPE, HOMOLYTIC-P, IDENTIFY-CHARGE-DISTRIBUTION), to assist the transformation process. After a molecular bond has been selected for cleavage, it is passed to GENERAL-SCISSION by $K_{ab-initio}$. Operations composing GENERAL-SCISSION identify the bonds parent-structure, perform the cleaving function, and manage fragments resulting from the cleavage process.

Fragment management is facilitated by ABSTRACT-GROUPING and ABSTRACT-ATOM-GROUPING. These methods partition in `abc`, given a set of starting points (e.g., atoms). Normally, these points are specified by the endpoints (e.g., connecting atoms) of the reaction center identified by $K_{\text{get-sites}}$. In `ethane` disassociation, the reaction center endpoints are the two carbon atoms associated with the carbon–carbon single bond. As a consequence, in the application of CLEAVE-BOND to `weakest-bond` (i.e., the element specified by *weakest bond**), two `abc`'s are identified. These are grouped by ABSTRACT-ATOM-GROUPING and become radicals (i.e., `methyl-radical-1` and `methyl-radical-2` on instantiation by `abc`).

The (`abc`) instantiation process evaluates, updates, and classifies the `abc` as an instance of radical. During instantiation, `abc` invokes MAP-OLD-ATOM-TO-NEW-ATOM, a method that maintains pointers within the chemical system so that atoms constituting a substrate know where they came from. This is accomplished using attribute's identifier and old-identifier. During the dissociation of `ethane` the value of `carbon-1` identifier is "C-T0764," whereas its old-identifier has value nil, reflecting external creation for `ethane` (i.e., database for user specification of the `abc`). However, the attribute value of old-identifier for `carbon-3`, the carbon making up new methyl radical, `methyl-radical-1`, reflects its progenitor `carbon-1`:

carbon-3		carbon-1	
identifier:	"C-T0790	identifier:	"C-T0764
old-identifier:	"C-T0764"	old-identifier:	nil
...	...	...	...

Since the value of old-identifier describing `carbon-3` has the same value as `carbon-1`'s identifier value, these pointers enable construction of a substrate's complete history together with the operators that enabled each transformation.

A history trace is constructed by chaining through the values of `abc` attribute progenitors. Similarly, the values of enabling-conditions, an attribute describing reaction, is traced to identify the various reactions constituting a particular pathway. These utilities, in combination, allow every attribute describing a modeling element (e.g., `abc`) to be logged and accessed, in chronological order. This schema is one way in which LCR affords multifaceted and multilevel description of a modeling element throughout its history, and allows, for example, a multilevel description of a particular chemical species and the reaction pathways in which it participates throughout its life cycle.

Once a composite operation has successfully been executed, FEASIBILITY-P returns true and ab initio operations making up the sequence of transformations contained in K_t return viable transformations; unique products identified by `reaction` are appended to the global queue denoted *potential-reactant-queue*. This queue contains a complete listing of potential reactants throughout the reaction cycle. Management of potential reactants in this manner allows the behavior of a system as it moves toward an equilibrium state, to be simulated.

Continued application of composite operations, invoked by FIND-ALL-PATHWAYS on the substrates contained in *potential-reaction-queue* identifies additional instances of `transfer`, `propagation`, and `termination reaction`. By tracing the history of these instances of `reaction`, `free-radical-reaction` (i.e., free-radical pathways), one can construct a free-radical reaction. Methods of `free-radical-reaction` provide the utility for searching through a set of chemical species, using the value of the progenitors attribute and MOLECULE-EQUIVALENT-P, to identify substrate loops within the propagation reaction network. Propagation reactions are then identified and the attribute value established.

However, with the myriad products capable of being formed in termination sequences, application of $K_{free-radical}$ to substrates in the *potential-reactant-queue* may not terminate. This potential exists because new reactants, resulting from `coupling-reactions` and `combination-reactions`, are constantly appended to *potential-reactant-queue*. For example, two `ethyl-radicals` can combine to form `butane`, and hydrogen abstraction of `butane` can form `butyl`, which may combine to form `octane`. This cycle can, in principle, continue indefinitely. To prevent such cycles, substrates are removed from *potential-reactant-queue* when they have been consumed, or when a homologous series of a reactant has been previously examined. The latter is accomplished using HOMOLOGOUS-SERIES-P, with the register function responsible for appending potential reactants to *potential-reactant-queue*.

Pathways of the overall reaction sequence are constructed using instances of `free-radical-reaction`. Information associated with the individual instances `initiation-reaction`, `transfer-reaction`, `propagation-reaction`, `branching-reaction`, and `termination-reaction` facilitates this task. With this information, `free-radical-reaction` is able to construct individual free radical pathways.

IV. MODEL.LA.: A Modeling Language for Process Engineering

MODEL.LA. is a high-level, special-purpose language, which was developed to support the modeling activities of a broad range of process engineering tasks; process design, simulation, operations planning, diagnosis, configuration of control systems, and others. In this regard, MODEL.LA. is quite distinct in both scope and capabilities from other modeling languages such as ASCEND, MODASS, and OMOLA. The technical details of MODEL.LA.'s structure and implementation can be found in Stephanopoulos *et al.* (1990a, b) and Henning and Leone (1990).

MODEL.LA. shares many common features with LCR. Both languages share a common set of semantic relationships and a common syntax. They differ in their modeling elements, which reflect the different vocabularies of chemistry and process engineering. Nevertheless, the modeling elements of each language are organized into subsets, which depict the *structural* and *behavioral* knowledge of the corresponding domains. In this section we will provide an overview of MODEL.LA.'s structure, and in Section V we will discuss the utilization of MODEL.LA. within the scope of engineering problems.

A. BASIC MODELING ELEMENTS

To account for all representational needs in process engineering, MODEL.LA uses six (6) basic classes of modeling elements, and fairly rich modeling hierarchies of subclasses, emanating from the basic modeling elements.

1. Elements Defining Structures

To represent topological structures of processing systems, MODEL.LA employs the following modeling elements:

a. Modeling-Element 1: `generic-unit`. This is used to capture a system at any level of detail. Thus, an overall plant, a plant-section, a processing unit, a part of a processing unit (e.g., tubes of a heat exchanger), a phase in a processing vessel (e.g., liquid 1 in a two-liquid phase reactor), are all represented as instances of `generic-unit` (or, its specialized subclasses, as we will see in subsequent section). Also, sensors, actuators, control loops, and information processing systems are all represented and instances of `generic-unit`. Each instance of a generic unit encapsulates its own boundary, internal structural components, modeling relationships, and associated assumptions.

b. Modeling-Element 2: `port`. These are special objects that the generic units use to pass information to each other. Thus, the boundary of a generic unit is in essence defined by the set of ports through which it communicates with the rest of the world. Flow of materials, energy, or information into or from a generic unit takes place *only* through a port.

c. Modeling-Element 3: `stream`. These objects relate the ports of connected generic units, and are the conduits through which material, energy, information, or other quantities pass from one generic unit to the next. The type of a specific stream is always the same as the type of the ports that is associated with it.

2. Elements Defining Behavior

The following three modeling elements are used to capture the behavioral characteristics of materials and processing systems.

a. Modeling-Element 4: `constraint`. This class is used to capture any piece of knowledge associated with a declarative relationship among variables and parameters, such as a logical, qualitative, or quantitative relationship. An instance of `constraint` (or, its subsidiary subclasses) contains information about the form of the relationship it represents, the terms and variables that compose it, the "meaning" and significance of the relationship, as well as the range of its applicability.

b. Modeling-Element 5: `generic-variable`. Instances of this class (or its subsidiary subclasses) constitute the building blocks for the construction of modeling relationships. They represent parameters and variables describing physical quantities, engineering terms, design or operating specs, etc. An instance of `generic-variable` encapsulates information about the physical significance, value, range of possible values, trends over time, units, and other parameters of the quantity it represents.

c. Modeling-Element 6: `modeling-scope`. This object is a list of consistent instances of `constraint`, which represent the declarative relationships that apply to all components of a model, i.e., assumptions, simplifications, conjectures expected to be true, and modeling relationships. It is clear that `modeling-scope` is a *redundant* modeling element (e.g., all information in an instance of `modeling-scope` is contained in a set of instances of `constraint`). The importance, though, of maintaining a high-level container of the context in which the model was developed is sufficient to warrant the element's elevation into a basic element of the modeling language, allowing easier and direct manipulation of the modeling context.

B. SEMANTIC RELATIONSHIPS

MODEL.LA. possesses the same set of 13 semantic relationships as LCR, with the same semantic implications:

1. *Semantic-Relationship 1: **is-attribute-of.*** Indicates that an entity is an attribute of a particular class of objects.

   "entity"   ***is-attribute-of***   `object-class`

2. *Semantic-Relationship 2: **is-method-of.*** Indicates that a particular algorithmic procedure is a method of a specific class of objects.

   PROCEDURE   ***is-method-of***   `object-class`

3. *Semantic-Relationship 3: **is-a.*** Relates a subclass to the mother class:

   object-subclass   ***is-a***   object-class

4. *Semantic-Relationship 4: **is-a-member-of.*** Relates an instance to the class that generated it:

   instance-of-class-x ***is-a-member-of***   class-x

5. *Semantic-Relationship 5: **is-composed-of.*** Relates a modeling object to its component parts:

   jacketed-CSTR   ***is-composed-of***   (jacket; stirred-tank)

6. *Semantic-Relationship 6: **:is-part-of.*** It is the semantic relationship symmetrical to the previous one.

7. *Semantic-Relationship 7: **is-attached-to.*** This relationship is used primarily to define the boundary of a generic unit, by defining the ports

attached to a specific unit:

$$\texttt{part-x} \qquad \textit{is-attached-to} \qquad \texttt{unit-y}$$

8. *Semantic-Relationship 8: **is-connected-by.*** It is the inverse of the previous one.
9. *Semantic-Relationship 9: **is-described-by.*** It is used to indicate the variables or relationships used to describe the behavior of a generic unit:

$$\text{unit-}x \qquad \textit{is-described-by} \qquad (\text{variable-}y; \text{equation-}z; \text{rule-}w)$$

10. *Semantic-Relationship 10: **is-describing.*** This is the inverse of the previous one.
11. *Semantic-Relationship 11: **is-characterized-as.*** It is used to specialize a modeling class by specifying a fixed, default value for a given attribute of the class; for example, the following statement

$$\texttt{Reactor-x} \qquad \textit{is-characterized-as} \qquad \text{``isobaric''}$$

puts a constant value in the attribute, "pressure", of the class `Reactor-x`. All instances of reactor-*x* will have constant pressure.
12. *Semantic-Relationship 12: **is-disaggregated-in.*** As in LCR, this semantic relationship associates a modeling element, located in a given context, with its constituent modeling components that are located at a different (more detailed) context. For example,

$$\text{plant-}x \quad \textit{is-disaggregated-in} \quad (\text{reaction-section; separation-section})$$

13. *Semantic-Relationship 13: **is-abstracted-by.*** This is the inverse of the previous one.

The semanic relationships of MODEL.LA. obey the same axioms as those of LCR, namely *transitivity, monotonicity, commutativity,* and *merging.* For more details, see Section III.D.

C. HIERARCHIES OF MODELING SUBCLASSES

Six basic modeling classes presented in Section IV.A are the root-classes for expanded trees of modeling subclasses. Whereas the six basic elements are designed to possess generic attributes and methods, independent of the particular domain they represent, the hierarchical trees emanating from

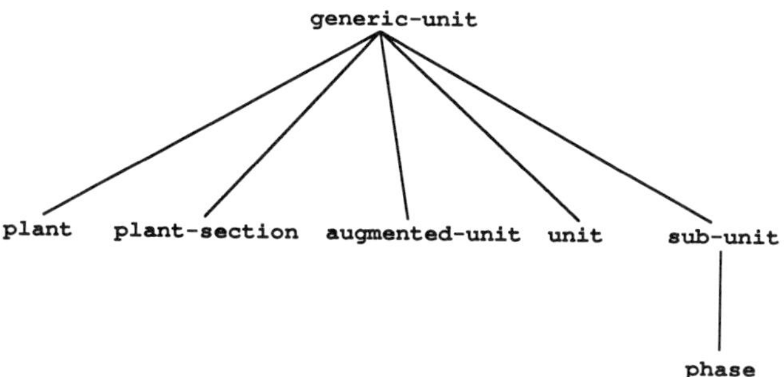

Fig. 17. The tree of `generic-unit` modeling subclasses.

them include modeling subclasses that are specialized to *reflect the vocabulary of process systems engineering*:

1. *The tree of* `generic-unit` *classes* (Fig. 17). The subclass `generic-processing-unit`, encompasses the various abstractions of processing systems: `plant`, `plant-section`, `augmented-unit`, `sub-unit`. The `sub-unit` has a specific subclass, `phase`, to capture the specialized attributes of a material with uniform thermodynamic state.

2. *The tree of* `port` *classes*. Four subclasses of ports emanate from the basic modeling element; `convective-port`, `material-port`, `energy-port`, and `information-port`. Each is specialized to express the characteristics of the quantity that flows through it.

3. *The tree of* `stream` *classes*. The number of stream subclasses is equal throughout and reflect similar structure of attributes as the `port` subclasses.

4. *The tree of* `modeling-scope` *classes*. It has two subclasses: `model` and `context`. The first captures all those relationships that reflect assumptions, hypotheses, and conjectures, all of which define the contextural scope of a model. The second captures the list of mathematical relationships describing the behavior of a generic-unit.

5. *The tree of* `constraint` *classes*. (Fig. 18). The `assignment` subclass is used to represent constraints, which are *solved* with respect to a particular variable. The `relationship` subclass is used to represent *unsolved* constraints.

6. *The tree of* `generic-variable` *classes*. This is composed of two subclasses, the `variable` and the `term`. The most important subclass is the `term`, which represents a compound quantity, made up from the

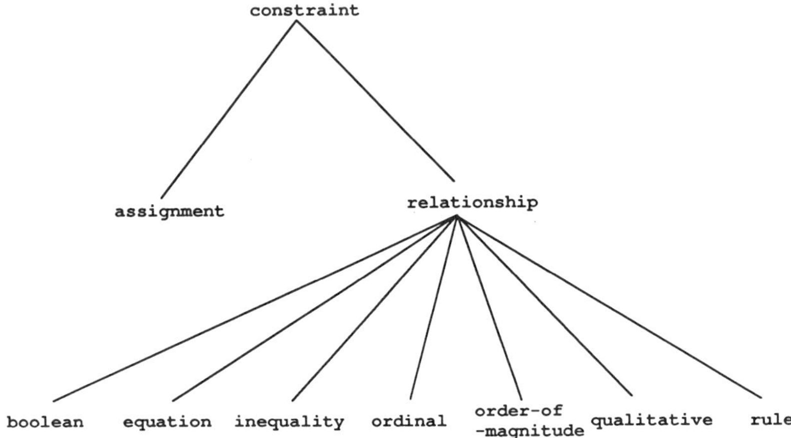

FIG. 18. The tree of constraint modeling subclasses.

combination of other elementary variables and parameters. It is used to model; terms appearing in modeling equations (e.g., enthalpy of stream-i, accumulation), dimensionless numbers [e.g., Re, Pr, Nu (Reynolds, Prandtl, Nusselt numbers)], design quantities [e.g., L/mG (liters per milligauss) of absorbers].

D. SYNTAX

MODEL.LA., like LCR, uses the BNF grammar rules for the construction of syntactically correct modeling sentences. For more details on the syntactic characterization of new models, the reader is referred to Stephanopoulos *et al.* (1990a, b), and MODEL.LA.'s manual (Henning and Leone, 1990).

V. Phenomena-Based Modeling of Processing Systems

MODEL.LA. is a language with infinite extensibility of its vocabulary, enabled by a fixed set of six modeling hierarchies and a fixed set of 13 semantic relationships. In Section IV.C, we discussed the hierarchies of modeling subclasses emanating from the six basic modeling elements. What is far more important for the modeling power of MODEL.LA. is its ability to capture all aspects of chemical engineering science, using the physical and chemical phenomena as the basis for the automatic generation of models for processing systems, using as input the declared physical and chemical characteristics of a processing system. For example, the user (or, another program) declares the following about a reactor-x:

Reactor-x is a two-liquid-phase jacketed CSTR.

MODEL.LA. is capable of parsing the sentence and of automatically creating the following "interpretation":

1. Reactor-x is composed of two subsystems: jacket and continuous stirred-tank.
2. The reacting mixture in the stirred-tank is made up of two liquid phases.
3. There is mass transfer between the two liquid phases.
4. There is heat transfer between reacting mixture and jacket.
5. Materials enter and leave the reactor through convective flows.
6. Enthalpy enters and leaves the reactor through convective flows.

Once this "interpretation" has been established, MODEL.LA. (a) generates all the requisite modeling elements and (b) constructs the modeling relationships, such as material balances, energy balance, heat transfer between jacket and reactive mixture, mass transport between the two liquid phases, equilibrium relationships between the two phases, estimation of chemical reaction rate, estimation of chemical equilibrium conditions, estimation of heat generated (or consumed) by the reaction, and estimation of enthalpies of material convective flows. In order to automate the above tasks, MODEL.LA. must possess the following capabilities:

1. Rich hierarchies of modeling elements, which can be used to represent any conceivable quantity of interest in chemical engineering science.
2. A series of procedures, which can automatically generate the complete set of modeling equations representing the behavior of a processing system.

A. The "Chemical Engineering Science" Hierarchies of Modeling Elements

The modeling hierarchies of `constraint` and `generic-variable` (see Section IV.C) have been expanded to a series of subclasses, which have been specialized to capture the knowledge of chemical engineering science, and use it in an explicit manner during the construction of models for processing systems.

1. Classes of Variables

Class `variable` is a subclass of the basic class `generic-variable`. From the class `variable` emanates the tree of subclasses, a partial view of which is shown in Fig. 19. Unlike other modeling approaches, MODEL.LA. does not represent variables through their values alone, but it provides an extensive structure that includes many additional attributes in the description of a variable. The additional attributes allow MODEL.LA. to reason about these variables and not just acquire their values. Thus, a set of methods in the class `variable` allow any of the subclasses to monitor their values, react with predefined procedures when the value of the variable changes, invoke values from external databases, and so on.

2. Classes of Terms

The class `term` is a subclass of the `generic-variable`, and one of the most important modeling elements in MODEL.LA. It is used to represent a very broad spectrum of compound physical quantities, which appear in modeling relationships. Figure 20 shows a partial view of that

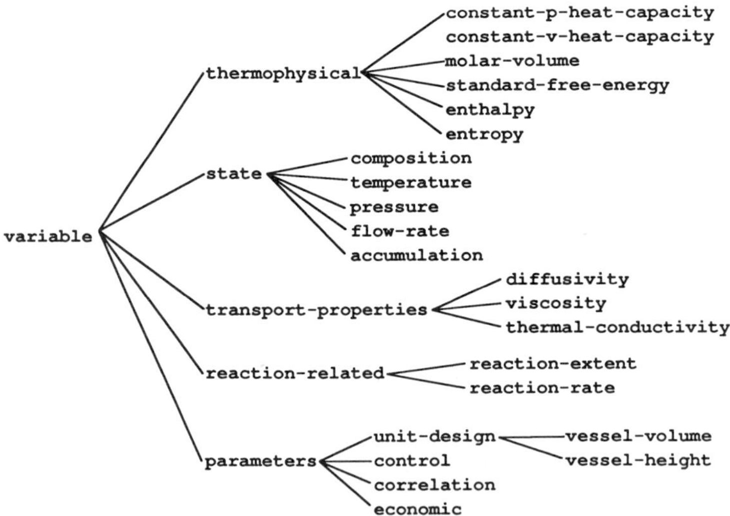

Fig. 19. Partial view of the `variable` hierarchy of modeling subclasses.

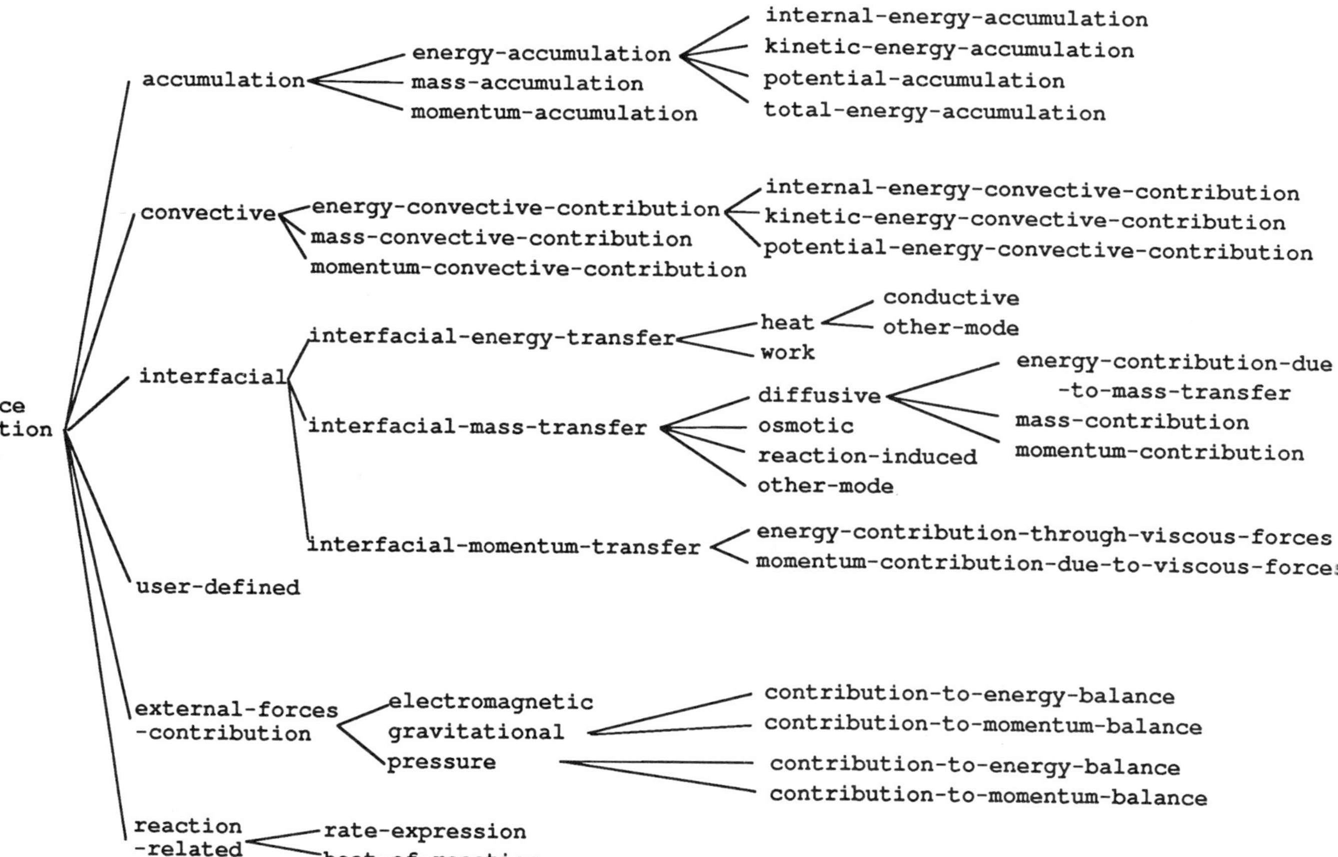

Fig. 20. Partial view of the modeling hierarchy of term class, which represent terms in balance equations.

part of the hierarchy of classes emanating from `term`, which are involved only in the definition of balance equations during the modeling of processing units. The complete structure of the hierarchy of terms is fairly extensive and covers the bulk of terms appearing in balance equations, transport phenomena, and equilibrium relationships.

3. Classes of Equations

All these modeling elements emanate from the class `equation` (Fig. 18), and a partial view is shown in Fig. 21. Each subclass captures all the information about a given equation, including its significance, preconditions for its correct use, and implications on the physics of the modeled system.

B. FORMAL CONSTRUCTION OF MODELS

Every model constructed through MODEL.LA. is represented through a MCDD (see Section III.B), with the modeling elements playing the role of nodes and the semantic relations among the modeling elements representing the edges of the digraph. Figure 22 shows the MCDD associated with the model of a continuous-stirred-tank reactor (CSTR) without a jacket. Note the nodes representing topological elements (e.g., the ports, `input1-vessel` and `input2-vessel`), variables (e.g., `pressure, temperature, composition`), modeling equations (e.g., `mass-lumped-balance, energy-lumped-balance, phase-equilibrium-equation`). The element `vessel-surroundings-heat-transfer` is not specified. If it is specified to be a jacket, then the MCDD of the jacket is appended to that of the CSTR and produces the composite MCDD of Fig. 23, which represents the model of a *jacketed CSTR*. This type of modular construction of processing models provides MODEL.LA. with infinite flexibility in representing any processing systems. Furthermore, predefined models can be used as components in future representations, making the next modeling effort no more difficult than previous ones.

C. MULTIFACETED MODELING OF PROCESSING SYSTEMS

Consider the three abstractions of the plant shown in Fig. 1. How could one use the capabilities of MODEL.LA. to generate consistent represen-

FIG. 21. Partial view of the hierarchy of modeling classes representing various types of equations in chemical engineering science.

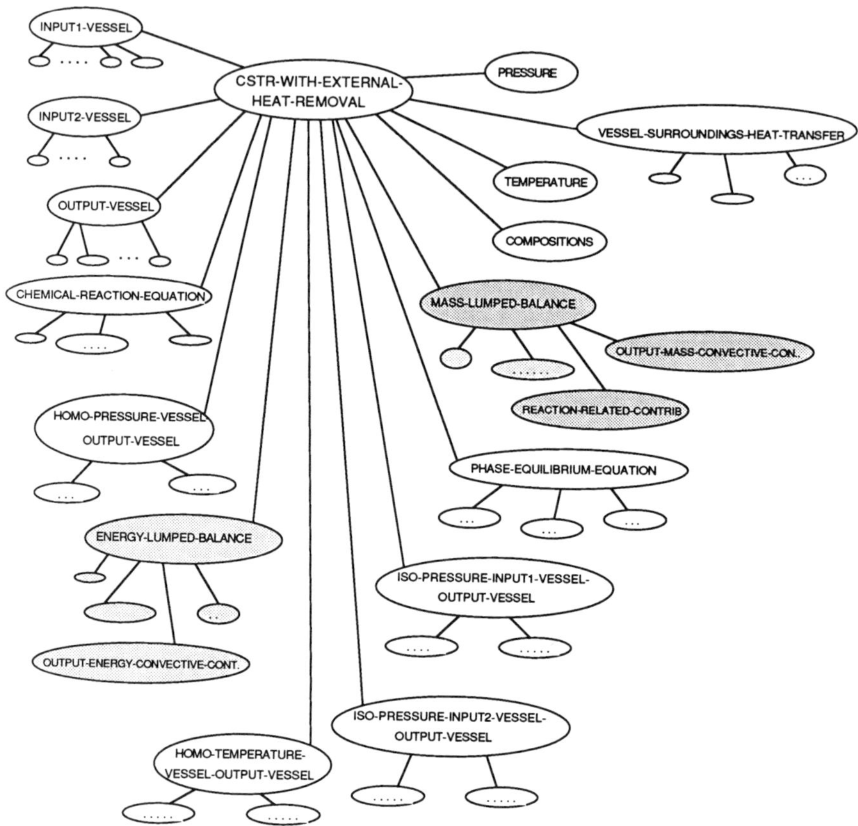

FIG. 22. The MCDD associated with the model of a CSTR without jacket. (Reprinted from *Comp. Chem. Eng.* **14**, Stephanopoulos, G., Henning, G., and Leone, H. MODEL. LA A modeling language for process engineering. Part I, Page 813, Copyright 1990, with kind permission from Elsevier Science Ltd, The Boulevard, Langford Lane, Kidlington 0X5 1GB, UK.)

tations of the three views of the same plant? On the other hand, the two versions of the plant shown in Fig. 2 have many modeling elements in common. How could one use MODEL.LA. to maintain two distinct facets of the same plant at the same level of detail? Such multiviewing of process models, along with the corresponding multilevel and multicontext requirements, define the scope of the so-called *multifaceted modeling of processing systems*.

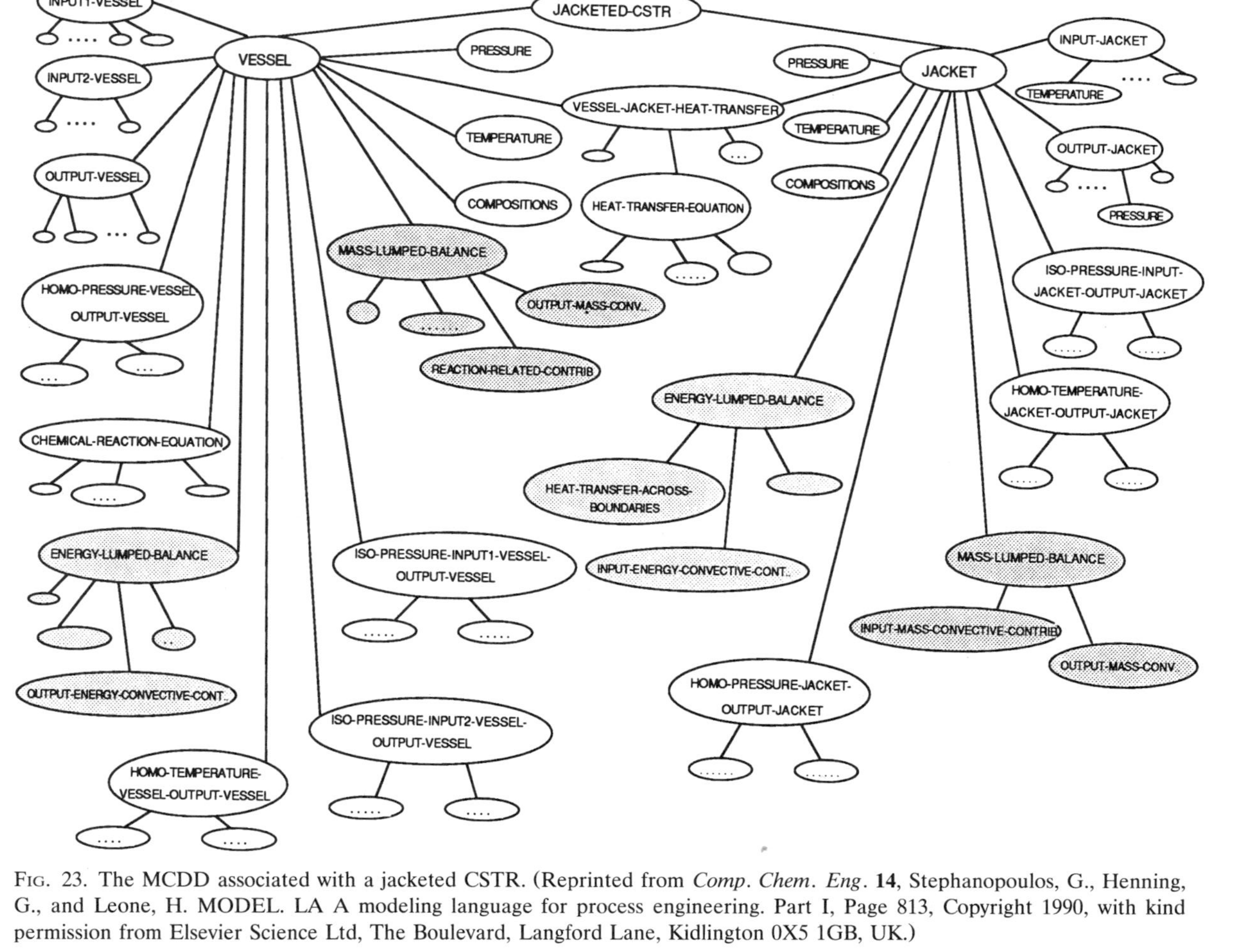

FIG. 23. The MCDD associated with a jacketed CSTR. (Reprinted from *Comp. Chem. Eng.* **14**, Stephanopoulos, G., Henning, G., and Leone, H. MODEL. LA A modeling language for process engineering. Part I, Page 813, Copyright 1990, with kind permission from Elsevier Science Ltd, The Boulevard, Langford Lane, Kidlington 0X5 1GB, UK.)

1. Concurrent Models at Multiple Abstractions

Consider again the "overall process" of Fig. 1a. It can be represented by an instance of the modeling element, `generic-unit` (or, `plant`, if we want to take advantage of its special attributes). MODEL.LA. can derive automatically the necessary modeling instances for the refined plant of Fig. 1b, through the invocation of the following semantic relationship:

overall-process   ***is-disaggregated-in***   (generalized-reaction-section;
generalized-separation-section)

If the `generalized-reaction-section` and `generalized-separation-section` are defined as instances of the class `plant-section`, then the correspondence between the various ports in Fig. 1a and 1b is automatically established. Thus, all the "external" ports `port-A`, `port-B`, `port-P`, and `port-BP` in Fig. 1b acquire the same information that their "external" counterparts possess in Fig. 1a. Furthermore, MODEL.LA. automatically establishes instances of the streams `Gas-Recycle` and `Liquid-Recycle` and associates them with the corresponding ports of the two plant sections in Fig. 1b. The automatic generation of models for the refined representation of the same plant as in Fig. 1b is carried out by a series of procedures that MODEL.LA. activates when an instance of the `generic-unit` (or its subclasses) is (are) refined through the ***is-disaggregated in*** semantic. Specifically, these procedures carry out automatically the following tasks:

1. *Establish structural compatibility* between the abstract and the set of refined generic units.
2. *Establish topological compatibility* between the ports and streams of the abstract entity and those of the refined entities generated after the disaggregation process.
3. *Establish compatibility between the physicochemical phenomena* occurring at the abstract and refined contexts of the plant.
4. *Propagate values of variables* from the abstract to the more detailed context (or, vice versa), using the compatibility links established above.

For the technical details of the procedures, carrying out the above tasks, the reader is referred to Stephanopoulos *et al.* (1990a, b).

2. Concurrent Models at the Same Abstraction

MODEL.LA. maintains concurrent models of the same process at the same level of abstraction using the notion of *context* (see also Section

III.D). For example, let CONTEXT-1 signify the set of (1) topological relationships characterizing the structure in Fig. 2a and (2) modeling relationships describing the functionality and behavior of the generalized reaction and separation sections. Then, we can create CONTEXT-2 as the child of CONTEXT-1, by changing a few of its relationships so that the new context can capture the "splitter" and the "purge-stream" of Fig. 2b. Since CONTEXT-2 (1) inherits all relationships from CONTEXT-1, (2) overrides or eliminates those that changed, and (3) adds new modeling elements and relationships, the two contexts share the common modeling elements (without duplication) and maintain their distinctive character through their private modeling elements or/ and relationships. For more technical details on the MAT, see Section III.D; and for the knowledge sharing of concurrent models at the same level of abstraction, see Stephanopoulos *et al.* (1990a, b).

D. Computer-Aided Implementation of MODEL.LA.

MODEL.LA. was implemented within IntelliCorp's KEE, using Common LISP (list processing language) as the low-level programming language. KEE's frame system provided a fairly flexible, object-oriented programming environment. To take advantage of a number of facilities and to be integrated with an advanced, computer-aided engineering framework, MODEL.LA. was incorporated into the DESIGN-KIT (Stephanopoulos *et al.*, 1987).

1. Organization

The different objects comprising the modeling system are grouped into several knowledge bases:

- The basic modeling elements of the language, and the subclasses that emanate from them, are stored in a knowledge base (KB) called NETWORK-MODEL.
- When the composite objects representing new model classes are generated, their parts constituent are created as more specialized subclasses of the modeling elements of the NETWORK-MODEL KB. However, all the specialized entities constituting model classes are stored in the LIBRARY KB.
- When a model class is instantiated to represent a given model situation, all the entities used to generate the composite object that represents the model instance are created in the WORKING KB.

The rationale behind this division is that managing large knowledge bases is difficult. As knowledge bases grow in size and complexity, they strain the capacities of software tools for knowledge editing and maintenance. This division is based on the different functionality of the several KBs. The NETWORK-MODEL KB contains the basic knowledge, and consequently, is almost fixed. It will only be changed if the language is upgraded or modified. The LIBRARY KB will change if the model classes are modified or new ones are defined. Finally, the WORKING KB contains the most volatile type of knowledge, that is, entities that are created to represent very specific situations.

The WORKING KB stores all the instance objects that are used to represent the models employed during a particular process engineering activity. As was shown in previous sections, however, a model may have multiple descriptions of its structure, topology, behavior, etc. These different modeling situations are encapsulated in CONTEXTS. The CONTEXT facility is built on top of KEE worlds. KEE worlds is a set of tools provided by the KEE system to allow the representation of alternative states of the knowledge base, and to facilitate the exploration and comparison of alternative scenarios. The information in a standard KEE knowledge base provides a starting point for modeling hypothetical situations represented as worlds. In our case this information, called the "background," is stored in the WORKING KB. This KB contains the assumptions that are true in every hypothetical situation, and perhaps the problem-specific facts describing the initial modeling situation.

2. Facilities for the Generation of Models

Model classes can be generated in one of two ways:

(a) By using the MODEL.LA. language, a natural language built on top of KEE's Tell And Ask. The Tell And Ask language is a special facility for interacting with a knowledge base, by telling it new facts to incorporate, and asking about the facts it already contains. It is basically a set of KEE functions and a set of linguistic forms that format information for KEE.

(b) By using the MODEL-EDITOR, that is a specially designed mouse-and-menu-style interface. The description of the MODEL-EDITOR interface can be separated clearly in two parts: one governing the appearance and properties of the display itself, and the other defining the logical relations between display and modeling objects.

Our application domain makes it natural to adopt menus as the underlying display metaphor. *Menus* are represented as structured objects organized in *panels*. Menus themselves are constructed of rectangular primitive regions called *boxes* and nonprimitives called *polyboxes*. Boxes (which need not have visible outlines) are used to accept input data (typed or mouse-selected data), to allocate regions for displaying text and annotations of various kinds, etc. The adoption of a menu-driven interface has several advantages, among which we note the following: (1) *error prevention*, which is achieved naturally, because menus give the user the complete list of assumptions that can be specified at any point during the definition of a model; (2) *memorization*, to avoid or minimize the need for reference to an external manual or calls to help facilities; (3) *immediate visual feedback*; and (4) *active menus*, which help ensure that all the assumptions that need to be set for the model to be defined properly are in fact set. The model editor interface has been designed according to a "conditional display" paradigm. The complete definition of a model class can in principle require the user to specify a very large number of individual assumptions. In practice, fortunately, most model classes can be defined supplying only a small set of the possibly relevant data.

3. Facilities for User-MODEL.LA. Interaction

Specially designed tools provide an integrated set of knowledge access facilities serving various purposes such as inspecting the current status of the several KBs, linking the available model classes and their associated assumptions, displaying the structure of models or just partial views of them, listing the assumptions associated with a model instance in several contexts where it is defined, listing the relations and variables associated with a model instance, and listing the variable values.

To the extent that the structure of a model reflects the logical relations among its several constituent entities, models can be verified in a straightforward way, by confirming that the necessary links exist. Thus, specific browsers are used to display the different specification semantic relations existing among the pieces of a model. It is obvious that this browser facility allows a selective access to the different views of a model.

Other types of browsers are employed to show different semantic relations between the components of the language, for example, to display *"is-a"* or *"is-a-member-of"* links, or to show the directed graph that is formed by the contexts and the parent/child relations between them.

Summarizing, we can say that the user-interaction facilities incorporated in this system are governed by the need to provide a smooth, natural, and efficient environment for accessing, analyzing, and debugging knowledge.

References

Andersson, M., An object-oriented modeling environment. *In* "Simulation Methodologies, Languages and Architectures and AI and Graphics for Simulation" (Iazeolla *et al.*, eds.), 1989 European Simulation Multiconference, Rome, pp. 77–82. The Society for Computer Simulation International, 1989.

Corey, E. J., *Pure Appl. Chem.* **14**, 19 (1967).

Dugundji, J., and Ugi, I., An algebraic model of constitutional chemistry as a basis for chemical computer programs. *Top. Curr. Chem.* **39**(19) (1973).

Evans, L. B., Boston, J. F., Britt, H. I., Gallier, P. W., Gupta, P. K., Joseph, B., Mahalec, V., Ng, E., Seider, W. D., and Yazi, H., ASPEN: An advanced system for process engineering. *Comput. Chem. Eng.* **3**, 319 (1979).

Henning, G. P., and Leone, H. P., "Design-Kit Users' Guide." Laboratory for Intelligent Systems in Process Engineering, Massachusetts Institute of Technology, Cambridge, MA, 1990.

Meyer, B., Reusability: The case for object-oriented design. *IEEE Software*, March, pp. 50–64 (1987).

Nagel, C., Identification of hazards in chemical process systems. Ph.D. Thesis, Department of Chemical Engineering, Massachusetts Institute of Technology, Cambridge, MA (1991).

Naur, P., Revised report on the algorithmic language ALGOL 60. *Commun. ACM* **6**, 1 (1963).

Nilsson, B., Object-oriented modeling of chemical processes. Doctoral Thesis, Department of Automatic Control, Lund Institute of Technology (1993).

Pantelides, C. C., SPEEDUP—recent advances in process simulation. *Comput. Chem. Eng.* **12**, 745–755 (1988).

Perkins, J. D., and Sargent, R. W., SPEEDUP: A computer program for steady-state and dynamic simulation and design of chemical processes. Selected topics on computer-aided process design and analysis. *AIChE Symp. Ser.* **214**, 78 (1982).

Piela, P. C., ASCEND: An object-oriented computer environment for modeling and analysis. Ph.D. Thesis, Carnegie-Mellon University, Pittsburgh, PA (1989).

Piela, P. C., Epperly, T. G., Westerberg, K. M., and Westerberg, A. W., ASCEND: An object oriented computer environment for modeling and analysis: The modeling language. *Comput. Chem. Eng.* **15**(1), 53 (1991).

Sørlie, C. F., A computer environment for process modeling. Dr. Ing. Thesis, Department of Chemical Engineering, Norwegian Institute of Technology (1990).

Stephanopoulos, G., Artificial intelligence: What will its contributions be to process control? *In* "The Second Shell Process Control Workshop" (D. M. Prett, C. E. Garcîa, and B. L. Ramaker, eds.), p. 591. Butterworth, London, 1990.

Stephanopoulos, G., Johnston, J., Kriticos, T., Lakshmanan, R., Mavrovouniotis, M., and Siletti, C., DESIGN-KIT: An object-oriented environment for process engineering. *Comput. Chem. Eng.* **11**(6), 655 (1987).

Stephanopoulos, G., Henning, G., and Leone, H., MODEL.LA. A modeling language for process engineering. I. The formal framework. *Comput. Chem. Eng.* **14**(8), 813 (1990a).

Stephanopoulos, G., Henning, G., Leone, H., MODEL.LA. A modeling language for process engineering. II. Multifaceted modeling of processing systems. *Comput. Chem. Eng.* **14**(8), 847 (1990b).

Vernin, G., and Chanon, M., eds., "Computer Aids to Chemistry." Ellis Horwood, Chichester, 1986.

Wipke, W. T., Heller, S. R., Feldmann, R. J., and Hydes, E., eds., "Computer Representation and Manipulation of Chemical Information." Wiley, New York, 1974.

Woods, E. A., The Hybrid phenomena theory: A framework integrating structural descriptions with state space modeling and simulation. Dr. Ing. Thesis, Department of Engineering Cybernetics, Norwegian Institute of Technology (1993).

Woodward, R. B., *in* "Perspectives in Organic Chemistry" (A. Todd, ed.), p. 155. Wiley (Interscience), New York, 1956.

Woodward, R. B., *in* "Pointers and Pathways in Research" (G. Hofteizer, ed.), p. 23. Ciba of India, Ltd., Bombay, 1963.

AUTOMATION IN DESIGN: THE CONCEPTUAL SYNTHESIS OF CHEMICAL PROCESSING SCHEMES

Chonghun Han and George Stephanopoulos

Laboratory for Intelligent Systems in Process Engineering
Massachusetts Institute of Technology
Cambridge, MA 02139

James M. Douglas

Department of Chemical Engineering
University of Massachusetts, Amherst, MA 01003

If you really know how to carry out an engineering task, then you can instruct the computer to do it automatically. This self-evident truism can

ADVANCES IN CHEMICAL ENGINEERING, VOL. 21

be used as litmus test of whether a human "really" knows how to, say, design an engineering artifact. Experience has shown that in very few instances, engineers have been able to automate the process of design, thus demonstrating the presence of serious flaws in (1) their understanding of how to do design or/and (2) their ability to clearly articulate the design methodology, both of which can be traced to the inherent difficulty of making the "best" design decisions. The pivotal element in automating the design process is *modeling the design process itself*, which includes the following modeling tasks: (1) modeling the *structure of design tasks* that can take you from the initial design specifications to the final engineering artifact; (2) representing the *design decisions* involved in each task, along with the assumptions, simplifications, and methodologies, needed to frame and make the design decisions; and (3) modeling the *state of the evolving design*, along with the underlying rationale.

In this chapter, we will show how one can use ideas and techniques from artificial intelligence, such as symbolic modeling, knowledge-based systems and logic, to construct a computer-implemented model of the design process. By using the Douglas hierarchical approach as the conceptual model of the design process itself, this chapter will show how to generate models of the structure of design tasks, design decisions, and the state of design, thus leading to automation of large segments of the synthesis of chemical processing schemes. The result is a *human-aided, machine-based* design paradigm, with the computer "knowing" how the design is done, what the scope of design is, and how to provide explanations and the rationale for the design decisions and the resulting final design. Such paradigm is in sharp contrast with the traditional *computer-aided, human-based* prototype, where the computer carries out numerical calculations and data fetching from files and databases, but it has no notion on how the design is done, i.e., knowledge resting exclusively in the province of the individual human designer. In addition, we will argue that the human-aided, machine-based design is the paradigm that will characterize future design systems, where rapid conceptualization and prototyping of engineering artifacts will be the source of competitive edge.

I. Introduction

The engineering design of products and/or processing systems is a dialectic process (Stefik *et al.*, 1982) between goals (i.e., what is desired) and possibilities (i.e., what is actually realizable), aimed at the satisfaction of functional and performance specifications. No general theory exists for the systematic and rigorous development of a procedure that leads to the design of the desired engineering artifacts. This is due to two inescapable facts: (1) any design is a knowledge-intensive task; the more and better-quality knowledge, the more efficient the design and better the quality of the designed artifact and (2) the knowledge required for a particular class of design problems is specific to the class of problems. Thus, various attempts to formalize the overall design procedure as a large combinatorial optimization problem, have not provided a *generic theoretical framework for design*, but have simply underlined the fact that every design problem is a complex decision process with real-valued and integer decisions. On the other hand, combinatorial optimization techniques based on branch-and bound strategies (Edgar and Himmelblau, 1988), when applied to narrowly defined specific domains, have been shown to be fairly effective in selecting good design solutions through the implicit enumeration of many alternatives (Grossmann, 1985, 1989; Kocis and Grossmann, 1989). Nevertheless, even in such cases, the successes are an indication of the efficiency of the implicit enumeration techniques rather than a manifestation of a theory for design.

In the absence of a general theory on how design is done, research in the filed of artificial intelligence has been addressing the following distinct but complementary areas of inquiry:

1. *Axiomatic theory of design*, with the objective to establish a theoretically firm ground for the definition of design and thus bring it into the realm of "science," rather than "art," where it presently stands. Efforts in this direction have led to a number of approaches for the representation of cognitive models of the design process, but no significant breakthroughs have been achieved.

2. *Engineering science of knowledge-based design* (Tong, 1987; Tong and Sriram, 1992), aiming at the development of a rational framework for organizing, evaluating and formulating knowledge-based models on how the design is done. Efforts in this direction have been more successful. They have led to a number of new representation schemes (e.g., frames, scripts, streams, actors; see Rich and Knight, 1991; Quantrille and Liu, 1991), which have broken previous limitations in articulating, representing, and utilizing all forms of available knowledge; a particularly important requirement for a successful design approach. In addition, theoretical work has led to the development of new mathematical frameworks, which can manipulate these forms of knowledge representation (e.g., algebra of approximations, logic inferencing through semantic networks, analysis of logic clauses). Consequently, today we can effectively use all forms of

available knowledge within a computer-based environment; this is a significant addition to the numerical algorithms of the traditional computer-aided design tools.

Computer-aided design environments should support the designer in reaching better solutions than in the past, but in shorter periods of time. *We cannot afford to have the designs of tomorrow take as long as the designs 20 years ago.* Here is where the new artificial intelligence (AI)-based programming styles (e.g. object-oriented programming), the development of design-oriented languages, and object-oriented databases have already made significant contributions to the design process, by allowing the development of highly complex, design-oriented software systems in remarkably short periods of time. The existing applications, especially in other areas of engineering, are very impressive and do not allow room for disputing this fact. In particular, the complexity of the so-called *cooperative design environments* (or equivalently, *concurrent engineering*) is of such scale that traditional procedural programming approaches would have rendered the development of such software systems impossible. Thus, *AI is enabling the development of new design environments*. In chemical process design, the corresponding effort in academic institutions is at early stages, although the articulation of the design needs by the industry has been expressed in more mature terms. Lack of education in the developments of modern computer science has been the primary reason for this delay.

A. Conceptual Design of Chemical Processing Schemes

Conceptual process design involves the series of tasks leading to the development of a process flowsheet from the given reaction information and product specifications. This process usually comes at the early stage of a design project. Since all other design activities depend on the results of the conceptual process design, it has been considered the most important stage during the development and deployment of a new process. According to Douglas (1988), while the cost of the conceptual design usually takes 10–20% of the total cost to develop a new commercial process, the decisions committed at this stage fix about 80% of the total project cost.

Conceptual process design is an underdefined problem. Only a very small fraction of the information needed to define a design problem is available from the problem statement. The design decisions of the process designer provide this missing information. For example, the designer makes design decisions about what kind of process units to use, how to interconnect those units, and so on. From this perspective, the conceptual design is a combinatorial problem. It has been reported (Douglas, 1988) that there are 10^4–10^9 ways that we might consider to accomplish the conceptual design. On the other hand, experience indicates that at the conceptual design level, less than 1% of ideas ever prove to be successful, so the emphasis should be placed on the quick screening of numerous process alternatives.

1. Previous Approaches and Their Limitations

The pivotal role of the conceptual process design has motivated significant levels of research on the automation of the conceptual process design. AIDES (Adaptive Initial DEsign Synthesis) (Siirola et al., 1971; Powers, 1972; Rudd et al., 1973) is the pioneering prototype in the automatic synthesis of conceptual process flowsheets. AIDES makes use of a heuristically modified version of the means–ends analysis technique (Quantrille and Liu, 1991, pp. 268–270); given information on raw materials and the desired products, it attempts to bridge the difference between raw material and products through a sequence of operations that are eventually transformed into unit operations. BALTAZAR (Mahalec and Motard, 1977) is another early example on the use of AI techniques in process synthesis. Using techniques from the area of automatic theorem proving, BALTAZAR proposes logic-based approaches to the synthesis of flowsheets.

Douglas (1988) has proposed a hierarchical decision procedure. This procedure decomposes the design problem into a series of design or decision levels, sequenced and solved in a hierarchical order. It first sketches a design that is complete but vague, and then refines the vague parts into more detailed designs until finally the design has been refined to a complete sequence of detailed unit operations. Heuristics are used at all levels to fix the structure of the flowsheet. Economic potential is computed at each level to reduce the number of alternatives that have to be considered. It should be noted that hierarchial decision procedure is a systematic design procedure, not a design automation system.

Synthesis of process flowsheets through a simultaneous structural and parametric optimization, has received significant levels of attention (Grossmann, 1985, 1989; Kocis and Grossmann, 1989). This approach requires a superstructure as an initial structure from which all the designs of interest can be derived. Branch-and-bound algorithms have been used to solve the mixed-integer nonlinear programming (MINLP) formulations, and find the optimal structure and parameters of the unknown flowsheet. The possible design alternatives need to be articulated and modeled, if one is to develop an MINLP formulation that, using an implicit enumeration algorithm, will solve the MINLP problem and find the optimal solution. Clearly, this approach does not give much attention on the design process that we want to automate. However, it is a very promising technique for selecting the best alternative, once the superstructure has been identified.

Knowledge-based expert-system development has been very active during the last decade. After a design methodology has been identified for a given class of design problems, the expert system can be constructed in such a way that it maps uniquely the engineering design methodology into a computer program. The human usually interacts with the program by monitoring the design process, providing decisions and guidance at critical junctures, or the values which the program requests. Example of such programs are *PIP* (Kirkwood, 1987; Kirkwood *et al.*, 1988), *BioSepDesigner* (Siletti, 1988; Siletti and Stephanopoulos, 1992), *ProDesigner* (Kritikos, 1991). The construction of PIP (a program for the synthesis of process flowsheets) led to a rather rigid program with a strict procedural model to emulate the design methodology. Currently, a new object-oriented version is under development with significantly improved flexibility, extensibility and maintenance. *BioSepDesigner*, on the other hand, was developed to be a flexible design environment for the synthesis of separation sequences for the recovery and purification of proteins. Object-oriented in character, it is easily extendible with new knowledge and provides integrated treatment of data, models, and design decisions. Nevertheless, all three of the above systems and several others do not contain an explicit model of the design process and thus they cannot easily accommodate the incorporation of new knowldege as it becomes available.

B. Issues in the Automation of Conceptual Process Design

Extensive efforts over the last 30 years have led to rich computer-aided design environments that support the various tasks during the synthesis, development, and engineering of processing schemes. Figure 1 shows the standard outlay of such a computer-aided design (CAD) environment. A series of tools (e.g., equation solvers, optimization routines, specific design methodologies for localized problems, physical property estimation techniques) are being invoked by the human designer to execute various design tasks, such as design of heat exchanger networks, synthesis of separation systems, design of individual processing units, sizing, and costing. Convenient graphic interfaces make the interaction between designer and computer very smooth and transparent. Sophisticated database management

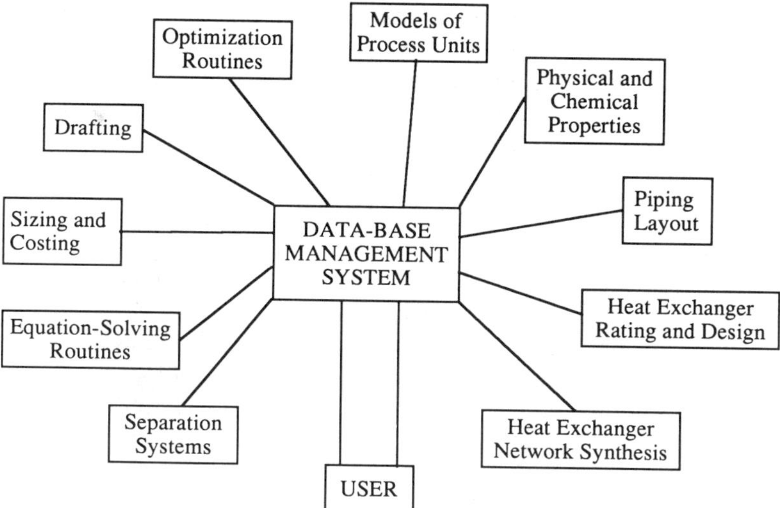

FIG. 1. Typical computer-aided design environment.

systems provide easy means for depositing and retrieving of data, intermediate results, previous designs, etc.

Despite the continuous enrichment of CAD environments, the character of the overall process design procedure remains the same (Mostow, 1985): "the human does the design and the computer provides the support tools, without understanding the design process, its rationale, or the design decisions." The drawbacks of the CAD paradigm are several and, at times, critical. They all stem from the fact that the design procedure is implicit in the designer's mind, for example

(a) Where does a design start from?
(b) What is to be done next?
(c) How are the assumptions, conjectures, and simplifications needed for the design to proceed, to be formulated?
(d) How are major design decisions made?

In other words, the CAD paradigm "knows" nothing about (1) the design agenda and (2) the "context" of the design activities.

Furthermore, the computer structure of tasks during the synthesis of a process flowsheet can be very large, detailed and complex for any human designer to document and mentally carry with him/her. To the extent that we can untangle and make explicit the design procedure, thus emulating the designer's own methodology, the process can be mechanized. But in this case, we are moving towards a *human-aided, machine-*

based design paradigm, where the computer, through human guidance, can carry out significant portions of a design by "knowing" the design procedure itself, its rationale and the reasoning behind a number of design decisions. This is the paradigm whose development and computer implementation has been significantly advanced by research in artificial intelligence, and which we will discuss in subsequent sections. The benefits from the availability of such mechanized models for design are many and diverse:

1. Rapid prototyping and evaluation of processing schemes.
2. Improvements in cost and reliability.
3. Explicit documentation of the design process itself: why certain goals were set during the design and how they were achieved, how design decisions were made, what assumptions and simplifications were involved, what models were used at the various stages of design, what alternative designs were examined, and why certain ones were selected over others.
4. Explicit documentation of the designed process itself, i.e., what are its components and their characteristics, how are they interconnected, what are their functional and performance characteristics, what are the critical design variables and the intrinsic tradeoffs.
5. Easy verification and modification of the resulting design. Having an explicit documentation of the intermediate design tasks, generated alternatives, rationale behind various design decisions, assumptions, conjectures, and simplifications, one can replay the design scenario and easily verify the validity of the derived processing scheme, or modify its design premises for further improvements.
6. The mechanized model of a design methodology offers an excellent depository for the organization of new empirical knowledge and/or the systematic incorporation of new theoretical results and analytic tools. Such inclusion of new knowledge will progressively increase the automation of the process synthesis procedure itself.

1. Modeling the Process of Design

How do you construct a software system, which emulates a specific methodology for the synthesis of processing schemes? Figure 2 shows a generic, conceptual model of the design methodology. It is composed of three distinct components; a *Planner*, a *Scheduler*, and a *Designer*.

a. Planner. Defines a top-level milestones through which the state of the processing scheme is expected to pass. For the synthesis of processing schemes, using the Douglas hierarchical approach, these milestones are (1)

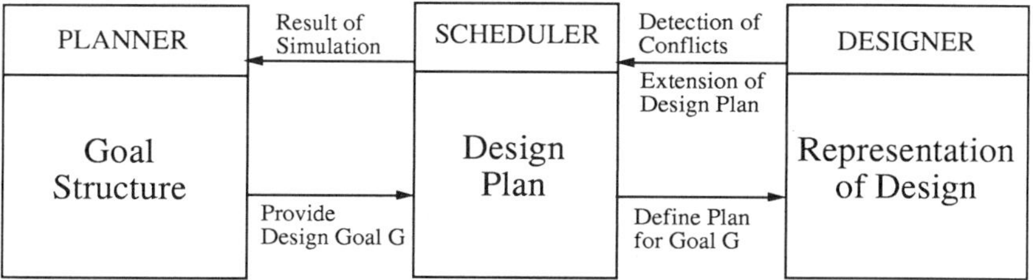

FIG. 2. A generic model for the design process.

design of the multiplant complex configuration, (2) design of the input/output structure for each plant, (3) design of the recycle structure, (4) design of the generalized separation structure, (5) synthesis of the separation subsystems (vapor, liquid, solid), (6) synthesis of heat exchanger networks, and (7) integration of processing subsystems.

b. Scheduler. Determines the sequence of design steps to be taken as one attempts to advance the current state of a processing scheme to the next milestone, defined by the Planner. It embodies a theory of how design goals and associated tasks are created, prioritized, and decomposed; how they interact, and how they are satisfied. As a result, it offers a fairly detailed design plan, by identifying all the requisite engineering design tasks. In subsequent sections, we will see that the *Scheduler* is represented by a tree of design tasks, which emulate the Douglas hierarchical design methodology (e.g., see Figs. 7–10, 12, 13).

c. Designer. It maintains the representation of the processing scheme being synthesized and other domain-specific knowledge. Thus, given a design step (from *Scheduler*), the *Designer* knows what must be done to execute the design step, e.g., reason with a specific set of rules, execute a design algorithm, carry out an optimization procedure, etc. It also updates the state of the evolving flowsheet, detects conflicts, and using domain-specific data, prescribes modifications to the design plan, which are communicated back to the *Scheduler*.

2. High-Level, Design-Oriented Languages

The conceptual model of the design process, discussed above (see also Fig. 2), should be turned into a logically well-defined computational process for the computer. There are high-level programming languages, such as FORTRAN, C, LISP, which can bridge the gap between the engineering model of conceptual design of Fig. 2, and its computer-based counterpart. However, as it has been pointed out by Newell (1982) and Chandrasekaran (1986), this gap is too wide for these languages to bridge in an expressive and efficient manner. Process designers cannot describe their design activities in their own terms, using FORTRAN, C, or LISP. Since the structure of language defines the boundaries of thought and expression, it is clear that the computer-based representation of a design methodology must rely on higher-level linguistic constructs, which allow us to do the following: (1) to represent the evolving structures of processing schemes in significant detail, including all the semantic relationships between the various representational components (processing units, streams, materials, modeling relationships, variables, decision-making procedures, numerical algorithms, etc.) and (2) to represent the design methodology itself, i.e., the intermediate design milestones, the goal structures taking the process from one milestone to the next, the procedures supporting various design tasks, etc.

3. Object-Oriented Representation of Knowledge

As a conceptual process design is a knowledge-intensive activity, its computer-based automation leads to a very large and complex computer program, which represents a very broad range of data models and procedural methodologies. Writing a large computer program is neither a trivial task, nor a straightforward exercise when all the design activities are already known.

The recent advances in object-oriented programming technology make this complex task much simpler by introducing the concepts of the *encapsulation* (or information hiding), *inheritance*, and *polymorphism*. In an object-oriented environment, the structure of the program can be designed in such a way that it represent a one-to-one mapping of the conceptual design methodology itself. For each object in engineering design methodology, a corresponding computational object can be constructed. For each engineering task, a symbolic operation possibly containing numerical computations in the computational model can be defined. With the use of this strategy, extending the model to accommodate new objects or new actions requires no strategic changes to the structure of the program itself. It only needs the addition of the new symbolic analogs of those objects or actions.

Many currently available CAD programs do not fully implement this concept. This is in part due to the unavailability of this technology when those systems were developed, and in part due to the lack of compatibility between the old codes and these new technologies. As a result, it is not a trivial task to modify the existing models of the CAD systems, or add a new model to the design system.

4. Human–Computer Interface

Given that the design process is such a complex process, the human–computer interface is very important for the user, in order to monitor the progress of the evolving design, provide decisions, and guide the direction of design in a user-friendly manner. It is not surprising that many computer-aided simulation environments, such as ASPEN Plus, PRO/II, DESIGN II, and HYSIM, have incorporated advanced graphic user interfaces into their environment. However, to help the process designer control the complexity of the design process, these user interfaces should be designed in a manner consistent to the model of the design process itself. The goal is to make the design process explicit to the user and allow the user to monitor the design process and communicate his/her ideas through user interfaces. When the user needs to make a design decision, the system should provide him or her with a context for the design decision. This context should have minimal information, yet still provide the key information necessary for the decisionmaking. After the user makes a design decision, the design decision should be recorded and made available to the user. The user should be able to see the effect of the design decisions on the process flowsheet, request and receive explanations, and solicit advice for the evolution of design. When the user interfaces satisfy these requirements, the user can take full control of the design process, while large segments of the design process are automated by the computer.

II. Hierarchical Approach to the Synthesis of Chemical Processing Schemes: A Computational Model of the Engineering Methodology

Douglas' hierarchical design approach (Douglas, 1985, 1988) was chosen to be the model of the engineering methodology, whose computer-based automation we seek to achieve. This approach has been used for the design of single-product continuous processes with single or multi-step reactions (Douglas, 1985, 1988, 1989), solid processes (Rossiter and Douglas, 1986a, b; Rajagopal *et al.*, 1992), polymer processes (McKenna and Malone, 1990), pharmaceuticals, and specialty chemicals

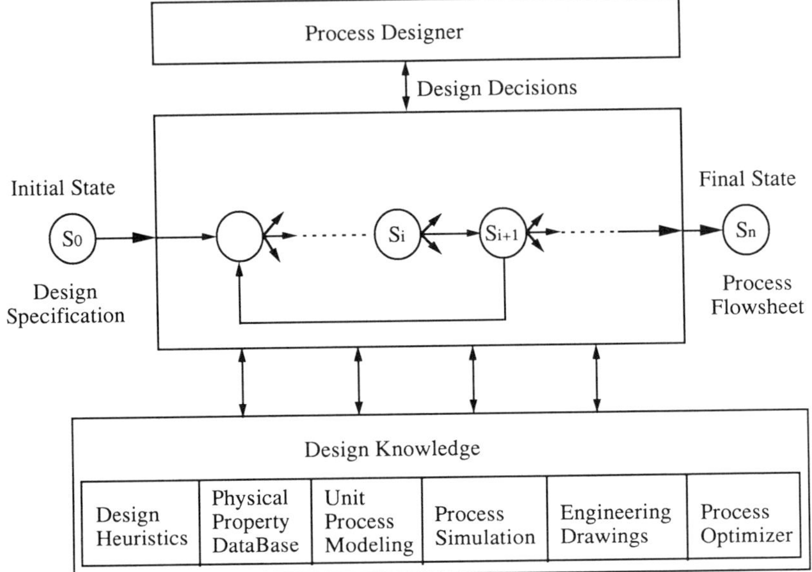

FIG. 3. Schematic diagram of the generic design process.

(Stephanopoulos *et al.*, 1994). It provides a well-defined structure of how the synthesis of conceptual processing schemes proceeds from very abstract, vague, and incomplete to fairly detailed, exact, and quite complete flowsheets. However, when it was developed, it was not intended to be a computational model for the design process itself. As a result, it lacks the requisite formalism and representational exactness needed for the construction of a computational process. In this section, we will provide a formalized restatement of the hierarchical procedure, in an effort to cast it more closely to a computational process for its computer-based automation.

A. HIERARCHICAL PLANNING OF THE PROCESS DESIGN EVOLUTION

It is common to characterize a design history as a series of transformations from specification to final design (Fig. 3). However, every design methodology does not handle the gap between "specifications" and "final design" in one sweeping structure of design steps. Instead, it identifies the abstract milestones through which the design is expected to go. In the case of the synthesis of processing schemes, the Douglas methodology has identified the following sequence of intermediate design milestones:

Beginning. Collect project specifications, e.g., chemistry to be used, feedstocks available, products to be produced, desired production specifications (amounts, purity), and economic constraints not to be violated (e.g., required capital $\leq a$, a product unit cost $\leq b$, return on investment $\geq c$, where a, b, and c are available bounding values).

Milestone 1. Synthesize and evaluate *plant complex structure*, i.e., identify the number of abstract "plants" each of which is formalized around a set of reactions, which take place under the same conditions (Fig. 4a).

Milestone 2. Design the *input / output structure* for each "plant," i.e., identify the set of *distinct* input and output streams associated with each "plant" of the overall structure (Fig. 4b).

Milestone 3. Select the *recycle structure* connecting the generalized abstract reaction and separation sections (Fig. 4c). During this phase of the design, the essential design alternatives of the generalized reaction section are also identified, i.e., the types of feasible reactors and their interconnections.

Milestone 4. Select the *generalized separation structure*, i.e., the elements of all-encompassing structure, shown in Fig. 4d, which are needed for the specific output from the reaction section of a given "plant."

Milestone 5. Synthesize the structure of the specific separation subsystems, e.g., vapor recovery, gas separation, liquid separation, or / and solid separation, which are relevant to the generalized separation structure of a given "plant."

Milestone 6. Integrate the process flowsheets, identified for each "plant" of the plant complex structure (Fig. 4a), and carry out consolidation of common processing tasks. The resulting design represents a flowsheet for the whole plant.

Milestone 7. Synthesize energy management systems for the consolidated flowsheet, derived at the last step.

End. Carry out sensitivity analyses of the process flowsheet's economic measures and revisit the critical design decisions, thus giving rise to improved designs.

It is clear from the above discussion that the Douglas approach to the synthesis of conceptual process flowsheets is based on the idea of *abstract refinement* (Mitchell *et al.*, 1981; Stephanopoulos, 1989), which is shown in Fig. 5. Since top-down abstract refinement does not handle goal coupling

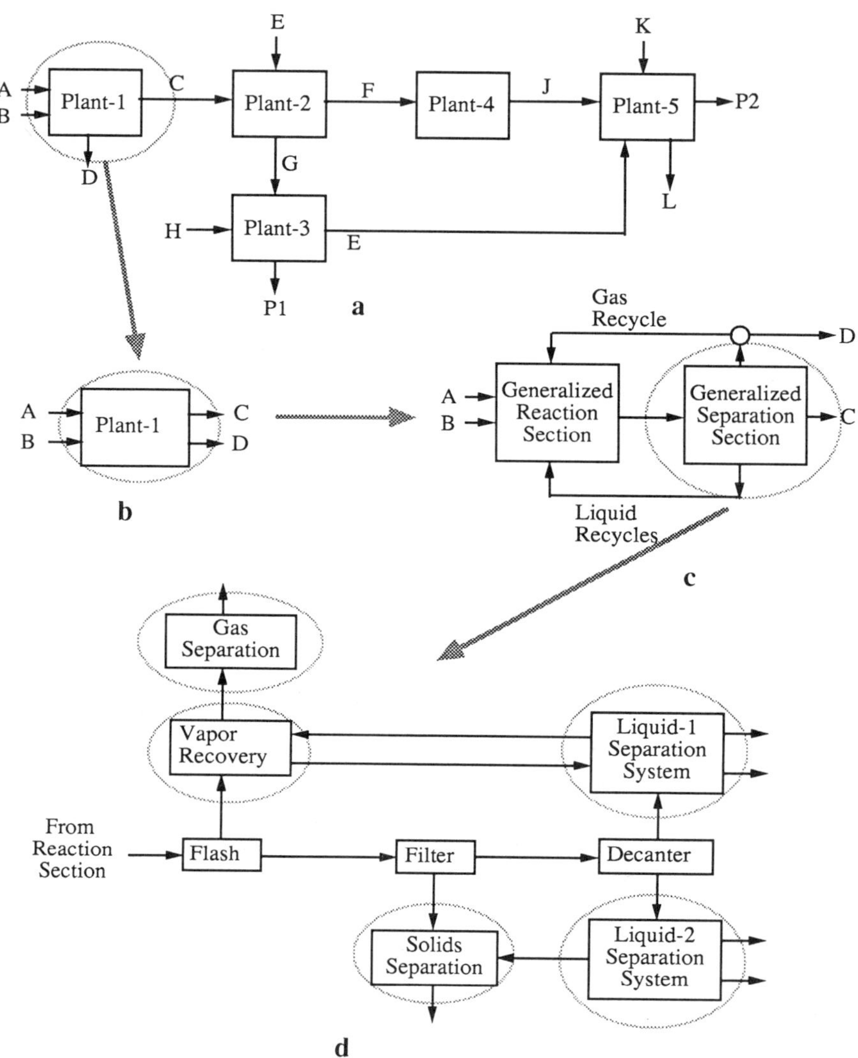

FIG. 4. Hierarchical evolution of conceptual processing schemes.

well, the approach must use, and in fact it does, *constraint propagation* to achieve consistency between the different parts of the overall processing scheme, as it will be discussed in a subsequent paragraph. However, for the time being, it is clear that Fig. 5 outlines the hierarchical planning stages for the conversion of specifications into one or more final process flowsheets.

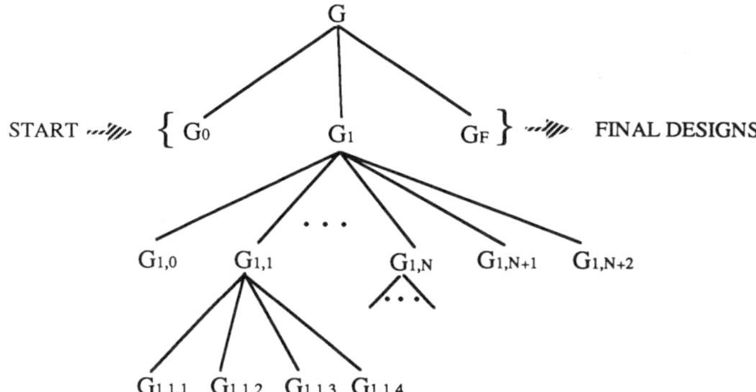

FIG. 5. Tree of goal structures in abstract refinement—(G—design the conceptual processing scheme for a multistep reaction plant; G_0—initialize the project's specifications; G_1—synthesize the process flowsheet for the whole-plant designs; G_F—evaluate process flowsheet and create improved designs; $G_{1,0}$—synthesize the plant complex structure; $G_{1,i}$—synthesize the flowsheet for plant-i; $G_{1,i,1}$—design the input/output structure for plant-i; $G_{1,i,2}$—design the recycle structure for plant-i; $G_{1,i,3}$—design the generalized separation structure for plant-i; $G_{1,i,4}$—design the liquid separation subsystem for plant-i; $G_{1,N+1}$—consolidate the flowsheets of all plant-i, $i = 1, 2, \ldots$, into one; $G_{1,N+2}$—synthesize energy management systems for consolidated process).

B. Goal Structures: Bridging the Gap between Design Milestones

Moving the synthesis of process flowsheets from one milestone to the next entails a series of design steps with progressively increasing amount of details. These design steps need, nevertheless, to be articulated very explicitly and modeled by specific entities, before they can be automated by a computational procedure. In this section, we will focus our attention on the articulation of design tasks or steps, limiting our view to the synthesis of processes with a single fluid product with no solids present anywhere in the process.

1. Goal Structures and the Transformational Model

While the model of abstract refinement is quite satisfactory for the hierarchical planning of the process synthesis methodology, it is quite cumbersome and possibly incorrect as a model to describe the intricate web of design actions from one milestone to the next. Instead, the *transformational model* (Balzer *et al.*, 1976; Scherlis and Scott, 1983; Stephanopoulos, 1989) provides a more appropriate vehicle. This model converts specifications into designs through a series of correctness-

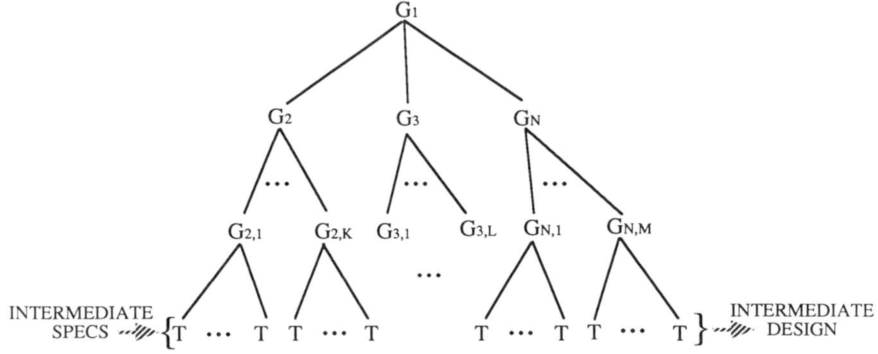

G: Goal   T: Transformation to achieve corresponding goal

FIG. 6. Structure of design tasks in the transformational model.

preserving transformations from one complete description to another (see Fig. 6). Thus, a single transformation may operate on several components of an evolving flowsheet at once. It is more general than the abstract refinement, which is limited to steps that replace a single component with a more detailed description of it. However, the generality comes at the potential cost of increased complexity, since a transformation on a complete description must deal with more information and must tackle all design decisions simultaneously. Nevertheless, since a transformation sequence can be viewed as an executable program for implementing the desired specifications, it is in agreement with the nature of the computational tools available in process engineering, i.e., branch-and-bound algorithms to solve mixed-integer constraint- or optimization-based design problems. Let us see now how we can model the transformation of one design milestone to the next, using Douglas' methodology for the synthesis of conceptual processing schemes. In doing so, in the subsequent paragraphs, we will generate the goal structures and requisite tasks following the general transformational model shown in Fig. 6, which can be easily converted into a computational process.

a. Goal-Structure 1: Initialize Project Definition. Looking back to Fig. 5, we notice that goal G_0, i.e., initialize the project's specifications, is the first to be accomplished. Figure 7 shows the refinement of G_0 into a series of data-acquisition subgoals, all of which are satisfied by user actions. Table I provides an example of project specifications at the outset of the process synthesis for a plant for the hydrodealkylation of toluence (Douglas, 1988).

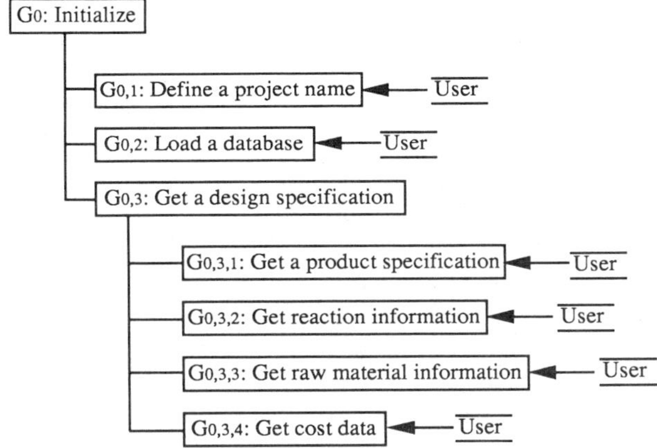

FIG. 7. Goal structure for the goal G_0, "Initialize the project specification."

b. Goal-Structure 2: Specify the Plant Complex Structure. For multistep reaction schemes, the synthesis of the process flowsheet starts by identifying the plant complex structure, i.e., achieving goal $G_{1,0}$ in Fig. 5. Figure 8 presents the transformational model of the goal structure, which identifies the individual "plants" defined around a set of reactions. From the design specification, the methodology can identify the number of simple plants. The reactions that take place at the same temperature, pressure, and catalysts are allocated to the same plant. The input and output streams of

TABLE I

DESIGN SPECIFICATIONS FOR HYDRODEALKYLATION (HDA) OF TOLUENE
TO PRODUCE BENZENE[a]

Product specification
    Product: benzene
    Production rate: 265 lb-mol/h
    Product purity: 0.9997 mole fraction
Reaction information
    Toluene + $H_2 \rightarrow$ benzene + CH_4 at gas phase, 500 psia, 1150°F
        Extent of reaction = $\text{prod}/[1 - 0.0036/(1 - X)^{1.5}]$
    2 Benzene $\rightleftharpoons$ diphenyl + H_2 at gas phase, 500 psia, 1150°F
        Extent of reaction = $\text{prod}[0.0036/(1 - X)^{1.5})/2(1 - 0.0036/(1 - X)^{1.5}]$
Raw material information
    Pure toluene at 15 psia, gas phase
    H_2: 95%, CH_4: 5% at 550 psia, 100°F, gas phase

[a]Key: prod represents the molar flowrate of the product; X represents the conversion of the primary reaction; psia = pounds per square inch absolute.

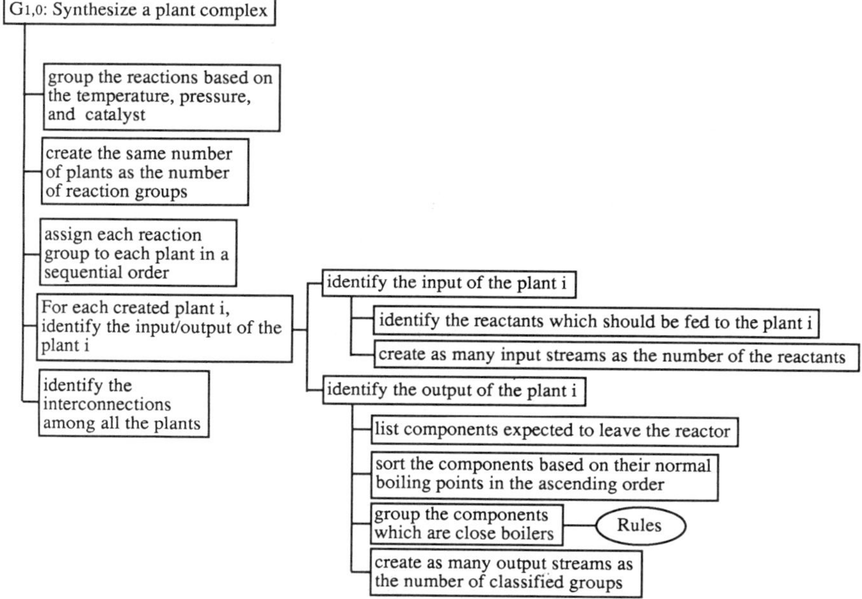

FIG. 8. Goal structure for the goal $G_{1,0}$, "Synthesize the plant complex structure."

each plant are identified. The rules (Douglas, 1988) for deciding the destination code are used to decide the number of output streams and the destinations of output streams. Once input and output streams of each plant have been identified, we can establish the interconnections among the plants using stream matching.

c. Goal-Structure 3: Design the Input-Output Structure. For each "plant-i" identified above, the hierarchical design of a process starts by selecting the structure of input and output streams. The design tasks at this stage are depicted in the transformational model of the goal structure shown in Fig. 9 and described below.

(i) *Synthesis.* When there are impurities in the feed streams, the methodology needs to make a decision for each feed stream about whether it needs feed pretreatment before the plant system.

(ii) *Analysis.* The design variable is a conversion for the primary reaction. If there are multiple reactions, then extents of reactions should be provided as functions of a conversion for the primary reaction, product

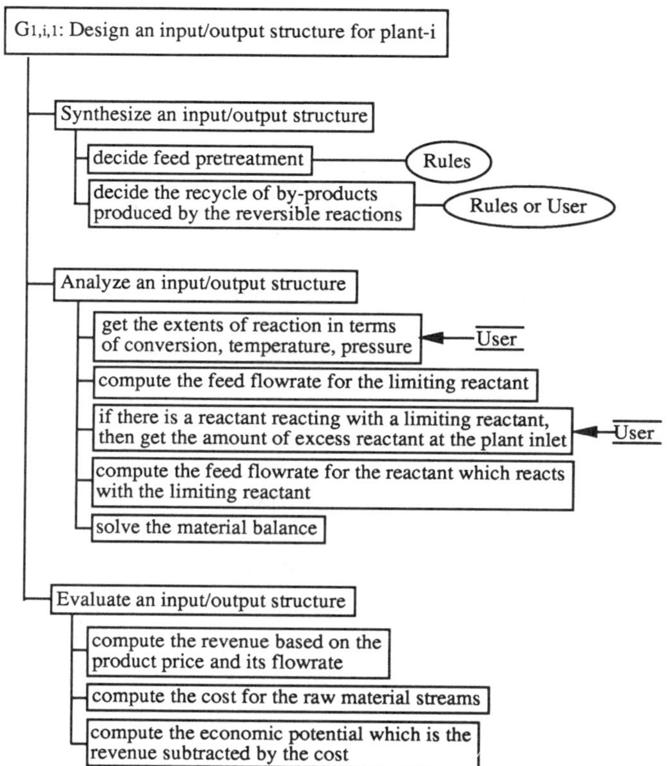

FIG. 9. Goal structure for the goal $G_{1,i,1}$, "Design an input/output structure for plant-i."

flowrate, temperature, and pressure. The extents of reactions represent the selectivity. The excess reactant ratio at the plant inlet is another design variable. It represents the ratio between the amount of the key reactant and that of the limiting reactant. A composition in the gaseous outlet stream can be substituted for the excess reactant ratio at the plant inlet.

(iii) *Evaluation.* Economic potential is computed by subtracting the costs of raw materials and feed pretreatment systems from the revenue from the product and the byproducts.

(iv) *Refinement.* The plant system identified at the input/output structure level is refined at the recycle structure level. The plant system is used as a boundary system at the recycle structure level.

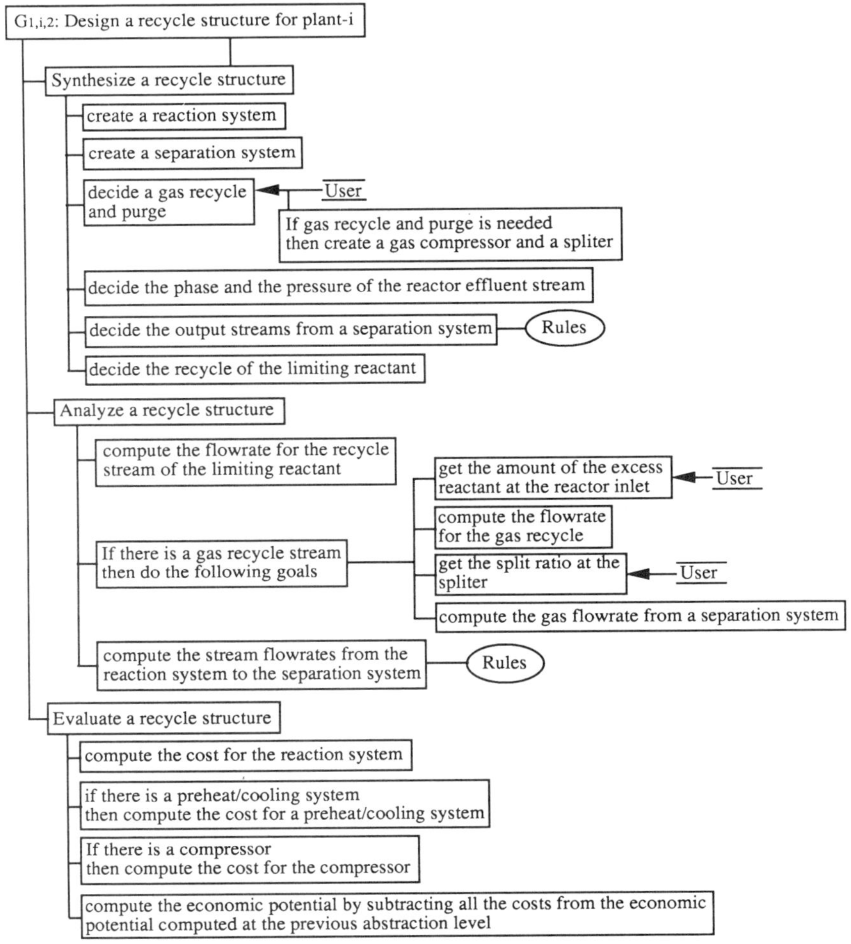

FIG. 10. Goal structure for the goal $G_{1,i,2}$, "Design a recycle structure for plant-i."

d. Goal-Structure 4: Design the Recycle Structure. The input/output structure of "plant-i" is refined into a more structured network of two sections; the generalized reaction and separation subsystems (e.g., see Fig. 4c). Figure 10 shows the goal structure, depicting the tasks needed for satisfying the top-goal, $G_{1,i,2}$, following again the transformational model in Fig. 6. The design tasks are discussed below.

(i) *Synthesis.* It is assumed that the recycle structure consists of a reaction section and a separation section. It is also assumed that 99% recovery is the same as 100% recycle. Thus, the unreacted limiting reactant will be recycled completely. The user needs to make a design decision about either just purging the gas stream that has reactants, or recycling a part of the gas stream and purging the remaining gas stream. As a result of this decision, a gas compressor and a spliter may be added to the flowsheet. When the plant system has a secondary reversible reaction, the user has to make a design decision about whether it is desirable to shift the equilibrium by recycling the byproduct.

(ii) *Analysis*. As the number of design alternatives is small, the material balance can be solved based on the structural decisions done at the synthesis phase. The ratio between the amount of the limiting reactant and that of the key reactant at the reactor inlet is needed as a parametric decision variable. This design variable indicates the excess amount of key reactant at the reactor inlet.

(iii) *Evaluation*. Economic potential is computed by subtracting the costs for the reaction system and gas compressor from the economic potential computed at the input/output structure level. To compute the costs for those units, several assumptions, such as the reactor type and the kinetic model, should be made. More detailed algorithms and example applications are available in Douglas' design text (Douglas, 1988).

(iv) *Refinement*. The separation system identified at the recycle structure level is refined at the general separation structure level. The separation system is used as a boundary system at the general separation structure level.

e. Goal-Structure 5: The General Separation Structure. The refinement of the generalized separation section leads to the superstructure of Fig. 4d, which accounts for all possible configurations of abstract separation subsystems. We will limit our attention to processes with fluids and single liquid phase only, thus eliminating the need for a filter, decanter and the associated subsystems of Fig. 4d. The resulting feasible options are shown in Fig. 11. Figure 12 shows the goal structure of the transformational model, with the corresponding tasks organized as follows.

(i) *Synthesis*. It is assumed that the general separation structure will be one of the three structures shown in Fig. 11. A structure is chosen on the basis of the phase of the reactor outlet stream. When the phase of the reactor effluent stream is not liquid, the user needs to make a design decision about whether it needs a vapor recovery system. If it does, the user should decide the location of the vapor recovery system. The location will be one of "on the recycle stream," "on the spliter outlet stream," or

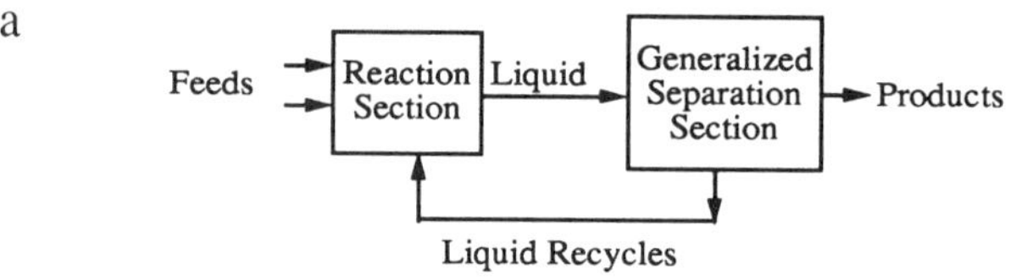

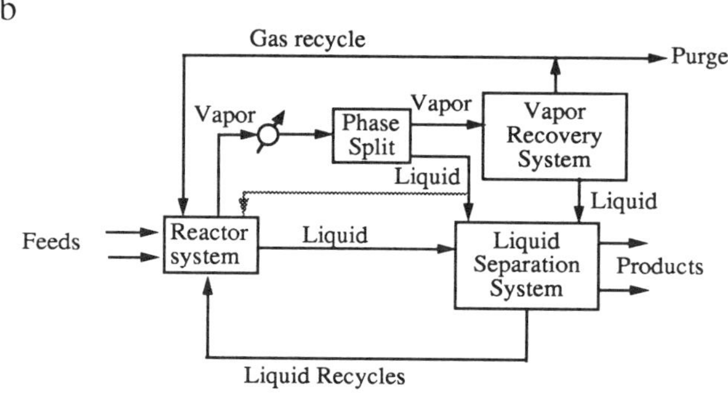

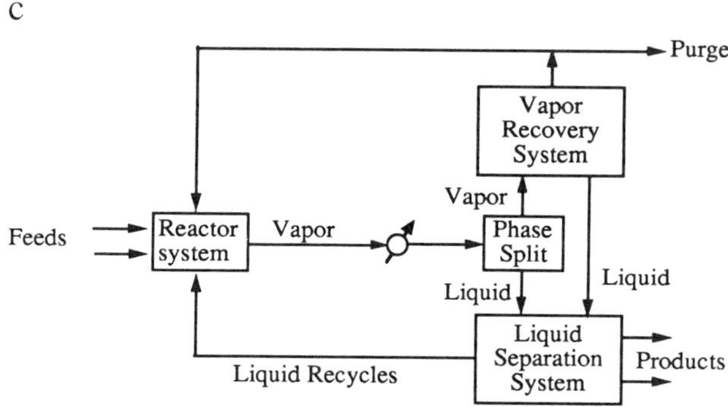

FIG. 11. Three possible general separation structures when reactor exit is (a) liquid (b) vapor, and liquid (c) vapor only [Reproduced from Hierarchical Decision Procedure for Process Synthesis, J. M. Douglas, *AIChE J.*, **31**, 353 (1985), by permission].

"on the purge stream." To help the user make this design decision, the system makes the flash calculation for the spliter and shows the flowrate and compositions of the vapor-phase outlet stream from the spliter. If the user decides to locate the vapor recovery system on the recycle stream to the reaction system, the control automatically goes back to the recycle structure level and locates the vapor recovery system on the recycle stream, and then the design procedure restarts at the general separation structure level.

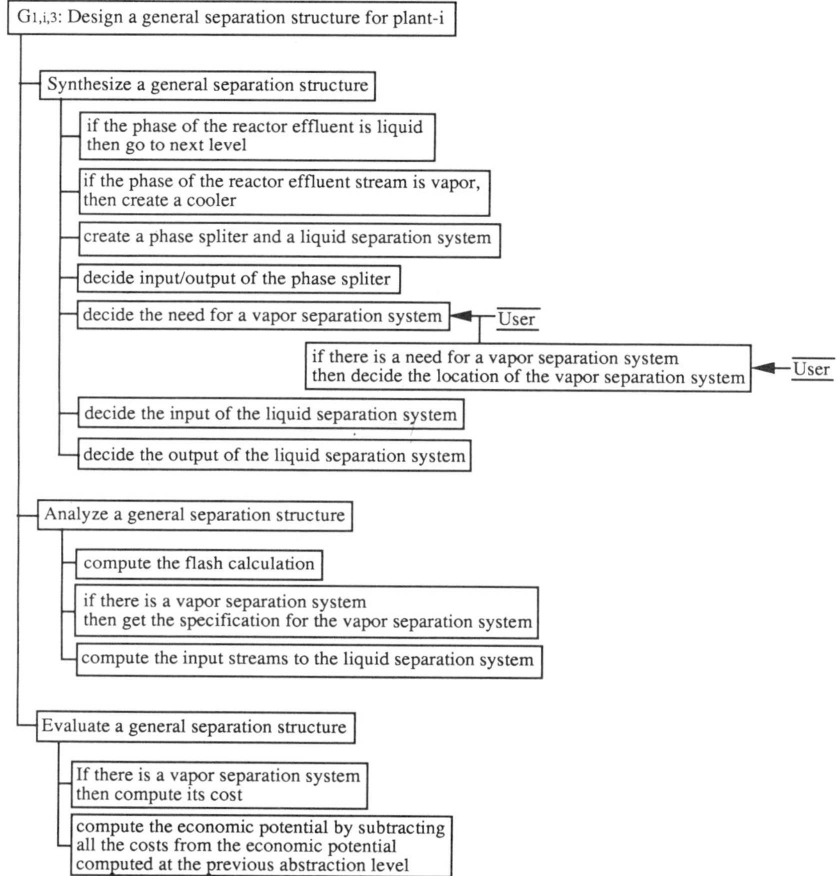

FIG. 12. Goal structure for the goal $G_{1,i,3}$, "Design a general separation structure for plant-i."

(ii) *Analysis.* If a vapor separation system is in a flowsheet, the user should give a specification for the vapor separation system, e.g., the existing components and their compositions at the liquid-phase outlet stream from the vapor separation system. As there are no more degrees of freedom, all the material and energy balance calculations can be done without any interaction with the user.

(iii) *Evaluation.* Economic potential is computed by subtracting the cost of vapor recovery system from the economic potential computed at the previous level. To compute the cost for a vapor separation system, the

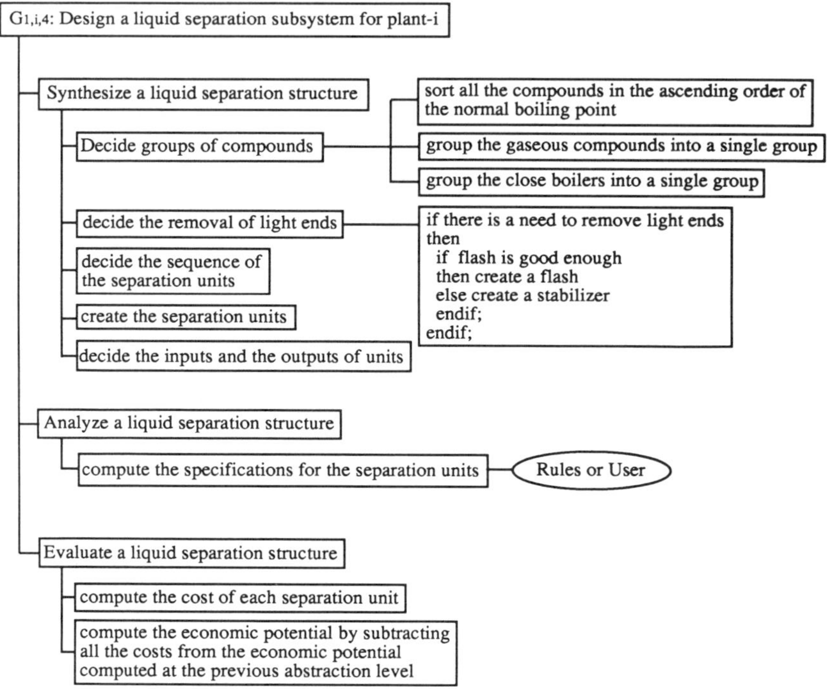

FIG. 13. Goal structure for the goal $G_{1,i,4}$, "Design a liquid separation subsystem for plant-i."

type of the system, e.g., absorption, condensation, should be specified. The algorithm for the absorber costing is available in Douglas' design text (Douglas, 1988).

(iv) *Refinement*. The liquid separation system identified at the general separation structure level is refined at the liquid separation structure level. The liquid separation system is used as a boundary system at the liquid separation structure level.

f. Goal-Structure 6: The Liquid Separation Subsystem. Figure 13 shows the goal structure and associated tasks during the design of the liquid separation subsystem.

(i) *Synthesis*. As the liquid separation system has only liquid feed streams, the only synthesis subproblem is that of the synthesis of the liquid separation sequence. If the input streams to the liquid separation system have light ends, then the methodology checks whether the light ends should be removed to satisfy the product specification. If light ends need to be removed, the methodology considers the flash as a means to remove the light ends. If the flash is not enough, then the design methodology decides to put a stabilizer column as a first column in the separation sequence. For the remaining components, the system decides the separation sequence. When the process has only three components to be separated, then, Eq. (1) is used to select either the direct or indirect sequence (Malone *et al.*, 1985). Whenever the left-hand side of Eq. (1) takes on a negative value, the indirect sequence is preferred.

$$\frac{\delta V}{F} = 1.2\left\{\left(\frac{x_B + x_C}{a_{BC}}\right)\left(\frac{x_A x_C}{1 + x_A x_C}\right) + \left(\frac{1}{a_{AC} - 1}\right)\left(\frac{x_C - f x_A + x_A x_C^2}{f(1 + x_A x_C)}\right) \right.$$

$$\left. - \frac{a_{BC}(x_A + x_B)(f - 1)}{(a_{AC} - a_{BC})f}\right\} - x_A, \quad (1)$$

where a_{xy} is a relative volatility between component x and component y, x_A is a mole fraction of component A in the feed stream, and f is a correction factor, $f = 1 + 1/100 x_B$.

When the process has more than three components, the selection is based on either heuristics (Douglas, 1988), or an implicit enumeration of the alternative sequences.

(ii) *Analysis.* It is assumed that in every separation unit, 99.5% of the light key is recovered in the overhead, and 99.5% of the heavy key in the bottom. It is also assumed that all the components lighter than the light key are taken overhead and that all components heavier than the heavy key leave with the bottom.

(iii) *Evaluation.* Economic potential is computed by subtracting all the costs of the distillation units from the economic potential computed at the previous abstraction level. To compute the cost of each separation unit, Fenske–Underwood–Gilland methods are used. To approximate the behavior of the distillation unit, a user can supply the average relative volatility for the column.

C. Design Principles of the Computational Model

Let us try to summarize the main principles (Han, 1994) on which a formalized computational model (outlined in the previous two sections) of the Douglas conceptual process design methodology was based (Douglas, 1988).

1. Principle 1: Hierarchical Planning of the Design Methodology

The overall design methodology was modeled as a hierarchical planning process. It starts with project specifications, goes through a predefined set of intermediate design milestones, and ends with a number of design alternatives.

2. Principle 2: Successive Refinement of Specifications into Implementations

The principle of hierarchical planning by itself cannot identify the design characteristics of the intermediate milestones. Successive refinement, on the other hand, does. Thus, at each stage of the hierarchical planning, we have a refinement of the specifications with more detailed description of the design characteristics of the next milestone. Figure 4 shows how the implementation specifications are successively refined from those of the input/output structure (first implementation), to recycle structure (second implementation), to generalized separation structure (third implementation), to separation subsystem structures (fourth implementation), etc. The generic *goal structure* of Fig. 5, based on the notion of *abstract refinement*, is the essential model that has been used to capture both the hierarchical planning and the successive refinement of the overall design methodology.

3. Principle 3: Propagation of Constraints

The top–down refinement of the abstract refinement model does not handle goal coupling well. Thus, the refinement of an abstract unit into a network of subunits, leads to design problems which could be solved independently; clearly, a gross violation of the overall system's intended functionality. Consequently, *constraint propagation*, to achieve consistency between the different parts of the design, is essential. Constraint propagation manifests itself through two distinct mechanisms; *induction of new constraints* and *propagation of constraint values*. Let us discuss two mechanisms in more detail since both appear during the synthesis of conceptual process design.

a. Propagation of Constraint Values. Consider the two abstractions of a process, shown in Fig. 14. At the input/output level (Fig. 14a), the input and output streams have flows as shown. At the recycle level (Fig. 14b), the input and output streams through the dashed boundary must have the same values, if the two abstractions are to be consistent. Thus, constraints on the flows of A, B, P, and (BP, A) have been propagated from one abstraction to the next.

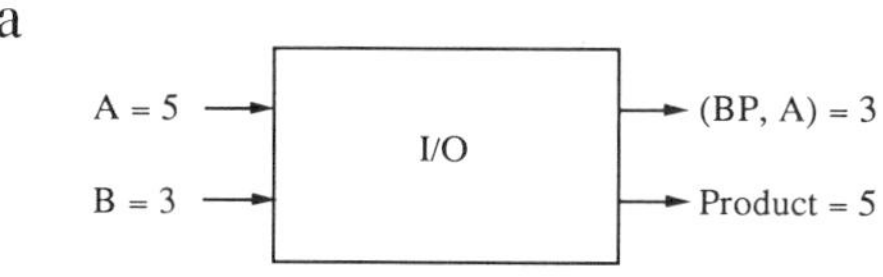

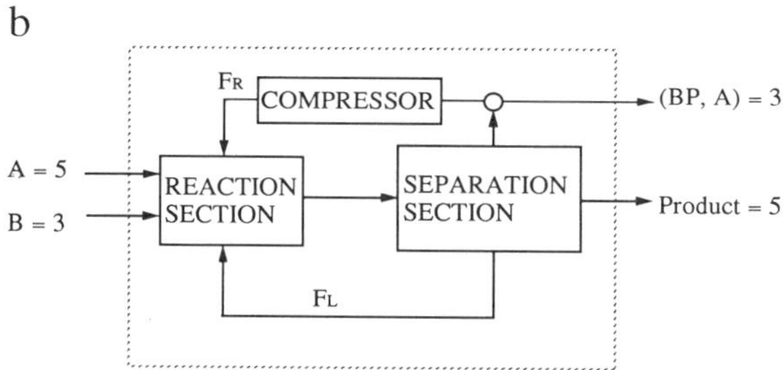

Economic Potential = 10 - Annualized_Compressor_Cost(F_R)
- Annualized_Reactor_Cost(F_R) > 0

Fɪɢ. 14. Schematic used to indicate propagation of constraints from the input/output structure (a) to the recycle structure (b).

b. Induction of New Constraints. The design abstraction of Fig. 14b has revealed the presence of a compressor on the gas recycle stream. Since the economic potential must remain positive, the following constraint on the value of F_R is induced:

$$\text{Economic potential} = 10 - \text{annualized\_compressor\_cost}(F_R)$$

$$- \text{annualized\_reactor\_cost}(F_R) > 0$$

Thus, as design constraints are propagated from abstraction i, to abstraction $i + 1$, they also induce new constraints that must be satisfied by the implementation at level $i + 1$.

4. Principle 4: Hierarchical Decomposition with Coordination

Each refinement consists of two steps: a decomposition and a coordination. Whenever a system is decomposed into a set of subsystems, a specific knowledge about how the current system should be decomposed into a set of subsystems is required. This knowledge should come from the problem context. For example, at the recycle structure level, the knowledge that a plant system will be decomposed into a reaction system and a separation system, is specific only for the decomposition at the recycle structure level. At the coordination phase, the inputs and outputs of all process units and their interconnections are identified. A coordination step is necessary because the decomposition of a system into a set of subsystems introduces additional details to the design. It should be emphasized that the coordination is accomplished through the two constraint propagation mechanisms discussed above.

5. Principle 5: Unified Transformational Design

The model of abstract refinement is satisfactory only for the depiction of the intermediate design milestones. It is unacceptable as the model to describe the design steps at each abstraction level. Instead, we have relied on the transformational model represented by the generic goal structure of Fig. 6. Therein we notice that the generation of an intermediate design is accomplished through a series of tasks, which preserve the correctness (i.e., constraint satisfaction and optimality) of the derived designs. It is important to note that the generic goal structure of Fig. 6 is an abstract notation of the intended transformations. Thus, it does take different forms depending on the specifics of the design methodology employed. For example, the goal structures of Figs. 9, 10, 12, and 13, all of which are manifestations of the transformational model, *are not unique* and simply signify the methodological design steps employed by the Douglas approach. One can envision an implicit enumeration algorithm, where the various tasks correspond to the computational steps of the particular algorithm.

In the present work, we have opted for an explicit description of the intended design goals along with the associated design tasks, thus giving rise to explicit transformational models at each level of abstraction, as shown by the detailed tasks in Figs. 9, 10, 12, and 13.

Furthermore, it is important to underline that the model of the unified transformational design has an internal structure that is fairly generic, and in all likelihood would be present in any specific implementation, namely, the cycle of *synthesis, analysis, and evaluation*. As can be seen from the goal structures in Figs. 9, 10, 12, and 13, synthesis, analysis and evaluation are the three common pivotal goals around which the specific design implementations are expressed.

Synthesis is the first and represents the activity of generating structural designs. When the design specifications are given as an input, the structure is synthesized by heuristic reasoning, algorithmic procedures, or based on the interaction with the user. During the synthesis phase, all the process units and the process streams are identified. Thus, the design is given to the analysis phase.

In the second phase, analysis implies the setting up and solving material and energy balances for the synthesized structure. When the degrees of freedom for the structure exceed zero, or the structure is underdefined, optimization search algorithms or the user supply as many values as the degrees of freedom. These variables are selected by the design heuristics.

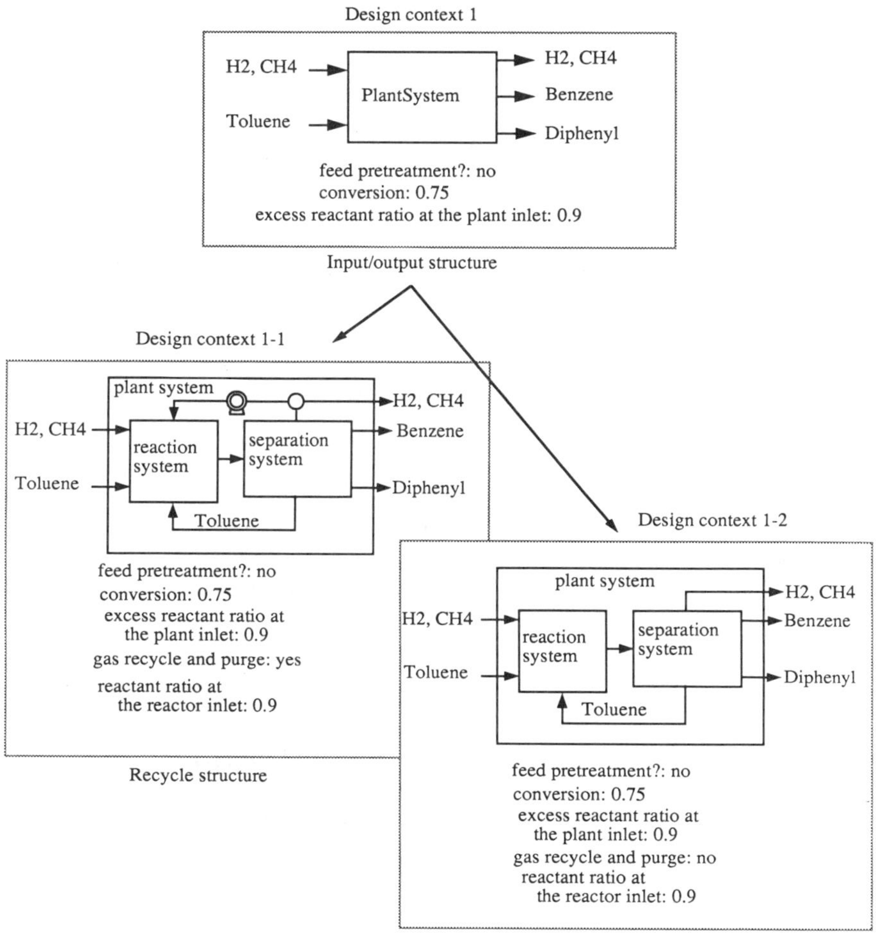

FIG. 15. Context-based design. This figure illustrates the scoping relationships among design contexts. The design decisions at the recycle structure are within the scope of the design decisions at the input/output structure level.

After the analysis phase, all process units and streams have quantitative values (flowrate, compositions, etc.).

Evaluation is the third phase and represents the activity of evaluating the structure in terms of economic measures, economic potential (*EP*), defined as follows:

$$
EP_i = \begin{cases} \text{revenue} - \text{cost}, & \text{at } i = 0, \\ EP_{i-1} - \text{cost} & \text{at } i > 0, \end{cases} \tag{2}
$$

where *i* represents the abstraction level.

Clearly, when implicit enumeration techniques are used, the parametric and structural optimization carries out synthesis, analysis, and evaluation in an integrated, unbroken cycle of simultaneous computations.

6. Principle 6: Context-Based Design

The design at the current abstraction level is used as a specification that will be implemented at the next abstraction level and defines the scoping relationship between the abstraction levels. This relationship is important to understand the mechanism for managing alternative designs. On the basis of this scoping relationship, we can identify the hierarchical dependency relationships among designs at all the abstraction levels. Figure 15 illustrates a design context generated from such relationships. The design decision "feed pretreatment?" has been made at the input/ output structure. This design decision affects not only the input/ output structure, but also all the designs refined from this input/ output. For example, a local decision, "gas recycle and purge," shown in Fig. 15, is within the scope of the global decision, "feed pretreatment".

III. HDL: The Hierarchical Design Language

In the previous section, we discussed the structure of a conceptual model that can be used to represent a design methodology. But the value of this model rests with the effectiveness of the representation schemes that one employs to describe the declarative and procedural components of the model in a way that the computer can "understand." Thus, we are led to the need of defining a design-oriented language for the description of the computational process.

From the discussion in Section II, it is clear that a design-oriented language must address the following two needs:

1. Multifaceted modeling of the various states of the evolving process design.
2. Modeling of the procedural design tasks, as these are described by the goal structures.

HDL (Han, 1994) has been designed to meet the above two classes of modeling needs. It has been influenced by other previous work on process-design-related languages, such as MODEL.LA. (Stephanopoulos *et al.*, 1990a, b; see also first chapter in this volume), ConStruct (Johnston, 1991), SYDERELA (Kritikos, 1991), and ASCEND (Advanced System for Computations in Engineering Design) (Piela, 1989). The structure of HDL conforms with the design principles discussed in the previous section, and the object-oriented modeling, and human–computer interface requirements discussed in Section I. HDL provides a fairly simple and powerful framework of modeling classes, which the user can extend to develop a customized design environment. It should be emphasized that HDL is a formal framework for the development of the computational process that emulates the Douglas methodology for conceptual process design. In the following paragraphs, we will discuss its specific characteristics.

A. MULTIFACETED MODELING OF THE PROCESS DESIGN STATE

As the process flowsheet of a chemical plant is designed, it goes through a series of design states with variable details, with each design state being a snapshot description of the process flowsheet. Such a description encompasses the physical entities that are parts of the design. In addition, it includes the relationships among these variables. Thus, the declarative representation defines what a design or partial design is. Clearly the rationale, expressed in the first chapter in this volume, for a language such as MODEL.LA., finds a perfect example in the needs of HDL. Therefore, it is not surprising that the multi-faceted modeling capabilities of HDL, have drawn heavily from the structure of MODEL.LA.

1. Basic Modeling Elements

When we describe the chemical processes, a model should fully highlight the problem at hand, although not at the expense of excluding other possibly coexisting models. For example, models of chemical processes have modeling primitives corresponding to process units (e.g., reactors, pumps, pipes). Models of chemical processes also have primitives for aggregated units (e.g., reaction system, separation system, plant systems). So, although both design tasks are in the same domain, and both designs ultimately are composed of the vessels, pumps, pipes, etc., different modeling elements are used to highlight the important features of the design. Designs using the appropriate modeling elements convey the necessary design information with greater comprehension, because modeling primitives match intuitive description elements. These primitives can then be used to create an effective declarative representation.

To account for all these requirements, HDL provides a set of basic modeling elements drawn and extended from MODEL.LA (see first chapter in this volume, Stephanopoulos *et al.*, 1990a), which are used to represent the states of evolving process flowsheets at varying levels of abstraction. They also provide an efficient vehicle for the representation of alternative process designs with consistent referencing to common design elements of the alternative processing schemes. The ability to provide hierarchical representations and consistent versioning, allows HDL a truly multifaceted modeling of processing schemes.

Let us present HDL's basic modeling elements. For a pictorial depiction of their character, see Fig. 16. The first three modeling elements are sufficient to represent the "structure" of any processing system. The following two elements are used to represent complex systems which consist of networks of instances of `GenericUnit`, `Port`, and `Stream`.

a. Modeling-Element 1. `GenericUnit`. This is identical in character to the MODEL.LA.'s modeling element with the same name. It is used to represent an isolated system by defining the boundary that separates it from the surroundings. Thus, any system can be modeled as a `GenericUnit`, if the system has a clear boundary between itself and the environment. For example, a plant can be modeled as a `GenericUnit`. A reactor can also be modeled as a `GenericUnit`. To distinguish a plant from a reactor, the `GenericUnit` has two subclasses that have more specific properties. They are the `GenericSystem` and the `GenericProcessUnit`. The `GenericSystem` is the unit that can be decomposed into subsystems. For example, within the context of the hierarchical decision procedure, the plant system at the input/output structure level is decomposed into a reaction system and a separation system at the recycle structure level. On the other hand, `GenericProcessUnit` is the unit that cannot be decomposed into subsystems without losing its physical identity. For example, when a reactor is decomposed into a vessel and a heating coil, a vessel does not have the function "reactor" any more.

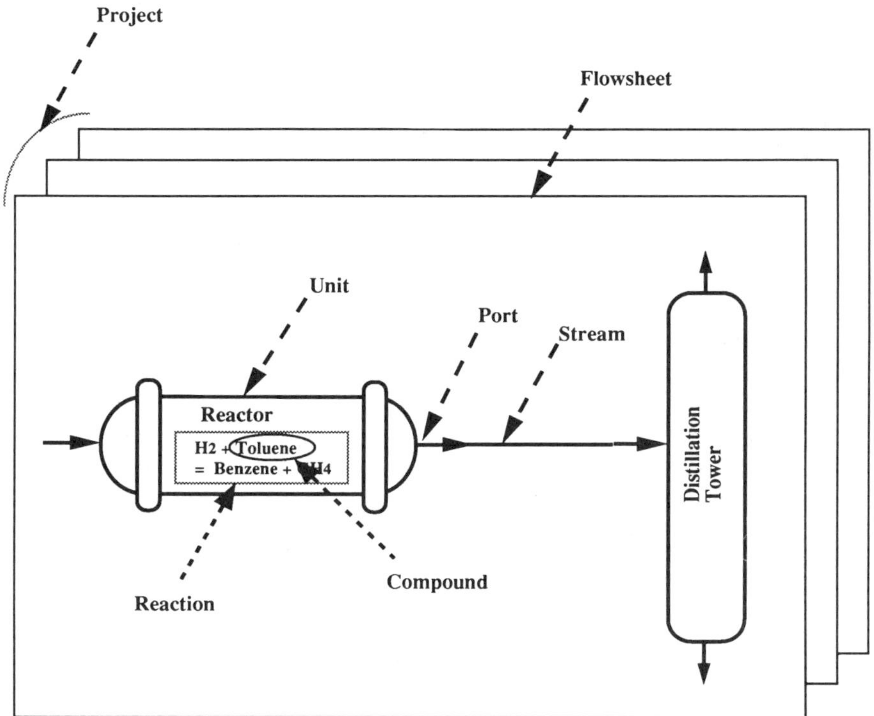

FIG. 16. Basic modeling elements in the HDL (hierarchical design language).

b. Modeling-Element 2: `Port`. As in MODEL.LA., instances of `Port` are used to define the boundaries of instances of `GenericUnit` and to provide the vehicles through which these instances transfer data to each other.

c. Modeling-Element 3: `Stream`. `Stream` is an abstraction that connects a port of an instance of `GenericUnit` to another port of another instance of `GenericUnit`. `Streams` know the `Ports` they are attached to, and symmetrically, each `Port` knows the stream to which it is attached. In logically consistent connections of `GenericUnits`, the flow through a stream from one port is connected to the another port of the same type.

d. Modeling-Element 4: `Flowsheet`. `Flowsheet` is an abstraction for a process flowsheet. It consists of process units and process streams. Except when the `Flowsheet` represents the design state at the highest abstraction level (least detail level), the `Flowsheet` has a boundary system which defines the design scope. The boundary system is transferred from the previous abstraction level. The user can monitor the design process through the `Flowsheet` and get the information which he/she wants to know from the `Flowsheet`. Each design alternative at each abstraction level is represented as an instance of `Flowsheet`.

e. Modeling-Element 5: `Project`. `Project` is an abstraction for a design project. An instance of `Project` keeps all the process flowsheets generated during the design project and handles the requests from a user, e.g., shows the `Flowsheet` at the recycle structure level. When a process alternative is generated, the alternative is also managed by the `Project`. Every instance of `Project` contains a tree-like data structure, in which it stores the state of the evolving `Flowsheet`. Specifically, as shown in Fig. 20, the root of the tree is the set of design specifications (empty `Flowsheet`), the first-level nodes represent alternative input-output structures, the second-level nodes the emanating recycle structures, etc. Clearly, the link between two connected nodes represents the design decision(s) generating one `Flowsheet` from the other.

f. Modeling-Element 6: `GenericVariable`. `GenericVariable` is an abstraction for a variable and is used to construct mathematical models for `GenericUnits`. It encapsulates the following information about a variable: variable name, variable value, possible range of values, and units. This additional information can be used to post constraints. These constraints are used to check whether the variable has a physically feasible value, e.g., a molar composition should be within the range from 0 to 1. The variable name and variable value type can be used to transfer information about the variable to the different environment, e.g. transfer the variable from HDL to Nexpert, which is a shell for the construction of expert systems, or copying a value from Nexpert to HDL.

g. Modeling-Element 7: `Compound`. `Compound` is an abstraction for a chemical compound. It has various elementary physical properties and the operations to access them from a database or to estimate derived properties through models and correlations. The elementary properties include normal boiling point, molecular weight, and critical temperature. The derived properties include the heat capacity at the given temperature or the vapor pressure at a given temperature.

h. Modeling-Element 8: `Reaction`. `Reaction` is an abstraction for a single reaction. It has information about the reactants, products and their stoichiometric coefficients, reaction temperature, reaction pressure, etc. It

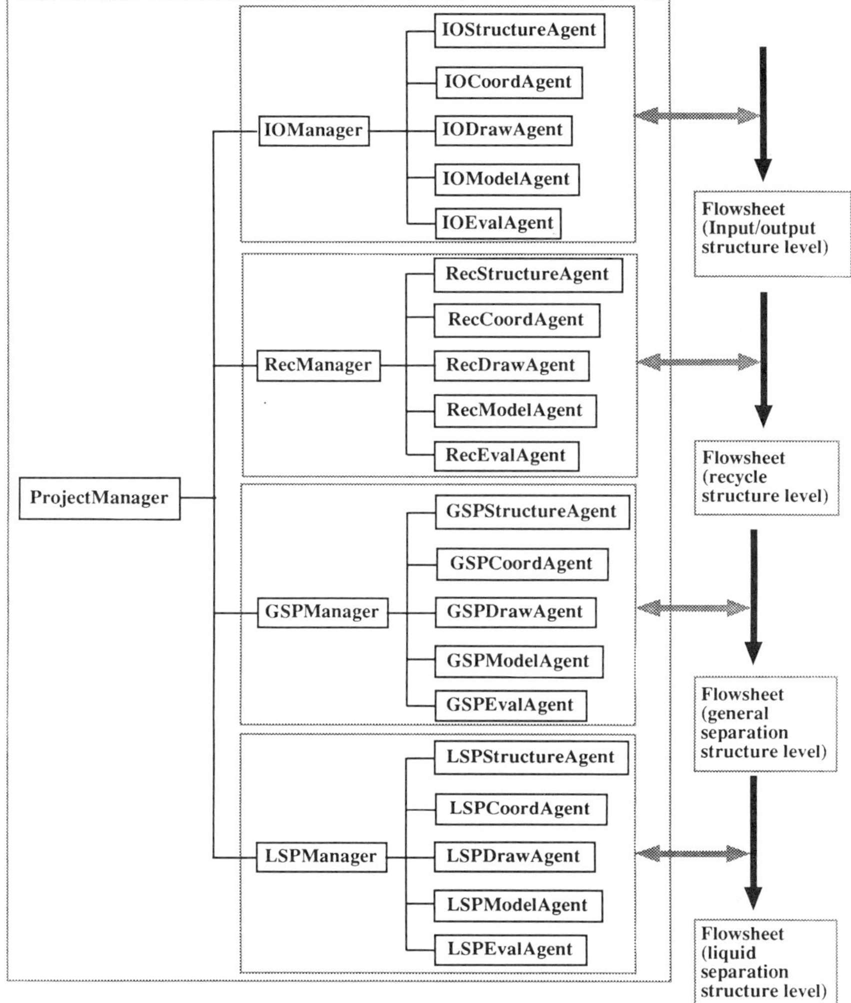

FIG. 17. Object model for design managers and design agents.

also has operations to estimate the properties such as the heat of reaction at a certain temperature, the equilibrium constant at a certain temperature and a pressure.

2. Semantic Relationships among Modeling Objects

Semantic relationships provide links between the various modeling elements, described above, and give rise to a network known as *semantic network*. Then, the description of a process is given by a semantic network. For more information on the nature of the semantic relationships and their role in modeling complex objects, the reader is referred to the more extensive discussion on the subject in the first chapter in the volume. For describing the structural aspects of a process plant, the following semantic relationships from MODEL.LA. are available in HDL. In the descriptions below, semantic relationships are written in boldface and italics. The class name starts with the uppercase letter, e.g., `Reactor`, and the instance name with a definite or indefinite article followed by its class name, e.g., `aReactor` or `theReactor`. A parenthesis represents a set, e.g., `(aReactor, aSeparator)`.

> *is-a*. The *is-a* relationship indicates the parent/child relationship of two classes, e.g., `Plant` *is-a* `GenericUnit`.
>
> *is-a-member-of.* The *is-a-member-of* relationship indicates the relationship between a class and its instance, e.g., `aReactor` *is-a-member-of* `Reactor`.
>
> *is-composed-of.* The *is-composed-of* relationship is used to identify the modeling objects that are parts of another object, e.g., `aHeatExchanger` *is-composed-of* `(aTube, aBundle, aShell)`.
>
> *is-part-of.* The *is-part-of* relationship is the inverse relationship of the *is-composed-of* relationship. For example, `aShell` *is-part-of* `aHeatExchanger`.
>
> *is-attached-to.* The *is-attached-to* relatinship provides the means for connecting process units and streams. The connection takes place through a linkage object called a port, e.g., `aPort` *is-attached-to* `aStream`.
>
> *is-connected-by.* The *is-connected-by* relationship is the inverse relationship of the *is-attached-to* relationship, e.g., `aGenericUnit` *is-connected-by* `aStream`.

In addition to the above, the following semantic relationships were found to be valuable and were introduced into the HDL.

> *is-a-model-of.* The *is-a-model-of* relationship indicates the relationship between a user interface and a model. An instance of a `GenericUnit` or its descendant classes represents the model of the process unit. For example, `ReactionSystem` has reaction information and has the operations to compute the heat of reaction and equilibrium constant, etc. A graphic unit associated with the reaction system represents the iconic structure of the reaction system, e.g., the number of input ports and the number of output ports. Figure 25 shows this relationship.

is-alternative-of. The ***is-alternative-of*** relationship relates a flowsheet to another alternative flowsheet. Both objects are of the same class and satisfy the same functions. However, they may have different structures, which result from different design decisions. Alternatives are used to keep track of the evolution of a flowsheet by recording the changes that are made to them.

B. Modeling the Design Tasks

In the previous section, we presented the modeling classes that allow one to describe the structure and behavior of chemical processing systems. Such a representation provides and structures the declarative knowledge about the current state of the design.

As a design is a transformation process of functional specifications into realizable physical objects, we need constructs that correspond to these transformations. A design task is a high-level construct that takes an action of transforming a design state to more detailed state, which satisfies the function of the design task. As a design task corresponds to the design action, the task has its own design goal and a design plan about how to accomplish its goal. Specific characteristics of design tasks are presented in this section.

1. Basic Task Elements

There are two kind of design tasks:design managers and design agents. A design manager is concerned primarily with organizing the computation, while the details of carrying out the steps are handled by design agents. Figure 17 shows the computational process for the various design tasks contained in HDL. The position of the task in the hierarchy shown in Figure 17 defines the role and the scope of the task during the design process. Figure 18 shows the inheritance relationships among design tasks.

a. Design-Task-Element 1: `GenericManager`. `GenericManager` can be considered a team leader who exercises control over subordinates, each of whom makes decisions within the context of its own bounded sub tasks. When subordinates encounter insurmountable difficulties in their sub tasks, some overall strategy may be introduced by the team leader to improve interactions between the subordinates.

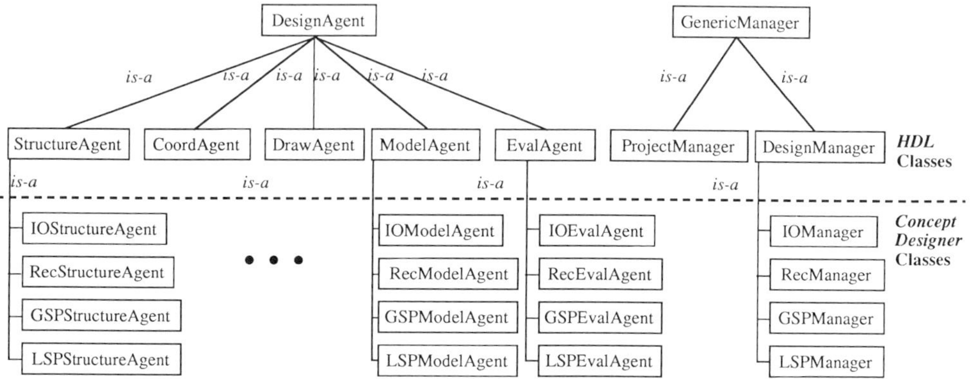

Fig. 18. The class tree of the design managers and the design agents in *ConceptDesigner*. The ancestor classes of `DesignAgent` and `GenericManager` exist, but not shown here. HDL classes are shown above the dotted line.

(i) *Design-Task-Element 1.1:* `DesignManager`. `DesignManager` is an abstract task that can be used to create a specific design manager. The classes that are descendants of the design manager can refine a process flowsheet from less detailed levels to more detailed levels. Each design manager carries out a functional transformation as shown below:

$$\texttt{aDesignManager(aFlowsheet}_i\texttt{)} \Rightarrow \texttt{aFlowsheet}_{i+1}$$

where subscript i denotes the abstraction level. For instance, `RecycleManager` refines a flowsheet at the input/output structure level into a flowsheet at the recycle structure level as follows:

$$\texttt{aRecycleManager(aIOStructure)} \Rightarrow \texttt{aRecycleStructure.}$$

(ii) *Design-Task-Element 1.2:* `ProjectManager`. `ProjectManager` is a task that organizes and coordinates the activities of design managers for a design project. Figure 17 shows `ProjectManager` at the top of the hierarchy among the design tasks. This hierarchy defines the roles of design tasks, while the class hierarchy defines the inheritance relationships.

b. *Design-Task-Element 2:* `DesignAgent`. `DesignAgent` is an abstract task that has a domain-specific knowledge. Specific design knowledge is represented as a design plan for each design agent. The following are subclasses of the `DesignAgent` class. Design agents 2.1–2.3 are sufficient to synthesize a structure and present it to the user in a very friendly manner. It should be noted that there is a certain order of executing these agents; `StructureAgent` first, `CoordAgent` second, and `DrawAgent`. The structural synthesis is completed by these agents. Agents 2.4 and 2.5 set up the material and energy balance and compute the economic potential of the flowsheet.

(i) *Design-Task-Element 2.1:* `StructureAgent`. `StructureAgent` is an abstract task that decomposes a given system into a set of subsystems and coordinates subsystems so that most of interconnections among sub systems are identified. The specific design knowledge for the decomposition and the coordination should be provided when the subclass is created for specific structure design. For instance, `aIOStructureAgent` shown in Fig. 17 has a knowledge and a design plan on how to synthesize an input/output structure. The `RecStructureAgent` shown in Fig. 17 has a knowledge and a design plan about how to sythesize a recycle structure. The interaction with the user is needed only when the agent does not have heuristic rules about the decision.

(ii) *Design-Task-Element 2.2:* `CoordAgent`. `CoordAgent` is an abstract task that establishes the connections among the `Ports` of the `GenericUnits` that do not have connections. Thus, this agent supplements the `StructureAgent` in terms of coordination.

(iii) *Design-Task-Element 2.3:* `DrawAgent`. `DrawAgent` is an abstract task that transforms a `GenericUnit` into an icon associated with the given `GenericUnit`. It also draws the streams that connect the ports of a generic unit to the other ports of generic unit based on the connection information of each port. It is a precondition that the connection information has been already identified by a `StructureAgent` and a `CoordAgent`.

(iv) *Design-Task-Element 2.4:* `ModelAgent`. `ModelAgent` is an abstract task that is in charge of the analysis. After the structure has been synthesized by the `StructureAgent`, a `ModelAgent` sets up and solves the material and energy balances, based on the synthesized structure. When design specifications are needed to finish the analysis, the agent may ask the user for these specifications.

(v) *Design-Task-Element 2.5:* `EvalAgent`. `EvalAgent` is an abstract task that is in charge of the evaluation. It computes the economic potential of the current process flowsheet.

2. *Semantic Relationships among Design Tasks*

The following list of semantics establishes the requisite relationships among the various design-task elements described above. The resulting semantic network (nodes are design tasks, edges are semantic relationships) models the overall design methodology.

decompose. The ***decompose*** relationship is used to decompose a boundary system into a set of subsystems. The syntax for this relationship is

`aStructureAgent` ***decompose*** `aBoundarySystem`

For example, `aRecStructureAgent` ***decompose*** `aPlantSystem`.

is-decomposed-into. The ***is-decomposed-into*** relationship is used to indicate the results of applying ***decompose*** relationship. The syntax is

`aGenericUnit` ***is-decomposed-into*** (a list of `aGenericUnits`)

For example, `aPlantSystem` ***is-decomposed-into*** (`aReactionSystem, aSeparationSystem`).

transform. The ***transform*** relationship is used to transform a list of generic units into a list of graphic units. Each graphic unit becomes associated with the corresponding generic unit. The syntax is

`aDrawAgent` ***transform*** (a list of `aGenericUnits`)

For example, `aIODrawAgent` ***transform*** (`aPlantSystem`).

is-transformed-into. The ***is-transformed-into*** relationship is used to indicate the results of applying transform relationship. The syntax is

(a list of `aGenericUnits`) ***is-transformed-into***
(a list of `aGraphicUnits`)

For example, (`aReactionSystem`) ***is-transformed-into*** (`aGraphicUnit`).

compute. The ***compute*** relationship is used to compute the material and energy balances for a flowsheet. A `ModelAgent` computes the material and energy balances for all the process units that exist on the flowsheet. The syntax is

`aModelAgent` ***compute*** `aFlowsheet`

For example, `aRecModelAgent` ***compute*** `aRecycleFlowsheet`.

evaluate. The ***evaluate*** relationship is used to evaluate the economic potential of a flowsheet. An `EvalAgent` evaluates an economic potential of a `Flowsheet` by sending the message "computeCost" to all the process units which exist on the flowsheet. The syntax is

`aEvalAgent` *evaluate* `aFlowsheet`

For example, `aRecEvalAgent` *evaluate* `aRecycleFlowsheet.`
refine. The *refine* relationship is used by a design manager to refine an instance of `aGenericSystem` at the current abstraction level into a more detailed flowsheet at the next abstraction level. The syntax is

`aDesignManager` *refine* `aGenericSystem`

For example, `aRecManager` *refine* `aPlantSystem` at the input/ output structure.
is-refined-into. The *is-refined-into* relationship is used to indicate the results of applying *refine* relationship. The syntax is

`aGenericSystem` *is-refined-into* `aFlowsheet`

For example, `aPlantSystem` at the input/ output structure level *is-refined-into* `aFlowsheet` at the recycle structure level.

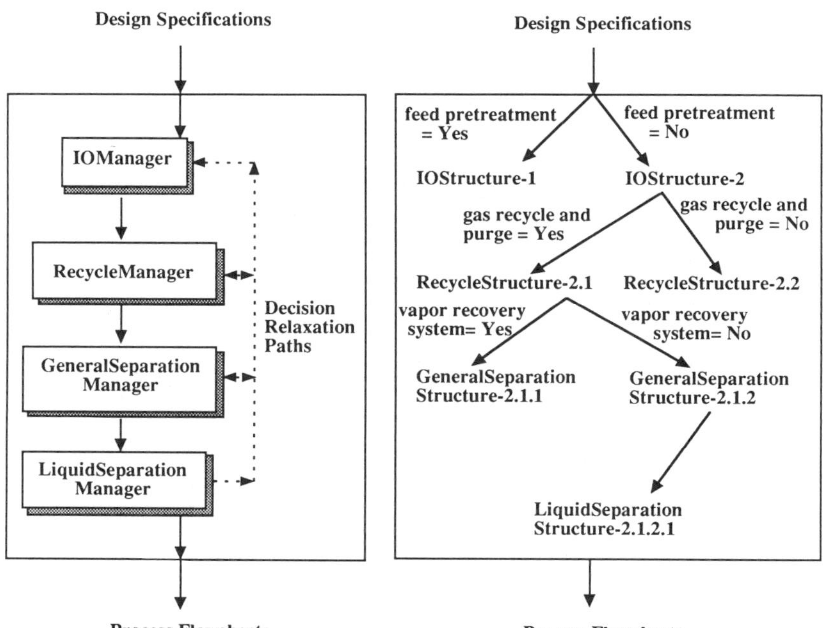

FIG. 19. Generation of design alternatives by relaxing design decisions.

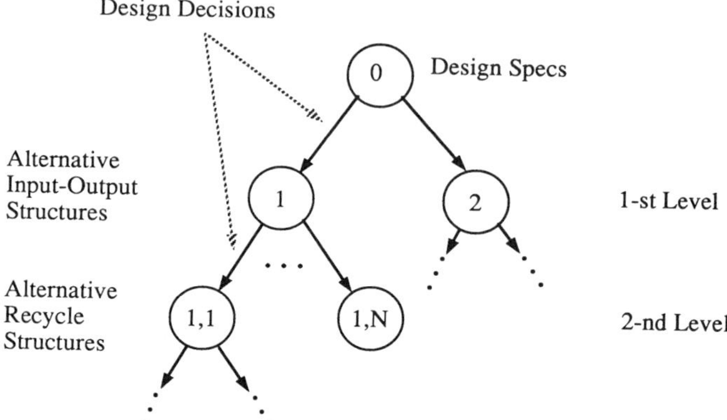

Fig. 20. The tree data structure maintaining design alternatives.

C. Elements for Human–Machine Interaction

In order to communicate with the user, a computational procedure should possess very informative user interfaces, such as those described in following paragraphs, which are part of the *ConceptDesigner* (see Section IV; see also Figs. 22–25).

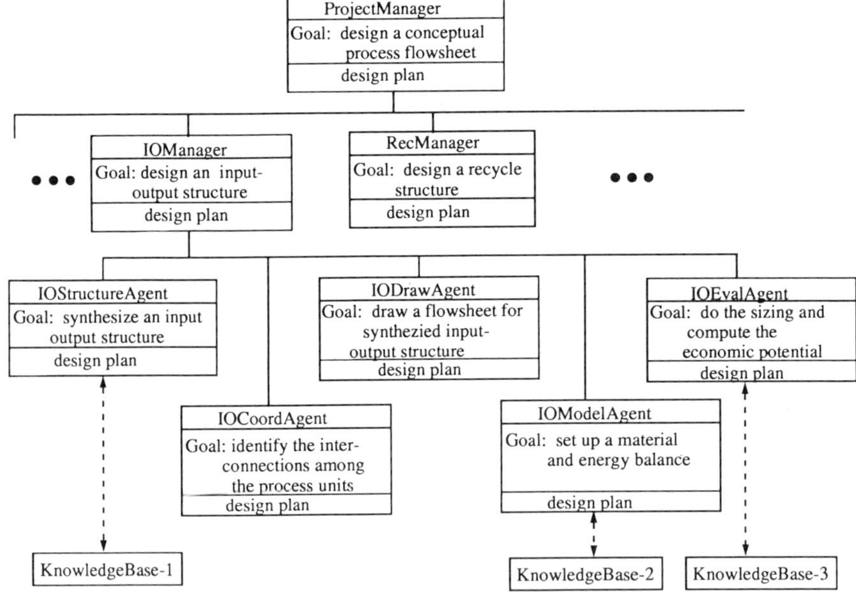

Fig. 21. Design task hierarchy in *ConceptDesigner*.

1. Human Interaction Element 1: `MainWindow`

This is an abstraction for the main interface. It provides several capabilities to define a design project. A user can define new project, load an old project, or save a current project into a file through `MainWindow`. The user can also define a new physical property database, load an existing database, modify the loaded database, and save the loaded database to a file. The user can specify the design mode: an automatic design mode, or an interactive design mode. In the automatic design mode, the design will be done by the system and the system will ask questions when they are needed. In the interactive design mode, the user can control the design process by executing each design task.

2. Human Interaction Element 2: `LayoutWindow`

This is an abstraction for the work space. A process flowsheet is developed and presented at the `LayoutWindow`. All design tasks are applied to an instance of `LayoutWindow` to transform a current process flowsheet. A snapshot of the transition in the process flowsheet is a design state at the moment. Figures 22 and 23 show the snapshots of a `LayoutWindow` during the design of hydrodealkylation-of-toluene process.

The elements described above are places where new modeling elements are designed and their values are displayed. The displayed elements are associated with the modeling elements and provide the user interface through which the user can access all the information generated during the design process. Those elements are presented below.

3. Human Interaction Element 3: `GraphicUnit`

This is an abstraction associated with a `GenericUnit`. `GraphicUnit`, displayed on the display area, is a means through which the user can communicate with the `GenericUnit`. A user can see the values of the design parameters. The user can also modify the graphic elements of the process unit, i.e., move the icon to the different place, rotate, or zoom. In Fig. 23b, `aGraphicUnit` displays the information of the associated distillation unit.

4. Human Interaction Element 4: `GraphicPort`

This is an abstraction associated with a `Port`. A user can modify the `GraphicPort` graphically, i.e., move, rotate, or zoom. The user can also see the values of `aPort` associated with the `GraphicPort`. In Fig. 23a, a `GraphicPort` shows the information of the associated port.

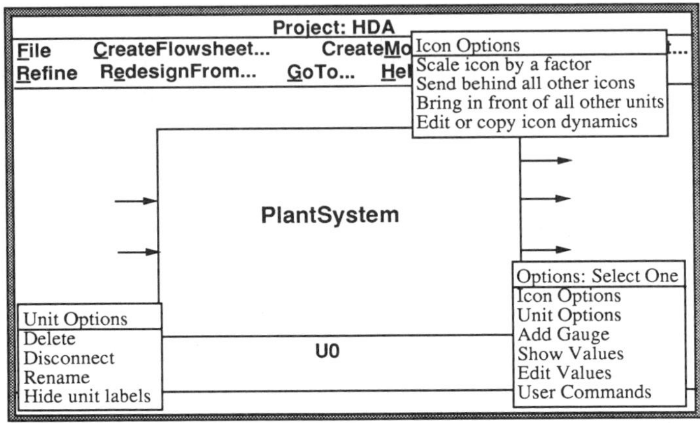

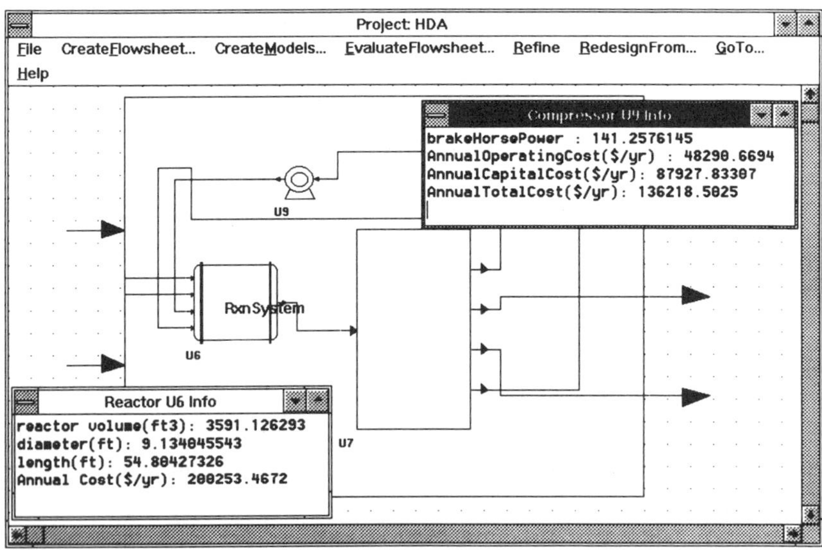

FIG. 22. User interfaces at (a) input/output structure level and (b) recycle structure level.

a

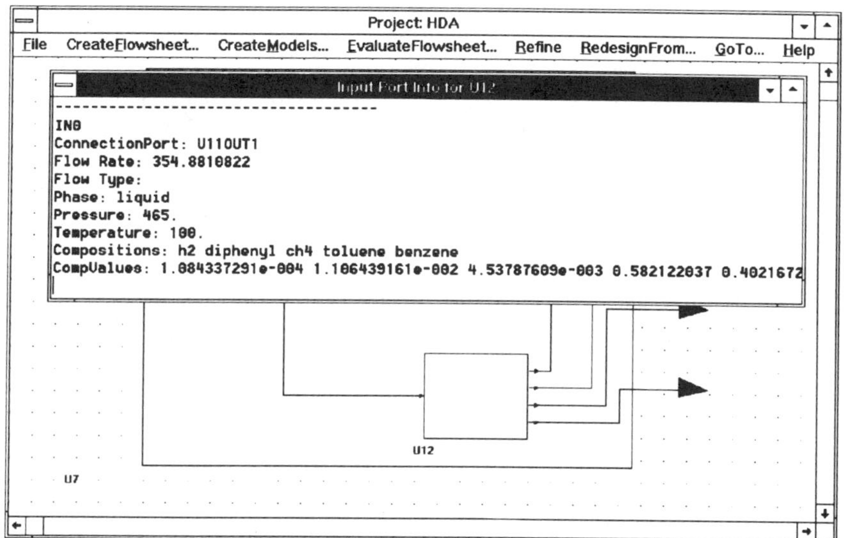

b

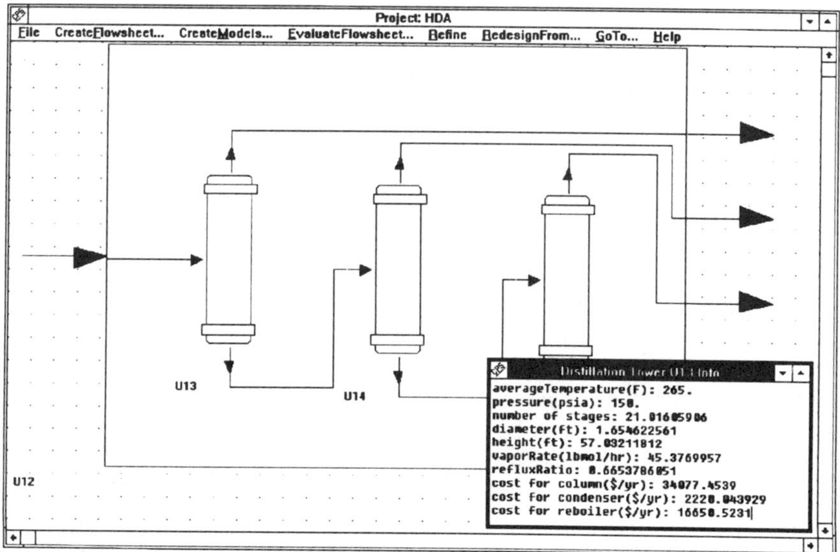

Fig. 23. User interfaces at (a) general separation structure level and (b) liquid separation structure level.

5. *Human Interaction Element 5:* `GraphicStream`

This is the abstraction associated with `Stream`. Like a `GraphicUnit`, the user can modify a `GraphicStream` graphically. The user can also see the values of a `aStream` associated with a `GraphicStream`.

D. OBJECT-ORIENTED FAILURE HANDLING

In producing a design, the software system will design a process flowsheet in a direction specified by its design plan. During the design process, the user may encounter obstacles or commit errors. Examples include violations of design constraints and numerical errors. These errors or failures are likely to occur more frequently in the conceptual design where HDL is intended to be used.

HDL models the way of producing a design as hierarchically structured design tasks. A hierarchically structured model leads to hierarchically structured failure handling. As shown in Fig. 17, each design agent has an explicitly defined role. The explicit structure of the model of the design process makes it possible to know what to do in the even of failure. When a failure occurs, the relevant information becomes immediately accessible due to the explicit localization of the failure. In some case, the system can guide the user to the point where the control should backtrack. Therefore, it is easy to correct the local error, and then to start a redesign from the agent which is responsible for the error.

E. MANAGEMENT OF DESIGN ALTERNATIVES

As the design decisions committed at the conceptual design fix 70–80% of the project cost, it is very important to design the economically optimal process flowsheet. A mechanism for managing alternatives has been developed, based on the characteristics of the conceptual process design itself: evolutionary, iterative decisionmaking process. The mechanism helps the user generate design alternatives by relaxing the previous design decisions.

The user can use the mechanism in the following way. First, he or she develops the base-case design which does not violate any process constraints. Then, the user improves the base-case design in the evolutionary manner by relaxing the decisions that look relatively promising in terms of contribution to the economic potential of the process flowsheet. Suppose we are working on the recycle structure level and we find one design decision about the gas recycle difficult to make because there are no heuristics available for the decision. We must then make a decision to recycle and purge gas components. After we complete the design of the recycle structure by deciding the values for the design variables and computing the economic potential, we can go back to the input/output structure level and restart the design. As all design tasks (including design managers and design agents) work as independent objects, it is a simple matter to go back to any abstraction level and start a redesign. The loaded design managers do the design in the same way except the decision about the gas recycle (e.g., we will choose not to recycle gas components). Figure 19 shows the design alternative generation process by relaxing design decisions. Figure 20 shows the tree structure which is dynamically generated by *ConceptDesigner* and is used to maintain design alternatives. The design decisions are recorded to a file and retrieved from the file along with design alternative. This helps a user identify the design decisions associated with a design alternative and relax them.

IV. ConceptDesigner: The Software Implementation

ConceptDesigner (Han, 1994) is the software system that implements the computational model, described in Section II, of the design process, using the linguistic constructs of HDL. The *ConceptDesigner* was designed to automate large segments of the design process with minimal interaction with the user. Nevertheless, a highly informative user-interface allows the human designer to monitor continuously the progress of the evolving design, and to override the automatic mode with a manual interactive mode of decisionmaking.

A. Overall Architecture

ConceptDesigner is an object-oriented software system, which was built entirely on the top of the HDL design language. Figure 18 shows the complete set of HDL objects that *ConceptDesigner* used to create its own customized design tasks. In the design of *ConceptDesigner*, there are one-to-one mappings between the design-oriented tasks, identified in Section II, and the modeling elements of HDL, discussed in Section III. In other words, the structure of the software objects in *ConceptDesigner* is in direct and explicit correspondence to the design methodology we have

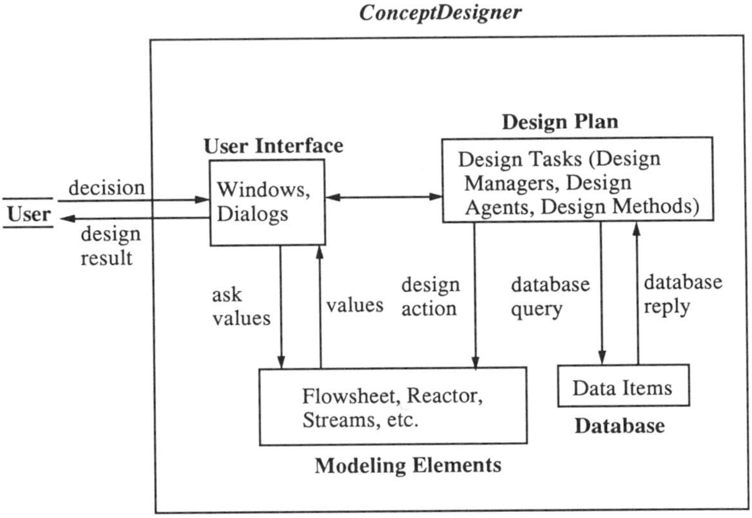

ConceptDesigner

FIG. 24. Functional modules of *ConceptDesigner*.

adopted for the synthesis of process flowsheets. For example, Fig. 21 shows the hierarchical interrelationship among various objects (provided by HDL) in *ConceptDesigner*. Notice the one-to-one correspondence of the objects in Fig. 21, to the design tasks of the goal structure in Fig. 5, which models the design methodology. For example, the object `Project-Manager` in Fig. 21 corresponds to the goal "design a conceptual chemical process" of Fig. 5. Similarly, we can see the following correspondence; objects, `IOManager`, `RecManager` in Fig. 21 correspond to goals "design an input/ output structure" and "design a recycle structure" of Fig. 5. Furthermore, the objects in Fig. 21, `IOStructureAgent`, `IOModelA-gent`, and `IOEvalAgent`, correspond to the goals of Fig. 9, "synthesize an input/ output structure," "analyze an input/ output structure," and "evaluate an input/ output structure."

Figure 24 shows the four major modules of *ConceptDesigner*. Figure 25 illustrates the semantic relationships among major objects in those modules.

1. Design Plan Module

This contains all the design-task modeling elements of HDL, and has structured them, through message passing methods, in such a way that

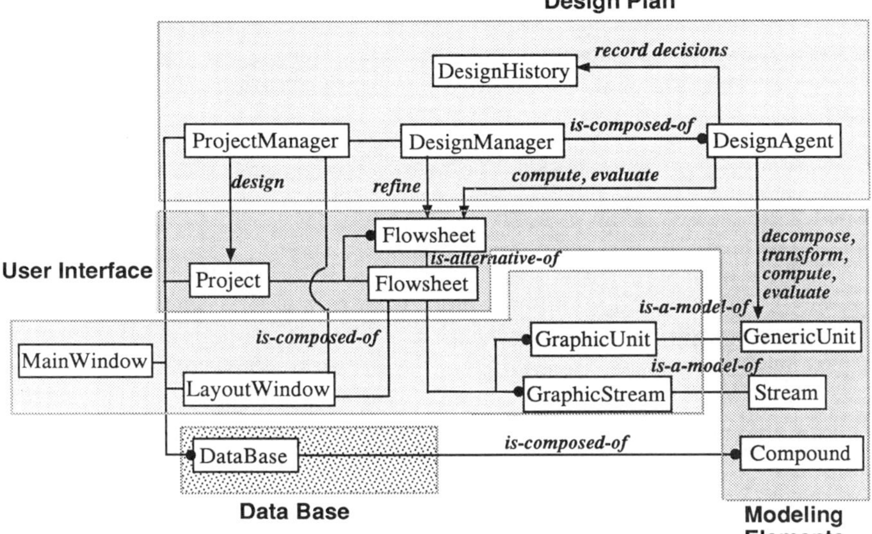

FIG. 25. Object model for *ConceptDesigner*.

they implement the goal structures of Figs. 8, 9, 10, 12, and 13, which model the design methodology. Thus, the `ProjectManager` can access information about the process design at any stage of its development, possess a "design plan" and calls on the `DesignManagers` to execute it. These managers are the softward objects: `IOManager`, `RecManager`, `GSPManager`, and `LSPManager` (see Fig. 17). Each of these managers possesses the data and methods to carry out the design of input/output structure, recycle structure, generalized separation structure, and liquid separation subsystem, respectively. They carry out their corresponding design tasks through the activation of five software objects, the `DesignAgents`, with common structure but different implementations. Figure 17 shows the structure of the specific `DesignAgents`, used at various stages of the design processed. The various `DesignAgents` contain specific methods, which are used to define the structure of processing systems, set up mathematical models, solve mathematical models, do sizing and costing of units, call on databases for physical properties, make decisions, etc. These methods are implemented through algorithms that contain symbolic manipulations, numerical procedures, or/ and production rule-based systems.

2. Modeling Objects Module

This contains all the modeling elements that HDL uses to provide contextual, hierarchical and multi-view representation of process flowsheets (see Section III, A).

3. User Interface Module

This contains a broad variety of classes that the *ConceptDesigner* acquired from *x-Kit* (a computer-aided environment for the expeditious tailoring of graphic-oriented software modules, developed at MIT-LISPE). These classes provide the following capabilities: construct icons for abstract or detailed processing units; construct iconic flowsheets; provide animation and dynamic views to the process flowsheets; dialog windows, tables, lists, control buttons, and other user interface elements; and graphs, and charts for the presentation of data. *x-Kit* has also provided the following facilities that can be used for linking *ConceptDesigner* with external programs and databases:

(a) *File-Linker*. A high-level editor to configure the mapping of data between the objects of *ConceptDesigner* and external ASCII (American Standard Code for Information Interchange) files. These mappings can be used to transfer data to input files for an external FORTRAN program, or retrieve data from output files that carry the results of the FORTRAN program.

(b) *C-function Linker*. A program that uses dynamic link libraries and allows any object-method to be implemented through an external C function.

(c) *Database-Linker*. A high-level editor to configure the mapping of data between the tables of external relational databases and the *ConceptDesigner*'s objects. Among the many graphic user interfaces, the *LayoutWindow* occupies a central position, since this is the main canvas where the flowsheet is shown. Various instances of the *LayoutWindow* depict the input/output, recycle, generalized separation, liquid separation, and other abstractions of the evolving flowsheet (see Figs. 22 and 23).

Figure 26 provides a different view of the *ConceptDesigner*'s structure. It shows that every project generates a series of distinct process flowsheets, all of which are handled by a file management system, which is organized in a hierarchical tree to reflect the evolution of the design and the various alternative designs. It also depicts the various sources of data

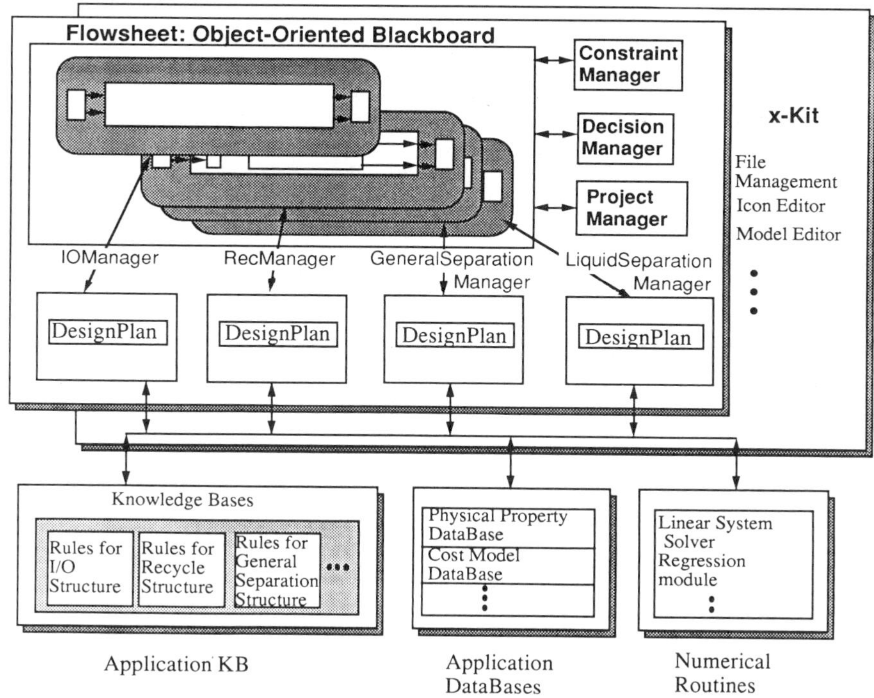

Fig. 26. Overall architecture of *ConceptDesigner*.

used and the types of knowledge bases and numerical procedures employed.

B. Implementation Details

HDL has been implemented in Actor, which is an object-oriented language. The user interfaces have been developed from Microsoft Windows 3.1. *ConceptDesigner* can be run in any PC (personal computer)-compatible computer.

Conceptual process design is a complex process. As the design goes through several abstraction levels and generates many design alternatives at each abstraction level, a huge space is needed to store all information which is not usually available in PC or workstation environment. *Concept-Designer* uses a dynamic memory allocation and a paging to resolve these memory problems. When an object is temporarily created, a certain amount of memory is allocated to the object. When the object is no longer needed, the allocated memory is immediately recovered by a garbage collection system. Another technique is a paging. After a process flowsheet is designed and the design process moves to the next abstraction level, the current process flowsheet is automatically saved to a file with a unique name. The memory consumed by the flowsheet is automatically recovered. These techniques enable *ConceptDesigner* to deal with large design cases.

V. Summary

In this chapter, we have tried to present a detailed approach for automating a complex design methodology. Three requirements are considered pivotal for the execution of projects, namely:

1. An explicit and formal model of the design methodology must be created. This model is the essence of the computational procedure to be developed.
2. A modeling language is needed to provide multi-faceted representation of the evolving design artifact, and represent the design tasks and their semantic relationships.
3. An object-oriented implementation of the software system is absolutely necessary.

The *ConceptDesigner* is a software system that was developed to automate large segments of Douglas' methodology for the synthesis of conceptual process designs. It satisfies all three of the above requirements. The most important lesson learned can be summarized as follows:

1. Hierarchical planning is an essential ingredient for modeling explicitly a design methodology. It sketches the general specs of the intermediate design milestones, without restricting the design activities too severely. Goal structures based on the concept of abstract refinement offer a very good model for describing the design plan.

2. Goal structures based on the notion of a unified, global transformation represent a very attractive generic model for the representation of the design methodology that transforms the artifact from an intermediate design milestone to the next. Such goal structures can accommodate equally well depth-first evolution of designs (such as the one suggested in Douglas' methodology), or exhaustive searches through implicit enumeration of design alternatives (such as those used in the MILP or MINLP formulation of process synthesis problems).

3. We cannot overstate the significance of modeling languages. A well-defined language captures explicitly, through its modeling elements,

domain-specific knowledge, and structures in a generic manner most of the design tasks.

4. The software that automates segments of a design activity should represent an explicit map of the design methodology. This is a necessary and sufficient condition in order to produce a software system which is (a) maintainable and (b) open to modifications and extensions with new knowledge.

References

Balzer, R., Goldman, N., and Wile, D., On the transformational implementation approach to programming. *Proc. Int. Conf. Software Eng.*, *2nd*, p. 337 (976).

Chandrasekaran, B., Generic tasks in knowledge-based reasoning: high-level building blocks for expert system design. *IEEE Expert* **1**, 3 (1986).

Douglas, J. M., As hierarchical decision procedure for process synthesis. *AIChE J.* **31**(3), 353 (1985).

Douglas, J. M., "Conceptual Design of Chemical Processes." McGraw-Hill, New York, 1988.

Douglas, J. M., Synthesis of multistep reaction processes. *In* "Foundations of Computer-Aided Process Design" (J. J. Siirola, I. E. Grossmann, and G. Stephanopoulos, eds.), p. 79. CACHE Corp., Austin, TX, and Elsevier, New York, 1989.

Edgar, T. F., and Himmelblau, D. M., "Optimization of Chemical Processes," pp. 413–422. McGraw-Hill, New York, 1988.

Grossmann, I. E., Mixed-integer programming approach for the synthesis of integrated process flowsheets. *Comput. Chem. Eng.* **9**, 463 (1985).

Grossmann, I. E., MINLP optimization strategies and algorithms for process synthesis. *In* "Foundations of Computer-Aided Process Design" (J. J. Siirola, I. E. Grossmann, and G. Stephanopoulos, eds.), p. 105. CACHE Corp., Austin, TX, and Elsevier, New York, 1989.

Han, C., Human-aided, computer-based design paradigm: The automation of conceptual process design. Ph.D. Thesis, Department of Chemical Engineering, Massachusetts Institute of Technology, Cambridge, MA (1994).

Johnston, J., Synthesis of control structures for complete chemical plants. Ph.D. Thesis, Department of Chemical Engineering, Massachusetts Institute of Technology, Cambridge, MA (1991).

Kirkwood, R. L., PIP-process invention procedure, a prototype expert system for synthesizing chemical process flowsheets. Ph.D. Thesis, Department of Chemical Engineering, University of Massachusetts, Amherst (1987).

Kirkwood, R. L., Locke, M. H., and Douglas, J. M., A prototype expert system for synthesizing chemical process flowsheets. *Comput. Chem. Eng.* **12**, 329 (1988).

Kocis, G. R., and Grossmann, I. E., A modelling/decomposition strategy for MINLP optimization of process flowsheets. *Comput. Chem. Eng.* **13**, 797 (1989).

Kritikos, T., A model for process design automation. Ph.D. Thesis, Department of Chemical Engineering, Massachusetts Institute of Technology, Cambridge, MA (1991).

Mahalec, V., and Motard, R. L., Procedures for the initial design of chemical processing systems. *Comput. Chem. Eng.* **1**, 57 (1977).

Malone, M. F., Glinos, K., Marquez, F. E., and Douglas, J., Simple, analytical criteria for the sequencing of distillation columns. *AIChE J.* **31**, 683 (1985).

McKenna, T. F., and Malone, M. F., Polymer process design. I. Continuous production of chain-growth homopolymers. *Comput. Chem. Eng.* **14**, 1127 (1990).

Mitchell, T., Steinberg, L., Reid, G., Schooley, P., Jacobs, H., and Kelly, V., Representations for reasoning about digital circuits. *Proc. IJCAI*-81 (1981).

Mostow, J., Toward better models of the design process. *AI Mag.* Spring, p. 44 (1985).

Newell, A., The knowledge level. *AI J.* **19**(2), 87 (1982).

Piela, P. C., ASCEND—An object-oriented environment for the development of quantitative models. Ph.D. Thesis, Department of Chemical Engineering, Carnegie Mellon University, Pittsburgh, PA (1989).

Powers, G. J., Heuristic synthesis in process development. *Chem. Eng. Prog.* **68**, 88 (1972).

Quantrille, T. E., Liu, Y. A., "Artificial Intelligence in Chemical Engineering." Academic Press, San Diego, CA, 1991.

Rajagopal, S., Ng, K. M., and Douglas, J. M., A hierarchical procedure for the conceptual design of solids processes. *Comput. Chem. Eng.* **16**, 675 (1992).

Rich, E., Knight, K., "Artificial Intelligence," 2nd ed. McGraw-Hill, New York, 1991.

Rossiter, A. P., and Douglas, J. M., Design and optimization of solids processes. Part 1. A hierarchical decision procedure for process synthesis of solids systems. *Chem. Eng. Res. Des.* **64**, 175 (1986a).

Rossiter, A. P., and Douglas, J. M., Design and optimization of solids processes. Part 2. Optimization of crystallizer, centrifuge and dryer systems. *Chem. Eng. Res. Des.* **64**, 184 (1986b).

Rudd, D. F., Powers, G. J., and Siirola, J. J., "Process Synthesis." Prentice-Hall, Englewood Cliffs, NJ, 1973.

Scherlis, W., and Scott, D., First steps towards inferential programming. *IFIP Congr. '83* (1983).

Siirola, J. J., Powers, G. J., and Rudd, D. F., Synthesis of system design. II. Toward a process concept generator. *AIChE J.* **17**, 677 (1971).

Siletti, C. A., Computer-aided design of protein recovery processes. Ph.D. Thesis, Department of Chemical Engineering, Massachusetts Institute of Technology, Cambridge, MA (1988).

Siletti, C. A., and Stephanopoulos, G., BioSep designer: A knowledge-based process synthesizer for bioseparations. In "Artificial Intelligence Approaches in Engineering Design" (C. Tong and D. Sriram, eds.). Vol. I, p. 295. Academic Press, San Diego, CA, 1992.

Stefik, M., Bobrow, D., Bell, A., Brown, H., Conway, L., and Tong, C., The partitioning of concepts in digital system design. *Proc. Conf. Adv. Res. VLSI*, p. 43 (1982).

Stephanopoulos, G., Artificial intelligence and symbolic computing. In "Foundations of Computer-Aided Process Design" J. J. Siirola, I. E. Grossmann, and G. Stephanopoulos, eds.), p. 21. CACHE Corp., Austin, TX, and Elsevier, New York, 1989.

Stephanopoulos, G., Henning, G., and Leone, H., MODEL.LA.:A modeling language for process engineering: Part I. The formal framework. *Comput. Chem. Eng.* **14**, 813 (1990a).

Stephanopoulos, G., Henning, G., and Leone, H., MODEL.LA.:A modeling language for process engineering: Part II. Multifaceted modeling of processing systems. *Comput. Chem. Eng.* **14**, 847 (1990b).

Stephanopoulos, G., Han, C., Linninger, A., Ali, S., and Stephanopoulos, E., Concept of ZAP (Zero Avoidable Pollution) in the synthesis and evaluation of batch pharmaceutical processes. Paper presented at *Ann. AIChE Meet.*, San Francisco (1994).

Tong, C., Toward an engineering science of knowledge-based design. *AI Eng.* **2**, 133 (1987).

Tong, C., and Sriram, D., "Artificial Intelligence in Engineering Design," Vol. I, p. 9. Academic Press, San Diego, CA, 1992.

SYMBOLIC AND QUANTITATIVE REASONING: DESIGN OF REACTION PATHWAYS THROUGH RECURSIVE SATISFACTION OF CONSTRAINTS

Michael L. Mavrovouniotis

Department of Chemical Engineering
Northwestern University
Evanston, IL 60208.

Given a fixed, predetermined set of elementary reactions, compose reaction pathways (mechanisms) that satisfy given specifications in the transformation of available raw materials to desired products. This is a problem encountered quite frequently during research and development of chemical and biochemical processes. As in the assembly of a puzzle, the pieces (available reaction steps) must fit with each other (i.e., satisfy a set of constraints imposed by the precursor and successor reactions) and conform with the size and shape of the board (i.e., the specifications on the overall transformation of raw materials to products). This chapter draws from *symbolic and quantitative reasoning* ideas of AI which allow the systematic synthesis of artifacts through a *recursive satisfaction of constraints* imposed on the artifact as a whole and on its components. The artifacts in this chapter are mechanisms of catalytic reactions and pathways of biochemical transformations. The former require the construction of *direct* mechanisms, without cycles or redundancies, to determine the basic legitimate chemical transformations in a reacting system. The latter are the chemical engines of living cells, and they represent legitimate routes for the biochemical conversion of substrates to products either desired from a bioprocess or essential for cell survival. The algorithms discussed in this chapter could be used in one of the following two settings: (1) synthesize alternative pathways of chemical or biochemical reactions as a means to interpret overall transformations which are experimentally observed and (2) synthesize reaction pathways in the course of exploring new, alternative production routes. In this chapter, we will discuss examples in both directions. Although we will be concerned only with constraints on the directionality and stoichiometry of elementary reactions, the ideas can be extended to include other types of constraints arising, for example, from kinetics or thermodynamics.

I. Reaction Systems and Pathways

Complex systems of chemical reactions, and the problems associated with their design and analysis, are omnipresent in process engineering applications. In biochemical processes, complex networks of biochemical reactions, catalyzed by enzymes, accomplish both the growth of cells in the bioreactor and the conversion of the raw materials to the target products. Catalytic processes with solid catalysts and fluid-phase raw materials and products are used for many inorganic chemicals as well as refining and petrochemics. Noncatalytic fluid-phase reaction processes involving complex feeds and products are also common.

1. Hierarchical Structure

In all of these systems, chemical reactions have a hierarchical structure. What we normally think of as one reaction is actually composed from several elementary steps, which are often called, collectively, the *mechanism* of the reaction. For processes with a solid catalyst, these steps describe the interaction of the catalyst's active sites with the fluid-phase reactants. For biochemical reactions, there are association and dissociation steps that involve complexes of the enzyme and the substrates. In all cases, unstable, short-lived intermediates may be formed and destroyed in a rapid succession of steps; the overall reaction we observe might not involve these unstable intermediates because it is a suitable composition of such elementary steps.

The hierarchical character of reaction systems does not end with decomposition of reactions into steps. At the next level, many reactions taken together make up pathways. In bioprocesses, such pathways form long chains, cycles, and branching structures that accomplish biologically identifiable functions. In other processes, a pathway is used to describe the sequence of transformations that are needed to obtain the desired product from the available raw materials. One might even envision, at the next level, pathways combined to describe entire chemical plants or families of processing technologies.

2. Synthesis of Pathways

This chapter concerns itself with the hierarchical view of reaction systems. We are not interested in the decomposition of a reaction into mechanism steps or a pathway into reactions, but rather the synthesis or design of these composite entities from appropriate building blocks: Construction of reactions and mechanisms from elementary steps, or construction of pathways from reactions.

The common feature is the imposition of constraints on the overall chemical transformation, and the assembly of building blocks so that these constraints are met by the construct (reaction mechanism or pathway) as a whole. The constraints refer to the stoichiometry of the transformations, i.e., the reactants and products that are involved, along with their proportions. As we will see, there are different ways to formulate these constraints, and each way is suitable for different applications.

In all applications, the core problem can be stated as follows. We have a set of constraints that specify what may or may not be used as a starting material and final product of the process, and we are given a set of transformations (steps or reactions) that can be used as building blocks. For each transformation, we know the participating compounds and their stoichiometric coefficients. We need to assemble the transformations into a composite transformation, so that the stoichiometry of this composite transformation satisfies the constraints. This composite transformation may utilize any compound it needs along the way, as long as in the end its net stoichiometry satisfies the constraints. We can think of the construction procedure as assembly of a device from components, or we can abstract it mathematically into the formation of suitable linear combinations of the building-block transformations. In the second view, we should keep in mind that these linear combinations must respect the initial direction of reactions, since some reactions may be permitted only in their forward direction.

3. Application Domains

Two types of applications that involve this core problem will be discussed in this chapter. In the synthesis of biochemical pathways, using enzymatic reactions as building blocks, it is useful to examine each chemical species and decide whether it is required or allowed to be a starting material for the pathway, and likewise whether it is required or allowed to be a final product of the pathway.

The second application is the construction of catalytic reaction mechanisms out of elementary steps, involving only on type of catalytic site. A more useful way to formulate the stoichiometric constraints for these systems is to classify every chemical species as either a terminal species or an intermediate. Terminal species represent stable compounds that can be produced or consumed in significant quantities, while intermediates are short-lived unstable species that participate in the mechanism but are neither raw materials nor final products of the process.

4. Interpretation of Pathway Design

There are various ways in which we can interpret the idea of meeting stoichiometric constraints by combining transformations or assembling steps. We alluded to these interpretations in earlier sections, in our informal statement of the problem. As with the formulation of the constraints, each application domain accepts more naturally a different interpretation.

One interpretation, well suited to the synthesis of biochemical pathways, is process design. In the development and design of a new bioprocess, there are specifications on the raw materials that may be used and the desired and permitted products. One must select a pathway to use and a microorganism that possesses this pathway. With the advent of genetic engineering techniques, pathways and microorganisms are not restricted to existing ones: It is possible to modify a pathway of an organism by introducing biotransformations from other organisms. Therefore, the design of pathways takes on a true synthetic character rather than mere selection among enumerated alternatives.

Another interpretation is related to the dynamic behavior of a closed chemical system and the quasi-steady-state assumption. It is common in the kinetics of reaction systems to assume that the concentrations of intermediates reach a quasi-steady state: The rate of production and the rate of consumption of the intermediate approximately balance each other, and their difference (the net rate that leads to accumulation or depletion of the intermediate) is small compared to each of the two. The quasi-steady-state assumption allows simplification of the equations, by setting the accumulation term in the mass balance for an intermediate to zero. The construction of pathways is tantamount to identification of quasi-steady-state patterns, because the pathway makes sure that the intermediates are consumed at the same rate as they are produced. We are specifically interested in the smallest possible pathways, which cannot be shortened through elimination of a step and cannot be reduced to a superposition of smaller subpathways.

A similar interpretation can be proposed for open systems if only some species are permitted to enter and leave the system: The remaining species are the intermediates, and the construction of pathways arranges the stoichiometries to reflect the fact that, at steady state, the production of intermediates must balance their consumption.

II. Catalytic Reaction Systems

Our discussion of pathway design in the context of catalytic reaction mechanisms will summarize the treatment presented by Mavrovouniotis (1992) and Mavrovouniotis and Stephanopoulos (1992). The interested reader may refer to these for mathematical details, analysis of computational complexity, and comparison to other approaches in the context of model reaction systems.

A. Basic Concepts, Terminology, and Notation

We will follow the terminology and symbols used by Mavrovouniotis (1992) and Mavrovouniotis and Stephanopoulos (1992), who in turn followed the nomenclature of Happel and Sellers (1982, 1983, 1989), Happel (1986), Sellers (1984, 1989), and Happel *et al.* (1990). For the rest of this chapter, H&S will denote the latter set of references and the entire approach of Happel and Sellers. In accordance with H&S, we use the term *mechanism* rather than *pathway* for this class of systems.

1. Intermediate and Terminal Species

Consider a given set of elementary reaction steps that are feasible in a system, and the species involved in these steps. Species can be classified as either *intermediates*, which occur in very small amounts, or *terminal* species that can occur in significant amounts and constitute the raw materials and products of the process. The intermediates are precisely those species for which the quasi-steady-state assumption could be made.

A chemical systems consists of A species, designated as $a_1, a_2, \ldots, a_A$, and S mechanism steps, designated as $s_1, s_2, \ldots, s_S$. Each step s_i accomplishes a specific elementary reaction or transformation, $r_i = R(s_i)$. Let α_{ij} represent the stoichiometric coefficient of species a_j in step s_i, with the usual convention that $\alpha_{ij} > 0$ if and only if a_j is a product of s_i, $\alpha_{ij} < 0$ if and only if a_j is a reactant of s_i, and $\alpha_{ij} = 0$ if and only if a_j does not participate in s_i. The stoichiometry of the transformation $R(s_i)$ can then be written as

$$r_i = R(s_i) = \sum_{j=1}^{A} \alpha_{ij} a_j.$$

The coefficients α_{ij} of each step s_i denote only ratios of species. Thus, the meaning of the reaction r_i is not affected if we multiply all corresponding α_{ij} by a positive constant.

The directionality of a step s_i will be denoted by a label, as either $\leftrightarrows s_i$ or $\rightarrow s_i$. The sign $\leftrightarrows$ denotes a *reversible* step, whose net rate may be either positive (i.e., in the forward direction) or negative (i.e., in the reverse direction). The sign $\rightarrow$ denotes an *irreversible* step that is either thermodynamically irreversible or known to proceed with a net positive rate (i.e., in the forward direction).

The first I species (a_j with $j = 1, \ldots, I$) are assumed to be the intermediates, whereas the remaining $T = A - I$ species (a_j, $j = I + 1, \ldots, I + T$) are terminal species. If intermediates are present in very low concentrations, and their high rate of production is balanced by a high rate of consumption, the quasi-steady-state assumption allows dropping the accumulation term from the mass balance of an intermediate.

2. Overall Mechanisms

An *overall* mechanism (Horiuti and Nakamura, 1967; Horiuti, 1973; Temkin 1973, 1979) is one that obeys the quasi-steady-state assumption: It must consist of steps combined in specific proportions, such that the net transformation involves only terminal species.

Although there is significant net production of some terminal species, and significant net consumption of others, within the quasi-steady-state assumption the concentrations of intermediates are low and the rate of production of each intermediate by some steps is approximately balanced by its rate of consumption by other steps. *Overall* reactions can thus be defined as the set of net transformations permissible under the quasi-steady-state assumption; *overall* mechanisms are the combinations of steps that accomplish this.

A *mechanism*, m_k, is a linear combination of steps:

$$m_k = \sum_{i=1}^{S} \sigma_{ki} s_i,$$

and belongs to an S-dimensional vector space. The coefficients σ_{ki} represent the proportional participation of the steps, and their signs must reflect the proper direction of steps: for irreversible steps, σ_{ki} must be nonnegative.

For clarity, the index i will hereafter be used for steps, the index j for species, and the index k for mechanisms. A reaction vector r_k, associated with each mechanism, describes the net transformation that is accomplished by the mechanism:

$$r_k = R(m_k) = \sum_{i=1}^{S} \sigma_{ki} R(s_i).$$

Through substitution of the expression of $R(s_i)$ the reaction $R(m_k)$ can be written as

$$R(m_k) = \sum_{j=1}^{A} \beta_{kj} a_j \quad \text{with} \quad \beta_{kj} = \sum_{i=1}^{S} \sigma_{ki} \alpha_{ij},$$

where β_{kj} is the stoichiometric coefficient of species a_j in the net reaction accomplished by mechanism m_k.

A reaction $r_k = R(m_k)$ that involves only terminal species (i.e., $\beta_{kj} = 0$ for $j = 1, \ldots, I$) is called an *overall reaction*, and the mechanism m_k is an *overall mechanism*, corresponding to a quasi-steady state.

3. Direct Mechanisms

Clearly, a linear combination of overall mechanisms is also an overall mechanism. We are interested only in the smallest possible overall mechanisms, which describe the most extreme modes of operation of the chemical system. To define these, we examine the set of steps participating in a mechanism—with nonzero coefficients σ_{ki}. We are interested in overall mechanisms whose set of participating steps is not a proper superset of some other overall prime mechanism m'_k. These mechanisms, called *direct* mechanism by H&S, are minimal with respect to inclusion (for the set of participating steps). One might equivalently state that a mechanism m_k is direct if and only if every other overall mechanism includes at least one step not participating in m_k. The definition used here, in contrast to the one given by H&S, makes no particular reference to the *reaction* accomplished by the mechanism, other than the requirement that only overall mechanisms be considered.

Reaction mechanisms in the literature are often direct, because physical intuition encourages avoidance of excess steps. In considering the mechanism of a reaction, one often postulates ahead of time an implicit quasi-steady state, i.e., a combination of steps that explains the observed net reaction and conserves all intermediate species. In our approach, however, we are interested in identifying all possible direct mechanisms that are consistent with the postulated set of steps.

Direct mechanisms are not necessarily linearly independent; it may be possible to express a direct mechanism as a linear combination of other direct mechanisms. However, direct mechanisms are chemically distinct, because each involves a unique combination of steps.

B. Previous Work on the Construction of Mechanisms

A related concept is that of *laminar* mechanisms, introduced by Sinanoğlu (1975), Sinanoğlu and Lee (1978), and Lee and Sinanoğlu (1981). These mechanisms can be constructed a priori (i.e., without regard to a specific set of mechanism steps) given the number of catalysts involved. Sellers (1971, 1972) also presented procedures for construction of a priori mechanisms for certain chemical systems. In both these instances of mechanism construction, the focus was the a priori enumeration of potential pathway shapes, without regard to the specific reactions permitted by the system. These techniques are thus not applicable to the synthesis of pathways or mechanisms from a given set of reaction steps.

Milner (1964) used the term *direct* mechanisms to describe the smallest possible overall mechanisms. Milner (1964) showed that, under certain conditions, a direct mechanism can involve at most $I + 1$ steps, where I is the number of intermediates; this allows the construction of the mechanisms by examining combinations of $I + 1$ steps at a time. Another algorithm for the construction of direct mechanisms was presented by H&S, under the fundamental assumption that all steps are reversible and the net reaction taking place is actually known. This chapter presents an algorithm characterized by much greater generality and efficiency. The advantages of the algorithms presented here over those of H&S will be pointed out in the description of the algorithms and the examples through-out this section.

C. STRUCTURE OF THE ALGORITHM

We will describe here the general operation of the algorithm for the construction of direct mechanisms for chemical systems. We will examine the algorithm only in its simplest form. Mavrovouniotis (1992) presents various enhancements and discusses computational implementation considerations. We will begin with two examples (ammonia and methanol synthesis) that illustrate the step-by-step operation of the algorithm; these should be particularly useful for readers that are not accustomed to the abstract description of algorithms, who may wish to study the examples first, or study the abstract algorithm and the examples simultaneously.

Initially, the algorithm considers each individual reaction step as a partial mechanism. Then, one intermediate after another are examined, and the set of partial mechanisms is modified so that the intermediate does not appear in the net stoichiometry; the modification of mechanisms is carried out in a way that preserves the correct direction of irreversible reaction steps. By processing all intermediates in this way, a set of overall mechanisms is constructed. This final set of mechanisms might include duplicate mechanisms and even indirect ones; these can be easily discarded. Similar action must be taken in the procedure of H & S. Mavrovouniotis (1992) discusses procedures for eliminating such redundant mechanisms in the end, or even preventing their construction.

The algorithm operates on a set of intermediate species N, the set of terminal species N_T, and a set of partial mechanisms M, which is iteratively modified. Other information maintained by the algorithm during its operation is arranged in a convenient format in Fig. 1, which also explains how the entries are initialized. The algorithm proceeds from this setup by successively eliminating each intermediate species from the system. In order to eliminate an intermediate, we consider all the mechanisms whose reactions involve the intermediate species at hand. We can create combinations of two mechanisms at a time, with combination coefficients such that the intermediate vanishes from the reaction. The new row for a combination of mechanisms has elements σ_{ki} (left portion of Fig. 1) and β_{kj} (right portion of Fig. 1) that are simply the linear combinations of the respective elements of the old mechanisms. To permit the best choice of intermediate to be made, the setup of the algorithm (Fig. 1) lists, below

s_1	s_2	.....	s_S		number of combinations $\Rightarrow$	n_1	n_2	.....	n_I	—	.....	—
ε_1	ε_2	.....	ε_S	mechanism $\Downarrow$	origin $\Downarrow$	a_1	a_2	.....	a_I	a_{I+1}	.....	a_A
σ_{11}	σ_{12}	.....	σ_{1S}	ε_1 m_1	—	β_{11}	β_{12}	.....	β_{1I}	$\beta_{1(I+1)}$	.....	β_{1A}
σ_{21}	σ_{22}	.....	σ_{2S}	ε_2 m_2	—	β_{21}	β_{22}	.....	β_{2I}	$\beta_{2(I+1)}$	.....	β_{2A}
$\vdots$	$\vdots$	$\vdots$	$\vdots$	$\vdots$ $\vdots$	$\vdots$	$\vdots$	$\vdots$	$\vdots$	$\vdots$	$\vdots$	$\vdots$	$\vdots$
σ_{S1}	σ_{S2}	.....	σ_{SS}	ε_S m_S	—	β_{S1}	β_{S2}	.....	β_{SI}	$\beta_{S(I+1)}$	.....	β_{SA}

FIG. 1. Initialized setup for the application of the algorithm for the construction of mechanisms. Rows correspond to mechanisms, and the operation of the algorithm adds and deletes rows. The left portion of the table contains the coefficients σ_{ki}; initially, each mechanism m_k contains only the corresponding step s_k (i.e., σ is the identity matrix). With each step symbol s_i we list the directionality label ϵ_i ($\rightarrow$ for an irreversible step, or $\leftrightarrows$ for a reversible step). Directionality labels are also listed for each mechanism m_k. The right portion of the table shows the reactions accomplished by the mechanisms, and groups intermediate species as a_1 to a_I, and terminal species as a_{I+1} to a_A. The algorithm deletes columns corresponding to intermediates as these are processed. Initially $\beta_{kj} = \alpha_{ki}$ when $i = j$, but this will not be the case for new mechanisms that are constructed. Above each intermediate species symbol the number of combination mechanisms that must be created to eliminate the species is listed; this will be explained further in the description of the algorithm. The column marked "origin" in the middle of the table is not essential; it is used for keeping track of how each new mechanism is constructed; one can list the intermediate that was eliminated and the mechanisms that were combined in the construction. [Reprinted with permission from Mavrovouniotis, M. L., and Stephanopoulos, G. "Synthesis of reaction mechanisms consisting of reversible and irreversible steps: I. A synthesis approach in the context of simple examples". *Ind. Eng. Chem. Res.* **31**, 1625–1637. (1992). Copyright 1992 American Chemical Society.]

each species, the number of combination mechanisms that must be created to eliminate the species. The description given below explains the individual operations involved in more detail.

1. Initialization

We initialize N to the set of all intermediates (assuming $I \neq 0$), i.e., $N := \{a_1, a_2, \ldots, a_I\}$. The set M is initialized with one-step mechanisms. Therefore, the initial σ is just the identity matrix, whereas the initial β matrix is equal to the α matrix (Fig. 1):

$$M := \{m_1, m_2, \ldots, m_S\} \quad \text{with} \quad \beta_{kj} := \alpha_{kj} \text{ and } \sigma_{ki} := \begin{cases} 1 \text{ if } k = i, \\ 0 \text{ if } k \neq i. \end{cases}$$

The initial directionalities are the same as those of the corresponding individual steps (Fig.1).

2. Number of Combinations for Each Intermediate

We will now define the sets of partial mechanisms in which each intermediate participates. We divide them into irreversible and reversible ones to take directionality restrictions properly into account. For each species $a_j \in N$, let Y_j be a subset of M, defined as follows:

A mechanism $m_k \in M$ belongs to Y_j if and only if it is irreversible and its net reaction contains a_j as a reactant, i.e., $m_k \in Y_j$ if and only if $m_k \in M$, $\rightarrow m_k$, and $\beta_{kj} < 0$.

Let y_j be the cardinality of Y_j, i.e., $y_j = |Y_j|$. Define similarly a subset Z_j of irreversible mechanisms as follows:

A mechanism $m_k \in M$ belongs to Z_j if and only if it is irreversible and its net reaction contains a_j as a product (i.e., $\beta_{kj} > 0$).

Let z_j be the cardinality of this set. For reversible partial mechanisms, it does not matter whether a_j is a net reactant or a net product, as long as it appears in the net transformation. Thus, we define another subset, X_j, as follows:

A mechanism $m_k \in M$ belongs to X_j if and only if it is reversible and its net reaction contains species a_j (i.e., $\beta_{kj} \neq 0$).

Let x_j be the cardinality of X_j, i.e., $x_j = |X_j|$. One can obtain the numbers y_j, z_j, and x_j by scanning the column of a_j in Fig. 1 for nonzero entries. A positive entry corresponding to an irreversible mechanism is counted into z_j. A negative entry corresponding to an irreversible mechanism is counted into y_j. A positive or negative (but not zero) entry corresponding to a reversible mechanism is counted into x_j.

We want to combine partial mechanisms, two at a time, so that a_j is not present in the net stoichiometry of the combination, always respecting directionality restrictions on mechanisms. We can combine a mechanism from X_j with one from Y_j, because the mechanisms in X_j can be reversed if necessary to make a_j a product (to balance with a mechanism from Y_j that uses a_j as a reactant); this gives rise to $x_j y_j$ combinations. Similarly, we can combine a mechanism from X_j with one from Z_j, reversing mechanisms in X_j if necessary to make a_j a reactant; this gives rise to $x_j z_j$ combinations. We can also combine a mechanism from X_j with any other mechanism from X_j in the same way; the number of combinations in this case is

$$\binom{x_j}{2} = \frac{x_j(x_j - 1)}{2}.$$

Finally, a mechanism from Z_j can be combined with any mechanism from Y_j, without any reversal, giving rise to $z_j y_j$ combinations.

The total number of possible combinations, n_j (placed above a_j in the layout of Fig. 1), necessary to eliminate a_j, is thus given by

$$n_j := \frac{x_j(x_j - 1)}{2} + x_j z_j + x_j y_j + z_j y_j.$$

Two limiting cases of this formula will be used later in the ammonia and methanol examples. If all steps are reversible, $n_j := [x_j(x_j - 1)]/2$, while if all the steps are irreversible $n_j := z_j y_j$.

3. Selection and Elimination of an Intermediate

The algorithm ultimately eliminates all intermediates, and one could process them in any order. However, to minimize computational effort in the current iteration, we select to eliminate the a_J that corresponds to the smallest n_J. For the intermediate a_J, the elimination constructs precisely those combinations that were counted into n_J in the previous phase of the algorithm. We put the new mechanisms in a set M_J. Specifically, the construction of M_J is governed by the following rules.

> *Rule 1.* For each distinct combination $m_k \in X_J$, $m_b \in X_J$, the combination mechanism is $m_c := \beta_{bJ} m_k - \beta_{kJ} m_b$ and it is reversible ($\leftrightarrows m_c$).
>
> *Rule 2.* For each $m_k \in Z_J$ and each $m_b \in X_J$, the mechanism $\rightarrow m_c$ must be constructed according to the sign of β_{bJ}, because the sign determines whether we need to reverse the mechanism m_b, and we must ensure that the coefficient that multiples m_k is positive. Thus
> (a) If $\beta_{bJ} > 0$, we form $m_c := \beta_{bJ} m_k - \beta_{kJ} m_b$.
> (b) If $\beta_{bj} < 0$, we form $m_c := \beta_{kJ} m_b - \beta_{bJ} m_k$.
>
> *Rule 3.* Similarly, for each $m_k \in Y_J$ and each $m_b \in X_J$, for $m \rightarrow m_c$ as follows:
> (a) If $\beta_{bJ} > 0$, we form $m_c := \beta_{bJ} m_k - \beta_{kJ} m_b$.
> (b) If $\beta_{bJ} < 0$, we form $m_c := \beta_{kJ} m_b - \beta_{bJ} m_k$.
>
> *Rule 4.* Finally, for each $m_k \in Y_J$ and each $m_b \in Z_J$, we form the mechanism $\rightarrow m_c$ as $m_c := \beta_{bJ} m_k - \beta_{kJ} m_b$. From the definition of the set Z_J it is easy to see that $\beta_{bJ} > 0$ and $-\beta_{kJ} > 0$, which guarantees that the irreversibility of mechanisms m_k and m_b is respected in the construction of m_c.

The combination mechanisms in all of these categories constitute the set M_J, whose cardinality is $n_J = x_J(x_J-1)/2 + x_J z_J + x_J y_J + z_J y_J$. The directionality of each new mechanism is $\leftrightarrows$ (reversible) if and only if both mechanisms used in the combination were reversible. The σ and β coefficients of each combination mechanism are obtained as linear combinations of the respective coefficients of the constituent mechanisms. For $m_c := \beta_{bJ} m_k - \beta_{kJ} m_b$, one obtains $\sigma_{ci} := \beta_{bJ} \sigma_{ki} - \beta_{kJ} \sigma_{bi}$ (for all i) and $\beta_{cj} := \beta_{bJ} \beta_{kj} - \beta_{kJ} \beta_{bj}$ (for all j). From the latter expression the coefficient for a_J can be computed by setting $j := J$, which gives $\beta_{cJ} = \beta_{bJ} \beta_{kJ} - \beta_{kJ} \beta_{bJ} = 0$, verifying that the choice of combination coefficients achieves the elimination of a_J, as intended.

A new mechanism is logged as a new row in Fig. 1, and its σ and β portions are computed as linear combinations of the rows of the mechanisms being considered. The "origin" entry in Fig. 1 is used for bookkeeping. In the examples presented later, we record the name of the intermediate eliminated and a symbolic linear expression of the combination of mechanisms (see Figs. 3–7 and 11–16).

4. Update of Active Sets

In order to eliminate a_J we can now modify the current set of partial mechanisms M by adding the new combination mechanisms M_J, and removing their ancestors X_J, Y_J, and Z_J, i.e., $M := (M \cup M_J) - (X_J \cup Y_J \cup Z_J)$. We can then remove the intermediate a_J, by setting $N := N - \{a_J\}$. Within Fig. 1, in this phase we remove all rows that have nonzero entries in the a_J-column and then remove the column a_J itself.

5. Termination or New Iteration

If Fig. 1 still contains active columns corresponding to intermediates (i.e., $N \neq \varnothing$), we go back to phase 2 of the algorithm for the next iteration. If no intermediate columns remain ($N = \varnothing$) the elimination is complete, and all direct mechanisms are in the set M.

D. FEATURES OF THE ALGORITHM

Our algorithm for the synthesis of direct mechanisms initially considers each reaction step as a partial mechanism. Then, one intermediate after another are examined, and the set of partial mechanisms is modified so that the intermediate does not appear in the net stoichiometry; the modification of mechanisms is carried out in a way that preserves the correct direction of irreversible reaction steps. By processing all intermediates in this way, we can construct a set of overall mechanisms.

The set of mechanisms M resulting from the application of the basic algorithm may contain duplicate mechanisms (as in the ammonia example, see next subsection) or even indirect ones (as in the methanol example, see next subsection). A simple procedure can be applied in the end to remove redundant (duplicate or indirect) mechanisms from the final set of mechanisms. Alternatively, incremental checking can be carried out for each new mechanism or each iteration of the algorithm. Redundancy of mechanisms is one of a number of issues that affect the operation of the algorithm, and variations of the basic algorithm can be defined to enhance either its computational efficiency or its conceptual simplicity. For example, when two or more intermediates entail the same number of combinations (n_J), the basic algorithm makes a completely arbitrary choice; a variation of the algorithm might instead make the selection in a manner that removes as many existing mechanisms as possible from the set M, increasing computational efficiency. In the direction of conceptual simplicity, instead of assembling and using the sets X_j, Y_j, and Z_j separately, one may modify the algorithm by dropping directionality labels from mechanisms, and lumping X_j, Y_j, and Z_j into a single set W_j; the construction of combinations that eliminate a selected intermediate a_J using every possible pair of mechanisms $m_k \in W_J$ and $m_b \in W_J$, and rejecting the ones that contain steps in the wrong direction. These and related issues are discussed by Mavrovouniotis (1992).

The algorithm presented here for the construction of direct mechanisms differs significantly in its operation from the method of H&S. First, the irreversibility of mechanism steps is taken into account by our algorithm as mechanisms are constructed; i.e., it is not necessary to reserve directionality considerations until the end. Second, the mechanisms do not have to be restricted a priori to a particular reaction. Third, direct mechanisms are constructed recursively through combination of steps or partial mechanisms, rather than as linear combinations from a basis set of linearly independent mechanisms.

E. EXAMPLES

The mechanism–construction approach is illustrated here through two examples, which have also been treated by H&S: an example involving reversible steps for the synthesis of ammonia, and an example with irreversible steps for the synthesis of methanol. We note, in particular, that in the methanol example H&S assumed that the steps that lead to ethane and dimethylether (which are byproducts) are irrelevant for methanol mechanisms; this assumption was motivated by the need to keep

$N_2+I \rightleftarrows N_2I$	$N_2I+H_2 \rightleftarrows N_2H_2I$	$N_2H_2I+I \rightleftarrows 2NHI$	$N_2+2I \rightleftarrows 2NI$	$NI+HI \rightleftarrows NHI+I$	$NHI+HI \rightleftarrows NH_2I+I$	$NHI+H_2 \rightleftarrows NH_3+I$	$H_2+2I \rightleftarrows 2HI$	$NH_2I+HI \rightleftarrows NH_3+I$	mechanism	species ⇒	N_2I	N_2H_2I	NHI	NI	HI	NH_2I	—	N_2	H_2	NH_3
s_1	s_2	s_3	s_4	s_5	s_6	s_7	s_8	s_9	⇓	number of combinations ⇒	1	1	6	1	6	1	28	—	—	—
⇆	⇆	⇆	⇆	⇆	⇆	⇆	⇆	⇆		origin ⇓	a_1	a_2	a_3	a_4	a_5	a_6	a_7	a_8	a_9	a_{10}
1									⇆ m_1	—	1						−1	−1		
	1								⇆ m_2	—	−1	1							−1	
		1							⇆ m_3	—		−1	2				−1			
			1						⇆ m_4	—				2			−2	−1		
				1					⇆ m_5	—			1	−1	−1		1			
					1				⇆ m_6	—			−1		−1	1	1			
						1			⇆ m_7	—			−1				1		−1	1
							1		⇆ m_8	—					2		−2		−1	
								1	⇆ m_9	—					−1	−1	2			1

FIG. 2. Application of the algorithm on the ammonia mechanism. The mechanism steps for the synthesis of ammonia are shown on the top left portion, above each step symbol s_i. The steps are those used by Happel and Sellers (1982, 1983) and Happel *et al.* (1990), and were originally derived from the work of Horiuti (1973) and Temkin (1973). The active site of the catalyst is denoted as "I," and all species that contain this active site are intermediates; N_2, H_2, and NH_3 are terminal species. The setup follows the description given in Fig. 1, with the entries that are equal to zero left blank. Directionalities are shown, but for this example all steps are considered reversible and the directionality does not matter. We start by taking partial mechanisms $m_1, m_2, \ldots, m_9$, each of which has length 1 and includes only the step with the same index (i.e., $\sigma_{ki} = 1$ if $k = i$; $\sigma_{ki} = 0$ if $k \neq i$). Naturally, these are only partial mechanisms, which will be gradually transformed to overall mechanisms. Since all steps are reversible, the number of combination mechanisms that must be created to eliminate each intermediate (listed below the species) is simply $x(x-1)/2$, where x is the number of mechanisms whose reactions involve the species. The algorithm proceeds by successively eliminating each intermediate from the system. If, for example, we pursue the elimination of NHI, there are four existing partial mechanisms, m_3, m_5, m_6, and m_7, to be considered; there are $4 \times 3/2 = 6$ pairwise combinations of these that must be constructed to eliminate NHI. The choice to eliminate the intermediate a_3 (NHI) first, however, is not computationally efficient; if we choose NH_2I (a_6) instead, we only need to construct one mechanism. In fact, a_6 has the minimum number of combinations and its column has been highlighted in Fig. 2, to show that this is the species which will be eliminated first. Intuitively, one can think of this elimination as follows. There are only two steps that involve NH_2I; if any mechanism uses one of the two steps, it must use the other, with a coefficient that would eliminate the intermediate NH_2I; hence, the two steps can be combined. [Reprinted with permission from Mavrovouniotis, M. L., and Stephanopoulos, G. "Synthesis of reaction mechanisms consisting of reversible and irreversible steps: I. A synthesis approach in the context of simple examples". *Ind. Eng. Chem. Res.* **31**, 1625–1637. (1992). Copyright 1992 American Chemical Society.]

the size of the problem small, since the H&S algorithm is not very efficient computationally. As we will see, our algorithm constructs one mechanism that involves simultaneous production of ethane and methanol (i.e., m_{35} in Figs. 16 and 17), showing that the omission of byproducts is misleading.

1. Synthesis of Ammonia

We first consider an example on ammonia synthesis that was used by Happel and Sellers (1982, 1983) and Happel *et al.* (1990) and contains only

s_1	s_2	s_3	s_4	s_5	s_6	s_7	s_8	s_9	mechanism	origin	N_2I a_1	N_2H_2I a_2	NHI a_3	NI a_4	HI a_5	NH_2I a_6	$\_$ a_7	N_2 a_8	H_2 a_9	NH_3 a_{10}
$N_2{+}I{\rightleftharpoons}N_2I$	$N_2I{+}H_2{\rightleftharpoons}N_2H_2I$	$N_2H_2I{+}I{\rightleftharpoons}2NHI$	$N_2{+}2I{\rightleftharpoons}2NI$	$NI{+}HI{\rightleftharpoons}NHI{+}I$	$NHI{+}HI{\rightleftharpoons}NH_2I{+}I$	$NHI{+}H_2{\rightleftharpoons}NH_3{+}I$	$H_2{+}2I{\rightleftharpoons}2HI$	$NH_2I{+}HI{\rightleftharpoons}NH_3{+}I$		number of combinations ⇒	1	1	6	1	3	##	21	–	–	–
1									⇌ m_1	—	1					##	−1	−1		
	1								⇌ m_2	—	−1	1				##			−1	
		1							⇌ m_3	—		−1	2			##	−1			
			1						⇌ m_4	—				2		##	−2	−1		
				1					⇌ m_5	—			1	−1	−1	##	1			
##	##	##	##	##	##	##	##	##	⇌ m_6	—	##	##	##	##	##	##	##	##	##	##
					1				⇌ m_7	—			−1			##	1		−1	1
						1			⇌ m_8	—					2	##	−2		−1	
##	##	##	##	##	##	##	##	##	⇌ m_9	—	##	##	##	##	##	##	##	##	##	##
					1			1	⇌ m_{10}	a_6: $m_6 + m_9$			−1		−2	##	3			1

FIG. 3. Setup for the application of the algorithm on the ammonia mechanism, after the elimination of a_6 (NH_2I). In the "origin" column, the construction of the mechanism m_{10} is documented as resulting from the elimination of a_6 through a specific linear combination of the mechanisms m_6 and m_9. The row of m_{10} has been obtained by carrying out the linear combination of the rows of m_6 and m_9. The rows for m_6 and m_9, as well as the column for a_6 have been crossed out from the table. The numbers of combinations have changed for a_5 and a_7 (but remain unchanged for other intermediates). Any one of the species a_1, a_2, and a_4, each entailing only one new combination, could be eliminated next. Both a_2 (N_2H_2I) and a_4 (NI) have been highlighted; their elimination can be carried out in parallel, because the two sets of mechanisms involved are disjoint; i.e., there is no mechanism whose reaction stoichiometry involves both a_4 and a_2. [Reprinted with permission from Mavrovouniotis, M. L., and Stephanopoulos, G. "Synthesis of reaction mechanisms consisting of reversible and irreversible steps: I. A synthesis approach in the context of simple examples". *Ind. Eng. Chem. Res.* **31**, 1625–1637, (1992). Copyright 1992 American Chemical Society.]

$N_2+I\rightleftarrows N_2I$	$N_2I+H_2\rightleftarrows N_2H_2I$	$N_2H_2I+I\rightleftarrows 2NHI$	$N_2+2I\rightleftarrows 2NI$	$NI+HI\rightleftarrows NHI+I$	$NHI+HI\rightleftarrows NH_2I+I$	$NHI+H_2\rightleftarrows NH_3+I$	$H_2+2I\rightleftarrows 2HI$	$NH_2I+HI\rightleftarrows NH_3+I$	mechanism	species ⇒	N_2I	N_2H_2I	NHI	NI	HI	NH_2I	_	N_2	H_2	NH_3
s_1	s_2	s_3	s_4	s_5	s_6	s_7	s_8	s_9	⇓	number of combinations ⇒	1	##	6	##	3	##	10	–	–	–
$\rightleftarrows$	$\rightleftarrows$	$\rightleftarrows$	$\rightleftarrows$	$\rightleftarrows$	$\rightleftarrows$	$\rightleftarrows$	$\rightleftarrows$	$\rightleftarrows$		origin ⇓	a_1	a_2	a_3	a_4	a_5	a_6	a_7	a_8	a_9	a_{10}
1									$\rightleftarrows$ m_1	—	1	##		##		##	-1	-1		
##	##	##	##	##	##	##	##	##	$\rightleftarrows$ m_2	—	##	##	##	##	##	##	##	##	##	##
##	##	##	##	##	##	##	##	##	$\rightleftarrows$ m_3	—	##	##	##	##	##	##	##	##	##	##
##	##	##	##	##	##	##	##	##	$\rightleftarrows$ m_4	—	##	##	##	##	##	##	##	##	##	##
##	##	##	##	##	##	##	##	##	$\rightleftarrows$ m_5	—	##	##	##	##	##	##	##	##	##	##
##	##	##	##	##	##	##	##	##	$\rightleftarrows$ m_6	—	##	##	##	##	##	##	##	##	##	##
						1			$\rightleftarrows$ m_7	—		##	-1	##		##	1		-1	1
							1		$\rightleftarrows$ m_8	—		##		##	2	##	-2		-1	
##	##	##	##	##	##	##	##	##	$\rightleftarrows$ m_9	—	##	##	##	##	##	##	##	##	##	##
						1		1	$\rightleftarrows$ m_{10}	a_6: m_6+m_9		##	-1	##	-2	##	3			1
	1	1							$\rightleftarrows$ m_{11}	a_2: m_2+m_3	-1	##	2	##		##	-1		-1	
			1	2					$\rightleftarrows$ m_{12}	a_4: m_4+2m_5		##	2	##	-2	##			-1	

Fig. 4. Application of the algorithm on the ammonia mechanism, after the elimination of a_2 and a_4, which yielded two new partial mechanisms, m_{11} and m_{12}, and eliminated m_2, m_3, m_4, and m_5. The number of combinations has been recomputed for each remaining intermediate species. The intermediate N_2I (a_1) will be eliminated next. [Reprinted with permission from Mavrovouniotis, M. L., and Stephanopoulos, G. "Synthesis of reaction mechanisms consisting of reversible and irreversible steps: I. A synthesis approach in the context of simple examples". *Ind. Eng. Chem. Res.* **31**, 1625–1637, (1992). Copyright 1992 American Chemical Society.]

reversible steps. The initial setup for the application of the algorithm is shown in Fig. 2, which also presents the actual elementary steps above the corresponding symbols s_i. The progression of the algorithm is explained in Figs. 2–7. Each figure shows one or more iterations through phases 2–5 of the algorithm; thus each figure eliminates one or more intermediates and flags those intermediates that will be eliminated next. Figure 7 provides the final results and compares them to the results obtained by Happel *et al.* (1990).

2. Synthesis of Methanol

Consider, as a second example, the mechanism for the synthesis of methanol (Fig. 8), analyzed by Happel *et al.* (1990). All the steps are

164 MICHAEL L. MAVROVOUNIOTIS

$N_2+I\rightleftharpoons N_2I$	$N_2I+H_2\rightleftharpoons N_2H_2I$	$N_2H_2I+I\rightleftharpoons 2NHI$	$N_2+2I\rightleftharpoons 2NI$	$NI+HI\rightleftharpoons NHI+I$	$NHI+HI\rightleftharpoons NH_2I+I$	$NHI+H_2\rightleftharpoons NH_3+I$	$H_2+2I\rightleftharpoons 2HI$	$NH_2I+HI\rightleftharpoons NH_3+I$	mechanism	species ⇒	N_2I	N_2H_2I	NHI	NI	HI	NH_2I	_	N_2	H_2	NH_3
s_1	s_2	s_3	s_4	s_5	s_6	s_7	s_8	s_9	⇓	number of combinations ⇒	##	##	6	##	3	##	6	–	–	–
⇌	⇌	⇌	⇌	⇌	⇌	⇌	⇌	⇌		origin ⇓	a_1	a_2	a_3	a_4	a_5	a_6	a_7	a_8	a_9	a_{10}
##	##	##	##	##	##	##	##	##	⇌ m_1	—	##	##	##	##	##	##	##	##	##	##
##	##	##	##	##	##	##	##	##	⇌ m_2	—	##	##	##	##	##	##	##	##	##	##
##	##	##	##	##	##	##	##	##	⇌ m_3	—	##	##	##	##	##	##	##	##	##	##
##	##	##	##	##	##	##	##	##	⇌ m_4	—	##	##	##	##	##	##	##	##	##	##
##	##	##	##	##	##	##	##	##	⇌ m_5	—	##	##	##	##	##	##	##	##	##	##
##	##	##	##	##	##	##	##	##	⇌ m_6	—	##	##	##	##	##	##	##	##	##	##
						1			⇌ m_7	—	##	##	−1	##		##	1		−1	1
							1		⇌ m_8	—	##	##		##	2	##	−2		−1	
##	##	##	##	##	##	##	##	##	⇌ m_9	—	##	##	##	##	##	##	##	##	##	##
				1				1	⇌ m_{10}	a_6: $m_6 + m_9$	##	##	−1	##	−2	##	3			1
##	##	##	##	##	##	##	##	##	⇌ m_{11}	a_2: $m_2 + m_3$	##	##	##	##	##	##	##	##	##	##
			1	2					⇌ m_{12}	a_4: $m_4 + 2m_5$	##	##	2	##	−2	##		−1		
1	1	1							⇌ m_{13}	a_1: $m_1 + m_{11}$	##	##	2	##			##	−2	−1	−1

FIG. 5. Application of the algorithm on the ammonia mechanism, after the elimination of a_1, with only five mechanisms (m_7, m_8, m_{10}, m_{12}, and m_{13}) remaining active. This reduction in the number of mechanisms as the first few intermediates are eliminated is quite common; however, the number of mechanisms tends to increase again at the end. The intermediate HI (a_5) will be eliminated next. [Reprinted with permission from Mavrovouniotis, M. L., and Stephanopoulos, G. "Synthesis of reaction mechanisms consisting of reversible and irreversible steps: I. A synthesis approach in the context of simple examples". *Ind. Eng. Chem. Res.* **31**, 1625–1637, (1992). Copyright 1992 American Chemical Society.]

irreversible, providing an opportunity to show how directionality of steps is taken into account as the mechanisms are constructed. In the beginning of their analysis, Happel *et al.* (1990) remove steps s_9 and s_{11}, because they lead to byproducts (ethane and dimethylether) rather than methanol. Here, we will keep all steps; it turns out that the products are not formed independently of each other, and isolation of the main product from the byproducts of the mechanism can be misleading.

The application of the algorithm is shown in Figs. 9–16, with explanation of the differences arising from the directionality of the steps; each figure eliminates one or more intermediates and flags those intermediates

$N_2+I\leftrightarrows N_2I$	$N_2I+H_2\leftrightarrows N_2H_2I$	$N_2H_2I+I\leftrightarrows 2NHI$	$N_2+2I\leftrightarrows 2NI$	$NI+HI\leftrightarrows NHI+I$	$NHI+HI\leftrightarrows NH_2I+I$	$NHI+H_2\leftrightarrows NH_3+I$	$H_2+2I\leftrightarrows 2HI$	$NH_2I+HI\leftrightarrows NH_3+I$	mechanism	species ⇒	N_2I	N_2H_2I	NHI	NI	HI	NH_2I		N_2	H_2	NH_3
s1	s2	s3	s4	s5	s6	s7	s8	s9		number of combinations ⇒	##	##	10	##	##	##	10	-	-	-
⇌	⇌	⇌	⇌	⇌	⇌	⇌	⇌	⇌		origin ↓	a1	a2	a3	a4	a5	a6	a7	a8	a9	a10
						1			⇌ m7	—	##	##	−1	##	##	##	1		−1	1
##	##	##	##	##	##	##	##	##	⇌ m8	—	##	##	##	##	##	##	##	##	##	##
##	##	##	##	##	##	##	##	##	⇌ m10	a6: m6 + m9	##	##	##	##	##	##	##	##	##	##
##	##	##	##	##	##	##	##	##	⇌ m12	a4: m4 + 2m5	##	##	##	##	##	##	##	##	##	##
1	1	1							⇌ m13	a1: m1 + m11	##	##	2	##	##	##	−2	−1	−1	
					1		1	1	⇌ m14	a5: m8 + m10	##	##	−1	##	##	##	1		−1	1
			1	2			1		⇌ m15	a5: m8 + m12	##	##	2	##	##	##	−2	−1	−1	
			1	2	−1			−1	⇌ m16	a5: m12 − m10	##	##	3	##	##	##	−3	−1		−1

FIG. 6. Application of the algorithm on the ammonia mechanism, after the elimination of a_5. We can generally, at any time, drop inactive (crossed out) mechanisms or eliminated intermediates, but we have so far maintained them so that one can easily keep track of the progress of the algorithm. At this point, in order to reduce the size of the setup, we drop from the figure all those mechanisms that had been crossed out in Fig. 5 or earlier. The two intermediates (a_3 and a_7) remaining in Fig. 6 involve the same number of combinations; in fact elimination of either one results automatically in the elimination of the other (Fig. 7), because the columns for a_3 and a_7 contained precisely opposite numbers; a_3 is arbitrarily chosen. [Reprinted with permission from Mavrovouniotis, M. L., and Stephanopoulos, G. "Synthesis of reaction mechanisms consisting of reversible and irreversible steps: I. A synthesis approach in the context of simple examples". *Ind. Eng. Chem. Res.* **31**, 1625–1637, (1992). Copyright 1992 American Chemical Society.]

that will be eliminated next. Figure 17 analyzes the results and compares them to those of Happel *et al.* (1990). With respect to the mechanisms m_{29} and m_{35}, which lead to byproducts, we observe (Figs. 16 and 17) that m_{29} can be thought of as unrelated to the synthesis of methanol, but omission of m_{35} may be misleading: This mechanism leads to simultaneous production of methanol and ethane, in stoichiometric proportions. Thus, the mechanism m_{35} should properly be viewed as one of the mechanisms that lead to methanol—with the drawback, of course, that it also leads to an equal number of moles of an undesired byproduct. The omission of byproducts is therefore risky simplification. We also note that the inclusion of directionality considerations appears particularly cumbersome in the treatment of Happel *et al.* (1990, p. 1061): The direct submechanisms for one of the basis reactions violate directionality restrictions; thus, an

166 MICHAEL L. MAVROVOUNIOTIS

$N_2+I\rightleftharpoons N_2I$	$N_2I+H_2\rightleftharpoons N_2H_2I$	$N_2H_2I+I\rightleftharpoons 2NHI$	$N_2+2I\rightleftharpoons 2NI$	$NI+HI\rightleftharpoons NHI+I$	$NHI+HI\rightleftharpoons NH_2I+I$	$NHI+H_2\rightleftharpoons NH_3+I$	$H_2+2I\rightleftharpoons 2HI$	$NH_2I+HI\rightleftharpoons NH_3+I$	mechanism	corresponding mechanism from Happel *et al.* (1990)	species ⇒	N_2I	N_2H_2I	NHI	NI	HI	NH_2I	—	N_2	H_2	NH_3
s_1	s_2	s_3	s_4	s_5	s_6	s_7	s_8	s_9	⇓	⇓	number of combinations ⇒	##	##	##	##	##	##	0	–	–	–
⇌	⇌	⇌	⇌	⇌	⇌	⇌	⇌	⇌			origin ⇓	a_1	a_2	a_3	a_4	a_5	a_6	a_7	a_8	a_9	a_{10}
1	1	1				2			⇌ m_{17}	m_4,Table V	a_3: $m_{13}+2m_7$	##	##	##	##	##	##		–1	–3	2
					1	–1	1	1	⇌ m_{18}	null,Table IV	a_3: $m_{14}-m_7$	##	##	##	##	##	##		0	0	0
			1	2		2	1		⇌ m_{19}	m_1,Table V	a_3: $m_{15}+2m_7$	##	##	##	##	##	##		–1	–3	2
			1	2	–1			–1	⇌ m_{20}	m_3,Table V	a_3: $m_{16}+3m_7$	##	##	##	##	##	##		–1	–3	2
1	1	1			2		2	2	⇌ m_{21}	m_5,Table V	a_3: $m_{13}+2m_{14}$	##	##	##	##	##	##		–1	–3	2
–1	–1	–1	1	2			1		⇌ m_{22}	null,Table IV	a_3: $m_{15}-m_{13}$	##	##	##	##	##	##		0	0	0
3	3	3	–2	–4	2			2	⇌ m_{23}	m_6,Table V	a_3: $3m_{13}-2m_{16}$	##	##	##	##	##	##		–1	–3	2
			1	2	2		3	2	⇌ m_{24}	m_2,Table V	a_3: $m_{15}+2\,m_{14}$	##	##	##	##	##	##		–1	–3	2
			1	2	2		3	2	⇌ m_{25}	m_2,Table V	a_3: $m_{16}+3m_{14}$	##	##	##	##	##	##		–1	–3	2
			1	2	2		3	2	⇌ m_{26}	m_2,Table V	a_3: $3m_{15}-2m_{16}$	##	##	##	##	##	##		–1	–3	2

FIG. 7. The final set of mechanisms produced here and their correspondence to the mechanisms constructed by Happel *et al.* (1990). As was explained earlier, the intermediate a_7 does not need to be eliminated because its column contains only zeros. All the direct mechanisms identified by Happel *et al.* (1990) have been produced. However, the last three mechanisms in Figure 7 are actually identical. Hence, the simple procedure discussed here does not preclude multiple occurrences of the same mechanism, or, as the next example will show, the occurrence of mechanisms that are not direct. For small studies this is not a significant drawback, because one can easily eliminate the redundancies in the end. The potential duplication of mechanisms and construction of indirect mechanisms are addressed by Mavrovouniotis (1992). [Reprinted with permission from Mavrovouniotis, M. L., and Stephanopoulos, G. "Synthesis of reaction mechanisms consisting of reversible and irreversible steps: I. A synthesis approach in the context of simple examples". *Ind. Eng. Chem. Res.* **31**, 1625–1637, (1992). Copyright 1992 American Chemical Society.]

alternative reaction is manually constructed which gives acceptable direct submechanisms.

3. *Importance of Reaction Directionality*

We provide here an abstract example (Fig. 18) to discuss the complexity characteristics of the method of H&S, and in particular that method's assumption at the outset that all reactions are reversible. This assumption might be insignificant for very simple reaction mechanisms or those that in fact do not contain irreversible steps. However, the assumption constitutes

s_1: $CH_4 + O_2 \rightarrow CH_3 + HO_2$

s_2: $CH_3 + O_2 \rightarrow CH_3O_2$

s_3: $CH_3O_2 \rightarrow CH_2O + OH$

s_4: $CH_3O_2 + CH_4 \rightarrow CH_3O_2H + CH_3$

s_5: $CH_3O_2H \rightarrow CH_3O + OH$

s_6: $CH_3O \rightarrow CH_2O + H$

s_7: $CH_3O + CH_4 \rightarrow CH_3OH + CH_3$

s_8: $OH + CH_4 \rightarrow CH_3 + H_2O$

s_9: $CH_3 + CH_3 \rightarrow C_2H_6$

s_{10}: $CH_3 + OH \rightarrow CH_3OH$

s_{11}: $CH_3 + CH_3O \rightarrow CH_3OCH_3$

s_{12}: $CH_2O + CH_3 \rightarrow CH_4 + CHO$

s_{13}: $CHO + O_2 \rightarrow CO + HO_2$

s_{14}: $CH_2O + CH_3O \rightarrow CH_3OH + CHO$

s_{15}: $CHO + CH_3 \rightarrow CO + CH_4$

FIG. 8. Mechanism steps for the synthesis of methanol, as used in an example by Happel *et al.* (1990). The mechanism was proposed by Yarlagadda *et al.* (1988) and assumes that all steps have a net rate in the indicated direction. The species CH_4, O_2, CH_3OH, CO, H_2O, C_2H_6, and CH_3OCH_3 are terminal and all others are intermediates; formaldehyde (CH_2O) is an intermediate because it is present in small amounts (Happel *et al.*, 1990). [Reprinted with permission from Mavrovouniotis, M. L., and Stephanopoulos, G. "Synthesis of reaction mechanisms consisting of reversible and irreversible steps: I. A synthesis approach in the context of simple examples". *Ind. Eng. Chem. Res.* **31**, 1625–1637, (1992). Copyright 1992 American Chemical Society.]

a significant drawback if many steps are irreversible and one starts with a very large reaction system, a large portion of which eventually does not participate in the final solution. In particular, the presence of many parallel irreversible routes leading from intermediates to other intermediates can have devastating consequences on efficiency. Figure 18 shows an idealized schematic of these features; there is actually only one mechanism that converts A to E, but a method that postpones consideration of reaction directionality until the final mechanisms have been constructed will produce an enormous number of possibilities. Thus, the procedure of H & S, as well as any method that ignores directionality, faces great difficulties for large systems of mostly irreversible reactions.

168   MICHAEL L. MAVROVOUNIOTIS

s1 →	s2 →	s3 →	s4 →	s5 →	s6 →	s7 →	s8 →	s9 →	s10 →	s11 →	s12 →	s13 →	s14 →	s15 →	mechanism ⇓	species ⇒	CH3	HO2	CH3O2	OH	CH3O2H	CH3O	H	CHO	CH2O
																combinations⇒	24	0	2	4	1	4	0	4	4
																origin ⇓	a_1	a_2	a_3	a_4	a_5	a_6	a_7	a_8	a_9
1															→m_1	—	1	1							
	1														→m_2	—	−1		1						
		1													→m_3	—			−1	1					1
			1												→m_4	—	1		−1		1				
				1											→m_5	—				1	−1	1			
					1										→m_6	—						−1	1		1
						1									→m_7	—	1					−1			
							1								→m_8	—	1			−1					
								1							→m_9	—	−2								
									1						→m_{10}	—	−1			−1					
										1					→m_{11}	—	−1					−1			
											1				→m_{12}	—	−1							1	−1
												1			→m_{13}	—		1						−1	
													1		→m_{14}	—						−1		1	−1
														1	→m_{15}	—	−1							−1	

FIG. 9. Initial setup for the application of the algorithm on the methanol mechanism. Terminal species are not shown in this figure; they will be incorporated in Fig. 10. The initial arrangement shows the mechanisms m_1 to m_{15}, each using only the step with the same index. The arrows below each step and to the left of each mechanism serve as indicators of directionality; here, all steps are irreversible ($\rightarrow$), and the algorithm that is proposed in this chapter can take directionality into account. In the formation of combinations of partial mechanisms to eliminate an intermediate species, it is no longer possible to take *any* combination of two mechanisms whose net reactions involve the species. If, for example, we attempt to combine m_1 and m_{13} to eliminate a_2 (HO$_2$), then the combination expression $\beta_{b2} m_k - \beta_{k2} m_b$ would lead to either $m_1 - m_{13}$ (for $k = 1$ and $b = 13$) or $m_{13} - m_1$ (for $k = 13$ and $b = 1$), violating the directionality of one of the mechanisms m_1 and m_{13}. This happens because a_2 participates as a net product in both mechanisms; it is simply not possible to eliminate a_2 if we insist on using both mechanisms in the forward direction. Thus, for irreversible steps and mechanisms, a legitimate combination that eliminates an intermediate must include one mechanism in which the species is a net product and one mechanism in which the species is a net reactant. This makes each intermediate's number of combinations equal to yz, where z is the number of mechanisms for which the species is a net product, and y is the number of mechanisms for which the species in question is a net reactant. The somewhat more complicated procedure for chemical systems that include both irreversible and reversible steps was given in the algorithm. As in the previous example, the intermediates with the smallest numbers of combinations are highlighted. The species a_2 (HO$_2$) and a_7 (H) are chosen for elimination, because they give rise to zero combinations (they occur only as products). [Reprinted with permission from Mavrovouniotis, M. L., and Stephanopoulos, G. "Synthesis of reaction mechanisms consisting of reversible and irreversible steps: I. A synthesis approach in the context of simple examples". *Ind. Eng. Chem. Res.* **31**, 1625–1637. (1992), Copyright 1992 American Chemical Society.]

S_1	S_2	S_3	S_4	S_5	S_6	S_7	S_8	S_9	S_{10}	S_{11}	S_{12}	S_{13}	S_{14}	S_{15}	mechanism ⇓	species ⇒	CH_3	CH_3O_2	OH	CH_3O_2H	CH_3O	CHO	CH_2O	CH_4	O_2	CH_3OH	CO	H_2O	C_2H_6	CH_3OCH_3
→	→	→	→	→	→	→	→	→	→	→	→	→	→	→		combinations⇒	18	2	4	1	3	2	2							
																origin ⇓	a_1	a_3	a_4	a_5	a_6	a_8	a_9	a_{10}	a_{11}	a_{12}	a_{13}	a_{14}	a_{15}	a_{16}
	1														→m_2	—	−1	1							−1					
		1													→m_3	—		−1	1				1							
			1												→m_4	—	1	−1		1				−1						
				1											→m_5	—			1	−1	1									
						1									→m_7	—	1				−1			−1		1				
							1								→m_8	—	1		−1					−1				1		
								1							→m_9	—	−2												1	
									1						→m_{10}	—	−1		−1							1				
										1					→m_{11}	—	−1				−1									1
											1				→m_{12}	—	−1					1	−1	1						
													1		→m_{14}	—						−1	1	−1		1				
														1	→m_{15}	—	−1					−1		1			1			

FIG. 10. Setup for the methanol mechanism, after the elimination of a_2 (HO_2) and a_7 (H). Mechanisms that have been abolished and intermediates that have been eliminated are immediately removed from the figures; thus, the rows of m_1, m_6, and m_{13}, and the columns of a_2 and a_7 are not present in Fig. 10 and subsequent figures. The terminal species, not shown in Fig. 9, have been included in the setup. The number of combinations has been recalculated for each species. The intermediates a_5 (CH_3O_2H) and a_8 (CHO), each of which gives rise to only two combinations, are next eliminated in parallel, since their sets of mechanisms (m_4 and m_5 for a_5; m_{12}, m_{14}, and m_{15} for a_8) are disjoint. [Reprinted with permission from Mavrovouniotis, M. L., and Stephanopoulos, G. "Synthesis of reaction mechanisms consisting of reversible and irreversible steps: I. A synthesis approach in the context of simple examples". *Ind. Eng. Chem. Res.* **31**, 1625–1637. (1992). Copyright 1992 American Chemical Society.]

III. Biochemical Pathways

The algorithm described for direct reaction mechanisms can also be used for the synthesis of biochemical pathways from basic bioreactions in the pursuit of quasi-steady-state behaviors of bioprocesses. More importantly, the algorithm identifies pathways leading from available raw materials to desired target products, enabling more informed decisions in the early stages of the design of a bioprocess. In translating the algorithm to biochemical systems, mechanism steps would map to individual bioreactions (usually catalyzed by enzymes), whereas overall reaction mechanisms correspond to acceptable biochemical pathways for a bioprocess.

Here, we will discuss a different formulation of the problem, which is better suited to the preliminary-design interpretation of the synthesis of biochemical pathways. This section will be based on the algorithm and

s1	s2	s3	s4	s5	s6	s7	s8	s9	s10	s11	s12	s13	s14	s15	mechanism ⇓	species ⇒ / origin ⇓	CH_3	CH_3O_2	OH	CH_3O	CH_2O	CH_4	O_2	CH_3OH	CO	H_2O	C_2H_6	CH_3OCH_3
→	→	→	→	→	→	→	→	→	→	→	→	→	→	→		combinations ⇒	18	2	4	3	2							
																origin ⇓	a_1	a_3	a_4	a_6	a_9	a_{10}	a_{11}	a_{12}	a_{13}	a_{14}	a_{15}	a_{16}
	1														$\to m_2$	—	−1	1					−1					
		1													$\to m_3$	—		−1	1		1							
						1									$\to m_7$	—	1			−1		−1		1				
							1								$\to m_8$	—	1		−1			−1				1		
								1							$\to m_9$	—	−2										1	
									1						$\to m_{10}$	—	−1		−1					1				
										1					$\to m_{11}$	—	−1			−1								1
			1	1											$\to m_{16}$	a_5: m_4+m_5	1	−1	1	1		−1						
											1			1	$\to m_{17}$	a_8: $m_{12}+m_{15}$	−2				−1	2			1			
													1	1	$\to m_{18}$	a_8: $m_{14}+m_{15}$	−1			−1	−1	1			1	1		

FIG. 11. Setup for the application of the algorithm on the methanol mechanism, after the elimination of a_5 (CH_3O_2H) and a_8 (CHO). Elimination of the intermediate a_9 (CH_2O), which involves two combination mechanisms, is carried out next. [Reprinted with permission from Mavrovouniotis, M. L., and Stephanopoulos, G. "Synthesis of reaction mechanisms consisting of reversible and irreversible steps: I. A synthesis approach in the context of simple examples". *Ind. Eng. Chem. Res.* **31**, 1625–1637. (1992). Copyright 1992 American Chemical Society.]

s1	s2	s3	s4	s5	s6	s7	s8	s9	s10	s11	s12	s13	s14	s15	mechanism ⇓	species ⇒ / origin ⇓	CH_3	CH_3O_2	OH	CH_3O	CH_4	O_2	CH_3OH	CO	H_2O	C_2H_6	CH_3OCH_3
→	→	→	→	→	→	→	→	→	→	→	→	→	→	→		combinations ⇒	18	3	6	3							
																origin ⇓	a_1	a_3	a_4	a_6	a_{10}	a_{11}	a_{12}	a_{13}	a_{14}	a_{15}	a_{16}
	1														$\to m_2$	—	−1	1				−1					
						1									$\to m_7$	—	1			−1	−1		1				
							1								$\to m_8$	—	1		−1		−1				1		
								1							$\to m_9$	—	−2									1	
									1						$\to m_{10}$	—	−1		−1				1				
										1					$\to m_{11}$	—	−1			−1							1
			1	1											$\to m_{16}$	a_5: m_4+m_5	1	−1	1	1	−1						
		1									1			1	$\to m_{19}$	a_9: m_3+m_{17}	−2	−1	1		2				1		
		1											1	1	$\to m_{20}$	a_9: m_3+m_{18}	−1	−1	1	−1	1			1	1		

FIG. 12. State of the algorithm, for the mechanism, after the elimination of a_9 (CH_2O). Similarly, elimination of the intermediate a_3 (CH_3O_2) leads to Fig. 13. [Reprinted with permission from Mavrovouniotis, M. L., and Stephanopoulos, G. "Synthesis of reaction mechanisms consisting of reversible and irreversible steps: I. A synthesis approach in the context of simple examples". *Ind. Eng. Chem. Res.* **31**, 1625–1637. (1992). Copyright 1992 American Chemical Society.]

s_1	s_2	s_3	s_4	s_5	s_6	s_7	s_8	s_9	s_{10}	s_{11}	s_{12}	s_{13}	s_{14}	s_{15}	mechanism ⇓	species ⇒	CH_3	OH	CH_3O	CH_4	O_2	CH_3OH	CO	H_2O	C_2H_6	CH_3OCH_3
															⇓	combinations⇒	10	6	3							
→	→	→	→	→	→	→	→	→	→	→	→	→	→	→		origin ⇓	a_1	a_4	a_6	a_{10}	a_{11}	a_{12}	a_{13}	a_{14}	a_{15}	a_{16}
						1									$\to m_7$	—	1		−1	−1		1				
							1								$\to m_8$	—	1	−1		−1				1		
								1							$\to m_9$	—	−2								1	
									1						$\to m_{10}$	—	−1	−1				1				
										1					$\to m_{11}$	—	−1		−1							1
	1		1	1											$\to m_{21}$	a_3: m_2+m_{16}		1	1	−1	−1					
	1	1									1			1	$\to m_{22}$	a_3: m_2+m_{19}	−3	1		2	−1		1			
	1	1											1	1	$\to m_{23}$	a_3: m_2+m_{20}	−2	1	−1	1	−1	1	1			

FIG. 13. State of the algorithm, for the methanol mechanism, after the elimination of a_3 (CH_3O_2). Next, elimination of the intermediate a_6 (CH_3O) yields Fig. 14. [Reprinted with permission from Mavrovouniotis, M. L., and Stephanopoulos, G. "Synthesis of reaction mechanisms consisting of reversible and irreversible steps: I. A synthesis approach in the context of simple examples". *Ind. Eng. Chem. Res.* **31**, 1625–1637. (1992). Copyright 1992 American Chemical Society.]

s_1	s_2	s_3	s_4	s_5	s_6	s_7	s_8	s_9	s_{10}	s_{11}	s_{12}	s_{13}	s_{14}	s_{15}	mechanism ⇓	species ⇒	CH_3	OH	CH_4	O_2	CH_3OH	CO	H_2O	C_2H_6	CH_3OCH_3
															⇓	combinations⇒	10	8							
→	→	→	→	→	→	→	→	→	→	→	→	→	→	→		origin ⇓	a_1	a_4	a_{10}	a_{11}	a_{12}	a_{13}	a_{14}	a_{15}	a_{16}
							1								$\to m_8$	—	1	−1	−1				1		
								1							$\to m_9$	—	−2							1	
									1						$\to m_{10}$	—	−1	−1			1				
	1	1									1			1	$\to m_{22}$	a_3: m_2+m_{19}	−3	1	2	−1		1			
	1		1	1		1									$\to m_{24}$	a_6: $m_{21}+m_7$	1	1	−2	−1	1				
	1		1	1						1					$\to m_{25}$	a_6: $m_{21}+m_{11}$	−1	1	−1	−1					1
	2	1	1	1									1	1	$\to m_{26}$	a_6: $m_{21}+m_{23}$	−2	2		−2		1	1		

FIG. 14. Setup of the methanol mechanism, after the elimination of a_6 (CH_3O). The intermediate a_4 (OH) will be eliminated next. [Reprinted with permission from Mavrovouniotis, M. L., and Stephanopoulos, G. "Synthesis of reaction mechanisms consisting of reversible and irreversible steps: I. A synthesis approach in the context of simple examples". *Ind. Eng. Chem. Res.* **31**, 1625–1637. (1992). Copyright 1992 American Chemical Society.]

s_1	s_2	s_3	s_4	s_5	s_6	s_7	s_8	s_9	s_{10}	s_{11}	s_{12}	s_{13}	s_{14}	s_{15}	mechanism ⇓	species ⇒ / origin ⇓	CH_3 a_1 (comb. 5)	CH_4 a_{10}	O_2 a_{11}	CH_3OH a_{12}	CO a_{13}	H_2O a_{14}	C_2H_6 a_{15}	CH_3OCH_3 a_{16}
→	→	→	→	→	→	→	→	→	→	→	→	→	→	→										
								1							→m_9	—	−2						1	
	1	1					1				1			1	→m_{27}	a_4: m_8+m_{22}	−2	1	−1		1	1		
	1		1	1		1	1								→m_{28}	a_4: m_8+m_{24}	2	−3	−1	1		1		
	1		1	1			1			1					→m_{29}	a_4: m_8+m_{25}		−2	−1			1		1
	2	1	1	1			2						1	1	→m_{30}	a_4: $2m_8+m_{26}$		−2	−2	1	1	2		
	1	1							1		1			1	→m_{31}	a_4: $m_{10}+m_{22}$	−4	2	−1	1	1			
	1		1	1		1			1						→m_{32}	a_4: $m_{10}+m_{24}$		−2	−1	2				
	1		1	1					1	1					→m_{33}	a_4: $m_{10}+m_{25}$	−2	−1	−1	1				1
	2	1	1	1					2				1	1	→m_{34}	a_4: $2m_{10}+m_{26}$	−4		−2	3	1			

FIG. 15. Setup of the methanol mechanism, after the elimination of a_4 (OH). Elimination of the only remaining intermediate, a_1 (CH_3), is carried out next. [Reprinted with permission from Mavrovouniotis, M. L., and Stephanopoulos, G. "Synthesis of reaction mechanisms consisting of reversible and irreversible steps: I. A synthesis approach in the context of simple examples". *Ind. Eng. Chem. Res.* **31**, 1625–1637. (1992). Copyright 1992 American Chemical Society.]

s_1	s_2	s_3	s_4	s_5	s_6	s_7	s_8	s_9	s_{10}	s_{11}	s_{12}	s_{13}	s_{14}	s_{15}	mechanism ⇓	species ⇒ / origin ⇓	CH_4 a_{10}	O_2 a_{11}	CH_3OH a_{12}	CO a_{13}	H_2O a_{14}	C_2H_6 a_{15}	CH_3OCH_3 a_{16}
→	→	→	→	→	→	→	→	→	→	→	→	→	→	→									
	1		1	1			1			1					→m_{29}	a_4: m_8+m_{25}	−2	−1			1		1
	2	1	1	1			2						1	1	→m_{30}	a_4: $2m_8+m_{26}$	−2	−2	1	1	2		
	1		1	1		1			1						→m_{32}	a_4: $m_{10}+m_{24}$	−2	−1	2				
	1		1	1		1	1	1							→m_{35}	a_1: $m_{28}+m_9$	−3	−1	1		1	1	
	2	1	1	1		1	2				1			1	→m_{36}	a_1: $m_{28}+m_{27}$	−2	−2	1	1	2		
	3	1	2	2		2	2		1		1			1	→m_{37}	a_1: $2m_{28}+m_{31}$	−4	−3	3	1	2		
	2		2	2		1	1		1	1					→m_{38}	a_1: $m_{28}+m_{33}$	−4	−2	2		1		1
	4	1	3	3		2	2		2				1	1	→m_{39}	a_1: $2m_{28}+m_{34}$	−6	−4	5	1	2		

FIG. 16. The final results for the methanol mechanism. [Reprinted with permission from Mavrovouniotis, M. L., and Stephanopoulos, G. "Synthesis of reaction mechanisms consisting of reversible and irreversible steps: I. A synthesis approach in the context of simple examples". *Ind. Eng. Chem. Res.* **31**, 1625–1637. (1992). Copyright 1992 American Chemical Society.]

s_1	s_2	s_3	s_4	s_5	s_6	s_7	s_8	s_9	s_{10}	s_{11}	s_{12}	s_{13}	s_{14}	s_{15}	mechanism	direct?	mechanism in Table X of Happel *et al.* (1990)
	1		1	1			1			1					m_{29}	√	not found, because s_{11} was omitted
	2	1	1	1			2						1	1	m_{30}	√	m_2/r_0; m_3/r_0 (lines 5 and 7)
	1		1	1		1		1							m_{32}	√	m_1/d_4; m_3/d_4; m_4/d_4; m_6/d_4 (lines 2,6,8,10)
	1		1	1		1	1	1							m_{35}	√	not found, because s_9 was omitted
	2	1	1	1		1	2				1			1	m_{36}	√	m_4/r_0 (line 7)
	3	1	2	2		2	2		1		1			1	m_{37}	$=m_{36}+m_{32}$	—
	2		2	2		1	1		1	1					m_{38}	$=m_{29}+m_{32}$	—
	4	1	3	3		2	2		2				1	1	m_{39}	$=m_{30}+2m_{32}$	—

Fɪɢ. 17. Analysis of the results for the methanol mechanism. The last three mechanisms that were produced (m_{37}, m_{38}, and m_{39}) are not direct; they can be formed from the direct ones. The last column shows the correspondence of the direct mechanisms to those of Happel *et al.* (1990). It is important to note that the same mechanism can be constructed many times in the procedure of Happel and Sellers (1983, p. 290)—as well as the simple algorithm presented here. It should also be noted that, in Table X of Happel *et al.* (1990), the mechanisms m_1/r_0 (line 1), m_2/d_4 (line 4), m_5/r_0 (line 9), m_5/d_4 (line 10), and m_6/r_0 (line 11) are all infeasible because they use either s_1 or s_5 in the wrong direction. The algorithm presented in this chapter never constructs mechanisms that violate the directionality of irreversible steps. [Reprinted with permission from Mavrovouniotis, M. L., and Stephanopoulos, G. "Synthesis of reaction mechanisms consisting of reversible and irreversible steps: I. A synthesis approach in the context of simple examples". *Ind. Eng. Chem. Res.* **31**, 1625–1637. (1992). Copyright 1992 American Chemical Society.]

examples presented by Mavrovouniotis (1989) and Mavrovouniotis *et al.* (1990, 1992). There are two differences between this view and the one adopted in the previous section. First, the classification of compounds as either intermediates or terminal species, introduced by H & S and used in Section II, does not distinguish between appearance of a compound as a reactant (raw material) and appearance as a product; in this section, we will permit separate specifications on net reactants and net products. Second, in Section II reversibility considerations were taken into account in the construction of mechanisms. Since most bioreactions are irreversible, we will treat reversible bioreactions as an exception, by splitting them into forward and reverse portions.

A. Fᴇᴀᴛᴜʀᴇs ᴏғ ᴛʜᴇ Pᴀᴛʜᴡᴀʏ Sʏɴᴛʜᴇsɪs Pʀᴏʙʟᴇᴍ

Biochemical pathway synthesis is the construction of consistent sets of enzyme-catalyzed bioreactions meeting certain specifications. One seeks to construct pathways which produce certain target bioproducts, under partial constraints on the available substrates (reactants), allowed byproducts, desired yield, productivity, etc. The pathway must include all reactions needed to convert initial substrates supplied to the bioprocess into final

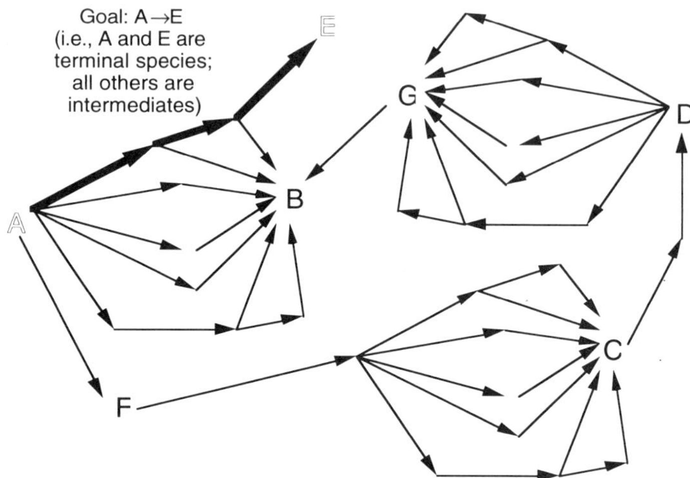

FIG. 18. The consequences of disregarding reaction irreversibility, in an abstract example. The species A and E are terminal and all others are intermediates. There are many parallel irreversible mechanisms that transform various intermediates to others (e.g., from A to B, from F to C, and from D to G). The direction of these internal mechanisms, however, is such that they do not participate in the final solution—which requires mechanisms only between A and E. There is actually only one mechanism (shown in thick arrows) that converts A to E in three steps. A method like that of H&S, which disregards directionality of steps and treats all steps as reversible initially, will come up with a very large number of mechanisms, going through the intermediates F, C, D, G, B (in that order). Our algorithm would *immediately* prune out the irrelevant portions of the network (starting with B) and identify the one feasible mechanism.

products that will be recovered from the process. Even if a computationally synthesized pathway is not present in any single known microorganism, it is of interest because of the possibilities offered by genetic engineering, which achieves the transfer of reactions and pathways from one microorganism to another. Through genetic engineering, a missing step can be inserted in a microorganism so that the synthesized new biochemical pathway (and hence a new bioprocess) is realizable.

Just like a direct reaction mechanism discussed in Section II, a pathway is not merely a set of reactions, because many distinct pathways can be constructed to include the same reactions but achieve different transformations. A pathway must include a *coefficient* (denoted by σ_{ki} in Section II) for each reaction, indicating the proportions in which the stoichiometries are combined. In this section, we will call these coefficients the *reaction stoichiometry* of the pathway. The overall transformation of net reactants to net products will be called the *molecular stoichiometry* of the pathway, and was denoted by β_{kj} in Section II.

Whereas in Section II we assumed that the participating steps will be given for a particular application, here it is useful to have a *database* with the most common enzymatic reactions and compounds, so that one only needs to add the specialized reactions for a given investigation.

B. Formulation of Constraints

A whole class of specifications in the synthesis of biochemical pathways can be formulated by classifying each compound and each reaction from the database according to the role it is required or allowed to play in the synthesized pathways' stoichiometries.

A given compound may occur in a pathway in any of three capacities: (1) as a net reactant or substrate of the pathway, (2) as a net product of the pathway, or (3) as an intermediate in the pathway. Constraints state whether a compound is allowed or possibly required to participate in the synthesized pathways in each of these three capacities.

For compounds as net reactants in the pathways, these conditions are inequality constraints on the coefficients β_{kj} (see Section II) that describe the participation of each species a_j in a pathway m_k. The specifications take the following forms:

(a) The species a_j is called a required reactant if it *must be consumed* by the pathway; this corresponds to a requirement that the coefficient β_{kj} be negative, i.e., $\beta_{kj} < 0$.

(b) A species is designated as an excluded reactant if it *cannot be consumed* by the pathway, and the corresponding restriction on the coefficient is $\beta_{kj} \geq 0$.

(c) Finally, the classification of a_j as an *allowed reactant* introduces no restriction on the coefficient β_{kj}.

In a realistic synthesis problem, the default constraint for compounds is that of excluded reactants, since most compounds in the database would not be available as economical raw materials.

The constraints for potential products of the pathways involve similar inequalities. A species a_j can be specified as (1) a required product, $\beta_{kj} > 0$, (2) an excluded product, $\beta_{kj} \leq 0$, which is the default, or (3) an allowed product, which involves no restriction on the sign of β_{kj}. Each compound a_j thus acquires one constraint from the reactant specifications and one from the product specifications. One of these combinations, however, the designation of the same compound as a required reactant ($\beta_{kj} < 0$) and a required product ($\beta_{kj} > 0$), is inconsistent. Of the remaining combinations, the designation of a compound as an allowed reactant and allowed product corresponds to the absence of any constraint on β_{kj}. The reader may compare these sets of constraints to the specification of intermediate and terminal species in Section II. There, we classified each species a_j using only two categories, as follows:

(a) If a species a_j is an intermediate, its coefficient in m_k is restricted to $\beta_{kj} = 0$. Clearly, this corresponds to the combination (conjuction) of an excluded reactant constraint ($\beta_{kj} \geq 0$) and an excluded product constraint ($\beta_{kj} \leq 0$).

(b) If a_j is a terminal species, its coefficient in m_k is not restricted at all. This corresponds to the combination of an allowed reactant and an allowed product.

Thus, in the case of biochemical pathways we have allowed a richer vocabulary for stoichiometric constraints but we can still reproduce the earlier specifications.

Notice the relation between the specification of a required reactant (negative stoichiometric coefficient—a strict inequality) and an excluded product (negative or zero stoichiometric coefficient—a loose inequality). In the operation of the algorithm, the first type of constraint is initially not fully satisfied; instead, the corresponding loose inequality constraint is satisfied in its place. Thus, for most of the phases in the algorithm, required reactants are treated merely as excluded products. As we will see, the distinction is eventually enforced, in the last phase of the algorithm. A similar observation holds for the strict inequality arising from required products, in relation to the loose inequality arising from excluded reactants.

A reaction can participate in pathways in either its forward or its reverse direction, giving rise to additional possible specifications: Reactions may be required, allowed, or prohibited to participate in the synthesized pathways in each of the two directions. Most reactions are likely to be *excluded* in the backward direction, because of prior knowledge about the (thermodynamic or mechanistic) irreversiblity of the reaction.

C. ALGORITHM

The algorithm for the synthesis of biochemical pathways follows closely the logic of the algorithm for the synthesis of catalytic reactions, i.e., it synthesizes biochemical pathways from a set of enzymatic reactions through an iterative satisfaction of constraints. A few minor differences reflect the richer vocabulary of specifications, given in the preceding section, and the biochemical context of compounds and enzymatic reactions.

Given a set of stoichiometric constraints and a database of biochemical reactions, the algorithm carries out iterative satisfaction of constraints, just like the algorithm in Section II. The algorithm proceeds as follows.

1. Initialization and Reaction Processing

The set M contains a one-step pathway for each individual enzymatic reaction available in the database, which has been compiled from known biochemistry. The set N contains the metabolites present in the enzymatic reactions of the set M. Naturally, these include the substrates (raw materials) that can be used, the desired products, and a large number of other compounds that occur in the bioreactions but will not serve as raw materials or desired products. This last set of compounds will carry the restrictions of excluded reactants and excluded products.

In order to account for the reversibility of some reactions, the algorithm decomposes reversible reactions into a forward and a backward step; the two directions are thereafter prohibited from participating together in the same pathway. This approach, unlike Section II, treats reversible reactions as the exception.

2. Number of Combinations for Each Intermediate

We note that the iteration through phases 2–4 was called compound-processing in the original description of this algorithm (Mavrovouniotis *et al.*, 1990). In these phases, partial pathways are gradually combined to satisfy the constraints (just as in Section II). This takes place as the algorithm tackles one compound at a time.

The objective of phase 2, in particular, is the selection of the most suitable compound for constraint satisfaction. We want to given priority to compounds that participate (as reactants or products) in only a small number of pathways that are active in the current state of the problem.

The computation of combinations (n_j as defined in Section II) facilitates this selection. Since all pathways are irreversible in our biochemical formulation, we only need to assemble (for each species a_j) the set Y_j, which contains all the partial pathways that include a_j as a net reactant, and the set Z_j, which contains all the partial pathways that include a_j as a net product. The cardinalities of these sets are y_j and z_j, respectively. It is clear that, in order to eliminate the metabolite a_j, we need to combine any pathway from Y_j, giving rise to $n_j = x_j y_j$ combinations.

3. Elimination of an Intermediate Metabolite

Partial pathways (which initially are just individual enzymatic reactions) are gradually combined in an effort to satisfy the constraints on the role of each metabolite. To minimize the computational effort, we choose in each iteration that metabolite that has the smallest number of combinations n_j. Let a_j be this metabolite. For the satisfaction of the constraints, a set M_J of new pathways is constructed through combinations of exactly one partial pathway from Z_J and exactly one pathway from Y_J. These pathways have the form

$$m_c = \beta_{bJ} m_k - \beta_{kJ} m_b,$$

where m_k are m_b are taken from the sets Y_J and Z_J, respectively (i.e., $m_k \in Y_J$ and $m_b \in Z_J$) and β_{bJ}, β_{kJ} are the net coefficients with which a_j participates in m_b and m_k, respectively. The net coefficient of a_j in the new pathway, m_c, is $\beta_{bJ} \beta_{kJ} - \beta_{kJ} \beta_{bJ} = 0$, verifying that, in the resulting new pathway, a_j has been eliminated.

A subtle precaution must be taken in the formation of the combinations, because of the way in which reversible reactions were decomposed into forward and reverse portions. Specifically, if m_k and m_p involve the same reaction (*any* reaction) *in different directions*, then the combination is rejected (or rather not formed at all).

4. Update of Active Sets

In order to modify the set of pathways such that the constraints on the compound a_j are met, we make use of the sets $M_J, Y_J,$ and Z_J. The set of active pathways is modified by selecting the first applicable modification from the following list:

1. If a_j is an excluded product and a required or allowed reactant, $M := M \cup M_J - Z_J$.
2. If a_j is an excluded reactant and a required or allowed product, $M := M \cup M_J - Y_J$.
3. If a_j is an excluded reactant and an excluded product, $M := M \cup M_J - Y_J - Z_J$.
4. If none of the previous conditions describe the constraints on a_j, then M remains unchanged.

Then, we can remove the metabolite a_j and set $N := N - \{a_j\}$.

5. Termination or New Iteration

At this point, if no metabolite remains in N (i.e., if $N \neq \varnothing$), we can proceed to the final phase. If metabolites remain, then we return to phase 2, with a new computation for the selection of the next metabolite to process. In effect, the procedure in phases 3–4 is eventually carried out for all compounds.

6. Pathway Processing

The necessity for this phase, which did not exist in the algorithm of Section II, is a consequence of the richer vocabulary of stoichiometric constraints used for biochemical systems. The set of active pathways resulting from the processing of all metabolites as above satisfies all the requirements, except the constraints designating required reactants, products, or reactions. For these, only the corresponding loose-inequality constraints are satisfied; for example, instead of a required reactant and prohibited product, a compound will have been treated, up to this point, as an allowed reactant and prohibited product. The pathways satisfying the original stoichiometric constraints are those combinations of pathways from the final set that use at least one constituent pathway satisfying each of the strict constraints. The combination must thus include at least one pathway consuming each required reactant, at least one pathway producing each required product, etc.

D. EXAMPLES

We will discuss briefly some examples, to illustrate the features of the pathway synthesis algorithm and the results it has produced.

1. Computational Efficiency

The structured character of biochemical reaction networks is exploited by the synthesis algorithm in early pruning and abstraction, with significant gains in computational efficiency. This happens because the algorithm processes first those compounds and reactions that lead to few or no new combinations. This is shown schematically in Fig. 19, where some irrelevant portions of the network are pruned and some linear chains of reactions are compacted. We should note that this kind of reaction sequences are very common in biochemical reaction systems.

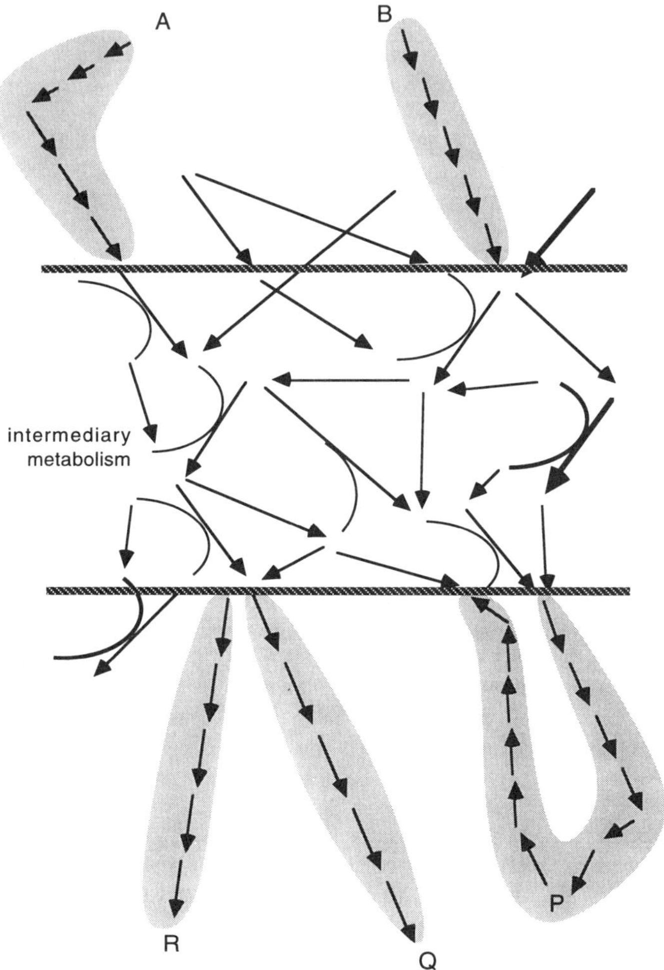

FIG. 19. An instance of efficient processing by the algorithm for the synthesis of biochemical pathways. Suppose that P, A, and Q are excluded reactants and excluded products; B is an allowed reactant; R is an allowed product; and all other compounds are excluded reactants and excluded products. In the early stages of the compound-processing phase, the algorithm will discard, one after the other, all reactions involved in the metabolism of B and R; it will construct the whole pathway that synthesizes and degrades P, discarding the individual reactions, and thus reducing the number of reactions/pathways that are being considered; it will also construct pathways for the consumption of A and the production of Q, discarding the individual reactions for further reduction of the pathway space.

2. Synthesis of Alanine

The synthesis of alanine from glucose will be discussed here. In the database of 250 reactions used by Mavrovouniotis *et al.* (1992), six reactions involve glucose, and four involve alanine (*alanine dehydrogenase, methylserine hydroxymethyltransferase, alanine aminotransferase,* and *β-alanine aminotransferase*).

The initial formulation of the problem has glucose as a required reactant, alanine as a required product, NH_3 as an allowed reactant, and CO_2 as an allowed product. Additionally, a set of compounds that serve as internal currencies in the cell are designated as allowed reactants and allowed products. These are likely to occur in the problem specification, and they include NAD, NADH, NADP, NADPH, ATP, ADP, AMP, coenzyme-A (CoA), phosphate, and pyrophosphate. This initial formulation is actually too tight to generate a manageable number of pathways. By examining the constraints that are difficult to satisfy, we can modify the formulation of the problem to permit a solution. Here, the problem is resolved by designating the compounds malate and acetyl–CoA as allowed reactants and allowed products, reaching a formulation that constructs 1446 useful pathways.

Figures 20 and 21 show two pathways, suppressing many of the side reactants and side products for simplicity. Figure 20 depicts the normal

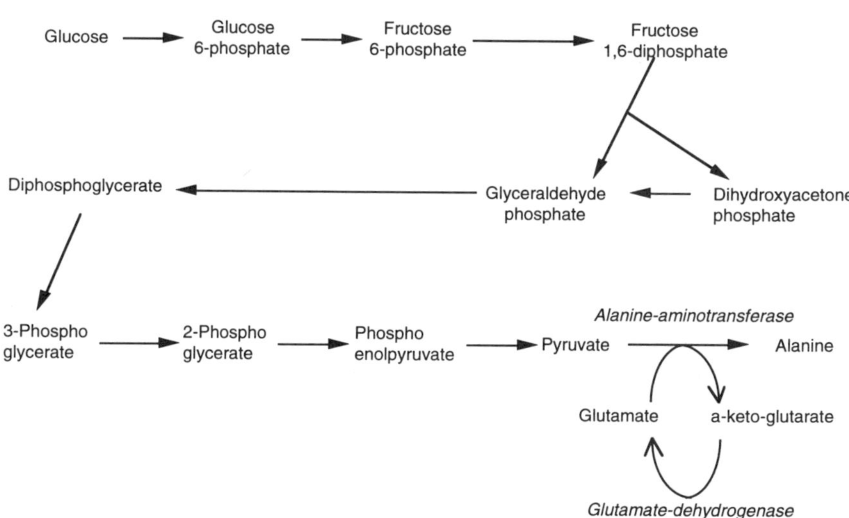

FIG. 20. Synthesis of alanine from glucose that follows the glycolysis pathway to pyruvate, which is in turn converted to alanine by *Alanine aminotransferase*. The glutamate required by this reaction is produced by *glutamate-dehydrogenase*.

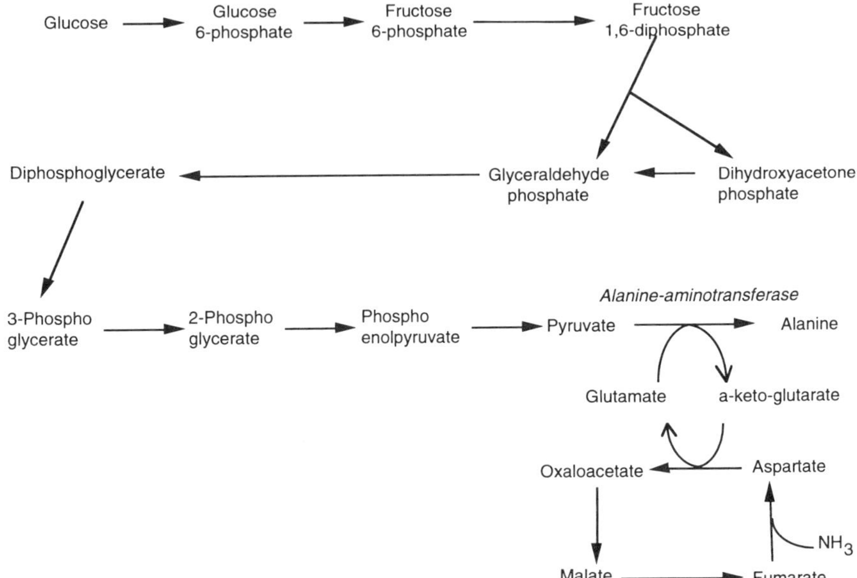

FIG. 21. Synthesis of alanine from glucose, with recovery of glutamate through a loop involving oxaloacetate, aspartate, fumarate, and malate.

pathway for the synthesis of alanine, with glucose as the main reactant. The pathway of Fig. 21 shows an alternative recovery of glutamate through a set of four reactions. The second pathway represents an interesting way for the cell to attach ammonia to an organic backbone; this is a variation that may be relevant for many aminoacid processes.

3. Synthesis of Lysine

In another study (Mavrovouniotis *et al.*, 1990), the biosynthesis of the aminoacid lysine was extensively investigated. In that case study, several reactions (shown in gray in Fig. 22) were assumed nonfunctional, and alternatives that avoid these steps were sought.

Figure 22 shows one of the resulting pathways, which involves the use of glyoxylate as an intermediate to bypass one of the nonfunctional steps. The pathway also includes conversion of malate to fumarate (which is the direction opposite to that normally used by microorganisms) and subsequent conversion of fumarate into aspartate and on to lysine. The use of the latter step is the key to the existence of this innovative pathway.

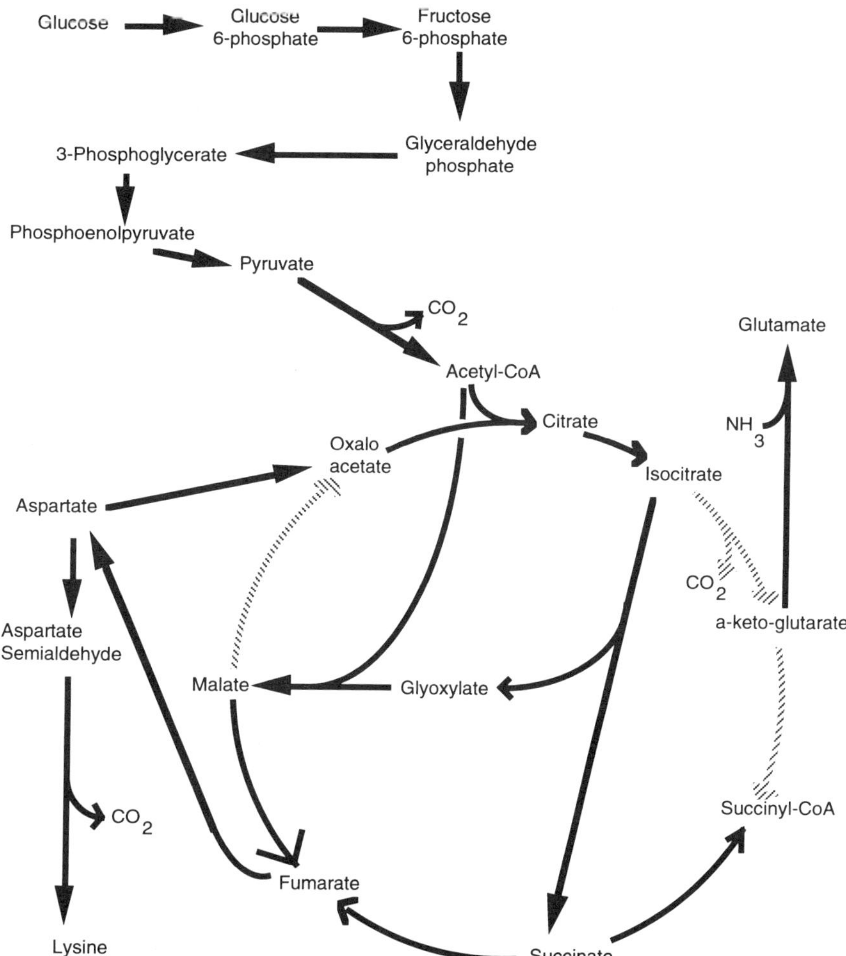

FIG. 22. A pathway for the production of lysine from glucose, drawn in a simplified form (many side reactants and side products are not shown, and many reaction sequences are lumped together).

IV. Properties and Extensions of the Synthesis Algorithm

Both forms of the algorithm (i.e., the basic version of Section II and the alternative of Section III) guarantee that each combination of pathways from the final set (derived by the synthesis algorithm) satisfies all the initial stoichiometric requirements. The correctness of the algorithm is based on the fact that each of the original requirements is satisfied in one of the iterations, and after a constraint is satisfied it is guaranteed to remain satisfied. As was pointed out earlier, constraints on required reactants or required products are satisfied in two phases. Initially, only the corresponding loose-inequality form is satisfied; the satisfaction of the strict-inequality version is ensured in the last phase (6) of the algorithm.

The second important property of the algorithm is its completeness. The synthesis algorithm creates a final set of pathways such that any pathway satisfying the original stoichiometry constraints is present in the final set. The basis of this property is that the elimination procedure is guaranteed to preserve legitimate pathways. A more detailed discussion of these properties is given by Mavrovouniotis *et al.* (1992).

We note that the algorithm may give rise to duplicate or indirect mechanisms (or pathways). Procedures can be applied in the end to remove such redundant mechanisms from the final set of mechanisms, or even to remove them incrementally during each iteration of the algorithm. These and other variations and implementational choices in the algorithm can be defined to enhance either its computational efficiency or its conceptual simplicity (Mavrovouniotis, 1992).

With respect to its computational complexity, the algorithm would require time, in the worst case, exponential in the number of reactions (Mavrovouniotis *et al.*, 1990). This worst-case complexity, however, is not encountered in practice, because the algorithm exploits the structure of biochemical networks, as was shown in Fig. 19. Another example, in Fig. 18, shows how the careful inclusion of directionality (irreversibility) considerations in the algorithm also serves to curb computational complexity.

A last feature of the algorithm that is worth discussing is its extendibility to other types of constraints. Additional specifications for the pathways or mechanisms under construction might arise from thermodynamics, kinetics, yield or productivity restrictions, biological regulation, etc. The question is whether such additional constraints can easily be incorporated into the synthesis procedure.

Some of these constraints are linear in nature. For example, a restriction on the endothermic or exothermic character of a pathway is a linear constraint on the enthalpy of reaction. If we take a linear combination of reactions (or pathways) the enthalpy of the resulting pathway is the corresponding linear combination of the enthalpies of the constituent reactions or pathways. For these linear situations, the quantity being constrained can be considered as merely another species, whose stoichiometry is subject to the type of constraints already tackled by the algorithm. Thus, the synthesis algorithm can deal with these specifications with little or no modification.

Other types of constraints, not corresponding to quantitative linear relations, can also be included, if they fulfill two requirements. The first requirement is that the constraint must possess the following convexity-like property: If the constraint is satisfied for two individual pathways, m_k and m_b, then it should be satisfied for any positive linear combination of these, $m_c = \mu_k m_k + \mu_b m_b$ (where $\mu_k > 0$, $\mu_b > 0$). The second requirement is that, given a set of partial mechanisms that do not satisfy the constraint, there must be a clear way to identify combination mechanisms that do satisfy it. These two requirements allow a constraint to be processed in much the same way that the stoichiometric constraints were, except that the details of construction of combinations would depend on the nature of the constraint. In one of the iterations, the constraint would be processed (through a procedure dictated by the second requirement), and it would thereafter remain satisfied (because of the first requirement).

V. Summary

In a chemical system we often discriminate between intermediate species and terminal species, and the latter are the only ones that appear in appreciable amounts in the stoichiometry of the overall transformation. Alternatively, we designate compounds as allowed or required to appear as reactants or products in the overall stoichiometry. These formulations are significant in the identification of quasi-steady-state behaviors or the synthesis of pathways to accomplish a desired transformation in a process.

We presented here a conceptual framework and algorithms for the synthesis of pathways or mechanisms given a set of steps. The algorithms have been applied to catalytic reaction systems and to biochemical pathways. The basic approach is based on successive processing and elimination of reaction intermediates that should not appear in the net stoichiometry of the overall reactions accomplished.

References

Happel, J., "Isotopic Assessment of Heterogeneous Catalysis." Academic Press, Orlando, FL, 1986.
Happel, J., and Sellers, P. H., Multiple reaction mechanisms in catalysis. *Ind. Eng. Chem. Fundam.*, **21**, 67–76 (1982).

Happel, J., and Sellers, P. H., Analysis of the possible mechanisms for a catalytic reaction system. *Adv. Catal.* **32**, 273–323 (1983).

Happel, J., and Sellers, P. H., The characterization of complex systems of chemical reactions. *Chem. Eng. Commun.*, **83**, 221–240 (1989).

Happel, J., Sellers, P. H., and Otarod, M., Mechanistic study of chemical reaction systems. *Ind. Eng. Chem. Res.* **29**˙ 1057–1064 (1990).

Horiuti, J., Theory of reaction rates as based on the stoichiometric number concept. *Ann. N.Y. Acad. Sci.* **213**, 5–30 (1973).

Horiuti, J., and Nakamura, T., On the theory of heterogeneous catalysis. *Adv. Catal.* **17**, 1–74 (1967).

Lee, L.-S., and Sinanoğlu, O., Reaction mechanisms and chemical networks—Types of elementary steps and generation of laminar mechanisms. *Z. Phys. Chem.* **124**, 129–160 (1981).

Mavrovouniotis, M. L., Computer-aided design of biochemical pathways. Ph.D. Thesis, Massachusetts Institute of Technology (1989).

Mavrovouniotis, M. L., Synthesis of reaction mechanisms consisting of reversible and irreversible steps: II. Formalization and analysis of the synthesis algorithm. *Ind. Eng. Chem. Res.* **31**, 1637–1653 (1992).

Mavrovouniotis, M. L., and Stephanopoulos, G., Synthesis of reaction mechanisms consisting of reversible and irreversible steps: I. A synthesis approach in the context of simple examples. *Ind. Eng. Chem. Res.* **31**, 1625–1637 (1992).

Mavrovouniotis, M. L., Stephanopoulos, G., and Stephanopoulos, G., Computer-aided synthesis of biochemical pathways. *Biotechnol. Bioeng.* **36**, 1119–1132 (1990).

Mavrovouniotis, M. L., Stephanopoulos, G., and Stephanopoulos, G., Synthesis of biochemical production routes. *Comput. Chem. Eng.* **16**, 605–619 (1992).

Milner, P. C., The possible mechanisms of complex reactions involving consecutive steps. *J. Electrochem. Soc.* **111**, 228–232 (1964).

Sellers, P. H., An introduction to a mathematical theory of chemical reaction networks I. *Archive for Rational Mech. Anal.* **44**, 23–40 (1971).

Sellers, P. H., An introduction to a mathematical theory of chemical reaction networks II. *Archive for Rational Mech. Anal.* **44**, 376–386 (1972).

Sellers, P. H., Combinatorial classification of chemical mechanisms. *SIAM J. Appl. Math.* **44**, 784–792 (1984).

Sellers, P. H., Combinatorial aspects of enzyme kinetics. In "Applications of Combinatorics and Graph Theory in the Biological and Social Sciences" (F. Roberts, ed.). Springer-Verlag, Berlin, 1989.

Sinanoğlu, O., Theory of chemical reaction networks. All possible mechanisms or synthetic pathways with given number of reaction steps or species. *J. Am. Chem. Soc.* **97**, 2309–2320 (1975).

Sinanoğlu, O., and Lee, L.-S., Finding all possible *a priori* mechanisms for a given type of overall reaction. *Theor. Chim. Acta* **48**, 287–299 (1978).

Temkin, M. I., The kinetics of transfer of labeled atoms by reaction. *Ann. N.Y. Acad. Sci.* **213**, 79–89 (1973).

Temkin, M. I., The kinetics of some industrial heterogeneous catalytic reactions. *Adv. Catal.* **26**, 173–291 (1979).

Yarlagadda, P. S., Morton, L. A., Hunter, N. R., and Gesser, H. D., Direct conversion of methane to methanol in a flow reactor. *Ind. Eng. Chem. Res.* **27**, 252–256 (1988).

INDUCTIVE AND DEDUCTIVE REASONING: THE CASE OF IDENTIFYING POTENTIAL HAZARDS IN CHEMICAL PROCESSES

Christopher Nagel[1] and George Stephanopoulos

Laboratory for Intelligent Systems in Process Engineering
Department of Chemical Engineering
Massachusetts Institute of Technology
Cambridge, Massachusetts 02139

All reasoning carried out by computers is *deductive*; i.e., any software system has all the necessary data, stored in various forms in a database,

---

[1] Present address: Molten Metal Technology, Inc., Waltham, Massachusetts.

ADVANCES IN CHEMICAL ENGINEERING, VOL. 21

and possesses all the necessary algorithms to operate on the set of data and *deduce* some results. Many researchers in the area of cognitive psychology make similar claims on the reasoning mechanisms of the human beings. The fact, though, remains that both humans and machines can use very simple "algorithms" on a small set of data and produce results, which could not have been visible by the "naked eye" of direct reasoning. In such cases, we tend to talk about the *inductive* capabilities of either of the two. These ideas are nowhere more prominent than in the area of *hazards identification and analysis*. One often hears, "if I knew that the conversion of A to B could be catalyzed by the presence of C then I would have forseen the last disaster, and have done something about it," with the speaker converting a problem of *inductive* identification (i.e., induce the possibility of a hazard from the list of chemicals) into an issue of deductive statement. In this chapter we try to demonstrate that the identification of hazards is essentially an interplay between inductive and deductive reasoning. Through inductive reasoning we attempt to generate all potential hazardous top-level events, which can be justified by the presence of a set of chemicals. We call the reasoning inductive because it has the potential to generate specific knowledge that was not "visible" ahead of time. Once the potentially harmful top-level events have been identified, deductive reasoning attempts to "walk" through the processing scheme and its unit operations and their design or operating characteristics (assumptions, or decisions), and generate the preconditions, which would enable the occurrence of a specific top-level event. The inductive reasoning procedures operate on a set of chemicals and create in an *exhaustive, bottom–up* manner many alternative reaction-pathways, some of which could lead to a hazard, e.g., release of large amounts of energy over a short period of time. On the other hand, the deductive reasoning procedures are *goal-directed* and operate in a *top–down* manner. In this chapter we will develop the detailed framework for the implementation of these ideas, which among other benefits offer the following advantages: (1) formalize the hazards identification problem and unify the methodological approaches at any stage of the design activities and (2) systematize the generation and evaluation of mechanisms for the prevention of hazards, or containment of their effects.

I. Introduction

When a process is transferred from the laboratory to the pilot or/ and commercial scale, a variety of hazards may appear that had earlier been well controlled under the relatively small scale of the discovery effort. Throughout the period of a process's development, two factors that may introduce new hazards and must be examined continuously are *change* and *scale-up* (Brannegan, 1985). Change complicates hazards evaluation by introducing new components that are associated with hazards that may be unknown. Scale-up, particularly initial scale-up, can generate significant potential hazards by escalating the magnitude of effects, initially thought to be benign. Since changes often occur throughout the life of the plant, the need to identify hazards as early as possible in the development stages does not imply that hazard identification ends when the design specifications have been approved. In fact, approval of a design means only "At the time of the study, the study team believes that, provided that the plant is constructed and operated in accordance with their recommendations, the plant will be acceptably safe" (Lowe and Solomon, 1983). The first uncontrolled change during construction, or the first unapproved modification during operation, negates this approval. Consequently, hazard identification is a continuing concern and a permanent ingredient of safe operations and should be applied, sometimes in a very simple form, to control any changes from the original intentions of the designers. Several approaches have been presented in the past decade to systematize hazard identification and hazard analysis, but procedural robustness is often constrained by the quality of information available and the expertise of the individuals involved.

No one doubts the importance of hazard identification, in advance of an unwanted event. However, the quality of the risk analysis results can be no better than the extent to which hazards are recognized in the first place. Furthermore, the analysis is no better than the analyst's understanding of the plant's design and its operations. Decisions about safety are made continuously. These decisions are made in light of all the uncertainties and are based on the understanding of the characteristics of the facility and the substances involved. Formal analytical processes may or may not be involved in the decision process. Studies recently indicated that design errors—a design error is deemed to have been committed when the design is changed after an incident—were rarely revealed before startup and accounted for 25% of all accidents (Haastrup, 1983). India's experience with design error is closer to 40%, and it has been suggested that if the definition is broadened to include management and organizational aspects of process design and engineering, design errors would account for nearly 90% of the recorded incidents (Batstone, 1987). Moreover, the percentage of precursors (leading to a hazardous incident), perceived to be known at the time of the incident, varies. It depends on the perceptions that an individual formed in his or/her particular capacity. Our survey suggests that plant personnel believe that 90–100% of the precursors leading to a hazard are known at the time of the incident, hazard specialists believe that 40–60% of those precursors are known at the time of the incident, and insurance analysts believe that only 20–30% of the precursors initiating or propagating the preconditions to a hazard are known at the time of the incident. Such data are exacerbating and they suggest, to some, that improved hazard identification methods are unnecessary, perhaps even unwanted. Yet, incidents continue to occur.

A. Predictive Hazard Analysis

The basic objective of hazard analysis is to identify and assess potentially hazardous situations, and their possible consequences and associated risk, in order to provide a rational basis for determining where risk reduction measures are needed. Hazard identification always has been an integral part of design and operational practice. However, it is to a large degree still an informal process depending on the experience of those directly involved.

Structured hazard identification methods can be classified roughly in two groups: (1) *comparative* methods that rely on systematic comparison of the process design against some recognized code or standard and (2) *fundamental* methods that can be applied in almost any situation (Boykin and Kazarians, 1987; Ozog and Bendixen, 1987). Individual experience is the essential ingredient of hazard identification for the first group of methods. However, such an approach requires that individual experience be collected, organized, recorded and standardized and become accessible information to those designing the equipment (e.g., through national and international codes and standards). Such codes and practices provide minimum standards against which deviations from safe practice can be identified and appraised. An important feature of these methods is that the experience gained through many years is incorporated in the company's practices and therefore available for use at all stages of design and construction. For new processes, the hazard identification procedure is strongly dependent upon information obtained a priori, and derived from the efforts of the research and development engineers and the hazard identification team members. This requires that hazard identification methods must be directed toward stimulating the team members to apply their own experience of safe and unsafe as the standard by which to appraise the design, mainly by raising a series of "what if" questions. Fundamental methods of hazard identification are aimed at two outcomes: to identify serious incidents that may result in personal injury or financial loss [known as "top-level events" or (TLES)]; and to identify the underlying root causes leading to top-level events.

In general, methods that identify actions that eliminate, avoid, or reduce the potential hazard of a particular design are referred to as *intrinsic*. This approach evolves the design technology toward inherently safer configurations through the use of codes, guidelines, and checklists. Intrinsic methodologies can be effective at the stages where the process scheme is conceived and the process flow diagram is developed. At these stages equipment changes are easily made without adversely affecting construction costs and schedules (see Fig. 1), but they lack formality and afford no means for complete and consistent hazard identification. Alternatively, methods that identify actions that reduce the likelihood of a hazard through control and safety devices are referred to as *extrinsic*. These methods tend to control an identified hazard, or the conditions leading to the hazardous state and often generate solutions that increase plant complexity and operational costs. The more formal methods, such as HazOp, FMECA (Failure Modes, Effects, and Criticality Analysis), fault trees, event trees, and cause–consequence analysis, require information that is often sparse in the early design stages (see Fig. 1) and the certainty of which may be indeterminate.

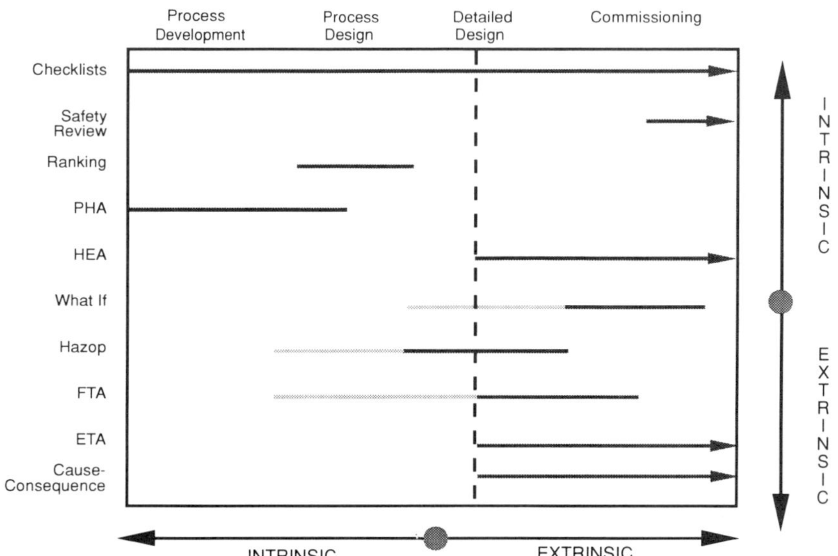

FIG. 1. Applicability of various hazards identification and analysis methodologies during various phases of the design process.

Lees (1980) stated that, "The safety of the plant is determined primarily by the quality of the basic design concept rather than by the addition of special safety features." This point cannot be overemphasized. The degree to which it is economic to eliminate, as opposed to control, a hazard is very dependent on when the hazard is recognized. By the time the design has reached the stage of sufficient documentation to allow a detailed hazard identification, the flexibility to eliminate hazards entirely is very reduced. If hazard/ error detection is to be shifted to predesign review, a less encumbered approach must be developed and integrated into the design process.

B. Incompleteness of Conventional Hazard Analysis Methodologies

The identification of potential hazards and the evaluation of their effects should be a continuing process from the conception of the processing scheme to plant shutdown and decommissioning. However, because conventional methods are limited by their scope and useful life span, it is difficult to integrate hazard analysis and evaluation into the design process. More importantly, all conventional methods are incomplete. Their ineffectiveness is inherent in their methodological approach and their limitations in capturing and utilizing all forms of available knowledge. These weaknesses result in two principal deficiencies: (1) the strength of analysis is dependent on the a priori identification of hazards and/ or events leading to these hazards, and (2) they are unable to offer a reliable estimate of the quality of the design technology. The same weaknesses prevent the available methods from reaching an adequate balance among the following non-commensurable objectives: (a) early identification of hazards to avoid costly redesign or construction modifications, (b) postponement of evaluation to await more detail, and (c) avoidance of costly duplication of effort. Currently, there is no single solution to the problem. Multiple methods are used over the extended time period of the design process (development, construction, operation). Each of these methods suffers from particular deficiencies, which are born out from the specific design context that they were to service and the character of the approach that they have adopted.

Intrinsic methods, for example, although they tend to increase the quality of the process design, must generally be employed in the early engineering stages. The window of opportunity for their application is very brief since the incorporation of intrinsic safety features at a late stage in the design will usually require major design changes with adverse consequences on the cost and the schedule for the commissioning of the plant's operation. Moreover, they do not provide a creative search for new hazards when experience is lacking, nor can they provide quantification as to the quality of a particular design. Heuristics have been proposed to fortify intrinsic methods (Dale, 1987; Kletz, 1985). They are ad hoc pieces of knowledge, offering no metric of sound origin to discriminate among alternative design concepts. However, they can be suggestive as to a design's quality provided they are intelligently applied; blind usage often leads to unforeseen difficulties.

Extrinsic methods attempt to quantify the implications of a hazard's occurrence by beginning with a detailed design. Unfortunately, by the time a detailed design is available, it is often too late to avoid hazards. The control of the hazards through external, safety devices is the only economic alternative. In principle, controlling the effect of hazards leads to an acceptable solution, although there is no assurance that the envisioned control scheme will effectively mitigate all eventual outcomes. In fact, the fundamental flaw in any method that is based on controlling hazards lies in the assumption that they have the ability to both identify accurately and pinpoint precisely the location of a future hazard. Virtually all experience suggests that the contrary is true. Indeed, wherever we suspect the possibility for the initiation of a hazard, we take added precautions to add control and safety devices in order to reinforce the assurance that the hazard will not occur. Consequently, conventional methods are approximations to hazard analysis and assessment. They are incomplete because the set of axioms they use and the deductive methodologies they employ are both incomplete. Furthermore, their performance and effectiveness is fundamentally limited due to their lack of expressive power. Conventional methods avoid the use of models and seek solutions from techniques that do not have firm chemical engineering foundations. What is needed is a formal, unified approach for systematically, automatically, and completely identifying hazards and pathways leading to hazards in design alternatives —at all stages of the design process. But how does one design such a methodology?

C. Premises of Traditional Approaches

The weaknesses of the traditional approaches, as discussed above, are all due to their inherent *lack of representational expressiveness*. This shortcoming can manifest itself in the following ways: (1) a methodology exhibits strength only in a particular design phase, where it captures and uses most of the knowledge available during that phase of the design process; (2) a methodology is incapable of transferring information derived at one phase to another phase of the design process, nor can it reason about the chemical process and its surrounding environment; (3) discrimination among design alternatives varies depending on the technique employed; and (4) complete identification of hazards cannot be guaranteed and therefore the quality of the analysis cannot be assessed. The fundamental premise of traditional approaches is that *"in hazards analysis there is no unifying theme"*; all hazards are different and that every unreliable design creates hazardous situations in its own way. The stand-alone nature of previous approaches is evidence of this statement.

We have found that the truth is somewhat different: *representation limitations of past efforts have prevented the exploitation of unifying themes*. Moreover, conventional methodologies do not include for use, or produce *explicit* information, such as (1) *underlying assumptions* on operational modes, phase equilibria, fluid mechanical aspects, reaction rates, mechanisms of mass and heat transfer, and selected approaches in estimating the value of thermodynamic properties; (2) *simplifications* made by the analyst to limit the model's validity over a given range of conditions or to underscore the relative importance of various physicochemical phenomena; (3) *missing relationships* including qualitative relationships, order-of-magnitude reasoning, and inequalities; and (4) *scope of the task*, that is, what was the process intended for. Difficulties are encountered in the identification and elucidation of root causes leading to the top-level event because their basis is neither concise nor explicit.

Efforts in engineering science reinforce these observations (Stephanopoulos, 1987). Many tasks are not achievable if representational expressivity isn't sufficiently rich to allow the description of the necessary concepts (Brachman and Levesque, 1985). Concepts must be manipulated directly if powerful reasoning is to be achieved. The success of any advanced computer-aided tool for enhancing the identification of hazards requires: (1) the development of a representational language sufficiently rich to embody advanced scientific concepts and (2) a means for manipulating these concepts and reasoning about them, directly.

D. OVERVIEW OF PROPOSED METHODOLOGY

In this chapter we will attempt to describe a concise framework for the development of hazards identification and analysis techniques. It is based on the following premises:

1. Every top-level hazardous event has been initiated by a physicochemical reaction.
2. A physicochemical reaction initiating a top-level event is determined solely by the type, chemical reactivity, and physical properties of the materials involved.

3. Every top-level event has been activated by the logical satisfaction of all its physicochemical precursor conditions.
4. The satisfaction of the precursor conditions in premise 3 depends on the structure and the design characteristics of the particular process, as well as its operating conditions.

The implications of these four stipulations are very important and determine the logic, character, and implementation of the hazards' identification and analysis approach discussed in this chapter. Specifically, we can conclude the following:

- Premises 1 and 2 (above) imply that an *inductive* generation of *all* physical and chemical reactions leading to potential releases or generations of uncontainable amounts of energy or mass, or both, per unit time is the essential foundation for the complete identification of all potential hazards. Such task *depends entirely on the chemical behavior of the materials/species involved and not on the structure of a processing scheme, or the operating conditions of the associated plant.*
- Premises 3 and 4 (above) imply that the design or operating modifications of a process leading to the elimination or containment of hazards (identified by the inductive approach) can be generated *deductively* from the knowledge of the plant and its operating conditions.

Observing the above logic, we have organized the material of the subsequent sections in this chapter as follows: Section II provides the essential foundation for the reaction-based identification of hazardous top-level events, while Section III describes the inductive reasoning used to identify these hazardous events from the set of available materials. Both of these two sections draw heavily from the material of the first chapter in this volume and specifically from the LCR, the modeling language developed by the authors to describe chemicals and their chemical reactivity. Section IV outlines the methodological framework for the deductive determination of the causes of hazards, using the available knowledge for the description of the plant's design and operating conditions. A fairly detailed example provides an illustration of the ideas and techniques discussed in Section IV.

II. Reaction-Based Hazards Identification

We present a new approach to hazard analysis and assessment, one that begins with the inductive determination of hazardous chemical or physical reactions and proceeds deductively through the network of processing steps to identify completely all causes initiating these hazardous reactions. This strategy is *more efficient* in the identification of reaction-based hazards than conventional methodologies. The approach ensures *completeness* and resolves uncertainty of design quality through first-principles-based quantification of the design risk. Furthermore, it enables a computer-aided automation. The methodology is based on two fundamental postulates that formalize and extend Bretherick's observation: "With the exception of releases of toxic or corrosive material, all accidents in storage, handling, or processing of chemicals involve the release of energy at rates too high to be dissipated in the immediate environment without damage." (Bretherick, 1990). These two postulates establish the theoretical framework that enables the design of a system for automatic identification and analysis of hazards in a decidable manner. By unifying the analysis of potential hazards, the framework can utilize the maximum knowledge available at every point in the design process. Regardless of what stage the design is in, a designer's attention can be focused at (1) earlier design stages and their vulnerable areas so that the associated hazards can be eliminated or (2) later stages, where the a priori sequence of events that leads to hazards can be used as an early warning structure (Lees, 1983).

The methodology employs domain-specific modeling languages (see first chapter in this volume) to describe

(a) Chemicals and their reactivity during the *inductive* identification of potential reaction-based hazards (see the modeling language LCR in first chapter in this volume).

(b) Processes during the *deductive* identification of their process-based causes (see MODEL.LA. in first chapter).

A. System Foundations

We now establish the *criterion for completeness* for any hazard identification methodology. [Herein, completeness refers to the ability of a methodology to identify all possible top-level events (Nagel, 1991.)] The methodology presented establishes how we use this criterion to completely and systematically identify hazards in an *efficient* manner.

Theorem. *Any hazard identification methodology must cover all subsets of sources present in a process network in order to guarantee completeness.*

In the discussion that follows we restrict our attention to hazards whose enabling path involves reactions among the chemical species, present in the process. This restriction excludes, for example, hazards (i.e., top-level events) created by falls, electrical shock, or impact with stationary objects, since there is no finite set of root causes leading to these hazards, thus making impossible the guaranteeing of completeness for hazards of this type. But, the preceding restriction does not exclude hazards initiated by these processes, provided they cause interactions among the chemical species present in the plant. For example, a static charge initiating a chemical reaction or the overheating of a tank's contents leading to its volatilization and subsequent over-pressuring, would be covered. Our approach is based on two fundamental postulates.

Postulate 1. *Hazards can only be created by the interaction of a system through its boundaries, or the altering of internal restraints of the system, such that the system proceeds toward a new equilibrium state.*

This postulate follows naturally from the First and Second Laws of Thermodynamics. Since states at equilibrium have no irreversible interactions, a state change is required to proceed toward equilibrium; these changes can only be brought about by interactions through a system's boundary or the altering of internal constraints. Therefore, a state change must accompany a hazard or the system would be at equilibrium. Although many nonequilibrium states may be transgressed in the development of a hazardous system, we need only consider the equilibrium states to identify potentially hazardous systems. This is possible because state changes can only be brought about by energy and entropy changes. Since these are state functions, an upper limit can be established on the potential of a hazardous state through the analysis of the equilibrium states leading to it. Our goal in inductive identification of hazards focuses on the identification of these states. Since any equilibrium state can be characterized completely by $(n + 2)$ variables, where n represents the masses of the particular chemical species initially charged and 2 represents two independent variable properties (e.g., temperature and pressure), our task is divided into

- The identification of all chemical species sets, which can potentially be present in the process.
- The identification of the independent variable properties that specify the environment of the chemical species.

The latter requires the identification of the processing environment, whereas the former establishes the need to identify the occurrence of all potential reactions. The inductive generation of all potential reactions will be discussed in detail in Section III. Whereas postulate 1 establishes the framework for inductively identifying hazardous states, postulate 2 advances the mechanism by which the pathway of events leading to a hazardous state can be identified deductively.

Postulate 2. *The degree of completeness of the set of internal restraints and exogenous factors specifying a hazardous system and its transformation, determines the degree of completeness with which the pathways leading to a hazardous state can be identified.*

Moreover, postulate 2 suggests that the deductive identification process is restricted by our understanding of mechanisms that allow states comprising the hazardous system to interact. Since any interaction results in an energy or entropy change of these states, the opportunities for preventing a hazard are limited by the incompleteness of our knowledge to determine the restraints and the external variables that specify these states. These in turn are more limited by the quality of the available details in describing the particular processing system (Battelle, 1985). However, there may not be a finite set of root causes leading to a hazardous state, or the identification of a finite set of causes may not be possible. As a consequence, the pathway of precursor events leading to a hazardous state is incomplete by definition. Despite this limitation, the constitutive equations that promote a hazardous state can be used to optimize the inherent safety of a particular design technology. The combined approach presented in this chapter allows the following:

1. Establishment of a systematic and formal methodology for increasing the inherent safety of a design technology.
2. A means for quantitatively assessing the inherent safety of design alternatives.
3. Establishment of a systematic and formal strategy for optimizing the inherent safety of a design technology through the selection of appropriate control points.
4. Guarantees on the correctness and completeness of the pathways leading to top-level events.

B. Modeling Languages and Their Role in Hazards Identification

Any methodology, whether it is applied in an automatic or manual manner, requires a rich representation of the process if hazards are to be effectively identified. This representation must have sufficiently expressive power to allow chains of precursor states or events to be easily identified. To satisfy these requirements, we have chosen to map the process description into a representational form that allows a multi-level, multi-faceted process description (see the modeling language MODEL.LA. in first chapter in this volume).

Research efforts outside the domain of hazards analysis have established that expressive, fully declarative, domain-rich modeling languages are indispensable, if complex integrated systems capable of synthetic tasks are to developed. Furthermore, they have shown that computer-aided systems with advanced reasoning capabilities require the satisfaction of the following three conditions:

- All declarative models should be fully articulated.
- Declarative knowledge should be completely decoupled from the procedural knowledge.
- A modeling language should be rich with domain-specific knowledge and thus allow the user to think about the task at hand in terms that are familiar (Stephanopoulos *et al.*, 1990a,b; Kritikos, 1991).
- For a computer-aided chemical reasoning system, we add a fourth condition. The logical structure of chemistry should be exploited by imbedding natural constraints into the declarative models describing chemicals and chemical reactivity.

The system described herein for the automatic identification of hazards and the pathways leading to them utilizes two modeling languages which satisfy the above requirements. Specifically, we have used

- LCR (language for chemical reasoning; see first chapter in this volume) to describe chemicals and their structure and atomic and molecular properties, as well as chemical reactions and their structure, directionality and contextual character.
- MODEL.LA. (modeling language; see first chapter in this volume) to describe processing systems and their unit operations and behavior, and to encapsulate the design decisions and operating conditions associated with any specific plant.

These languages are not ad hoc constructs but are based on clear formalisms and satisfy the essential premises of grammar, vocabulary, and semantics on which programming languages are founded. Both modeling languages allow multilevel description of processes, reactions, and materials with internal consistency (logical and quantitative) among the various models defining the corresponding objects at various levels of abstraction. Moreover, they support the representation of materials, reactions, and processes at multiple, coexisting contexts, an explicit comparison of the contextually alternative representations, and a systematic backtracking of decisions and assumptions that led the specific descriptions. Thus, different perspectives of the process, reactions, and materials—such as structural, topological, and physicochemical relationships—can be investigated independently. Both of these languages allow the following operations on processes, reactions, and chemicals:

- Multiple viewing in terms of structure, topology, and behavior
- Disaggregation of abstract descriptions to more detailed ones and aggregation of detailed descriptions to more abstract
- Contextual description of alternative models for the same process
- Controlled flow of information among the models at various levels or in various contexts, and detection of modeling conflicts
- Propagation of qualitative and quantitative knowledge through the defining models of behavior

Utilizing the specialized modeling language, MODEL.LA. (Stephanopoulos et al., 1990a, b; see also first chapter in this volume), we can construct a process description that is suitable for the tasks of hazards analysis or/ and identification. The process description we have employed is an abstraction of the conventional process representation (centered around the topology of a specific network of processing units). It is built on top of the conventional representation and thus provides complete access to the functionality contained in the base representation. This functionality comes with a set of tools that allow us to solve heat and mass balances, identify work interactions, evaluate phase partitioning, etc., and permits the propagation of qualitative and quantitative knowledge through the defining network. By focusing our representation on the thermodynamic state description of the process, we can facilitate the identification of pathways leading to potential hazards in the most efficient manner. The equipment state space (i.e., the process flowsheet representation) can be mapped to a thermodynamic state space, if we know the trajectory of thermodynamic states, comprising a process, combined with the transformations that connect these states. The mapping process that allows us to construct a thermodynamic state-based graph representation of a process flowsheet focuses on the state description of the process. The representation is composed of *streams* and *nodes* in accordance with the following definitions:

- *Streams* are idealized flows that connect nodes, and they have no accumulation or energy losses.
- *Nodes* are identified as points where discrete thermodynamic transformations occur, the result of changes in intensive or extensive variable values.

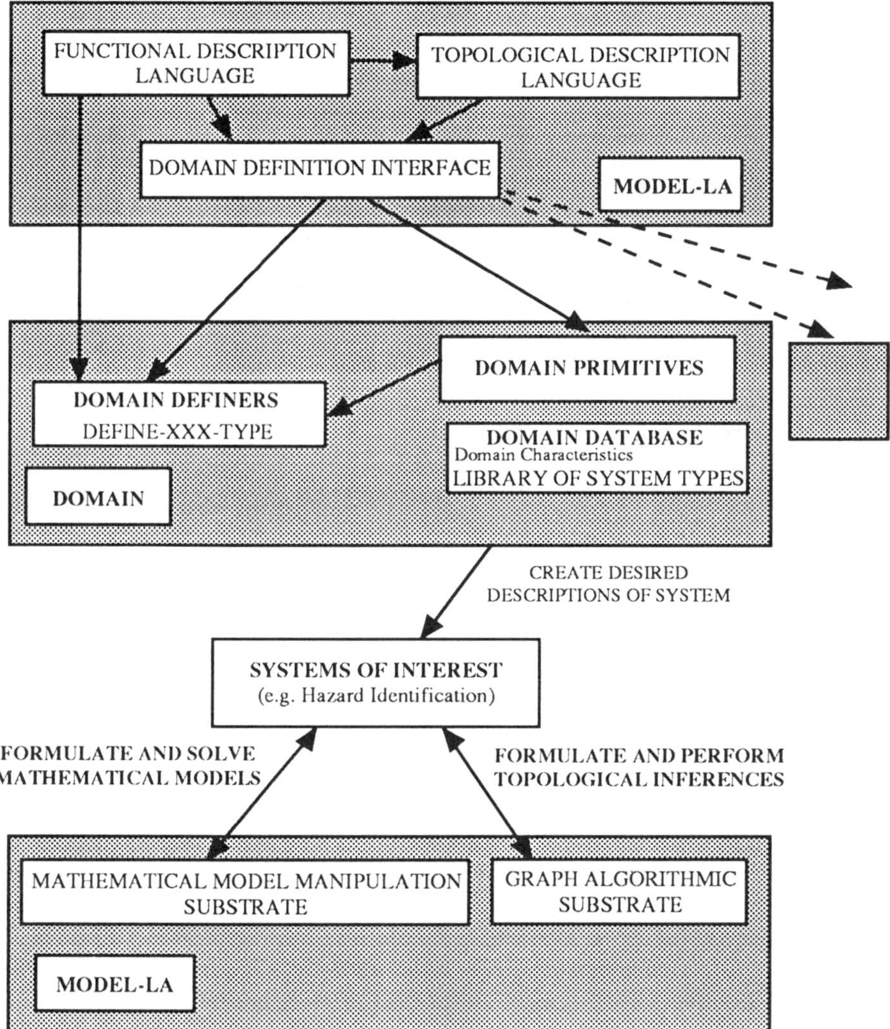

FIG. 2. Overall organization and use of the modeling language MODEL.LA.

These definitions are used to abstract the underlying process representation. As shown in Fig. 2, MODEL.LA. allows us to construct an abstract representation utilizing these definitions for our specific application from a conventional process description. This abstract representation has access to any utility, tool, or component that constitutes the base representation,

```
identifier:             unbound
element-name:           #<state-1>
element-type            composite-state
chemical-species-set:   unbound
operating-conditions:   #<op-cond-1>
boundary-elements:      (#<flow-port-3> ... #<heat-transfer-port-5>)
connecting-states:      (#<state-2>  #<state-3>)
system-description:     #<node-1>
system-volume:          unbound
```

Fig. 3. The attributes of the modeling object, *state*.

e.g.,

* The topology of the network is accessed through the graphical construction of the base representation.
* Mathematical relationships such as phase relations, energy balances, mass balances, and reaction or transport rates can be accessed through the content of the instances of objects of type **relationships** (see first chapter in this volume).

Knowledge about the abstract representation and the mapping process are contained in the modeling element, **state**. Figure 3 identifies several of the attributes that describe a state. Notice that the state is a composite object, i.e., an object comprised of objects. (Herein objects bound to attributes are denoted by $\#\langle \text{NAME} \rangle$ or will appear in bold.) This allows a multilevel, hierarchical representation of a process to be constructed. For example, the operating conditions that specify a state are accessed through the object **op-cond-1**, which is the attribute value of *operating-conditions*. Attributes of **op-cond-1** include the unit with which it is associated as well as the temperature, pressure, flowrate, and composition associated with that unit and their respective interval ranges.

The chemical species set of a state is the set of chemical constituents that are associated with the system description of that state. The values of the attributes *chemical-species-set*, *operating-conditions*, and *system-volume* provide the $(n + 2)$ independent variable quantities that are necessary to define a thermodynamic state. An important feature of this representation is that each state is described by a vector of *intensive* and *extensive* variables. The intensive vector defines the operational state of the process, while the extensive vector defines the maximum accumulation of mass and energy that can occur. This is bounded by flowrate, reaction rate and physical size of the process equipment. The values of these variables are accessed through the attributes: interval flowrate vector, interval accumulation, and system description. The values bounding an interval are dynamically set. (Intervals are defined as the minimum and maximum allowable value that the variable being described can achieve.)

Boundary elements and connecting states are also associated with the state description. Since a state is described as a system and a system has a boundary, boundary elements identify the vehicle for transforming the current state to a new state. The trajectory of connecting states is contained in the value of the attribute, *connecting-states*. This attribute allows a thermodynamic state representation of a process to be constructed from the individual states that compose it.

Alternatively, an equilibrium state in which there is no net effect on system boundaries can still be transformed to a new state when the internal restraints that specify the system are altered. This information, as well as the mapping that takes from an equipment based representation to a state-based representation, is contained in the value of the attribute *system-description*. An expansion of this state attribute value provides access to the base representation through the description of the modeling object, **node**. Figure 4 identifies several of the attributes describing the modeling element, **node**. This detailed description allows us to identify system boundary partitions: flow openings, heat transfer openings, work exchange openings, and mass transfer openings. Additionally, the system boundary type is specified as a thermodynamic boundary. The topological system type and topological connector type provide the links needed to connect the various nodes. Each of these elements has a description that allows us to evaluate its functionality. In addition, application specific handlers are also provided by the underlying modeling language. For example, the boundary element in the attribute, *flow-opening*, may have a description that requires an understanding of vapor-phase, liquid-phase, solid-phase, or multiphase transport. Each of these handlers in turn may access additional methods or handlers to facilitate the evaluation process.

The value of the attribute *system-interior-type* specifies the interior system description of the **node**. The value of this attribute allows us to map the node representation to the process equipment representation. Notice that it is the attributes of the **node** in combination with the features of the underlying specialized modeling language that afford a multilevel, multifaceted view of the process. Hence, we are able to move independently from the node representation to the equipment representation and pursue, as needed, analysis of momentum, mass, and energy interactions of the defining network. It is clear that using the elements of the modeling language, LCR, we also have access to the representation specifying the chemical constituents that are associated with the state. In the sections

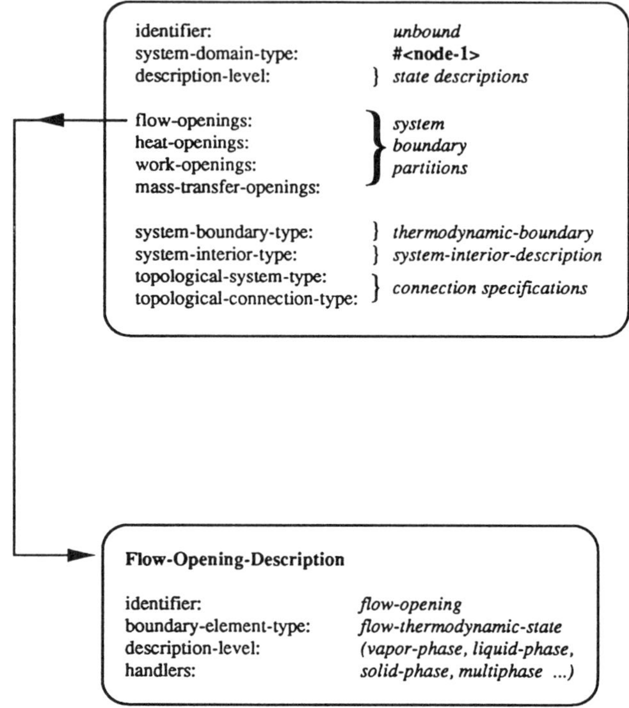

FIG. 4. Select attributes of the modeling objects, **node**, and **flow-opening-description**.

that follow, we will show how the information contained in the equipment-based description of a process through the use of MODEL.LA., is accessed by LCR and the chemical species in order to assist in the generation and evaluation of potential reaction alternatives.

An example of a mapping from the equipment representation to the thermodynamic state representation is shown in Fig. 5. It represents an isothermal vertical packed-bed catalytic reactor equipped with temperature and pressure sensors, an explosion vent, and a distributor plate. Notice that the equipment and sensors are not associated with the state representation. They are contained in the base representation and reside in the process description at the equipment level. As discussed earlier, flow, work, heat, and mass interactions are all modeled independently. This allows us to evaluate independently the effect of these processes. Independent evaluation assists in the identification, evaluation, and assessment of event pathways leading to hazardous states.

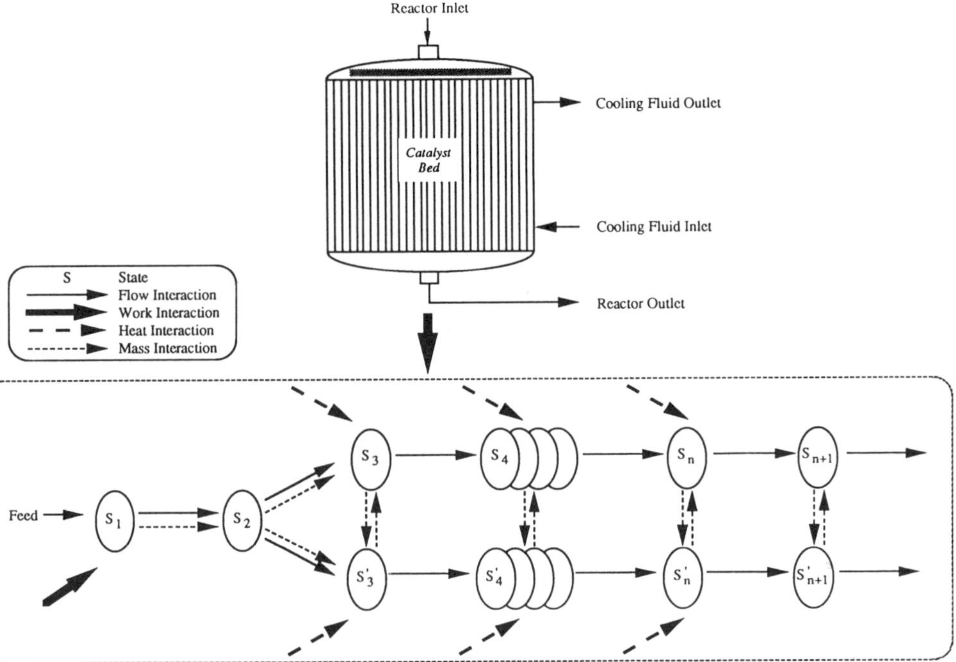

FIG. 5. Mapping of equipment into a thermodynamic state space.

C. GENERATION OF REACTIONS AND EVALUATION OF THERMODYNAMIC STATES

Since new thermodynamic states are created by reactions, it is required that we have a strategy for generating potential reactions between species resident in a particular node of the process description. To describe the reacting species and the reactions they undergo as well as the changes in their molecular structures, we use the language, LCR, which is presented in the first chapter in this volume (for a detailed description of the language, see Nagel, 1991). The language is used extensively in the inductive identification of reaction-based hazards. LCR provides a utility for generating reactions and identifying the resulting pathways, given a set of substrates and a reaction environment. The procedure that provides this utility and the keyword arguments (i.e., arguments with a colon prefix) accepted by this method are given below:

(FIND-ALL-PATHWAYS   *:substrates   :operators*
                            :override-environment :initiator-p)

Let us examine the character of the keywords:

- *:substrates*, is the argument that contains the list of chemicals, which are available at a particular process node.
- *:operators*, is the keyword argument that allows the user to specify the types of transformations to be used in order to focus the generation of reaction pathways. To investigate all theoretically feasible pathways subject to encoded preferences the user may supply the keyword argument *:operators* with the value, K^*, a modified composite operator, and allow the generation and evaluation of all pathways having prespecified features ($\Delta G < 10$ kcal/mol, stereocenters, etc.). Similarly, if there are no preferences, all theoretically feasible reactions can be generated by calling directly FIND-ALL-PATHWAYS with the argument *:operators* having the value $K_{ab\text{-}initio}$ (Nagel, 1991).
- *:override-environment*, is the argument that lists the operating conditions in which the alternative reaction pathways will be generated. It allows the user (or an automatic procedure) to set alternative operating environments and generate the corresponding alternative reaction pathways.
- *initiator-p*, determines whether an initiator of specific chemical reactions is present or not. It allows the user (or an automatic procedure) to investigate various reaction trajectories without knowing the specifics concerning the mechanisms for initiating reaction pathways.

Using this representation, we can now focus on the identification of the associated thermodynamic states. We assess the likelihood of reactions as well as the inherent instability of each species associated with a node using LCR. The generation of infeasible species is limitated by the following values (i.e., knowledge) embedded in the corresponding attributes;

:override-environment = **reaction-environment**

:operators = **K**, **K***, or $K_{ab\ initio}$.

Let us examine the generation of alternative reactions within the scope of the operating conditions associated with a particular node (or, its corresponding state). For example, suppose that the procedure FIND-ALL-PATHWAYS is applied to a chemical species set (CSS) (i.e., the set of chemicals bound to a specific **state**) composed of three chemicals, **A**, **B**, and **C**, each

Reaction Object

identifier:	*unbound*
name:	initiation-1
reactants:	(# <Cl$_2$>)
products:	(#<Cl> #<Cl>)
stoichiometry:	((#<Cl$_2$> . -1)
	(#<Cl> . 2))
reaction-environment:	#<reaction-environment-1>
enabling-conditions:	K$_f$
composing-transformations:	K$_t$
composing-reactions:	*unbound*
rate-expression:	#<rate-expression-1>
equilibrium-constant:	#<equilibrium-constant-1>
context:	*unbound*

FIG. 6. Description of the modeling object, ***reaction***.

of which is described by an instance of the modeling object, **atom-bond-configuration** (see first chapter). The procedure generates the following set of potentially reactive mixtures:

$$\mathbf{A} \rightarrow \, ? \tag{I}$$

$$\mathbf{B} \rightarrow \, ? \tag{II}$$

$$\mathbf{C} \rightarrow \, ? \tag{III}$$

$$\mathbf{A} + \mathbf{B} \rightarrow \, ? \tag{IV}$$

$$\mathbf{A} + \mathbf{C} \rightarrow \, ? \tag{V}$$

$$\mathbf{B} + \mathbf{C} \rightarrow \, ? \tag{VI}$$

$$\mathbf{A} + \mathbf{B} + \mathbf{C} \rightarrow \, ? \tag{VII}$$

Unique products generated by these reactions are added to the CSS

Pathway Object

identifier:	*unbound*
name:	pathway-1
reactants:	(# <C$_2$H$_6$O> #<Cl$_2$>)
products:	(#<C$_2$H$_5$C$_{10}$>)
stoichiometry:	*unbound*
competing-pathways:	(<pathway-2>
	#<pathway-3>)
composing-reactions:	(#<initiation-1>
	#<abstraction-1>
	#<combination-1> ...)
global-rate-expression:	#<composite-rate-exp-1>
global-equilibrium-constant:	*unbound*

FIG. 7. Description of the modeling object, ***pathway***.

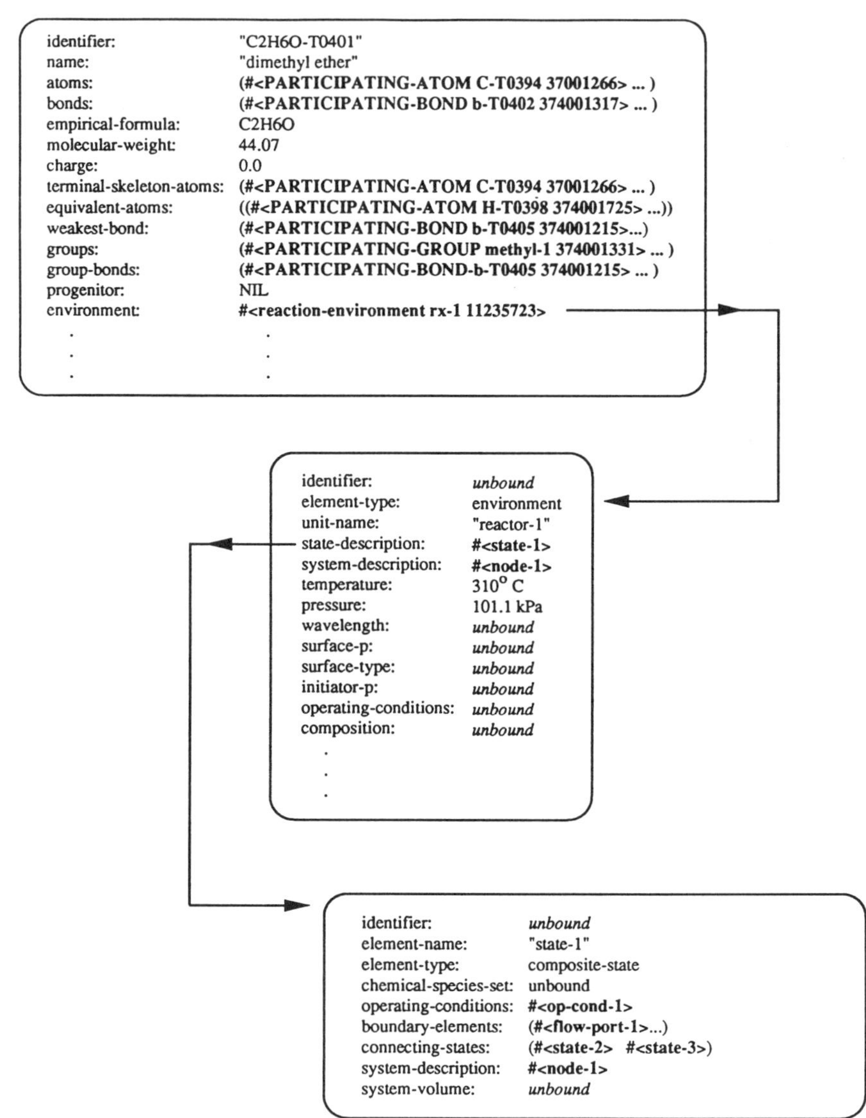

Fig. 8. Accessing process representation from chemical representation.

and the process is repeated. A single call to FIND-ALL-PATHWAYS

$$(\text{FIND-ALL-PATHWAYS} : substrates\ (\mathbf{ABC})\ :operators\ \mathbf{K*})$$

generates all possible pathways.

LCR provides modeling objects to contain the information generated by these chemical transformations. For example, the attributes describing the modeling object **reaction** are shown in Fig. 6. These attributes describe not only the reactants and products of the reaction but also contain information on the reaction environment, enabling conditions, composing transformations, reaction stoichiometry, equilibrium constant, rate expression, and identification of competing reactions. Similarly, Fig. 7 shows the attributes describing the object **pathway**, which is made of a network of one-step reactions. Whether any of the above potential reactions, e.g., (I) through (VII), will proceed and develop a hazardous event or not, depends on the value of the operating conditions that characterize the **state** of the corresponding **node**. Therefore, it is imperative that a link be established between the description of the process operations and the **atom-bond-configuration** encoding the information about the available chemicals.

The **atom-bond-configuration** object provides direct access to the **state** description of a **node** through the attribute, *:environment*. Figure 8 shows how the value of the "state-1," describing the conditions in "reactor-1," is transferred to the instance of the **atom-bond-configuration** describing the chemical reactant, "dimethyl ether." Such links allow the transfer of information among different modeling objects in LCR; a mechanism driven by the semantic relationships of LCR (see First chapter).

In a similar manner, the semantic relationships among the various modeling objects of MODEL.LA. allow the transfer of information among these objects. When used together, LCR and MODEL.LA. *allow functional and topological information derived at one point of the process to be accessible from any other point in the process.*

III. Inductive Identification of Reaction-Based Hazards

Using these tools and the representations that are built with them, potential hazards are inductively identified by combining various chemical and physical environments of the process in an attempt to generate new process states. The types and numbers of process states are determined by both the process configuration and the operating conditions. Enabling

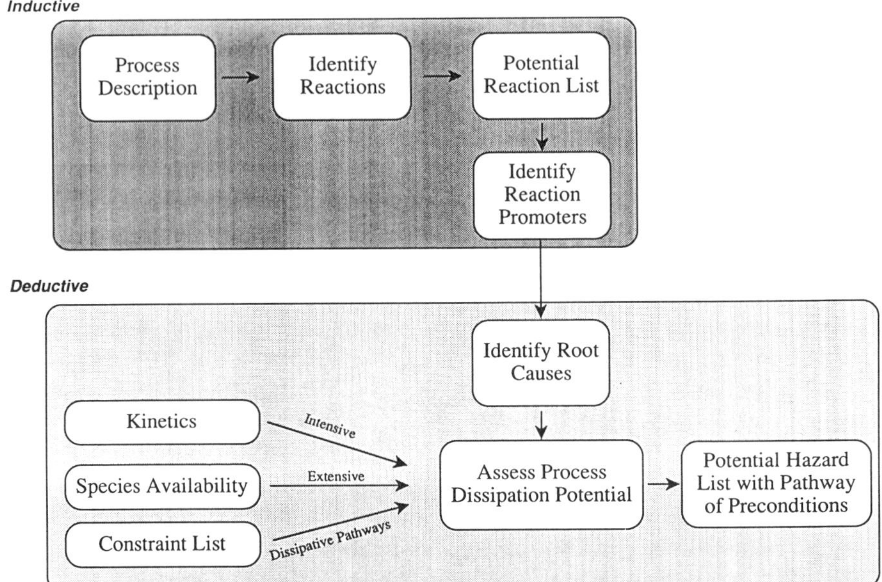

FIG. 9. Overview and implementational elements of the methodology for the identification and analysis of hazards.

conditions of a potential hazard are identified by establishing the conditions specifying the thermodynamic states that precede it. These conditions can be either intended or induced, resulting from changing boundary elements, system descriptions, or reacting states. Normally, they manifest themselves as procedural, material, or boundary changes.

A potential hazard is said to exist when the environment or a sequence of environments cannot be dissipated or prevented by the process. Using this environment as a goal state, we can then deductively identify the sequence of process and/or operational changes that led to its creation. Thus, the root causes and their temporal preconditions (whether they are equipment and/or operational failures) can be identified and associated with the corresponding hazard. Evaluation of the pathways leading to the specific hazard allows the quantification of risk and permits a concrete assessment of the potential hazard in the context of a given design technology.

Information flow during the execution of the hazards identification procedure is shown in Fig. 9. It is composed of two phases:

Phase 1. Inductive identification of reactive states and their assessment as potential hazards.

Phase 2. Deductive identification of pathways leading to those states.

The inductive phase begins with a process description from which reactions are identified using LCR, and the tools described earlier, which allow the identification of potential chemical reacting states. Once all the potential chemical transformations have been identified and the entire chemical species set has been elucidated, a **node** can be fully described and the **states** disseminating from it can be established (i.e., the $n + 2$ independent variable quantities are known). With this description potential physical reactions (e.g., overpressurizing, overheating, overcooling) can be elucidated using classical thermodynamic (open- and closed-system treatment) and transport phenomena analysis around the specific **node**.

Potential physical reactions can only be identified with any assurance of completeness *after* the complete specification of the chemical species set, CSS. This is a result of Postulate 1, $(n + 2)$ *defining variables must be known before a thermodynamic state can be specified*. Since incomplete specification of "n" leads to incomplete specification of potential states, a complete node description is required before physical reactions can be identified completely.

Once potential reactions are identified, they are posted on a potential reaction list. The promoters of each reaction are then identified by tracing the variable values that led to their creation. This is achieved through the links provided in the thermodynamic representation of the **state** and the **atom-bond-configuration** modeling objects. Similarly, by accessing the information contained in $\mathbf{K}$, $\mathbf{K}^*$ or $\mathbf{K}_{ab\ initio}$, we can also identify the conditions that promoted a particular chemical reaction. These conditions may be chemically induced, as in the case of nucleophilicity, or physically induced, as in the case of high-energy environments enabling radical or photochemistry, or both. These promoters are then used to establish the underlying pathways that enable the top-level event.

In this combined approach, the inductive identification of top-level events and the deductive identification of enabling pathways allow potential hazards and the sequence of events leading to them to be identified completely within the scope of the modeling effort. Moreover, since the top-level event is generated from its underlying states, the pathway of preconditions and the temporal ordering of those preconditions are explicitly spanning only the minimum cut set (see Nagel, 1991).

A. Hazards Identification Algorithm

The algorithms used in this approach are described below using the pseudocode conventions found in Cormen *et al.* (1990). Therein, a *terminal node* is defined as the "first node a feed enters or the last node a product or byproduct exits" in the process flowsheet. The routine calls for two loops, which is necessary to simulate multiple simultaneously occurring boundary failures.

Algorithm 1 ⟨GLOBAL-HAZARD-IDENTIFICATION⟩

   input: process-flowsheet
   initialize
      process-node-representation ← **apply** MAKE-PROCESS-TRANSFORMATION
            to process-flowsheet
      terminal-nodes ← **apply** FIND-ALL-TERMINAL-NODES to process-node-
            representation
      potential-hazard-list ← nil
   for each node in process-node-representation
      potential-hazard-list ← **append** (**apply** IDENTIFY-POTENTIAL-HAZARD
            to (node process-flowsheet))
            to potential-hazard-list
    return
   for each node in terminal-nodes
      expand node-scope
         until UNIQUE-STATE-IDENTIFIED-P   ;predicate test to identify
                                            unique states
           or EXPANSION-NOT-POSSIBLE   ;empty node-scope
                                            (i.e., terminal node
                                          is reached)
      then
        potential-hazard-list ← **append** (**apply** IDENTIFY-POTENTIAL-HAZARD
            to (node process-flowsheet))
            to potential-hazard-list
    return
  end
  return

The internal routine, IDENTIFY-POTENTIAL-HAZARD, called from GLOBAL-HAZARD-IDENTIFICATION, returns a potential hazard and the list of the enabling conditions. The algorithm for IDENTIFY-POTENTIAL-HAZARD, expressed in pseudocode, is given below:

Algorithm 2 ⟨IDENTIFY-POTENTIAL-HAZARD⟩

   input: node   ;starting point-hazards are associated
                                     with nodes
      process-flowsheet   ;context of the node
   initialize

```
potential-reaction-list ← nil
associated-sates ← nil
extended-CSS ← nil
potential-reaction-list ← apply FIND-ALL-PATHWAYS
                    TO CSS of node; identify
                                      potential
                                      chemical
                                      reactions
extended-chemical-species-set ← collect UNIQUE-CHEMICAL-
                                 CONSTITUENTS from
                                 potential-reaction-list
associated-states ← apply EVALUATE-STATES
     extended-CSS      to node and ;identify states
                                    associated with each
                                    node due to reaction
for each state in associated-states
   when UNIQUE-STATE-DESCRIPTION-P
      potential-reaction-list ← apply EVALUATE-PHYSICAL-
                                 REACTIONS to state
      return
   return
for each reaction in potential-reaction-list
   enabling-criteria ← collect FIND-ENABLING-CRITERIA
   classified-influence-paths ← apply CONSTRUCT-VARIABLE-INFLUENCE-
                                 PATHWAYS to (enabling-criteria process-
                                 flowsheet)
   root-causes ← apply IDENTIFY-NONDISSIPATIVE-PATHWAYS to classified-
                       influence-paths
   when root-causes
      potential-hazard ← list reaction enabling-criteria root-causes
      return
   return
end
return
```

Notice the interplay between the methodology and the specialized modeling languages and the importance of that interplay. These languages enable the methodological approach. For example, FIND-ALL-PATHWAYS applied to a chemical species set (CSS) utilizes the semantic relationships of LCR to construct the *potential-reaction-list*. Similarly, the representation utilized by LCR allows enabling-criteria to be explicitly associated with each reaction; these criteria can come from the **state** representation or from the inherent physicochemical properties associated with the chemical constituents themselves. Likewise, EVALUATE-PHYSICAL-REACTION utilizes functionality of the base representation to assess possible effects and to identify enabling criteria. These criteria come from the operating environment of the process equipment.

By interweaving the semantic relationships of the two modeling languages throughout the methodology, we are able to interplay the respective representations to satisfy the representational needs of the task at hand. For example, root causes are identified by propagating (i.e., solving) the enabling criteria through the network of equations lying in the base representation and defining the behavior of the overall process flowsheet. These criteria, however, are located in different representations, at different abstraction levels, to afford the resolution necessary to identify potentially hazardous conditions. The specialized modeling languages provide the tools to map between different representational forms and identify potential hazards with their enabling conditions and underlying root causes in an efficient manner.

Two additional features of Algorithm 1 are worth discussions:

- Interval values, describing ranges of operating conditions, are *dynamically* redefined whenever a condition is found to exceed the existing value range (values may be exceeded in the positive or negative direction).
- The description of a **node** (i.e., the scope of the defining system) can be expanded to simulate common boundary failure or process equipment malfunction.

B. PROPERTIES OF REACTION-BASED HAZARDS IDENTIFICATION

Let us now consider the relative efficiency of different methods for the identification of hazards (for details and proofs, see Nagel, 1991). Since it is meaningless to compare the efficiency of methods that are incomplete versus those that are complete, for the purposes of this section we consider only complete methodologies. But before beginning, we should point out one additional result that can be derived from the completeness analysis presented in Section II.B.

Corollary. *Equipment-based methodologies for hazards identification can be complete.*

An equipment-based methodology, e.g., HazOp Analysis, begins the identification of hazards with the postulation of specific process variables

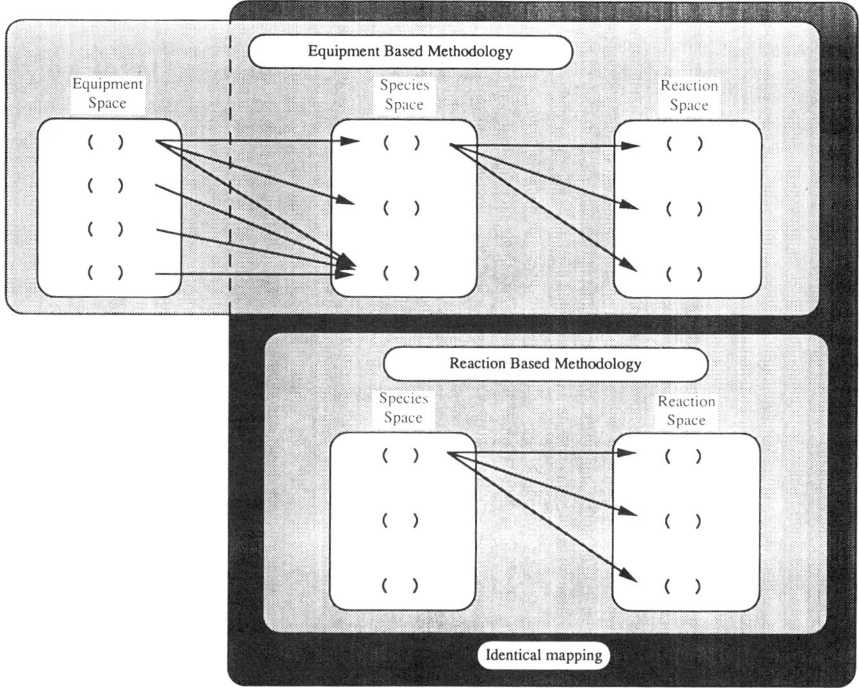

FIG. 10. Mapping of equipment space to species space.

deviations and/or equipment failures. From the point at which the complete chemical species set is generated, an equipment-based methodology is identical to a reaction-based methodology (see Fig. 10), such as the one described in this chapter. The implication of this statement is that the only information generated by the equipment analysis that is used in the identification of top-level events is the state space, which collapses to the species set since temperature and pressure are local variables.

Theorem. *No hazards analysis method exists that is both complete and whose running time is bounded by a polynomial in the number of initial species present in the plant.*

Theorem. *A reaction-based method is more efficient than any equipment-based method.*

The only difference between the equipment and reaction-based methods is in the generation of the chemical species set. Equipment-based methods generate the species set *indirectly* via the equipment state space; a reaction-based method generates it *directly*. Since both approaches must generate every element in order to supply a complete set of chemical species, the difference in efficiency between an equipment-based method and a reaction-based method must result from the fact that the former must generate the members of the chemical species set more than once, while the latter as we know does not. But, this assertion is trivially satisfied if the size of the equipment space is larger than the size of the species set space (see Nagel, 1991).

However, the equipment-based approach is also engineered to perform a second task, namely, *identification of root causes*. It can be shown (see Nagel, 1991) that if the number of chemical species present in the CSS is S_0, the possible number of subsets of species and thus the maximum number of possible reacting mixtures is 2^{S_0}. Consequently, when the complexity of an equipment-based methodology becomes larger than 2^{S_0}, the added complexity is an effect of the method's intention to also identify the root causes of hazards. Assuming that c_1 represents the number of additional guidewords in the selection of hazards, and c_2 the number of additional process parameters invoked by an equipment-based hazards identification method, then the resulting complexity of such method is given by

$$(2 + c_1)^{(S_0 + c_2)^E},$$

where E represents the pieces of equipment to be searched for the identification of hazards. For a mixture of 2 chemical species the corresponding search space of a method based on the chemical species (as the one presented in this chapter) is $2^2 = 4$. On the other hand, for a simple process with 2 pieces of equipment, $c_1 = 1$ and $c_2 = 2$, an equipment-based hazards identification technique must search a space of $3^{4^2} = 6561$ alternatives.

However, as shown in Section IV, completeness cannot be guaranteed in the identification of root causes by equipment-based techniques, since they cannot guarantee that all reacting mixtures can be identified, and thus not all possible thermodynamic states can be a priori identified. This is because guidewords and variable parameters (e.g., more, less) are designed to trace out *causes*, not generate thermodynamic states. This guarantee only comes when the resolution of the tracing mechanism completely defines the enabling state of the potential hazard.

For example, *to identify a potential hazard that is caused by changing the catalyst shape an equipment-based approach must include surface area as a process parameter to be searched; a piece of information embedded in the phenomenological kinetic rate expression and not explicitly available.* Thus, as a consequence of its formulation and the uncertainty of conditional operators, the equipment-based approach may not guarantee complete identification of potential hazards and cannot guarantee the complete identification of root causes leading to those hazards.

C. An Example in Reaction-Based Hazard Identification: Aniline Production

In this section, we will demonstrate how the methodology can be used to identify hazards inductively from the set of possible chemical reactions. We will focus on the detection of potential hazards arising from changing operating conditions rather than equipment malfunction or failure; the latter is discussed in Section IV.

The ability to elucidate competing chemical pathways when changes in control strategies occur is of particular importance in hazard identification. An understanding of the relationship between the reaction pathway topography and design and operating characteristics allows us to mitigate or constrain the offending reaction trajectory. Consider, for example, the catalytic production of aniline from nitrobenzene and hydrogen. Limiting our attention to the reactor, named **reactor-1**, we see that the initialization routine of the procedure GLOBAL-HAZARD-IDENTIFICATION transforms the reactor into a single terminal process node. The identification of hazards within this **node** is carried out by applying the procedure IDENTIFY-POTEN-TIAL-HAZARD, to this **node**. The procedure, IDENTIFY-POTENTIAL-HAZARD, begins by applying the procedure FIND-ALL-PATHWAYS to the chemical species set. This set, which is bound to the **node**, contains the known chemical species, i.e., CSS = {*nitrobenzene, hydrogen, Raney nickel*, and *phenol*}. Also, let **reactor-1** be bound to the **node**. The call to the procedure, FIND-ALL-PATHWAYS, and its application to the initial chemical species set, using composite chemical operators involving hydrogenation, is shown below:

$$(\text{FIND-ALL-PATHWAYS} \quad :substrates \text{ CSS} \quad :operators \ K_{\text{hydrogenation}})$$

Key reactions identified by the procedure, FIND-ALL-PATHWAYS, are

$$\varphi\text{-NO}_2 + H_2 \rightarrow \varphi\text{-NO} + H_2O, \qquad (\text{VIII})$$

$$\varphi\text{-NO} + H_2 \rightarrow \varphi\text{-NHOH}, \qquad (\text{IX})$$

$$\varphi\text{-NHOH} + H_2 \rightarrow \varphi\text{-NH}_2 + H_2O. \qquad (\text{X})$$

$$C_6H_5NO_2 + H_2 \longrightarrow C_6H_5NO + H_2O$$

$$C_6H_5NO + H_2 \longrightarrow C_6H_5NHOH$$

$$C_6H_5NHOH + H_2 \longrightarrow C_6H_5NH_2 + H_2O$$

$$C_6H_5NHOH + C_6H_5NO \longrightarrow C_6H_5NH_2 + C_6H_5NO_2$$

$$C_6H_5NHOH \longrightarrow C_6H_5N_2$$

$$\longrightarrow (C_6H_5)_2N_2$$

$$C_6H_5NO_2 \longrightarrow \text{decomposition products}$$

FIG. 11. Potential reactions of nitrobenzene.

These reactions combine to form the following overall reaction:

$$\varphi\text{-}NO_2 + 3H_2 \rightarrow \varphi\text{-}NH_2 + 2H_2O. \tag{XI}$$

A less restrictive implementation of the procedure, FIND-ALL-PATHWAYS, on the initial chemical species set, would employ the *:operators* **K** or **K*** and would generate several additional reactions of interest. Among these additional reactions of particular interest are, the decomposition reaction of nitrobenzene and the disproportionation of nitrobenzene with N-phenyl hydroxylamine forming aniline and nitrobenzene:

$$\varphi\text{-}NO_2 \rightarrow \text{products}, \tag{XIII}$$

$$AI\varphi\text{-}NO + \varphi NHOH \rightarrow \varphi NH_2 + \varphi - NO_2. \tag{XIV}$$

Several of the pathways that emanate from these reactions and are constructed by a recursive application of the procedure, FIND-ALL-PATH-WAYS, are shown in Fig. 11. Figure 12 shows how this information is managed by LCR using its modeling elements. This information is unavailable in conventional chemical synthesis programs (e.g., SECS and CYCLOPS) and provides an explanation (using the semantic relationships advanced by LCR) as to why the various reactions were activated and the specific intermediates were formed. For example, **pathway-1**, representing the transformation of nitrobenzene into aniline, "knows" that it *is-disaggregated-in* three separate individual reactions, denoted by **hydrogenation-1, hydrogenation-2**, and **hydrogenation-3**. Similarly, it "knows" that it *is-abstracting* the more general global reaction, **reaction-pathways**. It also "understands" which pathways are in competition; these in turn can be abstracted or disaggregated using the semantic relationships of LCR. In this way, the chemical sequence of events that led to a specific reaction can be traced all the way back to the reaction conditions that enabled it.

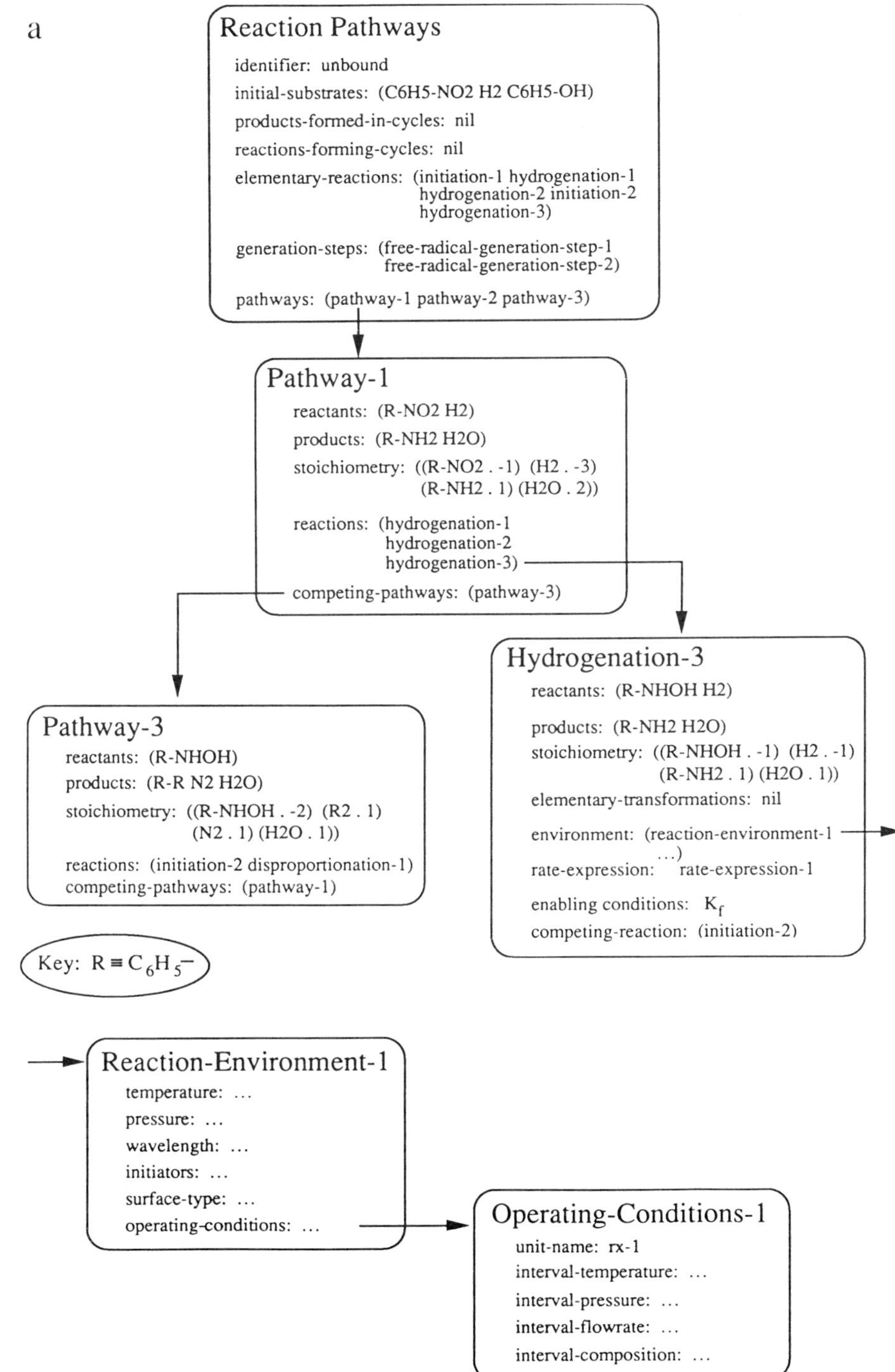

FIG. 12. Linkages between (a) reactions and pathways, (b) reactions and their operating conditions.

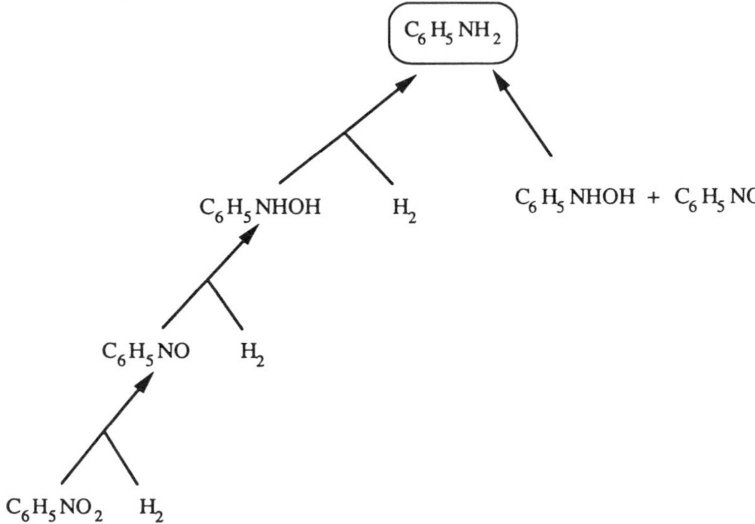

Fig. 13. Reaction path to aniline.

These enabling conditions can be related, in turn, to the operating conditions of the particular processing unit(s), using the *parent-equipment* attribute of *system-description* (an attribute of **node**). Through this mechanism the procedure FIND-ENABLING-CRITERIA traces out the sequences of events that lead to the reaction of interest. Figure 13 shows the path that leads to the formation of aniline. Because of the information contained in the LCR modeling objects that constitute this path, we know precisely the conditions that led to the creation of each element contained in the path. For example, by accessing the *heat-of-formation* attribute associated with the modeling element **reaction**, we identify that heat liberation is non-uniform during hydrogenation. This implies that an accumulation of the intermediate, φ-NHOH, could lead to an uncontrollable energy release.

More importantly, the tree spanning the generation pathways for aniline production (see Fig. 13) makes explicit the fact that hydrogen injection is an insufficient means for controlling reaction temperature. This results from the disproportion reaction identified by the procedure, FIND-ALL-PATHWAYS. As a consequence, a reduction in the hydrogen injection rate and regulation of the cooling fluid rate are necessary to control the heat released by the reaction. The realization that the heat released by the reaction cannot be controlled through hydrogen addition, alone, brings out an important parameter for the safe operation of the reactor, because the experimental detection of the *N*-phenyl hydroxylamine disproportionation with nitrosobenzene is difficult without prior knowledge of its occurrence (Stoessel, 1989).

Since LCR specifies the preconditions associated with each reaction generated by the procedure, FIND-ALL-PATHWAYS, the conditions that enabled a specific chemical transformation can be easily identified. This knowledge allows us to make explicit such information as the decomposition temperature of φ-NO_2, the disproportionation temperature of φ-NHOH, and the preconditions for the exothermic formation of such products as azobenzene and diazobenzene. By expanding the node description to include the atmosphere around the reactor (e.g., air), the initial CSS can be expanded to include *oxygen*. Then, the procedure, FIND-ALL-PATHWAYS, can identify additional reactions involving oxygen, e.g., combustion reactions with hydrogen, phenol, and Raney nickel as well.

The analysis of the inductively synthesized chemical pathways, described above, indicates that, if one started with only an understanding of the overall reaction, i.e., the hydrogenation of nitrobenzene to form aniline and water, the analyst may not have recognized the need to control reaction temperature, while moderating both cooling rates and hydrogen injection rates. Lacking this information, the loss of recirculation or the need to provide emergency cooling to satisfy additional duty requirements resulting from *N*-phenyl hydroxylamine accumulation, or potentially the need for redundant recirculation pumps may not have been understood and appreciated.

IV. Deductive Determination of the Causes of Hazards

The fundamental premise in all of the approaches that attempt to mitigate or control hazards lies on the assumption that they have the ability to both (1) identify accurately and (2) pinpoint precisely the location of a potential future hazard. We have shown that, although understanding the set of enabling conditions is essential for safe plant operation, the identification of the entire set of enabling conditions is an intractable task. It is impossible to completely identify the set of enabling conditions, leading to a hazardous state for the following reasons. The physicochemical conditions (e.g., presence of certain chemicals within predefined ranges of compositions, temperatures, pressures, flowrates within certain ranges of values) enabling the activation of particular chemical transformations form a set whose members possess variable values. When these variable-valued conditions are substituted into the equations representing the behavior of a chemical plant, the variables defining the conditions for the occurrence of a top-level event (TLE) (e.g., mixing of two chemicals forming an explosive mixture) achieve values necessary for the occurrence of a hazard. Since the variable-valued pre-conditions for a hazard are drawn from a range of, theoretically, infinite real-valued alternatives, we conclude that it is intractable to deduce all possibilities leading to a hazard. [*Note:* Expressed in different words, the analytic intractability in determining all conditions leading to hazards comes from the following weakness: There are no analytic (numerical) procedures that can take a set of equalities (e.g., the equations modeling the behavior of a plant) and a set of inequalities (e.g., preconditions for the activation of chemical pathways), solve them together, and produce a consistent set of new inequalities.]

Shifting to an interval representation, where we bracket the variable-valued preconditions into arithmetic intervals, over which the behavior is safe or unsafe, is only a partial solution for two reasons:

- First, the boundaries of the intervals, defining the conditions for the occurrence of a TLE are essentially functions of the values that the preconditions for the accomplishment of a reactive pathway can take on.
- Second, if the equations modeling the dynamic behavior of a plant display chaotic solutions (inherent in many non-linear systems), then we cannot be certain that the process behaves the same way when the preconditions defining a reactive pathway take on any values within given arithmetic intervals of values.

As a result, we have focused on the interpretation of the pathway *leading to a hazardous state* and its topography, and how these relate to the inherent safety of the process design technology rather than on the elucidation of pathways *leading to top-level events*. In the absence of a methodology for the complete identification of all conditions enabling the occurrence of a TLE, such an approach is essential if the safety of a chemical operation is to be enhanced.

A. Methodological Framework

In the previous section we discussed how an inductive approach can be used to generate all the chemical reaction pathways and the associated thermodynamic states, which lead to top-level hazardous events. A potential hazard is said to exist when the thermodynamic state or sequence of thermodynamic states leading to the hazard cannot be prevented, or the impact of these states cannot be dissipated by the design or the operating capabilities of the process. Using the precursor thermodynamic state(s) as a goal state(s), we can then deductively identify the sequences of changes in the structure of the processing system and/or its operation that led to the creation of the hazardous state. Consequently, root causes and their temporal preconditions, whether equipment and/or operational failures, can be identified and associated with the known hazard. Evaluation of the paths, leading to the hazard, permits the quantitative assessment of the top-level event and affords the risk assessment of the potential hazard in the context of the process design technology. In this section, we will show how the structure of the path itself provides a metric for assessing the inherent safety of the process design technology being evaluated.

The deductive identification of paths leading to a top-level event is accomplished through a *recursive tracing of the variable-influence links*, which describe the transition from state to state on the way to the hazardous event. These *variable-influence* links describe how the various physiochemical variables affect each other, and are generated from the network of modeling relationships describing the behavior of a plant. By propagating the values of the enabling/promoting conditions through the network of modeling relationships, one can assess the ability of the process to dissipate potentially hazardous effects. Obviously, the completeness and correctness of the results obtained through such an approach depend on the completeness and correctness of the modeling relationships. The modeling languages, LCR and MODEL.LA., are used to capture the requisite modeling relationships, which in addition to the first-principles-based equations, may include heuristics, empirical knowledge, experimental correlations, and design decisions. For example, LCR is used to capture the modeling of chemical kinetics, description of chemical structures for the computation of physical properties, whereas MODEL.LA. is used to capture the topology of the unit operations in a plant, the material and energy balances around the unit operations, and the description of operating variables (e.g., temperature, flow, pressure, composition).

Although the knowledge required to assess the potential for the prevention or dissipation of hazards is often held by different abstractions of the overall representation, the semantic relationships of the modeling languages afford efficient access to this information. The effect of protective processes, equipment restraints, sensors and control systems, emergency procedures, etc., are captured as constraints. Constraints may be embedded in the underlying representation as equations, or be associated directly to it via a constraint list, i.e., a collection of explicit process restraints. These restraints may be *passive*, such as materials of construction, or *active*, such as protection processes, sensors and control systems, and operating procedures. *Constraints limit the variable value of an enabling condition. They may achieve this task in one of two ways: demand mitigation or demand prevention.*

Mitigation requires no corrective action to limit a variable value. It is accomplished through the appropriate design of an equipment that (1) leads to the restraint of the top-level event, given an enabling condition, or (2) limits the value that an extensive (e.g., flow, total inventory of a material) or intensive variable (e.g., composition, temperature) can achieve. For example, the value of an intensive variable can be limited by the physical phenomena that are allowed to occur within a process. Orchestration of these phenomena, or where and how they occur, can place a restriction on the type of species that may be generated, the rates at which reactions occur, the separation of constituents into phases, etc. Similarly, extensive variable values such as equipment volume, maximum accumulation, or total mass, can be limited by process equipment design. Since the equations describing a process are a manifestation of these phenomena, select modification of the *variable-influence links* (i.e., of the cause-and-effect pathways) can maximize the inherent safety of the design technology. More importantly, *modification of the topology of the variable-influence links or limitation of the achievable variable values by the enabling preconditions are the only means by which the inherent safety of a process design technology can be altered.*

Preventing the occurrence of an enabling condition requires the undertaking of a corrective action. These actions are designed to limit the variable value of an enabling condition; however, they do so through active participation of protective equipment (e.g., release vents), control loops, or operating procedures. Like the mitigation, the prevention of enabling conditions necessitates an understanding of the *variable-influence links* (i.e., cause-and-effect links) and consequently, of the underlying physico-chemical phenomena. We begin the elucidation of the *variable-influence links*, leading to a potentially hazardous states, by first identifying the $(n + 2)$ independent variables that describe the potentially hazardous state. Since process design technologies are describable by the network of process modeling relationships (i.e., topology of the process flowsheet, material and energy balances, chemical and phase equilibrium relationships, kinetic and transport rate expressions, constitutive equations) that define the interactions of various variable quantities, we can trace out the influence of each variable that is associated with the state preceding the TLE. A trace of the variable-influence links defines the pathway of cause-and-effect interactions, which lead to the TLE.

B. Variables as "Causes" or "Effects"

The variable-influence pathway leading to the top-level event is constructed from the structure (Boolean) form of the incidence matrix, representing the network of process modeling relationships. But, the variable-influence pathways encompass certain information on directional causality. Consequently, it is important to identify the role of any variable in the set of modeling relationships: is a variable an input (cause), an output (effect), or of indeterminate directionality? In order to assign a role to each variable, we have developed a specific methodology that will be described in the following paragraphs. Consider the Boolean form of the incidence matrix that represents the modeling relationships, determining the behavior of a specific plant. These relationships capture all available knowledge about the plant; i.e., they are not limited to first-principles modeling equations, but include qualitative, order-of-magnitude, empirical correlations, etc. The fact that the matrix is in a Boolean form allows the simultaneous presence of relationships with inhomogeneous variables (i.e., real-valued, interval-valued, qualitative, or logical variables).

An *input-variable*, i.e., a variable that indicates the influence of the surrounding world on the process is clearly a potential *cause* and never an *effect* of the process' behavior. Typical examples of such input variables are design specifications, setpoints of control systems, characteristics of process feeds. The values of the input variables are established by factors external to the process. Input variables are generally extensive variables. An exception occurs when invariant intensive variables are associated with a feed. For example, the oxygen concentration in air may be considered invariant, when it is being used as a feedstock (e.g., formaldehyde production from air and methanol). Similarly feedstocks delivered to the process with concentration specifications can become inputs (e.g., 100% methanol). Manual settings by operators are input variables. These include manual valve manipulations, setpoints of controllers, structure of control loops, and starting or shutting of pumps. Furthermore, *all potential failures must be characterized as input variables*. This result occurs because potential failures are sources in the set of process equations that drive the causality in a particular direction. Therefore, we can use the following rules (see also Nagel, 1991) for the unambiguous characterization of certain variables as input variables (causes):

Rule 1. All influences of the surrounding world on a particular process are characterized as input variables, and are considered as causes of subsequent evolutions in the thermodynamic state of the process.

Rule 2. Process design specifications, and manual setting of operating variables and controller parameters are considered to be input variables, i.e., causes of subsequent events.

Once the set of input variables, associated with the influence of the surrounding world, has been identified, a systematic procedure examines the remaining variables in an effort to determine their unambiguous role as inputs or outputs. It should be clear at the outset of the subsequent discussion that *the unique and unambiguous characterization of all the process variables as inputs (causes) or outputs (effects) is in general impossible, and corresponds to an undecideable proposition.*

The assignment of a variable as *output* variable is always associated with a particular modeling relationship and implies that the variable takes on its value from the solution of the corresponding equation. Thus, it is identical to the concept of an output variable within the scope of the *input/output set assignment* for the solution of a set of nonlinear algebraic equations. But, unlike the solution of algebraic equations, an *output* variable within the scope of the deductive identification of hazards indicates a *physical consequence*, i.e., an *effect*, resulting from a specific set of causes (i.e., input variables). Consequently, we cannot use the variety of algorithms which have been developed for the identification of input/output set assignments, but we can employ some of the same ideas. Here are some of the rules which guide the selection of output variables (for detailed discussion, see Nagel 1991):

Rule 3. A variable occurring in a relationship, whose remaining variables have been characterized by Rules 1 and (or) 2 as input variables, is an output variable. It represents an effect of the process' behavior and can never be a cause.

Rule 4. A variable can be an output from only one relationship. Therefore, a variable already assigned as the effect (i.e., output variable) of a particular relationship, will be treated as cause (i.e., input variable) in subsequent relationships.

Rule 5. When a variable is the only variable in a particular relationship, then it must be characterized as output.

Rule 6. When a variable occurs in only one relationship and has not been characterized as input variable by Rules 1 and (or) 2, then it must be an output variable.

It is clear from the Rules 1–6 that the assignment of a variable as a *cause* or *effect* is guided by strict and unambiguous physical causality arguments. For example, if all the variables except one in a physical relationship have been characterized unambiguously as causes, the remaining variable *must* be the effect of the particular relationship. Also, it is obvious that if a particular variable has been characterized as the effect of a particular relationship, it can only be a cause in subsequent relationships. Finally, the variables that specify the state preceding a top-level event must be output variables.

It is also clear that Rules 1–6 do not resolve the character of all variables appearing in a set of modeling relationships. Thus, after a repeated application of Rules 1–6, two or more relationships with the corresponding unassigned variables, remain to be characterized. Such subsets of relationships should be "solved" simultaneously and can produce a number of alternative sets of variables, which could have been assigned as output. Clearly, any of these alternative assignments is arbitrary and does not reflect any unambiguous physical causality. What is more important is the fact that *the determination of unambiguous cause-and-effect links among the variables of the remaining relationships corresponds to an undecideable proposition, which can only be resolved through additional independent knowledge.*

Under conditions of incomplete assignment and ambiguity on the role of various variables in the cause-and-effect links, we have adopted a conservative attitude, exploring all potential causalities that may be produced from the modeling relationships.

C. Construction of Variable-Influence Diagrams

Let us now see how the ideas of the previous section can be used to construct the variable-influence diagram, that defines the paths leading to top-level event. Consider a set of four modeling relationships, represented by the structural matrix shown below.

	x_1	x_2	x_3	x_4	x_5	x_6	x_7
1			x		x		x
2		x			x	x	x
3	x	x	x	x		x	
4		x		x		x	

The columns represent variables in the defining process relationship and rows represent the relationships themselves. In accordance with the definitions given above, construction of the variable-influence path proceeds as follows:

1. Assignment of inputs:
    Step 1—assign constants as inputs (Rule 2)
    Step 2—assign invariant intrinsic properties of feeds as inputs (Rule 1)
    Step 3—assign setpoints as inputs (Rule 2)
2. Assignment of outputs
    Step 1—assign variables occurring in only one equation as outputs (Rule 6)
    Step 2—assign single variables in equations as outputs (Rules 3, 5)
    Step 3—eliminate assigned variables and corresponding equations (Rule 4)
    Step 4—assign dependent variables of scientific equations and definitions as outputs
    Step 5—repeat

In the structural incidence matrix given above, there are seven variables (*columns:* terms on abscissa, i.e., on horizontal axis; viz., x_1–x_7) and four relationships (*numbered rows:* terms on ordinate, i.e., on horizontal axis; viz., 1–4). Therefore, three variables must be assigned as inputs to specify the system. They are x_2, x_3, and x_5. Rewriting the structural matrix by moving the input variables to the right-hand side and delineating them with a vertical line, we have

	x_1	x_3	x_6	x_7	x_2	x_4	x_5
1		x		x			x
2			x	x	x		x
3	x	x	x		x	x	
4			x		x	x	

,

where the left-hand side of the structural matrix represents outputs and the right-hand side represents inputs. Using the definitions shown above, the following variables are characterized as effects (i.e., outputs): x_1 appears only in relationship 3 and therefore receives its value from that equation; similarly, equation 4 contains a single variable on the left hand side of the structural matrix, therefore x_6 takes its value from relationship 4. After the elimination of rows and columns associated with each assigned output variable, x_3 is identified as taking its value from relationship 1, and x_7 taking its value from relationship 2. Thus, x_3 and x_7 are characterized as effects (i.e., outputs) from relationships 1 and 2, respec-

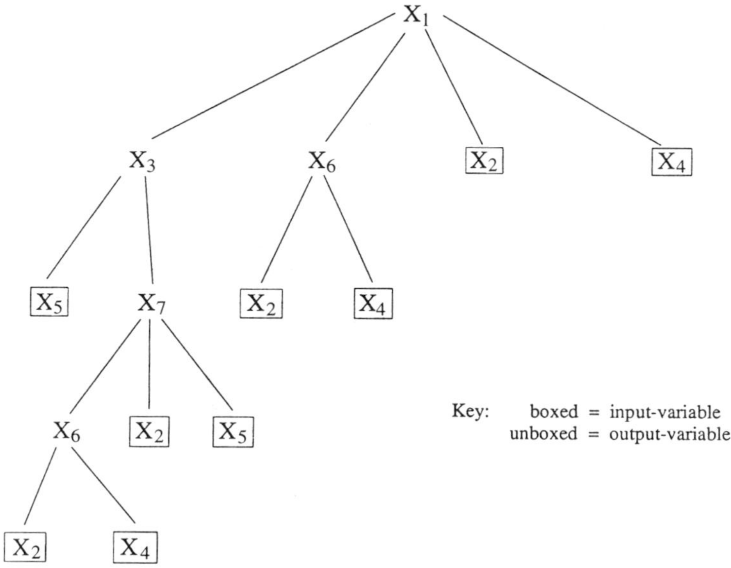

FIG. 14. Variable-influence pathway.

tively. The following structural incidence matrix indicates all assignments:

	x_1	x_3	x_6	x_7	x_2	x_4	x_5
1		$\underline{x}$		x			x
2			x	$\underline{x}$	x		x
3	$\underline{x}$	x	x		x	x	
4			$\underline{x}$		x	x	

The variable influence path leading to x_1 is shown in Fig. 14. As a consequence, x_1 can only be influenced by variables that appear on the path; causality is established by the input variables. This occurs because the output set assignment given to the set of relationships in the structural matrix is unique, as it is made evident by rearranging the tabulation order of relationships and variables comprising the structural matrix:

	x_6	x_7	x_3	x_1	x_2	x_4	x_5
4	$\underline{x}$				x	x	
2	x	$\underline{x}$			x		x
1		x	$\underline{x}$				x
3	x		x	$\underline{x}$	x	x	

Unique output set assignment occurs when the structural incidence matrix is triangular, such as the above. As a consequence, the variable-influence path leading to the top-level output variable (e.g., x_1) is also unique and can be affected only by inputs that occur along the path.

Alternatively, the network of relationships may not be represented by a triangle structural matrix. In this instance, the structural matrix may not have a unique output set assignment, and there may be alternative paths leading to the variable that describes the state preceding the top-level event. Alternative paths are created when loops exist in the set of process modeling relationships necessitating simultaneous solution. For example, in the structural matrix shown below there are three distinct output set assignments, leading to three distinct potential variable-influence paths. (*Note:* Underlined entries in bold indicate the assigned outputs to each relationship.)

	x_1	x_2	x_3	x_4	x_5	x_6	x_7	x_8	x_9	x_{10}	x_{11}	x_{12}
1		**X**	X		X				X			X
2					**X**		X		X	X		
3				**X**			X	X	X		X	
4			**X**	X		X		X	X		X	
5					X	**X**				X		X
6					X	X	**X**					X
7								**X**				
8									**X**	X		X
9	**X**	X		X	X		X				X	

	x_1	x_2	x_3	x_4	x_5	x_6	x_7	x_8	x_9	x_{10}	x_{11}	x_{12}
1		**X**	X		X				X			X
2					X		**X**		X	X		
3				**X**			X	X	X		X	
4			**X**	X		X		X	X		X	
5					**X**	X				X		X
6					X	**X**	X					X
7								**X**				
8									**X**	X		X
9	**X**	X		X	X		X				X	

	x_1	x_2	x_3	x_4	x_5	x_6	x_7	x_8	x_9	x_{10}	x_{11}	x_{12}
1		**X**	X		X				X			X
2					X		**X**		X	X		
3				**X**			X	X	X		X	
4			**X**	X		X		X	X		X	
5					X	**X**				X		X
6					**X**	X	X					X
7								**X**				
8									**X**	X		X
9	**X**	X		X	X		X				X	

Alternatively output set assignments, and consequently variable-influence paths, result from the block structure established by variables x_5, x_6, and x_7. Thus, variable x_5 can take its value from any one of the three relationships, since there is no way that x_5 can be uniquely and unambiguously assigned as an effect to any one of these three relationships.

Construction of the variable-influence diagrams, indicating the paths of the cause-and-effect links, which lead from the external causes to the variables describing the potentially hazardous state, is achieved by using the procedure CONSTRUCT-VARIABLE-INFLUENCE-PATHWAYS, a procedure called from the procedure, IDENTIFY-POTENTIAL-HAZARD. This method traces the influence of various variables defining the state preceding the top-level event. It constructs variable-influence paths from the set of process modeling relationships contained in the structural incidence matrix. The algorithm used for the generation of variable influence pathways is given below:

Algorithm 3: ⟨CONSTRUCT-VARIABLE-INFLUENCE-PATHWAYS ⟩.

input: process-flowsheet
    enabling-criteria   ;output from IDENTIFY-POTENTIAL-HAZARD
initialize
process-flow-equations ← **apply** IDENTIFY-PROCESS-FLOW-EQUATIONS
                  to process-flowsheet
VIM ← **apply** IDENTIFY-INFLUENCE-MATRIX to process-flow-equations
TLE-vars ← **apply** IDENTIFY-TLE-VARS to enabling criteria   ;identify top-
investigation-list ← TLE-vars   level event
for each tree in investigation-list   variables
if next-available-node ← **apply** IDENTIFY-NEXT-EXPANDABLE-NODE to tree
    then
    expansion-list
       ← **apply** EXPAND-NODE to VIM, tree, next-available-node
    endif
return
for each expansion in expansion-list
    apply CLASSIFY-BRANCH to next-available-node   ;classify
                              according to
                              technology-type
    if (apply CONTINUE-EXPAND-BRANCH-P to expansion,
      next available-node)
    then   ;predicate test to evaluate
                 branch expansion
    (**append** tree to investigation-list)
    endif
return

Whenever the procedure CONSTRUCT-VARIABLE-INFLUENCE-PATHWAYS has to generate a number of alternative variable-influence paths, it employs a breadth-first search strategy. The procedure, EXPAND-NODE, is used to establish the expansion list through the following algorithm:

Algorithm 4: ⟨EXPAND-NODE⟩.

> **input:** VIM, tree, var
> **initialize**
> > expansion-list ← nil
>
> **apply** CLEAR-ALL-ASSIGNMENTS to VIM
> **for** each node in (**apply** PATHWAY-NODES to tree, var)
> > **apply** ASSIGN-OUTPUT-VARIABLE to VIM, node
> > **return**
>
> **for** each equation in (**apply** EQNS-CONTAINING-VAR to VIM, var)
> > **if** (**apply** VALID-ASSIGNMENT-P to VIM, eqn, var)
> > > **then**
> > > > new-vars ← **remove** var from (**apply** VARS-IN-EQN to eqn)
> > > > **apply** CLASSIFY-VARS to new vars
> > > > new-tree ← **apply** APPEND-TREE to tree, node, new-vars
> > > > **append** new-tree to expansion-list
> > > **endif**
> > **return**
> **return**

D. Characterizing of Variable-Influence Pathways

The procedure CONSTRUCT-VARIABLE-INFLUENCE-PATHWAYS applies to each node of the variable-influence pathway the procedure, CLASSIFY-BRANCH, which in turn classifies each branch of the pathway according to the type of the technology, which can be used to control the variable-value specifying the node. The algorithm of this procedure is given below:

Algorithm 5 ⟨CLASSIFY-BRANCH⟩.

> **input:** node
> **for** each child-node in (**apply** CHILDREN to node)
> **select** case for child-node
> > terminal-variable
> **is** apply TAG-NO-EXPANSION to child-node   ;no node expansion
> > **endcase**
> > already-in-pathway

```
    is apply TAG-NO-EXPANSION to child-node
    endcase
endselect

select case for node
    inherent-controllability
    is apply TAG-TYPE-1-TECHNOLOGY to node        ;see definition below
    endcase
    Type-2-controllability and process-modification-applied
    is apply TAG-TYPE-2-TECHNOLOGY to node
    endcase
    Type-3-controllability and control-loop-applied
    is apply TAG-TYPE-3-TECHNOLOGY to node
    endcase
endselect
return node
```

Nodes are classified in this manner in order to make explicit the protective system responsible for mitigating a disturbance and its relationship to the top-level event. The procedure CONSTRUCT-VARIABLE-INFLUENCE expands each node of the tree and classifies each branch according to the technology type that could be used to control the variable-value, using the following definitions:

1. A *Type-1 technology* is capable of reducing the number of TLEs associated with the specific design, or is capable of modifying the topology of the path leading to the TLE through a change in process chemistry or the structure of the designed process flowsheet, provided design changes do not involve the introduction of *Type-2* or *Type-3 technologies*. Examples include reactant substitution, catalyst introduction or substitution, byproduct reduction, chemical character modification, solvent substitution, and intermediate chemicals elimination. Industrial processes with *Type-1 technology* changes include the following: (a) substitution of the reactant raw materials, hydrogen cyanide and acetylene, with propylene, ammonia, and air; (b) production of butyl lithium in dilute solutions, which prevents spontaneous combustion in the presence of air; and (c) direct hydroxylation of ethylene forming ethylene glycol, rather than oxidation of ethylene forming ethylene oxide, which in turn is hydrated to form ethylene glycol, thus eliminating ethylene oxide as an intermediate. Changes in the pathway topology, leading to the TLE, require design modification of the unit operations, changes in the piping, substitution of unit operations, or layout modification.

2. A *Type-2 technology* limits the value that a variable, which is involved in the pathway leading to a TLE, can achieve without requiring an action. Examples involving *Type-2 technologies* include the following:

(a) reduction of inventories, (b) minimization of operating (process condition) extrema, (c) reduction of intermediate accumulation. Industrial processes incorporating *Type-2 technologies* include the following: utilization of in situ reactions to limit intermediate accumulation (e.g., methyl isocyanate consumption via in situ reaction reduces the need for storage), utilization of continuous processes to minimize process hold up (e.g., continuous nitration processes eliminate the need for batch manufacturing of nitroglycerine), and reduction in the intensity of processing conditions can lead to smaller material and energy releases (e.g., catalyst improvements have enabled the Oxo process to produce aldehydes from syngas and olefins at lower pressure).

3. A *Type -3 technology* limits the value that a variable, involved in the path leading to a TLE can achieve *and require an action to do so. Type-3 technologies* involve active control of process variables in a direct or indirect manner, in order to limit the value that they can achieve. Examples include the following: (a) manipulation of feed rates, (b) use of cooling water, (c) temperature control and pressure control, (d) catalyst activity assessment through the monitoring of conversion and selectivity.

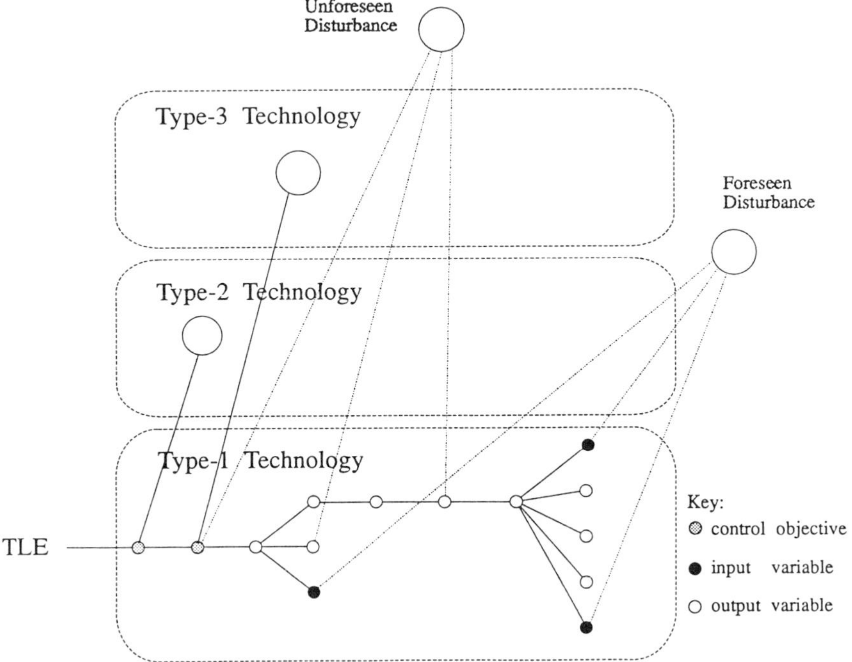

Fig. 15. Mitigation of disturbances through the specification of the control points.

The result of tagging the paths with the type of hazards-preventive technology is shown in Fig. 15. Having such a diagram one can then associate the different technology types with various parts of the variable-influence diagram, for the purpose of directly or indirectly controlling the variable values, composing the variable-influence pathway to the top-level event. Such an approach not only provides an understanding on how to control the values of each variable along the variable-influence path, but, also permit the designer to evaluate the feasibility of the process and its safety devices with respect to various operational criteria.

E. ASSESSMENT OF HAZARDS-PREVENTIVE MECHANISMS

The strategy we use for the specification of the most attractive hazards-preventive control objectives is based on *closeness*, i.e., hazards-preventing control objectives, which affect a variable that is at a minimum distance from the top-level event is preferred over those that affect more distant variables. However, the point on the variable-influence path, where the actual manipulation (e.g., design modification, controller or safety device) takes place, remains unspecified; it depends on whether the origin, type and intensity of disturbances are known ahead of time or not.

For disturbances that can be *foreseen*, it is preferred that the hazards-controlling objectives be placed near their origin (i.e., entry point of an input), in order to mitigate the effect of the disturbance before it has the opportunity to be amplified (see example in the diagram of Fig 15). *Foreseen disturbances* enter the process through the set of external variables or inputs. These include pump failures, valve failures, controller malfunctions, etc.

Notice, though, that the greater the distance between the location of the hazards-controlling objective and the top-level event, the greater the opportunity (i.e., the higher the probability) for the appearance of an unknown (*unforeseen*) disturbance along the pathway that connects the control objective to the TLE. *Unforeseen disturbances* can enter the process anywhere along the pathway and often change the pathway leading to the top-level event. Since the thermodynamic state preceding a top-level event is specified by the procedure, IDENTIFY-POTENTIAL-HAZARD, any disturbance, foreseeable or unforeseeable, must act as an input to these variables, if the TLE is to be enabled. Consequently, the only effective mechanism for the mitigation of unforeseen disturbances is the establishment of control objectives at *level 1*; the last defensive line before enabling the top-level event. This specification is translated into a series of conditional statements on the values of the variables at *level 1* of the variable-influence diagram (see Fig. 15). The methodology used by the procedure, GLOBAL-HAZARD-IDENTIFICATION, enables the identification of TLEs independently of the pathways that lead to them. Additional TLEs are identified by the procedure GLOBAL-HAZARD-IDENTIFICATION as it expands the scope of the process description in its search for potential hazards. Expansion of the process description includes the reformulation of the boundary defining the process, thus allowing the incorporation of disturbances (e.g., as new inputs) that were *"unforeseen"* by the earlier formulation of the process description.

Type-2 technologies can be particularly effective in mitigating unforeseeable disturbances, because they do not require active control of the disturbance. Nevertheless, whenever possible, *Type-1 technologies* offer the best mechanisms for the mitigation of unforeseeable disturbances. This is accomplished through the addition or subtraction of relationships from the structural incidence matrix; operations that describe the recommended changes in chemistry, materials, type of unit operations, or structure of the process flowsheet. On the other hand, control of foreseeable disturbances is better handled by *Type-3 technologies*. Each variable along the variable-influence pathway is subsequently characterized by the type of hazards-controlling technologies, i.e., *Type-1, Type-2,* or *Type-3 technologies*, which could be employed for controlling the value of the specific variable. Such characterizations allow fast and effective screening of the potential disturbance mitigation alternatives.

The procedure IDENTIFY-NONDISSIPATIVE-PATHWAYS constructs the set of enabling conditions that lead to a top-level event. It takes as its input the variable-influence pathways, which it obtains from the procedure, CONSTRUCT-VARIABLE-INFLUENCE-PATHWAYS. TLE variables are then identified, and each variable contained in the set is traced to identify potentially feasible roots for causing the TLE. When an achievable root is identified (i.e., an input disturbance, controlled or uncontrolled, which can enable the top-level event), it is collected into root causes and returned. The algorithm used by this procedure is shown below:

Algorithm 6: ⟨IDENTIFY-NONDISSIPATIVE-PATHWAYS⟩.

> **input:**　classified-influence-paths
> **initialize:**　TLE-variables ← (collect-TLE-variables)
> **for** each variable in TLE-variables
> 　　　　**when** (unique-path-p classified-influence-paths)　　;predicate test for
> 　　　　　　　　　　　　　　　　　　　　　　　　　　identifying unique
> 　　　　　　　　　　　　　　　　　　　　　　　　　　paths
> 　　　　　　　(**apply** CONSTRUCT-FEASIBLE-ROOTS　　　;constructs path-
> 　　　　　　　　　to TLE-variables classified-　　　　　ways or trees
> 　　　　　　　　　influence-paths)　　　　　　　　　　which support

```
                                                enabling criteria
            and collect into root-causes
            return
    else
            for each path in classified-influence-paths
            to (apply CONSTRUCT-FEASIBLE-ROOTS      ;constructs a
                to TLE-variables path)                feasible root to
            and collect into root-cause              the goal state
            return                                   (TLE variable)
                                                     path

        return goal state (TLE)
end
```

Figure 16 shows how the procedure IDENTIFY-NONDISSIPATIVE-PATHWAYS has identified a potential path, starting from a given TLE and deducing a specific (*foreseen*) disturbance as a potential cause for the given TLE. Furthermore, in Fig. 16 we can also see the type of the preventive technologies (as identified by the procedure, CLASSIFY-BRANCH) that are appropriate for each variable on the pathway, which leads from the foreseen disturbance to the specific TLE.

The procedure IDENTIFY-NONDISSIPATIVE-PATHWAYS calls the procedure CONSTRUCT-FEASIBLE-ROOTS, which assesses the feasibility of a particular variable trace to enable the TLE, given a disturbance as its input. The

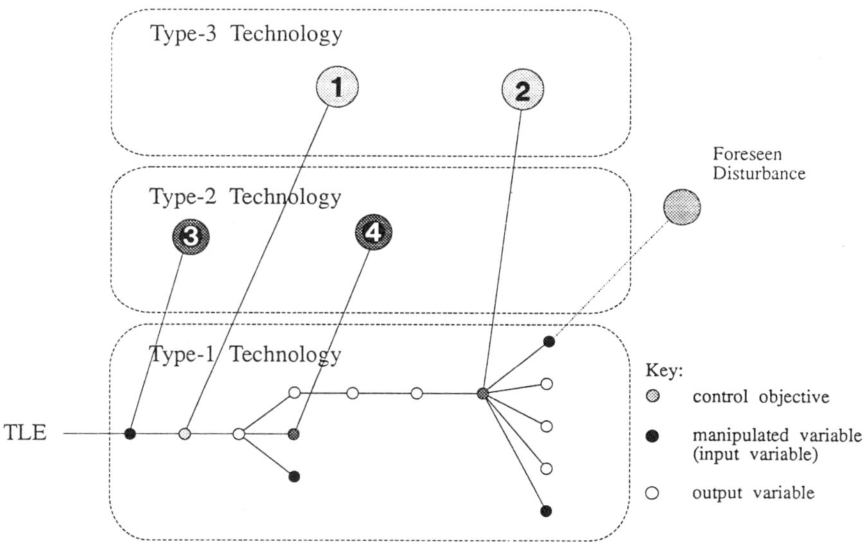

FIG. 16. Placement of technologies for the prevention of a specific TLE.

procedure CONSTRUCT-FEASIBLE-ROOTS also identifies the control mechanism responsible for the mitigation of the input disturbance. When a controlling mechanism is not associated with the input, the variable and its pathway are collected directly into roots. The algorithm defining the logic of the procedure, CONSTRUCT-FEASIBLE-ROOTS is shown below:

Algorithm 7: ⟨CONSTRUCT-FEASIBLE-ROOTS⟩.

```
      input: variable-set
             pathway
      when (enabling-feasibility-p variable-set pathway)
          for each variable in variable-set
              when (controlled-variable-p variable pathway)
                  controller ← (get-controller variable pathway)
                  collect variable, pathway, and controller into roots
                  return
              else
                  collect variable into roots
              return
          return
      return
      end
```

F. FAULT-TREE CONSTRUCTION

The construction of the variable-influence pathways and the classification of its nodes in terms of the type of technology that can be used to mitigate the propagation of a hazard, constitute the essential basis for the generation of topological fault trees, a tool that can guide the evaluation of hazards preventive mechanisms. The procedure CONSTRUCT-TOPOLOGICAL-FAULT-TREE is at the core of this construction. It takes as its input the top-level event, the associated hazardous state preceding the TLE, and the root causes for each variable defining the hazardous state. Using these inputs, it determines the reactions responsible for the hazardous state and the enabling criteria of the reaction. With this information, it constructs a *level-1-gate:* a logical gate immediately preceding the top-level event. This gate establishes the demands (i.e., inputs) of the top-level event. By looping through the set of demands associated with the *level-1-gate*, we can construct qualitative logical gates from a given demand input and its root causes. Since each qualitative gate has associated with it an input and an output, these can be linked together to form a tree. By linking the tree to the *level-1-gate* and in turn appending the *level-1-gate* to the TLE, a topological fault-tree can be constructed. The algorithm used is as follows:

Algorithm 8: ⟨CONSTRUCT-TOPOLOGICAL-FAULT-TREE⟩.

```
  input: TLE
         hazardous-state
         root causes
  initialize: hazard-variables ← (get-hazardous-variables hazardous-state)
              reaction ← (get-reaction hazardous-state)
              enabling-criteria ← (get-enabling-criteria reaction)
              level-one-gate ← (construct-level-one-gate
                                     hazard-variables enabling-criteria)
  for each input in level-one-gate
     gates ← (CONSTRUCT-TOPOLOGICAL-GATES input    ;gate construction
(get-root-causes ;input))
        tree ← (LINK-GATES input gates)             ;connects gates
        append tree to level-one-gate                via their logical
        return                                       outputs
     append TLE to level-one-gate
     return
     end
```

Basic gates are constructed using the procedure, CONSTRUCT-TOPOLOGI-CAL-GATES. This procedure builds *or-gates*, *and-gates*, and *special-gates*. *Special-gates are based on first-order predicate logic and require a certain amount of quantitative analysis prior to Boolean assignment (i.e., and-gates, or-gates, and special-gates).* The procedure receives an input variable and the root causes associated with that variable. Then, it uses the variable trace associated with the root causes to establish the pathway leading to the external input and the protective devices associated with that input. By collecting the controllers associated with the various controlled variables, the procedure identifies the necessary *and-gates*, *or-gates*, and *special-gates*. The algorithm used for the implementation of the procedure, CONSTRUCT-TOPOLOGICAL-GATES, is shown below:

Algorithm 9: ⟨CONSTRUCT-TOPOLOGICAL-GATES⟩

```
     input: variable
            root-causes
     initialize: and-gates ← nil
                 or-gates ← nil
                 special-gate ← nil
```

```
              initial-internal-input
                  ← (get-internal-input variable    ;network
                      root-causes)                   starting point
      for each variable in (get-output-assignment root-causes)
         up from initial-internal-input
         when (unbranched-node-p variable)          ;predicate test for
           controller                                determining if node
                  ← (get-controller variable         is branched
                      root-causes)
           when controller                          ;constraint
             collect (construct-and-gate            ;identification
                      variable controller            of variable
             (get-input variable))
                into and-gates
             and
             collect (construct-or-gate variable controller (get-input
                      variable))
                into or-gates
             return
           collect (construct-special-gate variable (get-inputs variable))
           into special-gates
           return
         gates ← append and-gates or-gates special-gates
         return
      return
      end
```

Complex topological fault trees are constructed using the gates identified by CONSTRUCT-TOPOLOGICAL-GATES. Since each logical gate has a set of inputs and an output, all the gates can be linked together by matching gate outputs to gate inputs.

Quantitative analysis of the pathway, leading to the TLE, is required to establish the logical basis for converting *special-gates* into structures (i.e., trees) containing *or-gates* and *and-gates*. By associating averaged probability and failure rate data with the resulting topological fault tree (i.e., with the variables and the technology type of preventive mechanisms), the latter can be transformed into a conventional fault-tree. Construction of the fault tree in this manner has particular utility because it is complete; all pathways leading to the TLE will be identified within the scope of the relationships describing the particular process design technology. This prevents incompleteness in the pathways leading to the TLE and minimizes errors in the frequence of that event, often by more than three orders of magnitude (ICI, 1988). (*Note*: ICI has shown that incompleteness can result in estimates of the TLE that are off by as much as three to five orders of magnitude; whereas errors in probability and failure rate data lead to estimates that are often within a factor of 2–5.)

G. An Example of Reaction-Based Hazard Identification: Reaction Quench

Consider a process that consists of a reactor used for the processing of a highly unstable chemical that is sensitive to small increases in temperature. The reactor is equipped with a quench tank to protect the system against a runaway reaction and is monitored by two temperature sensors (see Fig. 17): T_1 and T_2. Sensor T_1 automatically activates the quench tank outlet valve when it detects a temperature rise above the specified upper limit. Sensor T_2 sounds an alarm in the control room to alert the operator to the process upset. When the alarm sounds, the operator closes the reactor inlet valve. The operator also pushes a quench tank valve button in the control room in case the quench valve fails to open. Note that A is the reactant; B, the product; and C, the quench.

The analysis for the deductive determination of root causes begins by constructing the set of relationships that describe the process flowsheet.

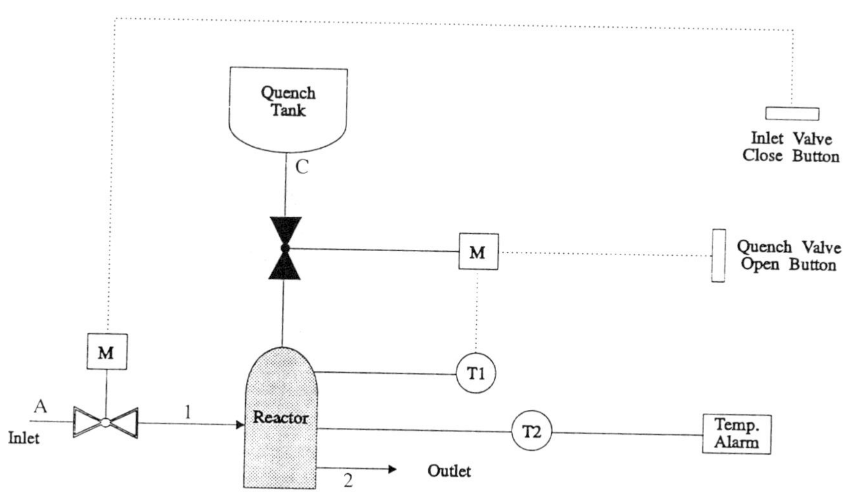

Fig. 17. The elements of the reactor-quench example.

These are shown below:

$$F_2 = f_2^A + f_2^B + f_2^C, \tag{1}$$

$$f_2^C = Q + F_1^C, \tag{2}$$

$$f_1^A = f_2^A + r_A, \tag{3}$$

$$f_1^B = f_2^B - r_A, \tag{4}$$

$$F_1 = f_1^A, \tag{5}$$

$$f_1^B = 0, \tag{6}$$

$$f_1^C = 0, \tag{7}$$

$$r_A = k[C_A]V_{RE}, \tag{8}$$

$$k = A * \exp(-E^*/RT_R) = f(T_R), \tag{9}$$

$$[C_A] = f(f_1^A, Q, r_A), \tag{10}$$

$$Q_r = f(\Delta h_r, r_A), \tag{11}$$

$$0 = f(Q, T_0, F_1, T_F, Q_r, F_2, T_R), \tag{12}$$

$$Q = f(SP_1), \tag{13}$$

$$SP_1 = f(T_R, T_1), \tag{14}$$

$$F_1 = f(SP_2), \tag{15}$$

$$SP_2 = f(Op_2), \tag{16}$$

$$OP_2 = f(T_R, T_2), \tag{17}$$

where F denotes a feed, f denotes a molar flowrate, Q denotes the quench feed, r_A is a reaction rate expression, V_{RE} is the volume of the reactor, SP denotes valve stem position, T denotes temperature, and Op denotes an operator. Complex relationships involving these variables show only the corresponding functional dependence [e.g., relationships (9) through (17)]. The *structural matrix* constructed from this set of equations and relationships is shown in Fig. 18. Each numbered row describes the relationship between the variables that are contained in it. The variables are represented by the columns of the matrix. The occurrence of a variable in a particular relationship is signified by a nonzero entry (e.g., x in Fig. 18) at the intersection of the column representing the variable with the row representing the relationship.

	F_1	F_2	Q	f_1^A	f_1^B	f_1^C	f_2^A	f_2^B	f_2^C	r_A	k	C_A	T_R	Q_R	V_{RE}	T_2	T_1	O_{p2}	SP_1	SP_2	Δh_r	T_Q	T_F
1		x					x	x	x														
2			x			x			x														
3				x			x			x													
4					x			x		x													
5	x			x																			
6					x																		
7						x																	
8										x	x	x			x								
9											x		x										
10			x	x						x		x											
11										x				x							x		
12	x	x	x							x	x											x	x
13		x																	x				
14																	x		x				
15	x																			x			
16																		x		x			
17													x			x		x					

FIG. 18. The structural incidence matrix for the reactor-quench example.

The process of constructing the variable-influence pathway, which leads to the state preceding the top-level event, begins with the identification of the potential input variables. Following the procedures outlined in Section IV.B, the assignment of the input variables is initiated. But, the identification of the *input specifications* is intricately related to the scope of the assumed *system's boundary*. Thus, by restricting our attention to the reactor only (see shaded unit in Fig. 17), examination of the process modeling relationships identifies two system constants: Δh_r (the heat of reaction) and V_{RE} (the reactor volume); these become input variables in the structural matrix. Similarly, if the scope of the process' boundary is expanded to include external feeds, then the boundary of the system encloses all the shaded area of Fig. 19. This expansion identifies the invariant intensive properties describing the inlet feed (T_F) and the quench feed (T_Q) as external or input variables. Step 3 of the input assignment procedure (see Section IV.B) assigns the setpoints T_2 and T_1 as input variables, thus further expanding the scope of the process boundary (see Fig. 20).

Once inputs are assigned, output assignment begins using the method described in Section IV.B. Using this procedure, f_1^B is identified as taking its value from Eq. (6) and f_1^C is identified as taking its value from Eq. (7).

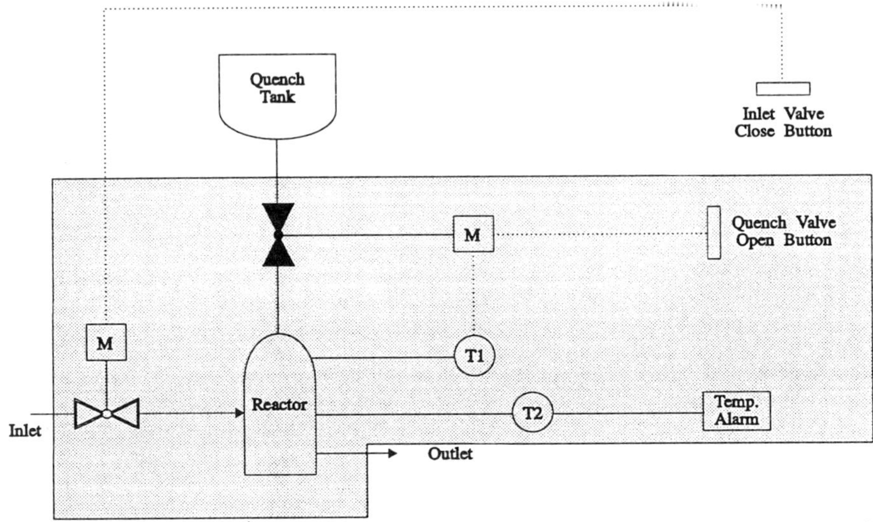

FIG. 19. The system boundary defining the expanded input specifications for the reactor-quench example.

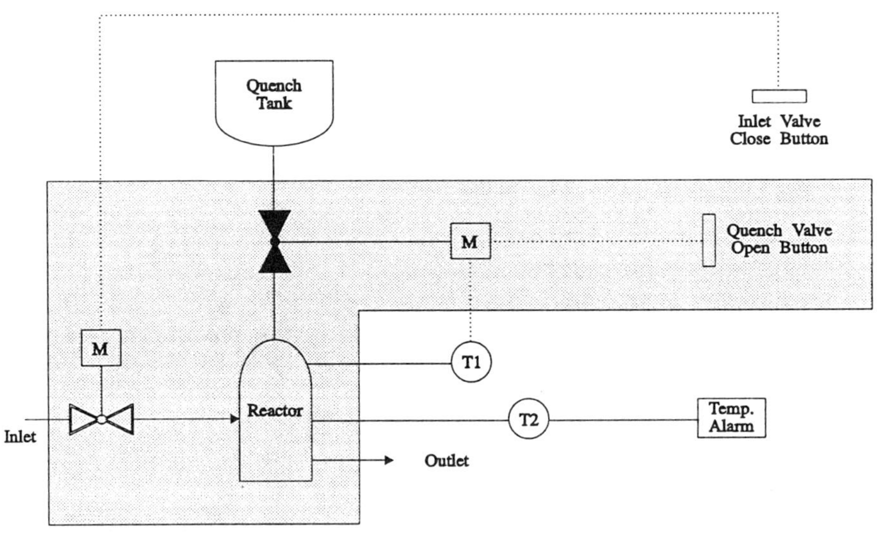

FIG. 20. The final system boundary indicating further expansion of the input specifications.

The respective rows and columns of the structural matrix are then eliminated. The value of the reaction rate constant can be given only by the definitional relationship (9), and thus it is assigned as output from Eq. (9). On elimination of the row and column corresponding to Eq. (9) and variable k, no other output assignments can be made. The block structure that results allows the remaining variables to take their output from any one of several equations.

To break the loop of cause-and-effect relationships, we make the unqualified assumption that Op_2, the variable denoting the human operator, obtains its value from relationship (17); i.e., the operator responds to the temperature in the reactor (T_R) and not vice versa. This assumption is then catalogued, as *Assumption-1*, so that it may be retracted at a later stage, as required. Such retraction allows the causality that emanates from the set of process equations to be modified. Once we have eliminated Eq. (17) and the column corresponding to variable Op_2, we can repeat the steps of the output set assignment (see Section IV.B), and find that F_1 obtains its value from Eq. (15). The set of input and output assignments, made so far, establish the causality between

- Reaction temperature and human operator
- Human operator and feed valve stem position
- Feed valve stem position and feed rate.

Elimination of the respective rows and columns identifies further that f_1^A takes its value from Eq. (5).

At the end of the preceding assignments, a block structure of *"simultaneous relationships"* still remains. Again, though, an unqualified assumption can be made regarding the causality of variables in Eq. (14). The assumption is that SP_1, *the variable describing the feed valve stem position, is driven by reactor temperature.* This assumption is catalogued and denoted as *Assumption-2*.

The variable-influence diagram resulting from the set of the two assumptions, i.e., *Assumption*-1 and *Assumption*-2, is shown in Fig. 21. This diagram makes explicit the pathways leading to T_R and illustrates how disturbances can propagate through the network of relationships and enable the occurrence of the TLE. If we now retract *Assumption*-1, and replace it by its opposite (which will be labeled *Assumption*-3), i.e., the operator does not react to the temperature of the reactor but *the value of the reactor temperature is defined by the actions of the operator*, then the variable-influence diagram changes and is now represented by that of Fig. 22. By collecting the entire set of graphs that describe the resulting set of cause-and-effect relationships, we can guarantee complete identification of pathways, leading to the TLE, within the scope of the modeling effort,

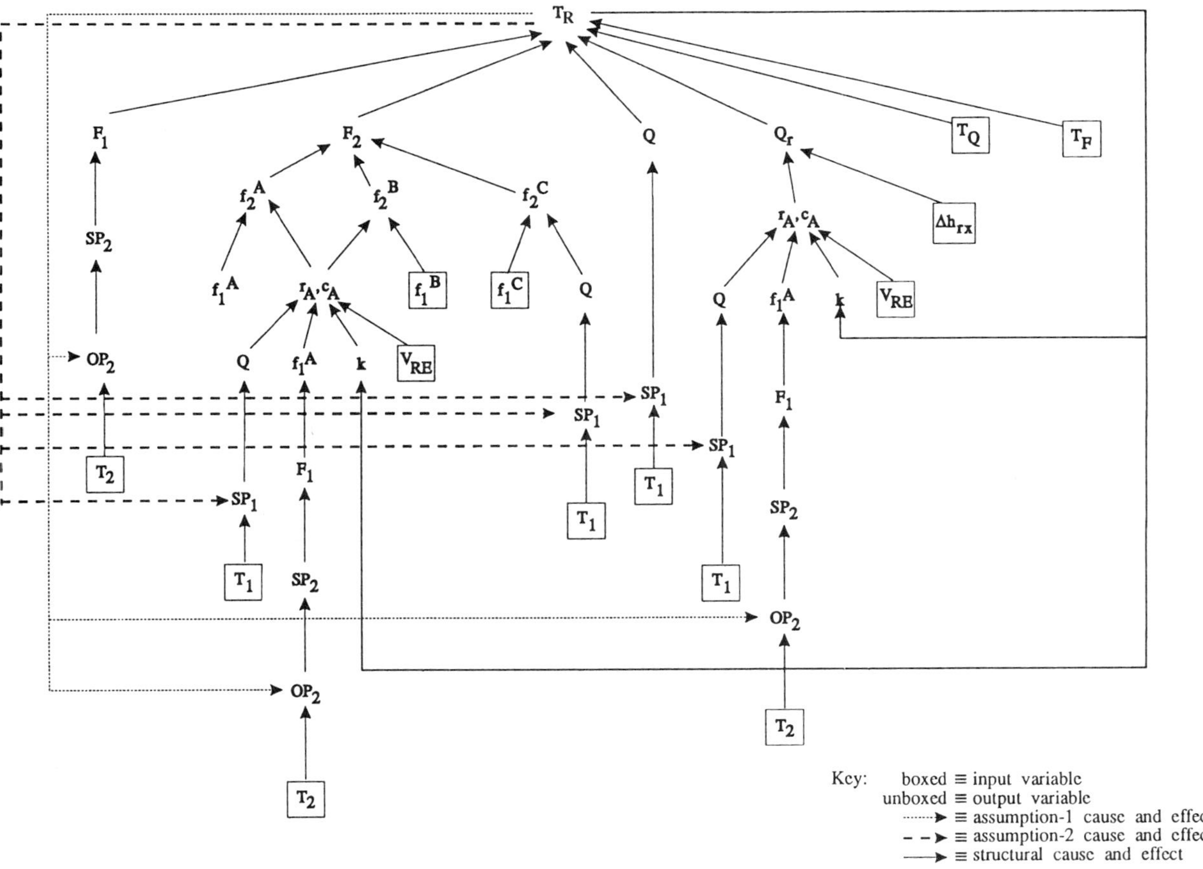

FIG. 21. The variable-influence diagram resulting from *Assumption-1* and *Assumption-2*.

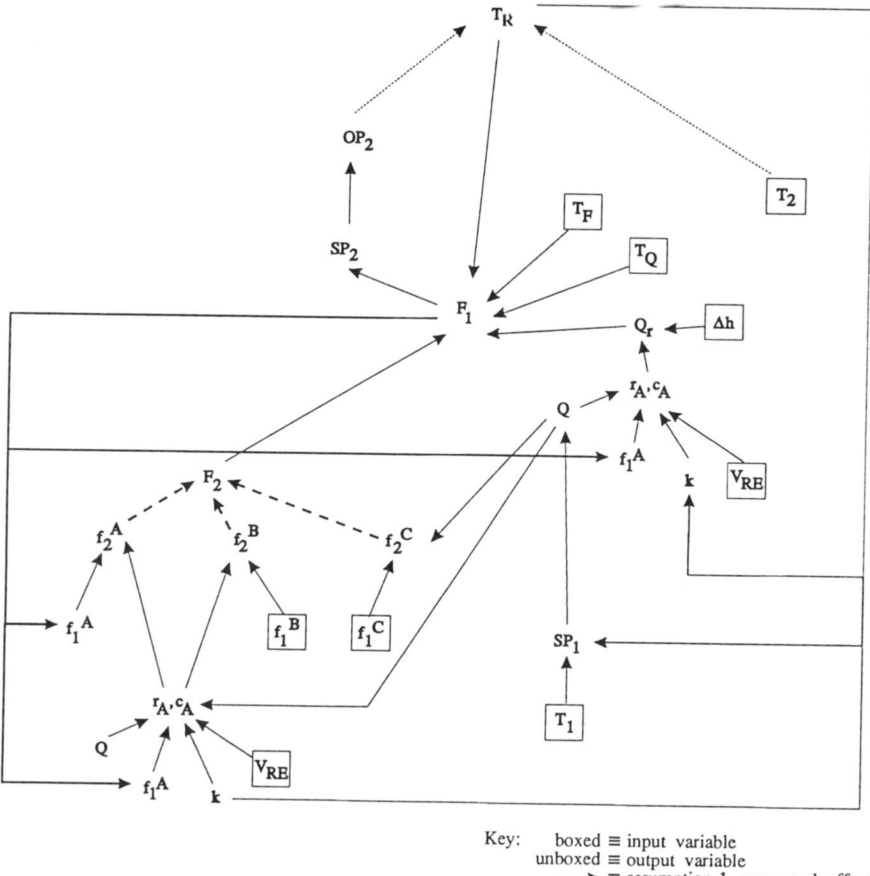

Fig. 22. The variable-influence diagram resulting from *Assumption-2* and *Assumption-3*.

in conjunction with the set of the assumptions made. By associating the variables, contained in the variable-influence diagram, with the appropriate types of hazards-preventing technologies (i.e., *Type-1, Type-2, Type-3 technologies*), the pathway of root causes is constructed. Figure 23 illustrates the association of the variables with specific *Type-1, Type-2,* and *Type-3 technologies*, for the case where the variable-influence diagram has been constructed assuming *Assumption-1* and *Assumption-2* to be true. Note that the heat of reaction Δh_r is the only Type-1 technology associated with the TLE. *This implies that a potential pathway for mitigating the*

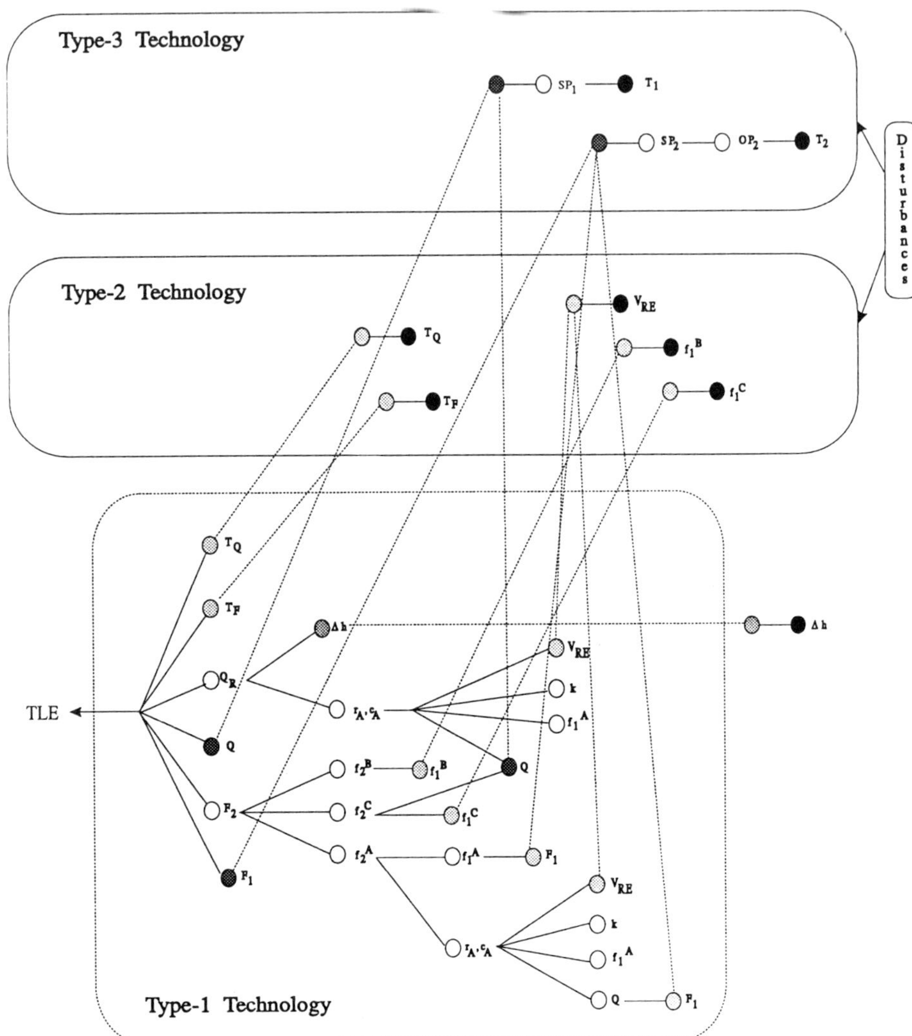

FIG. 23. Construction of root cause diagram.

TLE is to change the heat of reaction and obviously implies a change in chemistry. Furthermore, the variable-influence diagram makes explicit that top-level event prevention necessitates the placement of control objectives on every variable that immediately precedes the top-level, i.e., variables $T_Q, T_F, Q_R, Q, F_2, F_1$, should come under control.

Notice, also, that *the value of the variables described by Type-2 technologies are generally design specifications*; they affect the magnitude of the top-level event. For example, V_{RE} specifies the volume of the reactor. Similarly, feed and quench temperatures vary the rate at which a TLE is achieved; likewise, so may f_1^B and f_1^C as the values that describe the molar flowrates of the product and the quench, respectively, contained in the feedstream, vary. This is particularly so if the reaction described is autocatalytic. Trace amounts of product in the feed can catalyze the reaction leading to a thermal disturbance that ultimately could trigger a reaction runaway.

Type-3 technologies require active control to mitigate a disturbance and consequently are protective systems associated with the process flowsheet. Notice that certain control actions cannot be modeled easily without overspecifying the system such as the manual override of the quench valve by the operator. For this reason, the constraint list, as described earlier, is associated with each piece of process equipment. This allows us to associate in an *a posteriori* manner additional control structures that are available to the process.

How the different technology types are used to identify potential root causes that can enable the top-level event is illustrated in Fig. 23. This illustration makes explicit the fact that disturbances, which enter into the system, can enter only through the inputs identified by Type-2 and Type-3 technologies. These disturbances can either be known a priori (*foreseen*) or not (*unforeseen*). Modeled disturbances can affect the process only through its input variables and are therefore identified by the pathways leading to the top-level event. Unmodeled disturbances can act in one of two ways: they can affect the value of a variable through an unmodeled relationship, or they can change the casuality described by the pathway, negating the assumption set from which it was built. However, since various assumption sets are used to construct the alternative pathways leading to the top level event, the impact of unmodeled disturbances that enabled some pathways can be contained. Furthermore, because of the manner in which the top-level event has been identified, it can be enabled only through the set of variables that describe the state immediately preceding it. Since any disturbance must eventually pass through that state, if it is to enable the top-level event, control objectives designed around these variables can mitigate foreseen and unforeseen disturbances that lead to a top-level event. Using the knowledge associated with different technology types, and the pathways that lead to a top-level event, we can construct a topological fault tree. This is illustrated in Fig. 24. *Note the need for a special gate.* This type of gate is required because without quantitative assessment, it cannot be determined whether the gate is a single *and-gate*, a single *or-gate*, or a structure containing *and-gates* and *or-gates*.

The topological fault tree shown in Fig. 24 suggests that the top-level event can be enabled through a change in the input of T_Q (e.g., this could imply no quench). Using the algorithm presented in Section IV.C, the

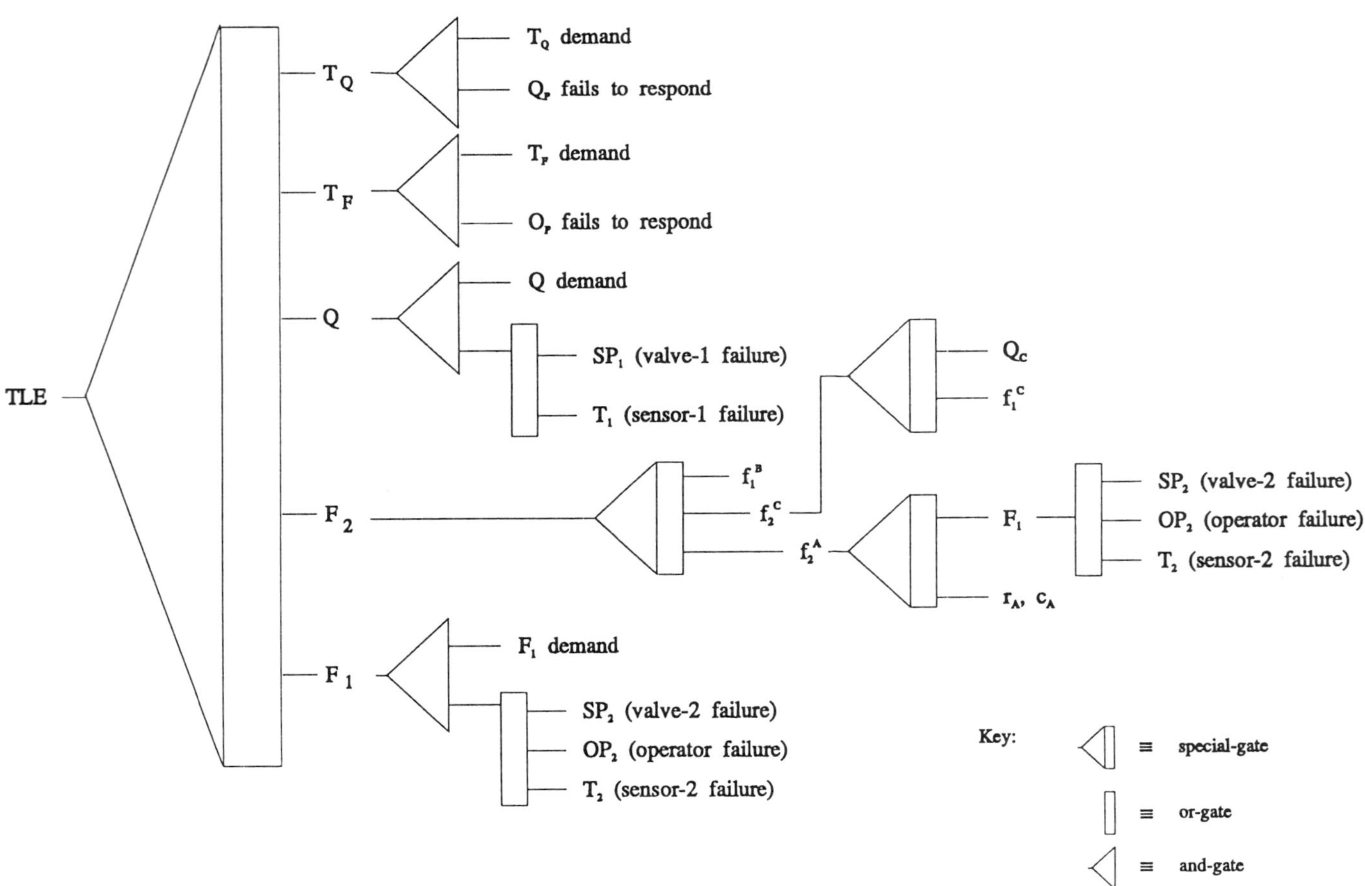

FIG. 24. The topological fault tree resulting from *Assumption-1* and *Assumption-2*.

logical gate constructed from this input variable is an *and-gate*, the result of a controlled variable involving an unbranched node. T_F, feed temperature, is the next variable with a pathway to a top-level event. Similarly, an *and-gate* is constructed for feed temperature, T_F, representing a thermal deviation demand in the feed and the failure of the operator to notice the upset, or take the required action. Quench flow rate, Q, also has a pathway to the top-level event (see Fig. 23). Following the algorithm proposed, construction of logical gates associated with this pathway illustrates that a demand by Q can be enabled in one of two ways: (1) the stem position of valve 1 fails to obtain the necessary position (i.e., the valve fails to close), or (2) the sensor setpoint T_1 is in error.

Unlike T_2, T_F, and Q, the outlet feed, F_2, requires a special gate to model the logical consequences of root causes passing through it. This gate takes as its input f_2^B, f_2^C, and f_2^A. Figure 23 illustrates that f_2^B obtains its input directly from f_1^B, a Type-2 technology. Consequently, any disturbance in f_1^B could potentially enable F_2. Since f_2^C and f_2^A are branched nodes, they in turn require construction of special-gates. The special-gate constructed from f_2^C takes as its input Q, and f_1^C. Similar to f_1^B, f_1^C is a Type-2 technology that takes its input directly from the

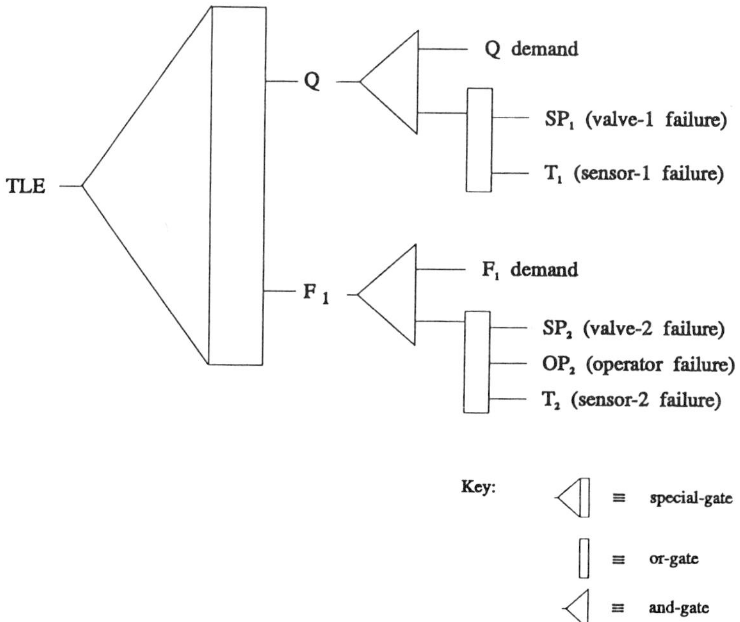

FIG. 25. The topological fault tree resulting from *Assumption-2* and *Assumption-3*.

external world. Therefore, any disturbance in the external surroundings that can lead to a change in the value of f_1^C could potentially also enable the top-level event.

Variable f_2^A, which is a branched node, also requires the construction of a special-gate. This gate requires as its input F_1 and $\{r_A, C_A\}$. Although the expansion for this node is not shown, what is important to recognize is the fact that with the exception of V_{RE}, each of the inputs leading to this set have been covered previously. The implications of changing V_{RE}, a Type-2 technology, is that a change in reactor volume could enable a top-level event. Although the mechanism for this process has not been modeled, it could be the result of an improper design or the accumulation of material internal to the reactor. *What is important is that the effect of V_{RE} has been made explicit.*

The final variable with a pathway leading to the top-level event is F_1, an unbranched node. Since F_1 has associated with it a Type-3 technology as its protective system, a demand on F_1 can be mitigated by the proper positioning of the stem controlling valve 2, proper action by the operator, and a proper setpoint (i.e., T_2) on the temperature controller. A con-

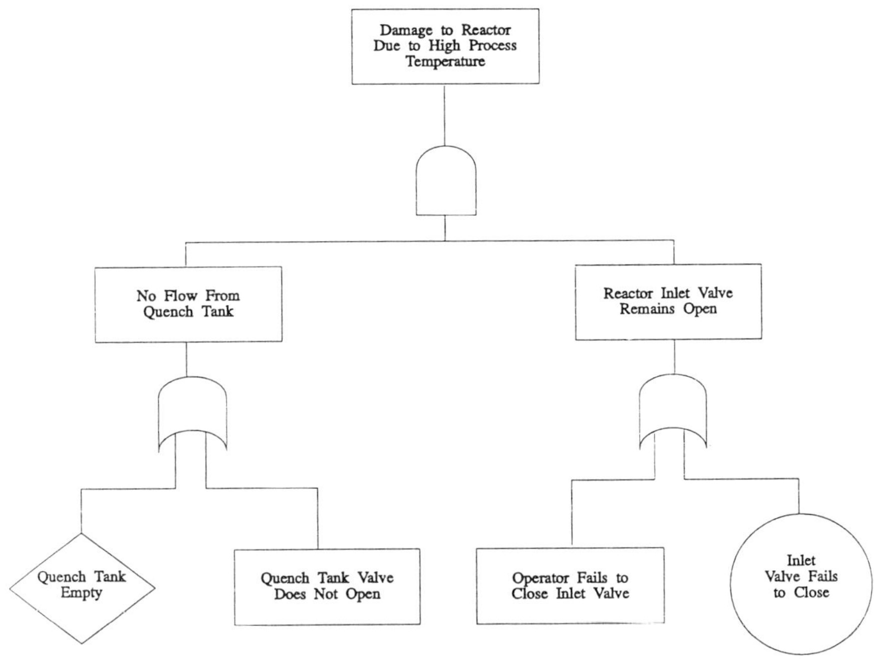

FIG. 26. The industrial implementation of a fault tree for the reactor-quench example.

densed version of the topological fault tree shown in Fig. 24, involving only Type-3 technologies, is shown in Fig. 25. It identifies two principal pathways to the top-level event: Q and F_1. Notice that a special-gate is still needed to connect these inputs to the output, T_R. The necessity of the special-gate will remain until *quantitative* analysis can determine whether the top-level event can be enabled by either Q or F_1, or whether both are required. This depends on the dynamics of the system: reaction rates, time requirements for runaway, heat release rates, heat removal rates, reactor volume, quench rates, etc.

The fault tree cited in literature for this process is shown in Fig. 26 (Battelle, 1985). Notice the similarity between Figs. 26 and 25, particularly in the structure of the two trees, and recognize that as a result of quantitative analysis, Fig. 26 has an *and-gate* as its top-level gate. More importantly, recognize that without complete quantification of the root causes, the fault tree given in Figure 26 may be incomplete.

V. Conclusion

By using the domain-specific modeling languages, LCR and MODEL.LA., we have developed a process-based methodology to identify potential hazards in a chemical process and generate mechanisms for the prevention or mitigation of their effects, through the identification and correction of inherent design weaknesses, or the adaptation of the operating procedures. The methodology is based on an interplay of inductive and deductive reasoning. *Inductive* reasoning has been used in order to identify (1) all potential chemical reactions, which could lead to a hazard, and (2) the requisite conditions that would enable the occurrence of these reactions. *Deductive* reasoning has been used to convert the enabling conditions needed for a reaction into process design or operational "faults" that would constitute the causes for the occurrence of a hazard. The methodology is more efficient, complete, and cost-effective than are current hazard analysis approaches. It contributes three important achievements: (1) formalization of the hazards identification problem, (2) systemization of hazard analysis through all phases of the design process, and (3) construction of a methodology that completely identifies all potential hazards within the scope of the modeling efforts. Furthermore, it establishes a formal strategy for the integration of safety into a design technology at any point in the design process and provides a means for discriminating among design alternatives with respect to disturbance mitigation. Finally, the methodology also provides the basis for the optimization of a design technology with respect to the parameters that describe its inherent safety.

References

Abelson, H., Sussman, G. J., and Sussman, J., "Structure and Interpretation of Computer Programs." MIT Press, Cambridge, MA, 1985.

Atallah, S., Assessing and managing industrial risk. *Chem. Eng.*, September, p. 8 (1980).

Batstone, R., *in* "Proceedings of the International Symposium on Preventing Major Chemical Accidents," February, p. 5.126. AIChE, Washington, DC, 1987.

Battelle, "Guidelines for Hazard Evaluation Procedures." AIChE Press, Washington, DC, 1985.

Boykin, R. F., and Kazarians, M., Quantitative risk assessment for chemical operations. *In* "Proceedings of the International Symposium on Preventing Major Chemical Accidents," February, p. 1.87. AIChE, Washington, DC, 1987.

Brachman, R. J., and Levesque, H. J., "Readings in Knowledge Representation." Morgan Kaufmann Publishers, Los Altos, CA, 1985.

Brannegan, D. P., "Hazards Evaluation in Process Development," Chemical Process Hazards Review, p. 18. American Chemical Society, Washington, DC, 1985.

Bretherick, L., "Bretherick's Handbook of Reactive Chemical Hazards." Butterworth, London, 1990.

Carson and Mumford, Analysis of incidents involving major hazards in the chemical industry. *J. Hazard. Mat.* **3**, 149 (1979).

Cormen, T. H., Leiserson, C. E., and Rivest, R. L., "Introduction to Algorithms." MIT Press, Cambridge, MA, 1990.

Cox, R. A., An overview of hazard analysis. *In* "Proceedings of the International Symposium on Preventing Major Chemical Accidents," February, p. 1.37. AIChE, Washington, DC, 1987.

Culbertson, T. L., and Searson, A. H., "Exxon Facility Design Assessment and Control of Hazards," Exxon internal publication. Exxon, 1983.

Dale, S. E., Cost effective design considerations for safer chemical plants. *In* "Proceedings of the International Symposium on Preventing Major Chemical Accidents," p. 3.79. AIChE, Washington, DC, 1987.

de Groot, J. J., and Van der Elst, F. H., Thermal properties of peroxides. *Inst. Chem. Eng. Symp. Ser.* **68**, 3/V:1 (1981).

Haastrup, P., Design errors in the chemical industry. *Inst. Chem. Eng. Symp. Ser.* **80**, J15 (1983).

Hendrikson, J. B., *J. Am. Chem. Soc.* **108**, 6748 (1986).

Hoffmann, J. M., Chemical process hazard review. *Am. Chem. Soc.*, p. 1 (1985).

ICI, "Process Safety," Course Notes. ICI, 1988.

Kletz, T. A., Make plants inherently safe. *Hydrocarbon Process.*, September, p. 72 (1985).

Kritikos, T., A model for process design automation. Ph.D. Thesis, Massachusetts Institute of Technology, Cambridge, MA (1991).

Lakshmanan, R., and Stephanopoulos, G., Synthesis of operating procedures for complete chemical plants. *Comput. Chem. Eng.* **12**, 985 (1988).

Lees, F. P., "Loss Prevention in the Process Industries." Butterworth, London, 1980.

Lees, F. P., Hazards warning structure: Some implication and applications. *Inst. Chem. Eng. Symp. Ser.* **80**, J1 (1983).

Lowe, D. R., and Solomon, C. H., Hazards identification procedures. *Inst. Chem. Eng. Symp. Ser.* **80**, G8 (1983).

Maher, M. L., *in* "Expert Systems in Engineering" (D. T. Phan, ed.). IFS Publications/ Springer-Verlag, Berlin, 1988.

Mosleh, A., Bier, V. M., and Apostolakis, G., A critique of current practice for the use of expert opinion in probabilistic risk assessment. *Reliab. Eng. Syst. Saf.* **20**, 63 (1988).

Nagel, C. J., Identification of hazards in chemical process systems. Ph.D. Thesis, Massachusetts Institute of Technology, Cambridge, MA (1991).

Ozog, H., Hazard identification analysis and control. *Chem. Eng.* (*N.Y.*). February 18, p. 161 (1987).

Ozog, H., and Bendixen, L. M., Hazard identification and quantification. *Chem. Eng. Prog.*, April, p. 55 (1987).

Perkins, J. D., and Barton, G. W., Modelling and simulation in process operation. *In* "Foundations of Computer-Aided Process Operations" (G. V. Reklaitis and H. D. Spriggs, eds.). CACHE Corp., Austin, TX, and Elsevier, New York, 1987.

Sheil, B., Power tools for programming. *Datamation*, February, p. 131 (1983).

Sheil, B., The artificial intelligence tool box. *In* "Artificial Intelligence Applications in Business" (W. Reitman, ed.), p. 113. 1984.

Slater, C., and Pitblado, "Major Industrial Hazards Project Report." The Warren Centre for Advanced Engineering, University of Sidney, Sidney, Australia, 1987.

Sriram, D., and Maher, M. L., *in* "Applications of Artificial Intelligence in Engineering Problems" (D. Sriram and R. Adey, eds.), Vol. 1. Southhampton University, UK 1986.

Stephanopoulos, G., The future of expert systems. *Chem. Eng. Prog.*, September, p. 44 (1987).

Stephanopoulos, G., Johnston, J., Kritikos, T., Lakshmanan, R., Mavrovouniotis, M., and Siletti, C., Design-kit: An object-oriented environment for process engineering. *Comput. Chem. Eng.* **11**, 655 (1987).

Stephanopoulos, G., Johnston, J., and Lakshmanan, R., An intelligent system for planning plant-wide process control strategies. *Journal A* **29**(3), 81 (1988).

Stephanopoulos, G., Henning, G., and Leone, H., MODEL.LA. A modeling language for process engineering. Part I. *Comput. Chem. Eng.* **14**, 813 (1990a).

Stephanopoulos, G., Henning, G., and Leone, H., MODEL.LA. A modeling language for process engineering. Part II. *Comput. Chem. Eng.* **14**, 847 (1990b).

Stoessel, F., Experimental study of thermal hazards during the hydrogenation of aromatic nitro compounds. *Proc. Int. Symp. Loss Prev. Saf. Promot. Process Ind.*, *6th*, p. 77-1 (1989).

Walling, C., "Free Radicals in Solution." Wiley, New York, 1957.

SEARCHING SPACES OF DISCRETE SOLUTIONS: THE DESIGN OF MOLECULES POSSESSING DESIRED PHYSICAL PROPERTIES

Kevin G. Joback[1] and George Stephanopoulos

Laboratory for Intelligent Systems in Process Engineering
Department of Chemical Engineering
Massachusetts Institute of Technology
Cambridge, MA 02139

Strings of letters form words. From words to verses and stanzas, a poet composes a work with its own dynamic behavior, such as emotional impact on the reader, that transgresses the character of its components. In an analogous manner, atoms form functional groups, and these, in turn, yield molecules with distinct behavior, e.g., physical properties. It takes a

---

[1]Present address: Molecular Knowledge Systems, Inc., Nashua, New Hampshire, USA.

Homeric or Shakespearean genius to convert letters to an epic with a predefined desired impact. It suffices to efficiently search a space of combinatorial alternatives, in order to identify the molecules that satisfy the desired constraints on a set of physical properties. Often, the requisite scientific knowledge is fragmented, dispersed, and nonformalized, making the deductive search for the desired molecules inefficient or impossible. The inductive "genius" of a scientist or engineer is needed to break the impasse in such cases. By evolution or revolution one needs to respond to tighter and shifting product specifications and identify new solvents, pharmaceuticals, imaging chemicals, herbicides and pesticides, refrigerants, polymeric materials, and many others. In this chapter we will sketch the characteristics of an intelligent, computer-aided tool to support the synthetic search for the desired molecules. With functional groups as the "letters" of an alphabet, automatic and interactive procedures compose and screen classes of potential molecules. The automatic synthesis algorithm defines and searches the space of discrete solutions (molecules) through a hierarchical sequence of the space's representations. At each level of detail, a set of explicit constraints is used to depict restrictions on the structure of molecules that can be generated from various combinations of functional groups. In addition, interval arithmetic is employed to test the satisfaction of design specifications, leading to the elimination of large classes of infeasible molecules. One, though, should never overestimate the effectiveness of search algorithms in locating the desired solution(s). Quite frequently we need to resort to human-driven, abductive jumps. In this chapter we will also describe how the automatic search can become interwoven with effective human–machine interaction. Thus, the resulting computer-aided tool, the *Molecule-Designer*, constitutes a paradigm of an intelligent system with two distinct but integrated and complementary capabilities. Examples on the synthesis of refrigerants, solvents, polymers, and pharmaceuticals will illustrate the logic and features of the design procedures in the *Molecule-Designer*.

I. Introduction

Physical properties have a major impact on the economics of many processes and the viability of many products. The refrigerant in a refrigeration cycle, the working fluid in a power cycle, and the solvent used in an azeotropic distillation all determine the physical and economic feasibility of the corresponding processes. Chemical products such as artificial sweeteners, lubricants, and textiles all must exhibit specific physical properties if they are to be accepted.

Until recently, the identification of compounds possessing desired physical property values required extensive experimental search through vast numbers of candidate molecules. Estimates indicate that 3000–5000 compounds need to be tested before a new, useful pharmaceutical is identified, and 5000–8000 to find a new pesticide (Verloop 1972). Computational estimation of the physical property values for candidate molecules can reduce significantly the need for experimentation. In this spirit, Horvath (1992) published a tremendous thesaurus of techniques and approaches for molecular design, by focusing on the estimation of physical properties from molecular structures. Numerous techniques are available for the computational estimation of thermodynamic properties (Reid *et al.*, 1987), environmental properties (Lyman *et al.*, 1982), polymer properties (van Krevelen 1976), biological activity (Martin, 1978; Hansch and Leo, 1979; Franke, 1984), phase equilibria (Fredenslund *et al.*, 1977), and others. However, these analytic methods can not be directly used in a synthetic manner for the design of molecules, as the following example illustrates:

Consider the design of a molecule whose physical properties, PP_1, PP_2, and PP_3, should have the values α, β, and γ, respectively. If functions f_1, f_2, and f_3 relate the three physical properties to the molecular structure, then

$$PP_1 = f_1(\,molecular\ structure\,) = \alpha,$$

$$PP_2 = f_2(\,molecular\ structure\,) = \beta, \qquad (1)$$

$$PP_3 = f_3(\,molecular\ structure\,) = \gamma.$$

The molecular structure of the unknown chemical could be found by inverting these three relationships. However, an explicit inversion is not analytic (the molecular structure is described by integer variables denoting the presence or absence of specific atoms and bonds), and it accepts multiple solutions (there may be several molecules satisfying the constraints). Implicit inversion of Eqs. (1) is possible through the formulation of appropriate optimization problems. However, in such cases the complexity and nonlinear character of the functional relationships used to estimate the values of physical properties in conjunction with the integer variables description of molecular structures, yield very complex mixed-integer optimization formulations.

Thus it is not surprising that the first efforts in systematically designing molecules possessing desired physical properties were heuristic in character, focusing on specific classes of chemical products. Godfrey (1972) used an empirical miscibility scale to determine whether two liquids were miscible. Francis (1944) used critical solution temperatures to choose solvents for the selective extraction of hydrocarbons. Berg (1969) developed a hydrogen bond classification scheme used to identify azeotropic distillation solvents. In recent years, the sophistication of the methods has improved (Venkatasubramanian *et al.*, 1994; Constantinou *et al.*, 1994; Gani *et al.*, 1991; Gani and Fredenslund, 1993; Nielsen, *et al.*, 1995), but the basic character of the various approaches has remained the same, specifically, all design-oriented techniques, (1) are *problem-specific*, i.e., for solvents, polymers, or pharmaceuticals; (2) are based on the *generate-and-test* paradigm with experiential heuristics employed to reduce the search space of potential alternatives; and (3) cannot utilize efficiently all available knowledge in a given area of molecular design.

A. Brief Review of Previous Work

Research findings from the areas of physical property estimation and chemical products selection are applicable to the design of molecules. A brief review of research in these areas is presented along with previous work in molecular design.

1. Estimation of Physical Properties

Every approach developed for the design of molecules with desired properties requires the estimation of physical properties. Some property estimation techniques lend themselves easily to the identification of the requisite molecular structures, starting from the desired design specs, and others do not. Depending on how various approaches attempt to relate molecular structure to physical properties, they can be grouped into five categories: pattern recognition, topological, group contribution, equation-oriented, and molecular-modeling-based techniques.

a. Pattern Recognition. Discriminant analysis and classification are the two statistical techniques used most often in pattern recognition. Both are multivariate techniques concerned with *separating* distinct sets of objects into classes and *allocating* new objects to previously defined classes. A discriminant function is developed from a set of experimental data called the "training set." This function is then used to classify new compounds. In many applications the number of classes equals 2; e.g., carcinogenic or noncarcinogenic, toxic or nontoxic.

Pattern recognition techniques lend themselves to synthetic designs of molecules for those problems for which the design specs require that a molecule is a member of a certain class.

b. Topological Techniques. These techniques ignore the actual three-dimensional shape of a molecule, the nature and lengths of the chemical bonds connecting its atoms, the angles between the bonds, and sometimes even atom types (Rouvray, 1986). Typically only the number of atoms and their interconnections are considered. This information is reduced to an *index* such as the Wiener Path Number (Wiener, 1947), Alternburg Polynomial Index (Alternburg, 1966), Gordon–Scantlebury Index (Gordon and Scantlebury, 1964), Hosoya's Z Index (Hosoya and Murakami, 1975), or Randić's Branching Index (Randić, 1975). These indices are then used to correlate the values of physical properties. The applicability of these techniques for property estimation has been extensively discussed in Kier and Hall (1986). By their nature, topological techniques require detailed information about the molecular connectivity of a compound and are difficult to incorporate into synthetic design procedures.

c. Group Contribution Techniques. These assume that each fragment of a molecule contributes a certain amount to the value of its physical properties. Contributions for each group are statistically regressed from large sets of experimental data. Techniques can become very complex, including nonlinear effects and interactions among groups. They are very appropriate for the design of molecules with desired physical properties and they constitute the basis for both *interactive* and *automatic* design procedures described in this chapter. Group contribution techniques have been used by other researchers for molecular design, as we will see in the next section.

d. Equation-Oriented Techniques. These techniques correlate estimated physical properties to properties more easily available or measured using empirical or theoretical models. Not relating a compound's molecular structure to its properties, these techniques cannot be used directly for molecular design. However, used in conjunction with group contribution techniques, rendering the values of the correlated physical properties, they broaden the list of physical properties which yield the specifications of the desired molecule.

e. Molecular-Modeling-Based Techniques. These techniques start with an atomic model of a molecule and use quantum and statistical mechanics to estimate its physical properties. Many of these estimates are more accurate than those obtained by any other estimation technique. Molecular-modeling-based techniques offer fairly complex and implicit relationships between molecular structure and physical properties. A straightforward generate and test is the only way such techniques are employed.

2. Selection of Desired Chemicals

Selecting a chemical product from a set of candidates is a two-step procedure. The first step is the most critical and involves the identification of those physical properties that are important to the performance of the chemical product and their values that give optimal performance. The second step involves a search through a database for existing compounds that possess these physical properties values. Unknown property values must be estimated. Such an approach has the advantage of being fast. Additionally, compounds in the database are, typically, commercially available or can be readily synthesized. The drawback of such an approach is that new compounds cannot be found.

3. Design of a Desired Chemical

This is also a two-step procedure similar to that of selecting a desired chemical from a list of candidate molecules. Unlike the selection from an existing set of compounds, the design of compounds implies the synthetic stipulation of molecules for which there are no experimental data of their physical properties. Therefore, using some of the available estimation techniques, different approaches have been proposed in the past and will be discussed in the following paragraphs.

a. Design of Solvents. Gani and Brignole (1983) and Brignole *et al.* (1986) used the UNIFAC (Fredenslund *et al.*, 1977) group contribution method to synthesize molecular structures with specific solvent properties for separation processes. Their synthesis procedure is divided into three steps:

1. Select the groups considered to be suitable building blocks for the molecular structures.
2. Combine the groups into candidate molecules according to specified combination rules.
3. Screen the candidate molecules using UNIFAC to evaluate their usefulness for a particular separation task.

To reduce the number of potential group combinations to a tractable number, several additional constraints are placed on the candidate solvents. For example, a high boiling point is required in order to facilitate simple separation of the solvent by distillation. In similar spirit are the works of Gani *et al.* (1991) and Gani and Fredenslund (1993), but the efficiency of search has improved with heuristic knowledge, while techniques for discrete optimization have been used to design optimal solvent mixtures. The works of Macchietto *et al.* (1990) and Odele and Macchietto (1993) have focused on the selection (rather than design) of optimal solvents for extractive separation processes.

b. Design of Polymers. Derringer and Markham (1985) proposed a generate and test methodology for designing polymers possessing desired physical properties, using the van Krevelen (1976) group contribution estimation techniques. Recognizing that the number of candidate polymers may be large, Derringer and Markham devised a ranking procedure that includes a desirability measure for each predicted property.

c. Design of Polymer Coatings. Tortorello and Kinsella (1983a, b) used the solubility parameter concept to design high-performance aircraft coatings resistant to water, fuels, hydraulic fluids, and lubricating oils. Additional design specs included resistance to high temperature and flexibility at low temperature.

d. Design of Drugs. Drug design has been the most active area for the development of systematic procedures to identify new chemical products. Beginning with a small set of experimental data on the efficacy of candidate compounds, one derives a statistical relationship between the drug's potency and a set of physicochemical properties such as Hammett's constant (1935), Taft's (1956) steric parameter, and the octanol–water partition coefficient, whose use was made popular by Hansch and co-workers (1963). These physicochemical properties are then related to structural characteristics. Such statistical relationships are called *quantitative structure activity relationships* (QSARs).

QSARs provide great insight into the drug design problem. Examination of the derived relationships often indicates how a drug's potency is being affected by reactive, transport, and steric considerations. To improve the potency, a drug designer can search for substituents, which when added to the candidate molecule will reduce steric hindrances or increase the rate of transport. Extensive tabulations exist (Hansch and Leo, 1979) listing the effect that specific substituents have on the various physical-chemical properties, typically used in drug designs.

B. General Framework for the Design of Molecules

The previous works on selecting and designing molecules with desired properties share certain common characteristics. Beginning with these characteristics, a general methodology was developed for designing molecules (Joback and Stephanopoulos, 1990). The overall philosophy of the design methodology is described in the following six paragraphs.

1. Problem Formulation

The first step in any design is to identify the target (Stephanopoulos and Townsend, 1986). A molecular design target consists of physical property, chemical, and structural constraints. Physical property constraints are typically concerned with the performance of the chemical product, such as its ability to perform as an aircraft coating. Chemical constraints are often related directly to molecular structure, restricting or requiring the occurrence of functional groups, such as the desire to design a diol with desired properties. Structural constraints are required when a molecule is constructed by assembling functional groups. The groups must have the correct type and occurrence of bonds so that they can be assembled into feasible molecules. Brignole *et al.* (1986) developed an extensive set of rules to constrain the choice of groups and ensure the structural feasibility of the resulting molecule. The effective formulation of the molecular design problem is crucial to the success of the design.

2. Target Transformation

For the computer to evaluate the performance of a candidate molecule, it must be able to estimate the values for those physical properties identified during problem formulation. The target transformation step develops *estimation procedures*, which enable the evaluation of the target constraints' physical properties in terms of the values of new physical properties (i.e., the transformed target). For example, if we want to design a molecule with vapor pressure P_{vp}, in a given range of values, we can use the Riedel–Plank–Miller correlation [Eq. (2)] and transform the target into these new properties, as shown by Equations (3a), (3b), and (3c):

$$P_{vp} = P_{vp}(T_b, T_c, P_c) \qquad \text{(correlation by Riedel–Plank–Miller)}, \quad (2)$$

where

$$T_b = T_b(\text{molecular structure})$$

$$\text{(group-contribution technique by Joback)}, \quad (3a)$$

$$P_c = P_c(\text{molecular structure})$$

$$\text{(group-contribution technique by Lydersen)}, \quad (3b)$$

$$T_c = T_c(\text{molecular structure})$$

$$\text{(group-contribution technique by Fedors)}. \quad (3c)$$

Starting with a compound's molecular structure, the T_b, P_c, and T_c are estimated first, using the three group contribution techniques indicated above. These values are then used in an equation-oriented technique to yield the final estimate for P_{vp}.

3. Design Procedure

The previous design approaches are based on the generate-and-test paradigm. This paradigm consists of two parts: the *generator* and the *tester*. The generator enumerates candidate solutions, whereas the tester evaluates each candidate and either accepts or rejects it. When the number of candidates becomes very large, exhaustive enumeration becomes impractical. Although not all previous molecular design approaches have explored them, several strategies are available to manage the search space (Hayes-Roth *et al.*, 1983), such as (1) move the tester into the generator, (2) prune partial solutions, and (3) abstract the search space. The last strategy is extremely powerful in managing the combinatorics of the design problem and constitutes the basis of the automatic design methodology to be discussed in Section II of this chapter.

4. Representation and Enumeration of Molecules

Designing molecules through the use of group-contribution estimation techniques results in candidate molecules which are represented as a collection of functional groups. To form complete molecules, it is necessary to connect these groups together. At times more than one way of connecting the groups is possible. For example, the following collection of groups

$$-CH_3 \quad {>}C{<} \quad -F \quad -F \quad -Cl \quad {>}C{=} \quad {=}CH_2$$

TABLE I

FOUR ENUMERATED MOLECULES

$$
\begin{array}{cc}
\underset{\displaystyle\underset{Cl}{|}}{CH_3-\overset{\displaystyle\overset{F}{|}}{C}-}\overset{\displaystyle\overset{F}{|}}{C}=CH_2
&
\underset{\displaystyle\underset{F}{|}}{CH_3-\overset{\displaystyle\overset{Cl}{|}}{C}-}\overset{\displaystyle\overset{F}{|}}{C}=CH_2
\\[3em]
\underset{\displaystyle\underset{F}{|}}{CH_3-\overset{\displaystyle\overset{F}{|}}{C}-}\overset{\displaystyle\overset{Cl}{|}}{C}=CH_2
&
\underset{\displaystyle\underset{Cl}{|}}{F-\overset{\displaystyle\overset{F}{|}}{C}-}\overset{\displaystyle\overset{CH_3}{|}}{C}=CH_2
\end{array}
$$

can be combined, ignoring stereoisomers, to form the four different molecules shown in Table I. Molecule enumeration has been extensively investigated by researchers doing work in structure elucidation (Gray, 1986). Before more rigorous estimation techniques, such as molecular modeling, or chemical constraints can be used, it is necessary to provide techniques for the representation and enumeration of all potential molecular structures.

5. Screening of Molecules

Chemical constraints are applied once the satisfactory candidate molecules have been enumerated. Typically, chemical constraints prevent the generation of unstable substructures within the structure of generated molecules. For example, if the substructure $-O-O-$ occurs in a compound desired to be stable, that compound is pruned. For design procedures based on group-contribution techniques, the application of chemical constraints must occur after enumeration, since the relative locations of the various groups within a molecule remain unspecified at earlier stages.

6. Final Evaluation

It is sometimes necessary to modify the physical properties estimation techniques employed by generate-and-test design procedures. Often this modification is introduced in order to remove computational steps in the estimation techniques that require knowledge of the global molecular structure. Using groups as the design basis, only the partial and local structure of a molecule is known during the design. However, once the candidate molecules are enumerated and screened, global molecular structure is known and more accurate estimation techniques can be used to further prune the candidates.

II. Automatic Synthesis of New Molecules

This section presents an algorithmic strategy for the automatic generation of molecules and their screening against a set of constraints, which represent the physical properties' values that the desired molecules should satisfy. *Functional groups* are the essential building blocks for the construction of molecules, allowing the use of *group-contribution estimation techniques* for the testing of physical property constraints. These techniques are simple, fast, and yield estimates of sufficient accuracy for preliminary screening purposes. The use of more complicated estimation techniques, such as molecular modeling, is unwise at this stage of design. The probability of an ab initio automatic identification of satisfactory molecules at this stage is low, implying that the effort expended for the examination of each candidate molecule should be kept to a minimum. Thus, a hierarchical approach has been adopted for the synthesis of desired molecules:

Phase 1. Simple estimation techniques are used to rapidly screen large number of candidate molecules, generated by a hierarchical search algorithm.

Phase 2. Candidate molecules satisfying the design constraints are reported to the human designer, who orders them using subjective preferences.

Phase 3. The retained molecules are evaluated through the use of more detailed estimation techniques, e.g. molecular modeling, complex equations of state.

In this chapter we will deal only with phase 1.

A. THE GENERATE-AND-TEST PARADIGM

The generate-and-test search paradigm, used by the automatic design algorithm for the synthesis of molecules, is composed of two modules. The first module, the *generator*, enumerates candidate molecules. The second, the *tester*, evaluates each molecule, estimating its physical properties and checking for structural feasibility, and either accepts or rejects it. We represent molecules as collections of groups, e.g., chloropropane is represented as ($-$Cl $-$CH$_3$ $-$CH$_2-$ $-$CH$_2-$). The generator simply constructs candidate molecules by selecting a collection of groups from an initial set. The representation of groups allows the tester to use group-contribution techniques to estimate physical property constraints and check the design constraints.

The combinations of groups that can be selected is infinite. However, from practical considerations, molecules for a typical application fall within some size range, which can be translated into an upper limit on the number of groups chosen. For example, refrigerants are generally of a small molecular weight. Placing a limit of 15 on the number of groups that can be used to form a molecule is a reasonable bound. A lower limit is established from the fact that at least two groups must be used to form a structurally feasible molecule. With limits on the minimum and maximum number of groups that can be chosen, the generator selects collections of groups beginning with all combinations of two groups, then all combinations of three groups, and so on until the upper limit is reached.

1. Design Constraints

The tester module checks each candidate molecule for satisfaction of the design constraints. Three types of constraints are used: (a) physical property constraints, (b) structural constraints, and (c) chemical constraints.

a. Physical Property Constraints. Estimation procedures are established for each physical property used in the design constraints. An estimation procedure is a collection of estimation techniques that determine physical property values when only the molecular structure is known. It is composed of group-contribution and equation-oriented estimation techniques. For example, using the constraint on vapor pressure, we obtain

$$P_{vp}(273 \text{ K}) > 1.01 \text{ bar.}$$

We need to employ an estimation procedure for P_{vp} and the associated independent physical properties, such as those described in Section I.B [see Eqs. (2) and (3a–c)], in order to determine each candidate molecule's vapor pressure at 273 K. Starting with a compound's molecular structure, T_b, T_{br}, and P_c are first estimated using the three group-contribution techniques [Eqs. (3a), (3b), and (3c), respectively]. These values are then used in an equation-oriented technique to yield the final estimate for P_{vp} [Eq. (2)].

b. Structural Constraints. Structural constraints determine whether a collection of groups can be connected in some manner to form a feasible molecule. Three requirements define the conditions for structural feasibility:

1. All groups in the collection should be able to be joined into a single connected component. The collection of groups $(-F \quad -F \quad -F \quad -F)$ is not feasible because it does not form a single molecule.
2. The single connected molecule formed from a set of groups cannot have any unconnected bonds. Connecting the groups $(-CH_2- \quad -F)$ gives us a single structured entity with one single bond unconnected.
3. The connections made in the single connected entity, formed from a set of groups, must all be between bonds of the same type. Single bonds may only connect with single bonds, double with double, etc.

Given a set of groups, it is possible to enumerate all ways in which they could be connected verifying that at least one candidate satisfies the three requirements. However, a graph theoretic examination of molecular structures provides a set of structural constraints that are much easier to apply. Such structural constraints have been developed and are shown in Table II.

c. Chemical Constraints. Chemical constraints are heavily dependent on the global connection of atoms within a molecule. Representing molecules as collections of groups does not provide knowledge about global connectivity. Chemical constraints are thus better used at later stages of the search methodology, where the complete structure of a molecule is known.

2. Combinatorial Explosion

Given a reasonable number of groups from which molecules can be constructed, the number of candidates that can be generated is extremely large. Allowing repetition of groups and ignoring the order of selection, the number of candidate molecules that can be generated by selecting n groups from a set of k groups is given by Eq. (4):

$$C^R(k,n) = \frac{(k+n-1)!}{n!(k-1)!}. \tag{4}$$

The total number of candidate molecules that can be selected from a set of k groups in which each candidate molecule has between 2 and n_{max}

TABLE II

STRUCTURAL CONSTRAINTS ON FORMING FEASIBLE MOLECULES

1. If G is a collection of n groups, then $n \geq 2$.
2. If G is a collection of n groups with n_c cyclic groups, n_m mixed groups, and n_a acyclic groups, and $n_a > 0$ and $n_c > 0$, then $n_m > 0$.
3. If G is a collection of n groups with n_c cyclic groups, n_m mixed groups, and n_a acyclic groups, then $n_m > 0$ implies that $n_a > 0$ or $n_c > 0$.
4. If G is a collection of n groups with n_c cyclic groups and n_m mixed groups, then either $n_c + n_m \geq 3$ or $n_c + n_m \geq 2$.
5. If G is a collection of groups, then the number of groups having an odd number of free bonds must be even.
6. If G is a collection of n groups with b free bonds, then $\frac{b}{2} \geq n - 1$.
7. If G is a collection of n groups with b free bonds, then $\frac{b}{2} \leq \frac{1}{2}n(n-1)$.
8. If a collection of groups contains more than one bond type, then there must be a transition group containing each bond type. A transition group is one that contains more than one bond type.
9. If G is a collection of groups with $n_{a,i}$ denoting the number of acyclic groups with a valence i and v_{mj} denoting the valence of some jth mixed group, then $n_1 \leq \Sigma_{\text{mixed}}(v_{m,j} - 2) + n_{a,3} + 2n_{a,4} + \cdots + (i-2)n_{a,i} + \cdots$.
10. If G is a collection of n groups with n_i denoting the number of groups with a global valence i and all n groups are acyclic, then $n_1 = 2 + n_3 + 2n_4 + \cdots + (i-2)n_i + \cdots$.
11. The number of occurrences of each bond type in a collection of groups must be even.

groups is given by Eq. (5):

$$\text{Total candidates} = \sum_{n=2}^{n_{\max}} C^R(k,n) = \sum_{n=2}^{n_{\max}} \frac{(k+n-1)!}{n!(k-1)!}. \tag{5}$$

Table III shows how this total number of candidates can quickly grow to very large values. Managing this combinatorial explosion is the major focus of the automatic design algorithm.

TABLE III

COMBINATORICS OF GROUP SELECTION
($k = 40$ GROUPS)

$n_{\max}$	# Molecules
4	135,710
5	1,221,718
6	9,366,778
7	62,891,458
8	377,348,953
9	2,054,455,593

TABLE IV
INITIAL SET OF GROUPS

$>\!CH_3$	$-CH_2-$	$>\!CH-$	$>\!C\!<$
$=CH_2$	$=CH-$	$=C\!<$	$=C=$
$\equiv CH$	$\equiv C-$	$-F$	$-Cl$
$-Br$	$-I$	$-OH$	$-O-$
$>\!CO$	$-CHO$	$-COOH$	$-COO-$
$=O$	$-NH_2$	$>\!NH$	$>\!N-$
$-CN$	$-NO_2$	$-SH$	$-S-$

B. THE SEARCH ALGORITHM

The magnitude of the combinatorial problem, resulting from a large number of functional groups, can be reduced only by reducing the number of groups. This reduction is done by abstracting the groups into families of groups, called *metagroups*. Table V shows the groups of Table IV clustered into four metagroups.

Instead of generating molecules by choosing from an initial set of groups, we choose from an initial set of metagroups. The candidate molecules formed from a collection of metagroups are called *metamolecules*, which are sets of molecules. Using the metagroups from Table V, the metamolecule (2 1 0 0) is the set of all molecules that can be formed by taking any two groups from *metagroup 1* and any one group

TABLE V
EXAMPLE METAGROUPS

Metagroup 1:
$\{-CH_3,\ =CH_2,\ \equiv CH,\ -F,\ -Cl,\ -Br,\ -I,\ -OH,\ -CHO,\ -COOH,\ =O,\ -NH_2,\ -NO_2,\ -CN,\ -SH\}$

Metagroup 2:
$\{>\!CH_2,\ =CH-,\ =C=,\ \equiv C-,\ >\!CO,\ -COO-,\ -O-,\ >\!NH,\ -S-\}$

Metagroup 3:
$\{=C\!<,\ >\!CH-,\ >\!N-\}$

Metagroup 4:
$\{>\!C\!<\}$

from *metagroup* 2. The number of molecules contained in metamolecule (2 1 0 0) is

$$C^{\mathrm{R}}(15,2) \times C^{\mathrm{R}}(9,1) = \frac{(15+2-1)!}{2!(15-1)!} \times \frac{(9+1-1)!}{1!(9-1)!},$$

or 1080 molecules.

1. Evaluation of Metamolecules

Abstracting groups into metagroups reduces the combinatorics of candidate molecule generation. However, we must be able to efficiently evaluate whether a metamolecule satisfies the design constraints.

a. Structural Constraints. As long as all the groups within each metagroup have a consistent molecular characteristic such as global valence, the structural constraints are still applicable. Metagroups 1 and 2 are consistent in ring class and global valence. Structural constraint 10 of Table II is thus applicable. Applying the constraint to metamolecule (2 1 0 0) yields

$$2 = 2 + 0 + 2(0) = 2,$$

i.e. the constraint is satisfied. This implies that each of the 1080 molecules contained in (2 1 0 0) satisfies the constraint.

b. Physical Property Constraints. Associated with each of the groups in a metagroup is a contribution toward a particular physical property. The contribution of a metagroup is called a *metacontribution* and is defined by a set of values. Table VI shows the contributions toward the value of the boiling point, T_{b}, for each of the groups in metagroup 2 (see Table V) toward T_{b}. The metacontribution of metagroup 2 toward T_{b} is thus the following set of values:

$$(22.42 \quad 22.88 \quad 24.96 \quad 26.15 \quad 27.38 \quad 50.17 \quad 68.78 \quad 76.75 \quad 81.10).$$

To use metacontributions in the calculation of physical properties, it is necessary to find a representation that can capture the set value of the metacontributions and can be manipulated by mathematical operators. Interval numbers were chosen as the representation.

TABLE VI

T_b Group Contributions for Metagroup 2
(Acyclic Groups)

Groups	Contribution
$-CH_2-$	22.88
$=CH-$	24.96
$=C=$	26.15
$\equiv C-$	27.38
$-O-$	22.42
$\diagdown CO \diagup$	76.75
$-COO-$	81.10
$\diagdown NH \diagup$	50.17
$-S-$	68.78

2. Interval Arithmetic and Meta-Contributions

The generalization of ordinary arithmetic to closed intervals is known as *interval arithmetic*. An interval is defined as a closed bounded set of real numbers (Moore, 1979):

$$X = \left[\underline{X} \quad \overline{X} \right] = \left\{ x \mid \underline{X} \le x \le \overline{X} \right\}. \tag{6}$$

Thus, intervals have a *dual* nature as both a number and a set. The basic interval arithmetic operations are

$$\left[\underline{X} \quad \overline{X} \right] + \left[\underline{Y} \quad \overline{Y} \right] \equiv \left[\underline{X} + \underline{Y} \quad \overline{X} + \overline{Y} \right],$$

$$\left[\underline{X} \quad \overline{X} \right] + \left[\underline{Y} \quad \overline{Y} \right] \equiv \left[\underline{X} - \overline{Y} \quad \overline{X} - \underline{Y} \right],$$

$$\left[\underline{X} \quad \overline{X} \right] * \left[\underline{Y} \quad \overline{Y} \right] \equiv \left[\min\left(\underline{X} * \underline{Y}, \quad \underline{X} * \overline{Y}, \quad \overline{X} * \underline{Y}, \quad \overline{Y} * \overline{Y} \right), \right.$$

$$\left. \max\left(\underline{X} * \underline{Y}, \quad \underline{X} * \overline{Y}, \quad \overline{X} * \underline{Y}, \quad \overline{X} * \overline{Y} \right) \right],$$

$$\left[\underline{X} \quad \overline{X} \right] \div \left[\underline{Y} \quad \overline{Y} \right] \equiv \left[\underline{X} \quad \overline{X} \right] * \left[1/\overline{Y} \quad 1/\underline{Y} \right] \quad \text{iff } 0 \notin \left[\underline{Y} \quad \overline{Y} \right].$$

The metacontribution of the metagroup 2 (see Table VI) in interval representation is [22.42 81.10]. Thus, we can construct Table VII, which shows the metacontributions for each metagroups, displayed in Table V for boiling point T_b, reduced boiling point T_{br}, and heat of vaporization ΔH_{vb} (Joback and Reid, 1987). Using the group-contribution estimation

TABLE VII

METACONTRIBUTIONS

Metagroup	T_b		T_{br}		ΔH_{vb}	
1	[−10.50	169.09]	[0.0027	0.0791]	[−0.670	19.537]
2	[22.42	81.10]	[0.0020	0.0481]	[2.205	9.633]
3	[11.74	24.14]	[0.0117	0.0169]	[1.691	2.138]
4	[18.25	18.25]	[0.0067	0.0067]	[0.636	0.636]

models

$$T_b = 198.18 + \sum_{\text{all groups}} n_i \Delta_{i,T_b}, \tag{7}$$

$$T_{br} = 0.584 + 0.965 \sum_{\text{all groups}} n_i \Delta_{i,T_{br}} - \left(\sum_{\text{all groups}} n_i \Delta_{iT_{br}} \right), \tag{8}$$

$$\Delta H_{vb} = 15.30 + \sum_{\text{all groups}} n_i \Delta_{i,\Delta H_{vb}}, \tag{9}$$

we can estimate the values of T_b, T_{br}, and ΔH_{vb} for metamolecule (2 1 0 0) as follows:

$$T_b = 198.18 + 2[-10.50 \quad 169.09] + [22.42 \quad 81.10]$$

$$= [199.6 \quad 617.46]\text{K}, \tag{10}$$

$$T_{br} = 0.584 + 0.965(2[0.0027 \quad 0.0791] + [0.0020 \quad 0.0481])$$

$$- (2[0.0027 \quad 0.0791] + [0.0020 \quad 0.0481])^2$$

$$= [0.549 \quad 0.783], \tag{11}$$

$$\Delta H_{vb} = 15.30 + 2[-0.670 \quad 19.537] + [2.205 \quad 9.633]$$

$$= [16.165 \quad 64.007] \text{ kJ/mol.} \tag{12}$$

The intervals given by Eqs. (10)–(12) span the range of physical property values possessed by each of the 1080 molecules in metamolecule (2 1 0 0).

Interval values for these fundamental properties can be used in equation-oriented estimation techniques. The Watson relation (Watson, 1943)

$$\Delta H_v = \Delta H_{vb} \left(\frac{1 - T/T_c}{1 - T_{br}} \right)^{0.38} \tag{13}$$

is used to estimate the enthalpy of vaporization at 250 K for metamolecule

$(2\ 1\ 0\ 0)$, where T_c is obtained from

$$T_c = \frac{T_b}{T_{br}} = \frac{[199.6 \quad 617.46]}{[0.549 \quad 0.783]} = [254.9 \quad 1124.7].$$

Inserting the value of T_c into Eq. (13), we obtain the interval value of the enthalpy of vaporization:

$$\Delta H_v = [16.165 \quad 64.007]\left(\frac{1 - 250/[254.9 \quad 1124.7]}{1 - [0.549 \quad 0.783]}\right)^{0.38}$$

$$= [0.688 \quad 229.40]\ kJ/mol.$$

3. Searching through Successive Molecular Abstractions

The generate-and-test search paradigm described earlier must now be modified to deal with the abstractions introduced by the metamolecules. Instead of generating and testing individual molecules, we generate and test metamolecules. Those metamolecules satisfying the test are reduced in abstraction, by dividing a metagroup into more meta-groups. This refinement produces a new generation of metamolecules that are retested.

Let us demonstrate the logic of the procedure using the metagroups of Table V and the meta-contributions of Table VII. Consider the following constraint on boiling point: $T_b > 500$ K.

Limiting the number of groups contained in a molecule between 2 and 4, the following 65 metamolecules are generated:

$$
\begin{array}{ccccc}
(2\,0\,0\,0) & (0\,2\,0\,0) & (0\,0\,2\,0) & (0\,0\,0\,2) & (1\,1\,0\,0) \\
(1\,0\,1\,0) & (1\,0\,0\,1) & (0\,1\,1\,0) & (0\,1\,0\,1) & (0\,0\,1\,1) \\
(3\,0\,0\,0) & (0\,3\,0\,0) & (0\,0\,3\,0) & (0\,0\,0\,3) & (2\,1\,0\,0) \\
(2\,0\,1\,0) & (2\,0\,0\,1) & (0\,2\,1\,0) & (0\,2\,0\,1) & (0\,0\,2\,1) \\
(1\,2\,0\,0) & (1\,0\,2\,0) & (1\,0\,0\,2) & (0\,1\,2\,0) & (0\,1\,0\,2) \\
(0\,0\,1\,2) & (1\,1\,1\,0) & (1\,1\,0\,1) & (1\,0\,1\,1) & (0\,1\,1\,1) \\
(4\,0\,0\,0) & (0\,4\,0\,0) & (0\,0\,4\,0) & (0\,0\,0\,4) & (3\,1\,0\,0) \\
(3\,0\,1\,0) & (3\,0\,0\,1) & (0\,3\,1\,0) & (0\,3\,0\,1) & (0\,0\,3\,1) \\
(1\,3\,0\,0) & (1\,0\,3\,0) & (1\,0\,0\,3) & (0\,1\,3\,0) & (0\,1\,0\,3) \\
(0\,0\,1\,3) & (2\,2\,0\,0) & (2\,0\,2\,0) & (2\,0\,0\,2) & (0\,2\,2\,0) \\
(0\,2\,0\,2) & (0\,0\,2\,2) & (2\,1\,1\,0) & (2\,1\,0\,1) & (2\,0\,1\,1) \\
(0\,2\,1\,1) & (1\,2\,1\,0) & (1\,2\,0\,1) & (1\,0\,2\,1) & (0\,1\,2\,1) \\
(1\,1\,2\,0) & (1\,1\,0\,2) & (1\,0\,1\,2) & (0\,1\,1\,2) & (1\,1\,1\,1) \\
\end{array}
$$

TABLE VIII

T_b VALUES FOR FOUR METAMOLECULES

Metamolecule	T_b	
(2 0 0 0)	[177.18	536.36]
(2 1 0 0)	[199.60	617.46]
(3 0 1 0)	[178.42	729.59]
(2 2 0 0)	[222.02	698.56]

Recall that the metamolecule (1 0 1 2) represents the set of all molecules that can be formed by taking any one group from metagroup 1, no group from metagroup 2, any one group from metagroup 3, and any two groups from metagroup 4.

Structural constraint 10 of Table II is used to prune the candidate metamolecules. The maximum valence any group has is 4. Therefore, each metamolecule is checked to ensure that it satisfies the constraint $n_1 = 2 + n_3 + 2n_4$; 61 metamolecules are pruned using this constraint. The remaining four metamolecules are

$$(2\,0\,0\,0) \quad (2\,1\,0\,0) \quad (3\,0\,1\,0) \quad (2\,2\,0\,0).$$

Physical constraints are applied next. The boiling point value T_b was estimated using the metacontributions of Table VII and Eq. (7). Table VIII shows these estimates for each of the remaining four metamolecules. These values show that all four metamolecules satisfy the constraint on boiling point T_b.

The next step of the search algorithm is to reduce the level of abstraction. Groups were abstracted into metagroups to reduce the combinatorics of the problem. However, this same abstraction reduced the effectiveness of property constraints. As the abstraction is reduced, this effectiveness is regained. Metagroup 1 is divided into two new metagroups:

$$\begin{matrix} -CH_3 & =CH_2 & =CH & -F & -Cl & -Br \\ -I & -OH & -CHO & =O & -NH_2 & -SH \end{matrix}, \quad \text{(metagroup 1,1)}$$

$$\{-COOH \quad -NO_2 \quad -CN\}. \quad \text{(metagroup 1,2)}$$

This division of metagroup 1 is propagated to the metamolecules. Metamolecule (2 0 0 0) was the set of all molecules that could be formed by taking any two groups from metagroup 1. With metagroup 1 divided into

TABLE IX

T_b Values for 13 Metamolecules

[(2 0) 0 0 0]	[177.18 385.86]
[(0 2) 0 0 0]	[449.50 536.36]
[(1 1) 0 0 0]	[313.34 461.11]
[(2 0) 1 0 0]	[199.60 466.96]
[(0 2) 1 0 0]	[471.92 617.46]
[(1 1) 1 0 0]	[335.76 542.21]
[(3 0) 0 1 0]	[178.42 503.84]
[(0 3) 0 1 0]	[586.90 729.59]
[(2 1) 0 1 0]	[314.58 579.09]
[(1 2) 0 1 0]	[450.74 654.34]
[(2 0) 2 0 0]	[222.02 548.06]
[(0 2) 2 0 0]	[494.34 698.56]
[(1 1) 2 0 0]	[358.18 623.31]

two new metagroups there are three possibilities:

1. Take any two groups from metagroup 1,1.
2. Take any two groups from metagroup 1,2.
3. Take any one group from metagroup 1,1 and any one group from metagroup 1,2.

These possibilities correspond to an expansion of the metamolecule (2 0 0 0) into three new metamolecules:

$$[(2\ 0)\ 0\ 0\ 0] \quad [(0\ 2)\ 0\ 0\ 0] \quad [(1\ 1)\ 0\ 0\ 0].$$

Table IX displays the 13 new meta-molecules resulting from the expansion of all four metamolecules.

The metacontributions toward T_b are also divided; e.g.

$$T_b \text{ of metagroup } 1, 1 = [-10.50 \quad 93.84],$$

$$T_b \text{ of metagroup } 1, 2 = [125.66 \quad 169.09].$$

T_b is estimated for each meta-molecule and the property constraint is applied. Table IX shows estimated T_b values for the 13 metamolecules.

Applying the property constraint prunes metamolecules [(2 0) 0 0 0] [(1 1) 0 0 0] and [(2 0) 1 0 0]. Additionally, the estimate of T_b for metamolecule (0 3 0 1 0) shows that all the molecules it contains have T_b values that satisfy the property constraint. None of the metamolecules resulting from further expansion of metamolecule [(0 3) 0 1 0] need to be checked.

The search continues with the expansion of metagroups until all metagroups contain only one group. At that point the abstraction has been removed, and the metamolecules generated represent individual molecules.

4. Strategies for the Formation of Molecular Abstractions

Metagroup division can be accomplished in many ways. Given a set of k objects, the number of ways these can be portioned into p sets is given by

$$S(k, p),$$

which is the Stirling number of the second kind. Starting with a set of hypothetical groups, $(a\ b\ c\ d)$, and dividing them into two metagroups, yields $S(4, 2) = 7$ possibilities. These are $[(a)(bcd)]$ $[(b)(acd)]$ $[(c)(abd)]$ $[(d)(abc)]$ $[(ab)(cd)]$ $[(ac)(bd)]$ $[(ad)(bc)]$. For a reasonable number of groups, the possible choices of metagroups is very large.

Two approaches can be used to add back detail: *expansion* and *division*. Expansion adds back knowledge about the metagroups, which is used by the structural constraints. Division focuses on reducing the width of metacontributions, thus improving the screening power of physical property constraints.

a. Division In Half. Dividing a metagroup in half is the simplest strategy. However, division without regard to the metacontributions could prove inefficient. Given the following set of groups with the corresponding contributions

Groups	$[g_1\ \ g_2\ \ g_3\ \ g_4]$,
Contributions	$[10\ \ \ 60\ \ \ 11\ \ \ 61]$,

our initial meta-group $[g_1\ \ g_2\ \ g_3\ \ g_4]$ would have a metacontribution of $[10\ \ 61]$. Dividing the meta-group in half would result in the two metagroups $[g_1\ \ g_2]$ and $[g_3\ \ g_4]$. The metacontributions for these new metagroups would be $[10\ \ 60]$ and $[11\ \ 61]$. These metacontributions are almost identical to the original. The division thus added to the combinatorics without improving the possibility for pruning.

Meta-contributions should be considered when dividing metagroups in half. The midpoint of the initial metacontribution is $(61 - 10)/2 = 25.5$. All groups whose contributions are less than $25.5 + 10 = 35.5$ are collected into the first new metagroup, and all those with contributions greater than 35.5 are collected into the second new metagroup. The resulting metagroups have tighter interval representation of their contributions, leading to more efficient screening.

b. Division by Largest Gap. The interval representation of metacontributions ignores their discrete nature. Dividing a metagroup at the largest gap in its contributions attempts to take advantage of the distribution of contributions and produce two new metagroups whose metacontributions are distributed over a much more narrow range than the original metacontribution. Using the same example set of groups and contributions given above, the initial metacontribution, [10 61], has a width of 51. Dividing the metagroup at the largest gap in the contributions produces two new metagroups with metacontributions [10 11] and [60 61]. The total width of these two intervals is 2, a considerable reduction from 51.

c. Division to Isolate Groups. Extreme values of the contributions by some groups can greatly affect the interval value of the calculated properties. The contribution toward the boiling point T_b, from Joback's method (Joback and Reid, 1987) for the group $=O$, is -10.5. If we are searching for low values of T_b, then it is desirable to have many $=O$ groups in our molecules. However, from chemical considerations it is unlikely that a molecule with a large number of $=O$ groups would be stable. Isolating $=O$ onto its own metagroup enables the designer to pose constraints on the maximum number of occurrences of the metagroup in any meta-molecule.

5. Evaluation of the Search Algorithm: Taming the Combinatorial Explosion

Tables X and XI summarize the results of the application of the search algorithm in two case studies. Let us look at the highlights of each one of them:

Case 1. In this case study we want to synthesize molecules that have a vapor pressure, at 273 K, larger than 1.0 bar. The molecules are to be composed from a set of 44 functional groups, and they can contain up to 3 functional groups, i.e., $k = 44$ and $n = 3$. There exist 15,180 molecules that can be created from various combinations of the functional groups (Table X). The search algorithm generates 460 meta-molecules and rejects 104 of them. The refinement of the metagroups and the pruning of infeasible molecules is guided by a series of constraints, as indicated in the footnotes of Table X.

TABLE X

PRUNING RESULTS FOR $k = 44$, $n = 3$ AUTOMATIC DESIGN
[CONSTRAINT $= P_{vp}(273\text{ K}) > 1.0$ BAR]

# Metagroups	# Metamolecules	Kept	Pruned
1	1	1	0
3^a	10	4	6
4^b	7	4	3
10^c	51	1	50
11	3	1	2
12^d	3	2	1
13	5	3	2
14	6	3	3
15	6	4	2
16	8	7	1
17	11	11	0
18	12	11	1
19	21	18	3
20	28	26	2
21	33	29	4
22	44	44	0
27	109	85	24
28^e	102	102	0
Total^f	460	356	104

^{a}Expanded by ring class.
^{b}Isolated $-$ COOH, $-$ NO$_2$, and $-$ CN.
^{c}Expanded by global valence.
^{d}Isolated $=$ O. Restricted $=$ O occurrences to 1.
^{e}12 metagroups never occurred in any metamolecules.
^{f}There are 15,180 molecules contained in the search.

Case 2: The premises are the same as in case 1, but here we allow the
formation of molecules with up to five functional groups. The total
number of potential molecules is 1,712,304. The search algorithm has
generated 4131 metamolecules and rejected 2,094 of them (Table XI).
See footnotes of Table XI for constraints guiding the pruning of
infeasible metamolecules.

From both these examples is clear the advantage of using a search with
successive molecular abstractions; *the number of metamolecules needed to
be evaluated is far smaller than the number of individual molecules.*

The algorithm was also analyzed in order to identify some of its
"bounding" properties. Assume that an initial metagroup is divided into
two children metagroups. The metamolecules formed from these metagroups are tested, and all those that contain occurrences of the second

TABLE XI

PRUNING RESULTS FOR $k = 44$, $n = 5$ AUTOMATIC DESIGN
[CONSTRAINT $= P_{vp}(273 \text{ K}) > 1.0$ BAR]

# Metagroups	# Metamolecules	Kept	Pruned
1	1	1	0
3[a]	21	8	16
4[b]	23	8	15
5[c]	23	12	11
6[d]	30	27	3
7	58	27	31
8	58	32	26
9	37	35	2
10	53	35	18
11	71	42	29
12	87	86	1
13	110	86	24
14	165	107	58
15	206	179	27
24[e]	1675	185	1490
30[f]	625	479	146
37[g]	888	688	200
Total[h]:	4131	2037	2094

[a] Expanded by ring class.

[b] Isolated $-\text{COOH}$, $-\text{NO}_2$, and $-\text{CN}$.

[c] Isolated $=\text{O}$. Restricted $=\text{O}$ occurrences to a maximum of 1.

[d] Isolated $-\text{F}$.

[e] Expanded by global valence.

[f] Expanded several metagroups containing two or three groups in half.

[g] Expanded all nonzero occurring metagroups to individual groups. Ten metagroups never occurred in any metamolecule.

[h] There are 1,712,304 molecules contained in the search.

group are pruned away. This scenario is repeated with the surviving metagroups, and so on.

Dividing a metagroup (MG) containing k groups into two metagroups, MG_1 and MG_2, containing k_1 and k_2 groups, respectively, allocates the

$$\frac{(k + n - 1)!}{n!(k - 1)!}$$

possible molecules into

$$\frac{(2 + n - 1)!}{n!(2 - 1)!} = n + 1$$

metamolecules. One metamolecule contains only occurrences of MG_1, one metamolecule contains only occurrences of MG_2, and the remaining $n - 1$ metamolecules contain occurrences of both metagroups. If no metamolecules containing MG_2 survive the testing, then the percentage of molecules pruned is given by

$$\left[1 - \frac{(k - 1)!}{(k_1 - 1)!} \frac{(k_1 + n - 1)!}{(k + n - 1)!} \right] * 100\%.$$

Repeating this expansion and pruning process r times, until the final metagroup contains only one group, requires the generation and testing of

$$r(n + 1) + 1$$

metamolecules. The advantage of abstraction, as measured by the total number of molecules contained in the search divided by the number of metamolecules needed to be generated and tested, is given by

$$\text{Advantage of abstraction} = \frac{1}{r(n + 1) + 1} \frac{(k + n - 1)!}{n!(k - 1)!}.$$

Considering the worst-case scenario, in which MG_2 and all subsequent second metagroups contain only a single group, we have $r = k - 1$ leading to

$$\text{Advantage of abstraction} = \frac{1}{(k - 1)(n + 1) + 1} \frac{(k + n - 1)!}{n!(k - 1)!}.$$

Table XII shows this advantage of abstraction for several values of k and

TABLE XII

ADVANTAGE OF ABSTRACTION

$k \setminus n$	2	3	4	5	6
15	5.6	11.9	43.1	136.8	391.5
20	7.2	20.0	92.2	369.6	1,321.6
25	8.9	30.2	169.2	819.0	3,513.5
30	10.6	42.4	280.3	1,590.0	7,956.7
35	12.2	56.7	431.7	2,808.6	16,060.2
40	13.9	73.1	629.6	4,621.3	29,726.5
45	15.6	91.6	880.5	7,195.8	51,426.2
50	17.2	112.2	1,190.3	10,720.4	84,272.3

n. In particular, Table XII shows that if we have an automatic design involving 40 groups with an occurrence value of 5, then the number of metamolecules needed for exhaustive searching will be 4621.3 times smaller than the number of potential molecules.

C. CASE STUDY: AUTOMATIC DESIGN OF REFRIGERANTS

Automotive air conditioners are a major source of refrigerant emissions, which contribute to the depletion of the Earth's protective ozone layer. This case study generates replacement refrigerants for automotive air conditioners.

Identifying the target set of constraints is the first step of the methodology. Constraints are derived from performance considerations and in an evolutionary manner attempting to find a compound with properties better than refrigerant 12. The design temperatures between which the refrigerant must operate are 110°F (43.3°C) maximum and 30°F (-1.1°C) minimum (Langley, 1986). The following constraints form the design target:

- $P_{vp}(T = -1.1°C) > 1.4$ bar
  The lowest pressure in the cycle should be greater than atmospheric (Dossat, 1981). This reduces the possibility of air and moisture leaking into the system. Douglas (1988) recommends a safety factor of 5 psig (pounds per square inch gauge).
- $P_{vp}(T = 43.3°C) < 14$ bar
  A high system pressure increases the size, weight, and cost of equipment (Dossat, 1981). A pressure ratio of 10 is considered to be the maximum for a refrigeration cycle (Perry and Chilton, 1973).
- $\Delta H_v(T = -1.1°C) > 18.4$ kJ / g-mol
  The value for refrigerant 12's enthalpy of vaporization at -1.1°C is 18.4 kJ / g-mol [American Society of Heating, Refrigerating and Air-Conditioning Engineers (ASHRAE), 1972]. A higher value reduces the amount of refrigerant required.
- $C_{p_L}(T = 21.1°C) < 32.2$ cal / g-mol · K
  It is desirable to have a low liquid heat capacity to reduce the amount of refrigerant that flashes on passage through the expansion valve (Dossat, 1981). Refrigerant 12's liquid heat capacity at 21.1°C is 32.2 cal / g-mol · K (ASHRAE, 1972).

Estimation procedures were developed for each physical property used in the target constraints: P_{vp}, H_v, C_{p_L}. These estimation procedures were based on correlations that require the evaluation of seven physical properties, given by group-contribution techniques:

(a) Reduced boiling point, T_{b_R}
(b) Normal boiling point, T_b
(c) Critical pressure, P_c
(d) Coefficients for cubic feet of ideal gas heat capacity with temperature, $C^{\circ}_{p,a}, C^{\circ}_{p,b}, C^{\circ}_{p,c}, C^{\circ}_{p,d}$

In the automatic design, 44 functional groups were used; molecules containing 2–7 groups were allowed; the number of group occurrences was limited to 7 to account for the fact that most refrigerants are of small molecular size, and 47 molecules were designed that satisfy the four physical property constraints and the structural constraints listed in Table II. Table XIII shows some of these 47 molecules along with their estimated values for P_{vp}, ΔH_v, and C_{P_L}.

Molecules 19 and 20 are of particular interest. These are two ringed compounds that possess physical properties satisfying the design constraints. Although the chemical stability of these compounds still needs to be verified, it is considered a success that the automatic design was able to design such not obvious compounds.

D. Case Study: Automatic Design of Polymers As Packaging Materials

In this second example we demonstrate the applicability of the automatic design methodology to the design of polymers with desired properties. The problem is to design polymers for use as integrated-circuit (IC) encapsulants.

Electronic packages are sealed to prevent gross contamination, handling damage, and the entry of corrosive gases (Mih, 1984). To package microelectronic circuitry so that it is useful and functions properly under various environmental conditions, it is essential to select the correct packaging materials (Fogiel, 1972). Polymeric coatings are widely used in the electronics industry because of their excellent properties and low cost (Goosey, 1985). Polymers used for semiconductor encapsulation must protect against moisture, chemical agents, wide temperature variations, and mechanical shock. The polymeric material must be able to satisfy these requirements with a minimal effect on device parameters over an

TABLE XIII

Refrigerant Design—Automatic Results

Molecule	$P_{vp}(272.05)$	$P_{vp}(316.45)$	$H_v(272.05)$	$C_{p_L}(294.35)$
1. $1(-CH_3)1(-Cl)$	1.59	6.20	21.58	20.00
2. $2(-F)1(\diagdown NH)$	2.69	10.96	19.06	22.26
3. $1(-Cl)1(-F)1(-CH_2-)$	1.65	6.61	20.71	22.89
4. $1(\equiv CH)1(-CH_3)1(\equiv C-)$	1.67	6.30	21.44	24.06
5. $2(-F)2(\equiv C-)$	2.09	7.87	19.61	21.97
6. $1(\equiv CH)1(-F)1(-CH_2-)$ $1(\equiv C-)$	1.74	6.72	20.56	26.95
7. $1(=O)2(-CH_3)1(=C\diagup)$	1.71	7.29	27.14	30.06
8. $1(-Cl)2(-F)1(\diagdown N-)$	2.65	10.41	19.04	24.83
9. $1(-Cl)2(-F)1(\diagdown CH-)$	1.74	6.96	19.45	25.28
10. $2(-CH_3)1(-F)1(\diagdown N-)$	1.81	7.29	20.42	30.04
11. $3(-F)1(\diagdown NH)1(\diagdown N-)$	1.70	8.12	20.88	29.98
12. $1(-CH_3)2(-F)1(-O-)$ $1(\diagdown N-)$	1.96	8.35	19.70	31.60
13. $1(=CH_2)2(-F)1(=CH-)$ $1(\diagdown N-)$	2.15	8.61	18.73	30.15
14. $1(=CH_2)2(-F)1(-CH_2-)$ $1(=C\diagup)$	1.41	5.78	19.59	30.78
15. $1(=CH_2)2(-F)1(=CH-)$ $1(\diagdown CH-)$	1.42	5.82	19.12	30.62
16. $1(\equiv CH)2(-F)1(\equiv C-)$ $1(\diagdown N-)$	2.77	10.54	18.90	28.88
17. $1(\equiv CH)2(-F)1(\equiv C-)$ $1(\diagdown CH-)$	1.83	7.07	19.31	29.34
18. $3(-F)2(=CH-)1(\diagdown N-)$	1.66	7.13	18.92	31.79
19. $2(\diagdown^r CH-)1(=C\diagup^r)1(=O)$ $2(-F)$	1.53	7.11	26.08	31.93
20. $3(\diagdown^r CH-)3(-F)$	1.40	5.80	18.67	31.11

extended period of time, and be relatively inexpensive and easy to process. Some of the important physical properties of packaging material are (Dillinger, 1988)

1. Permeability to water vapor at high temperatures.
2. Thermal conductivity.
3. Outgassing in plastics at elevated temperatures and the resulting impact on water vapor permeability.
4. Thermal expansion coefficient and mismatch between expansion coefficients of package and chip interconnect.

The following constraints are the design specifications of a good encapsulant:

- $T_g > 400°C$
  The glass transition temperature must be $> 400°C$. The encapsulant must keep its structural integrity during use. The high temperatures at which microelectronic circuits operate place a restriction on T_g.
- $R > 10^{16}$ Ω-cm
  The volume resistivity of the solid polymer must be $> 10^{16}$ Ω-cm. Since the packaging material will make contact with the metal leads of the microelectronics device, it is essential that the compound have a high-volume resistivity.
- $\lambda > 0.16$ w/ m $\cdot$ K
  The thermal conductivity of the solid polymer is desired to be greater than the thermal conductivity of the currently used polyimide. A high thermal conductivity is desirable allowing the microelectronic circuitry to be cooled more effectively.
- $P(O_2) < 1.0$ cm³-mil 100 in.$^{-2}$ day^{-1} atm^{-1}
  The permeability of the polymer to oxygen should be < 1.0 cm³-mil 100 in.$^{-2}$ day^{-1} atm^{-1}. Diffusion of oxygen and water through the polymer to the microelectronic circuitry could cause corrosion and is thus undesirable. To establish a value for this physical property constraint, we examined polymers used as barriers. Polymers with a permeability to oxygen of ≤ 1.0 cm³-mil 100 in.$^{-2}$ day^{-1} atm^{-1} are considered high-barrier materials.

Estimation techniques from van Krevelen (1976) and Salame (1986) were combined into estimation procedures for the properties of each target constraint: $T_g, R, \lambda, P(O_2)$. These procedures are detailed in the following paragraphs:

1. *Thermal conductivity.* This is estimated by the following correlation (van Krevelen, 1976)

$$\lambda(298 \text{ K}) = \lambda\left(C_p^S, V, U\right),$$

where

$C_p^s = C_p^s$ (groups)—specific heat by the van Krevelen

group-contribution technique

$V = V(\text{groups})$—specific volume by the modified van Krevelen

group-contribution technique

$U = U(\text{groups})$—Rao function by the van Krevelen

group-contribution technique

2. *Electrical resistivity.* This is estimated by the following correlation (van Krevelen, 1976)

$$R = R(P_{LL}, V),$$

where

$P_{LL} = P_{LL}(\text{groups})$—dielectric polarization by the van Krevelen

group-contribution technique,

$V = V(\text{groups})$—specific volume by the modified van Krevelen

group-contribution technique.

This estimation procedure requires two properties to be estimated by group-contribution techniques: P_{LL} and V.

3. *Glass transition temperature.* This is estimated by the following correlation (van Krevelen, 1976)

$$T_g = T_g(Y_g, M),$$

where

$Y_g = Y_g(\text{groups})$—glass transition function by the van Krevelen

group-contribution technique,

$M = M(\text{groups})$— repeat unit weight by the van Krevelen

group-contribution technique.

This estimation procedure requires two properties to be estimated by group-contribution techniques: Y_g and M.

4. *Permeability to oxygen.* This is estimated by the following correlation

TABLE XIV

Polymer Design—Automatic Results

Molecule	T_g	R	L	Pi
1. 1(—C6H4—CH2—C6H4—)	479.5	20.1	0.164	2.40e—03
2. 1(—OCONH—)	423.5	122.6	0.214	3.25e—15
3. 1(—CONH—)2(—C(CH3)(C6H5)—)	445.7	19.6	0.172	6.42e—02
4. 1(—CONH—)1(—C(CH3)(C6H5)—)1(—CH(C6H5)—)	408.8	19.4	0.181	1.74e—02
5. 1(—O—)1(—CHF—)2(—C6H4—CH2—C6H4—)1(—OCONH—)	453.7	19.6	0.163	1.33e—04
6. 1(—CH2—)1(—CHF—)2(—C(CH3)(C6H5)—)1(—OCONH—)	442.6	19.8	0.161	1.62e—01
7. 3(—CH2—)1(—CONH—)2(—C6H4—CH2—C6H4—)	429.9	19.8	0.166	7.73e—02

#	Structure				
8.	$2(-O-)1(-CH_2-)1(-CONH-)2\left(-\bigcirc-CH_2-\bigcirc-\right)$ 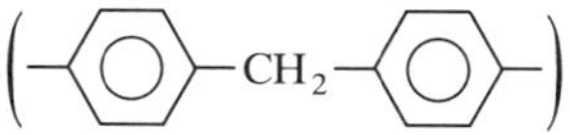	432.0	19.6	0.165	9.92e—03
9.	$1(-O-)1(-CH(CH_3)-)4\left(-\bigcirc-CH_2-\bigcirc-\right)$ 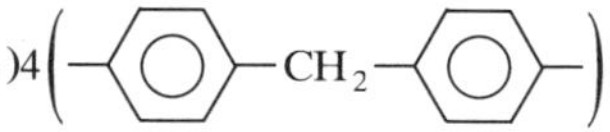	466.6	20.2	0.160	7.59e—02
10.	$2(-C(CH_3)C_6H_5-)3\left(-\bigcirc-\right)1(-OCONH-)$ 	445.9	19.8	0.168	2.24e—01
11.	$1(-C(CH_3)(C_6H_5)-)2(-CH(C_6H_5)-)2\left(-\bigcirc-\right)1(-OCONH-)$ 	421.7	19.8	0.174	1.73e—01
12.	$1\left(-CH_2-\bigcirc-CH_2-\right)3\left(-\bigcirc-CH_2-\bigcirc-\right)2(-OCONH-)$ 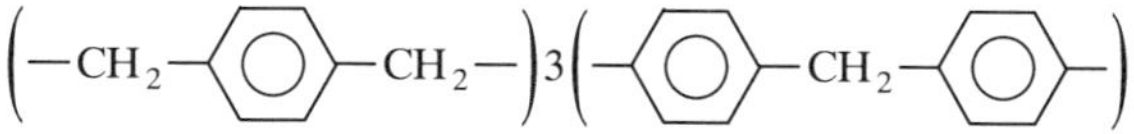	453.0	19.5	0.168	6.15e—07
13.	$3(-CH(C_6H_5)-)1\left(-\bigcirc-CH_2-\bigcirc-\right)2(-OCONH-)$ 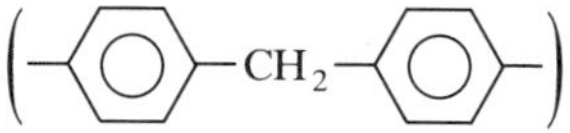	423.2	19.3	0.180	5.74e—05
14.	$2(-C(CH_3)(C_6H_5)-)2(-CH(C_6H_5)-)2(-OCONH-)$	434.3	19.4	0.176	1.27e—03

(Salame, 1986)

$$P = P(\pi),$$

where

$$\pi = \pi(\pi_i, N_b) \text{—is given by the Salame correlation,}$$

with

$$\pi_i = \pi_i(\text{groups}) \text{—permachor by the Salame}$$
$$\text{group-contribution technique,}$$

$$N_b = N_b(\text{groups}) \text{—number of backbone groups by the}$$
$$\text{Salame group-contribution technique.}$$

This estimation procedure requires two properties to be estimated by group-contribution techniques: π_i and N_b. These procedures result in seven physical properties that are estimated by group-contribution techniques: C_p^s, V, U, P_{LL}, Y_g, M, and π.

In the automatic design of new polymers, 21 groups were used. Polymers having between one and six group occurrences were allowed. The design procedure produced over 18,000 feasible polymers. Table XIV shows several polymers randomly selected from the set of feasible candidates along with the estimated values of their important physical properties. The large number of retained molecules indicates that the design specs are fairly "loose" and could be tightened by either tightening the bounding values of the physical property constraints or introducing additional physical properties constraints.

III. Interactive Synthesis of New Molecules

The automatic synthesis of molecules, described in the previous section, attempts to generate feasible candidates without resorting to the detailed, fragmented, often informal, but nevertheless sound and valuable knowledge possessed by the human designer. The search for new molecules is carried out efficiently, but it is entirely built on a limited amount of knowledge, which is represented by the sets of constraints on (1) physical properties, (2) the feasibility of molecular structures, and (3) chemical stability and other chemical properties of the resulting molecules.

The interactive design attempts to support the articulation of and incorporate the designer's informal knowledge and abductive capabilities in postulating "promising" molecular structures. In this manner, the resulting computer-aided tool can capture and utilize the best of two worlds: (a) the computer's ability to locate feasible solutions after extensive search of the solution space and (b) the human designer's "intelligence" in expanding and guiding the search in an efficient manner. Therefore, although this section will deal only with the features of the interactive design procedures, the *Molecule-Designer*, the computer-aided tool that implements both automatic and interactive procedures, has integrated both in a seamless manner (see Section IV).

The human-driven character of interactive design allows the use of (1) problem-specific subjective preferences, (2) informal, qualitatively stated scientific knowledge, (3) rapid evaluation of alternatives, and (4) evolutionary design of new molecules, starting from known and existing alternatives. It provides the designer with the following facilities:

1. Visualization of the target constraints, helping the designer in the selection of the most "promising" functional groups to be incorporated in a molecule.
2. Extensive databases for the extraction of patterns in the structural evolution of known molecules.
3. Interactive definition of new property estimation techniques and qualitative scientific rules, which are to be incorporated in the automatic design algorithm.

A. ILLUSTRATION OF INTERACTIVE DESIGN

The framework of interactive design is entirely built on the premise of *additivity of group contributions* for the estimation of physical properties. Let us look at some typical illustrations. Table XV shows the estimation of normal boiling points T_b, and normal melting points T_m, using Joback's group contribution techniques (Joback and Reid, 1987), along with the

TABLE XV

EXAMPLE OF LINEAR GROUP CONTRIBUTION
ESTIMATION TECHNIQUES

Groups	Contributions[a]	
	Δ_{i,T_b}	Δ_{i,T_m}
$-CH_3$	23.58	-5.10
$-CH_2-$	22.88	11.27
$>CH-$	21.74	12.64
$>C<$	18.25	46.43
$-F$	-0.03	-15.78
$-Cl$	38.13	13.55

[a] $T_b = 198.18 + \Sigma n_i \Delta_{i,T_b}$; $T_m = 122.5 + \Sigma n_i \Delta_{i,T_m}$.

contributions of a few select groups. The additivity of group contributions allows the molecules to be assembled group by group, while offering a "partial" estimate of the desired physical properties. Given a constraint, such as, $T_b > 300$ K, a group is added to the molecule and then the constraint is evaluated. Choosing one $-CH_3$ group results in $T_b = 221.76$ K. The constraint is not satisfied. Choosing a second $-CH_3$ group results in $T_b = 245.34$. The constraint is still not satisfied. Adding three $-CH_2-$ groups results in $T_b = 313.98$, which satisfies the constraint. Adding a second constraint, $T_m < 250$ K, makes the monitoring of the numerical values difficult. Graphical representations can simplify this problem.

Figure 1 shows a two-dimensional design space formed from T_b and T_m. The two imposed constraints form a feasible region denoted by the shaded area. The contributions of each group toward T_b and T_m form a two-dimensional (2D) vector. These are called *group vectors*. The intercepts of the group contribution models shown in Table XV establish the starting point for the first group vector. Beginning at the intercept point an appropriate set of group vectors are selected that terminate in the feasible region and produce a structurally feasible molecule. Figure 1 shows the group vectors for chloropropane.

Complex constraints on T_b and T_m are easily handled by the interactive design procedure. All that is required is to identify a feasible region or regions. The constraints need not be linear or convex.

The following heuristics on group selection are often helpful in selecting groups that satisfy the physical property constraints and are structurally feasible.

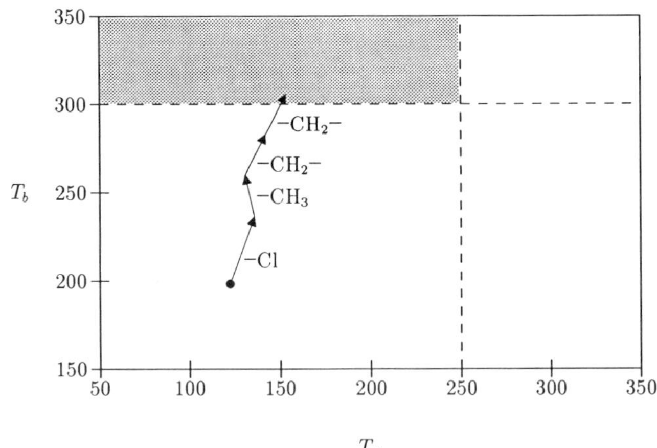

Fig. 1. An interactively designed molecule: chloropropane.

Rule 1. Separate the groups into three sets:

1. *Terminators*: all groups having one free bond.
2. *Extenders*: all groups having two free bonds.
3. Branchers: all groups having more than two free bonds.

Rule 2. Select the following initial group sets:

1. If a pure acyclic molecule is to be designed, choose two terminators.
2. If a pure cyclic molecule is to be designed, choose two cyclic extenders.

Rule 3. Continue by first considering extenders. If a brancher is to be added, follow it immediately with terminators.

These heuristics ensure that when a molecule reaches the feasible region, it is either structurally feasible or requires the addition of only one or two groups for feasibility.

1. Reduction in the Dimensionality of the Search Space

The dimension of the design space is equal to the number of fundamental physical properties needed to evaluate the property constraints. Using the estimation procedure for the vapor pressure P_{vp}, given by the Riedel–Plank–Miller correlation

$$P_{vp} = P_{vp}(T_b, T_{br}, P_c),$$

where

$T_b = T_b(\text{groups})$ is given by Joback's group contribution,

$T_{br} = T_{br}(\text{groups})$ is given by Lydersen's group contribution,

$P_c = P_c(\text{groups})$ is given by Ambrose's group contribution,

we obtain three fundamental properties: T_b, T_{br}, and P_c. Designing for constraints on P_{vp} thus requires a three-dimensional (3D) design space. This is unfortunate because representing and manipulating 3D objects graphically is more complex than 2D manipulation.

The dimensionality of the design space can become more than a mere complication. One estimation procedure for the liquid heat capacity requires seven fundamental physical properties. To interactively design for constraints on C_{p_L} would require a seven-dimensional physical property space to display each of the seven fundamental properties.

Studies on *factor analysis* (Cramer, 1980a, b; Klincewicz, 1982; Joback, 1984) show that a number of physical properties are highly intercorrelated.

TABLE XVI

PHYSICAL PROPERTY–FACTOR RELATIONSHIPS

$$
\begin{aligned}
1/\sqrt{P_\mathrm{c}} &= 0.157 - 0.019 F_1 \\
V_\mathrm{c} &= 296.1 - 89.66 F_1 - 36.68 F_3 \\
n_\mathrm{A} &= 14.50 - 5.35 F_1 - 1.18 F_3 \\
C^o_{\mathrm{p},298} &= 25.70 - 8.72 F_1 - 2.44 F_3 \\
T_\mathrm{c} &= 545.9 - 24.65 F_1 - 87.92 F_3 \\
T_\mathrm{b} &= 358.4 - 25.26 F_1 - 64.94 F_3 \\
\Delta H_\mathrm{vb} &= 7686.6 - 432.3 F_1 - 1614.3 F_3
\end{aligned}
$$

High correlations between two properties indicate the possibility of replacing one with a function of the other. This would enable us to reduce the dimensionality of a design space. One study (Joback, 1984) found that nine physical properties were well approximated by three new properties called *factors*. Table XVI shows several of the derived physical property estimation techniques, using essentially two of these factors, F_1 and F_3. Group contribution estimation techniques were developed for both factors, F_1 and F_3.

Incorporating these equation-oriented estimation techniques into a new estimation procedure for P_vp yields the following correlation, proposed by Riedel, Plank, and Miller

$$
P_\mathrm{vp} = P_\mathrm{vp}(T_\mathrm{b}, T_\mathrm{c}, P_\mathrm{c}),
$$

where the following properties can be estimated from their correlations to factors F_1 and F_3 (Table XVI):

$$
T_\mathrm{b} = T_\mathrm{b}(F_1, F_3),
$$

$$
T_\mathrm{c} = T_\mathrm{c}(F_1, F_3),
$$

$$
P_\mathrm{c} = P_\mathrm{c}(F_1, F_3),
$$

with F_1 and F_2 computed by group-contribution correlations established by Joback:

$$
F_1 = F_1(\text{groups}),
$$

$$
F_3 = F_3(\text{groups}).
$$

The fundamental properties are F_1 and F_3. Two fundamental properties enable design in a 2D physical property space.

2. Utilization of Interactive Design

The graphical representation of physical property constraints allows the designer to quickly gain insight into the feasibility of the problem and the relative importance of each constraint. Excessively large or small feasible regions, redundant constraints, or open feasible regions may indicate a need to respecify design constraints.

Once satisfied with the feasible region, design may proceed in one of the following three ways:

a. Evolutionary Design. Starting from an existing molecule, interactive design allows the evolution of the molecule to new altenatives through guided structural modifications. For example, Fig. 2a shows the feasible region defined by constraints on three physical properties and how two functional groups of the initial (infeasible) molecule were replaced by two

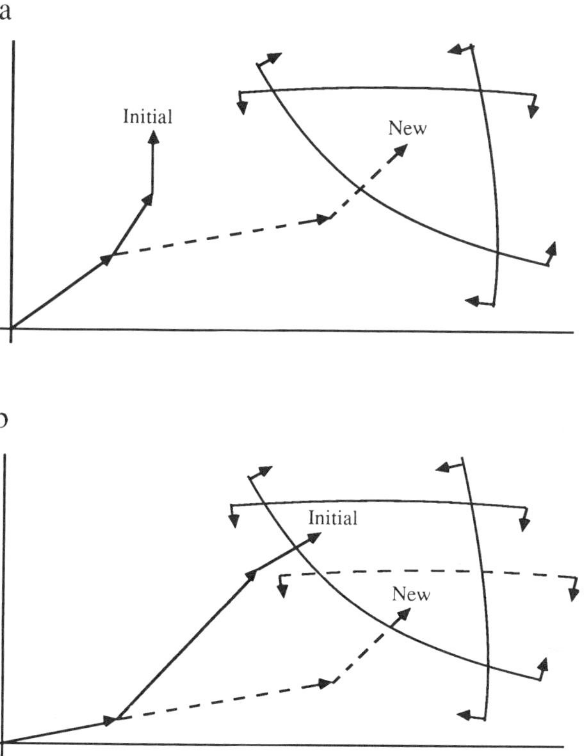

FIG. 2. Interactive design approaches: (a) evolutionary design of improved molecules; (b) evolutionary tightening of design constraints.

other groups to yield a feasible molecule. Figure 2b, on the other hand, shows how interactive design can be employed to tighten the specifications of a physical property constraint (e.g., by moving the location of a constraint; see dashed-line constraint), requiring the evolution of the initial molecule to a new one (satisfying the new set of constraints).

b. Grass-Roots Design. Building the molecule from scratch, group by group, using pure interactive or in conjunction with automatic search.

c. Combination of Evolutionary and Grass-Roots Designs. A molecule is designed from scratch so that it meets the initial specifications. Subsequently, improvements are searched for by tightening the design specifications and carrying out an evolutionary design. Figure 2b depicts such a situation.

B. CASE STUDY: INTERACTIVE DESIGN OF REFRIGERANTS

The specifications of this case study are identical to those described in Section II.C and will not be reproduced here.

To design in a 2D space, the factor relationships shown in Table XVI are used to reduce the fundamental properties to the factors F_1 and F_3. Each physical property constraint, which is a function of F_1 and F_3, is plotted in a 2D $\{F_1-F_3\}$ design space shown in Fig. 3. The region in which all constraints are satisfied is shaded. The displayed symbols correspond to the constraints as follows:

Symbol	Constraints
Circle	$P_{vp}\ (T = -1.1°C) > 1.4\ \text{bar}$
Triangle	$P_{vp}\ (T = 43.3°C) < 14\ \text{bar}$
Square	$\Delta H_v\ (T = -1.1°C) > 18.\text{kJ}/\text{g-mol}$
Circle	$C_{pL}\ (T = 21.1°C) < 32.2\ \text{cal}/\text{g-mol} \cdot \text{K}$

The graphical representation used in the interactive design procedure immediately provides insights into the design problem. The first insight is that the chosen constraints yield a feasible solution space. Although each constraint was justified on its own, there was no guarantee that the set of four constraints would produce a feasible space. If the feasible region were too large or small, this would indicate the need to respecify the design target. Additionally, it is seen that for a major portion of the design space the $P_{vp}\ (T = 43.3°C) < 14$ bar constraint is redundant. It is superseded by the $\Delta H_v > 18.4$ constraint.

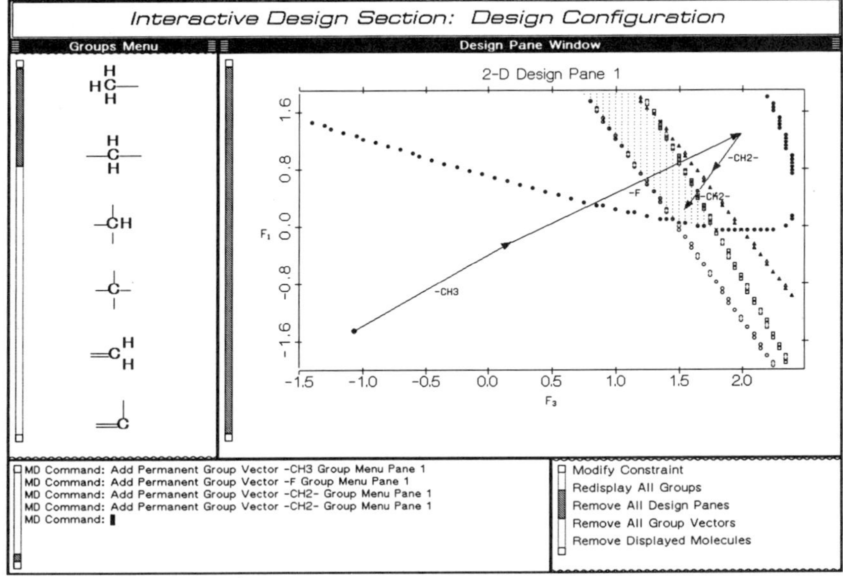

FIG. 3. Refrigerant design: a solution.

TABLE XVII

ESTIMATED PROPERTY VALUES FOR DESIGNED REFRIGERANTS

	P_{vp}		ΔH_v	C_{pL}
Compound	272.05 K	316.4 K	272.05 K	294.3 K
1. $CH_3 - CH_3$	2.710	9.486	18.67	22.49
2. $CH_3 - Cl$	1.590	6.193	21.58	19.99
3. $CH_3 - NH_2$	0.375	2.152	29.78	23.50
4. $CH_3 - CH_2 - CH_2 - F$	1.195	5.013	21.21	31.10
5. $CH_2 = CH - CH_3$	1.310	5.195	21.24	25.33
6. $CH_2 = CH - Cl$	0.750	3.340	24.14	23.01
7. $CH_2 = CH - O - CH_3$	0.543	2.648	24.83	29.83
8. $F - CH_2 - CH_2 - CH_2 - F$	1.236	5.330	20.34	34.03
9. $F - CH = CH - CH_2 - F$	1.045	4.573	20.54	29.89
10. CCl_2F_2	0.440	2.213	24.70	27.87
11. $CBrClF_2$	0.122	0.837	28.10	29.02
12. $CHBrF_2$	0.538	2.855	22.96	26.04
13. $CH(COOH)F_2$	0.002	0.041	42.56	37.58
14. $CH(HCO)F_2$	0.417	2.421	25.91	30.42
15. $NH_2 - NH_2$	0.027	0.311	41.41	26.03
16. $CH \equiv C - Cl$	0.968	4.084	24.33	21.72
17. $CH \equiv C - CH_3$	1.670	6.297	21.44	24.06
18. $CH \equiv C - NH_2$	0.214	1.353	32.56	25.42

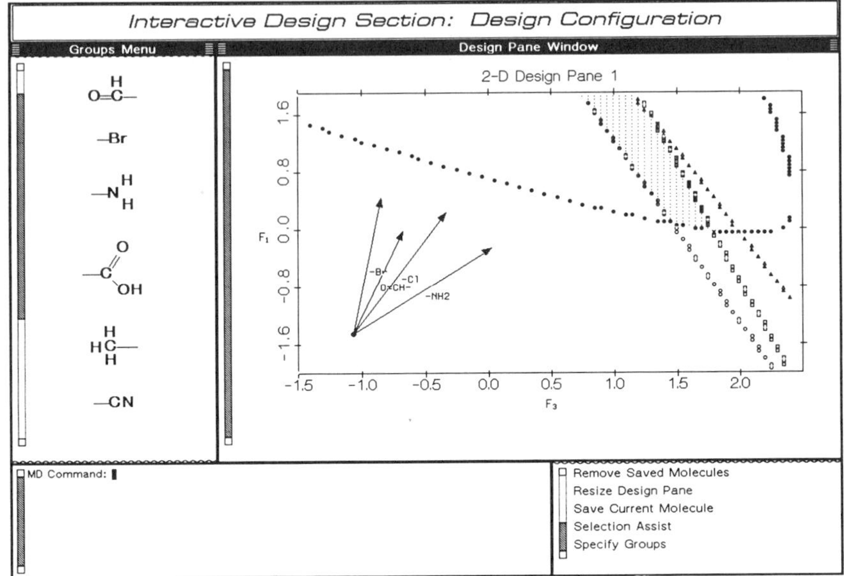

FIG. 4. Candidate groups for chlorine replacement.

Figure 3 also shows the group vectors for a designed molecule. The location of the vectors' end point can be directly interpreted into physical property information. For example, the molecule satisfies all constraints but has a liquid heat capacity near the constraint. Table XVII shows several interactively designed molecules with estimated values for their physical properties. Some of the molecules may be chemically unstable. At this stage of the design methodology only physical property and structural constraints are being explicitly considered. When choosing the groups, the designer implicitly considers chemical constraints. Once molecules are designed, the next steps of the methodology, molecule enumeration and screening, would identify and remove chemically unstable compounds.

1. Replacing Chlorine

Removing chlorine from current refrigerants would seem to reduce the hazard of ozone depletion. The interactive design procedure is well suited for searching for group replacements. The group vector for the chlorine group is shown in Fig. 4. The target is to find one or more groups making nearly equivalent contributions to that of chlorine.

The search for single group replacements is begun by restricting the possible groups to those with one single free bond. These groups are then sorted with respect to distance. The three closest groups, $-CHO$, $-Br$, and $-NH_2$, are displayed in the design space shown in Fig. 4.

Again the graphical representation used by the interactive design enables us to evaluate the effect of any substitution. Replacing, $-Cl$ by $O=CH-$, would result in a compound having reduced vapor pressure, an increased enthalpy of vaporization, and an increased liquid heat capacity. These alterations of physical properties are derived from the relative positions of the group vectors and the locations of the constraints.

C. Case Study: Interactive Design of an Extraction Solvent

The success of a liquid–liquid extraction process depends on the selection of the most appropriate solvent (Lo *et al.*, 1983). This case study examines the design of a solvent for the extraction of acetic acid from water. Lo *et al.*'s (1983) procedure for solvent selection was adapted for interactive design use.

The physical properties of the solvent used to facilitate separation in liquid–liquid extraction have major impact on process performance. Two important physical properties are (1) solvent selectivity for the solute and (2) the solute's partition coefficient between the solvent and the parent liquor.

Lo used the three-term solubility parameter (Barton, 1983) and a graphical procedure to identify solvents for liquid–liquid extraction. In a 2D space constructed from the polar component of the solubility parameter δ_p and the hydrogen bonding component of the solubility parameter δ_H, the distribution coefficient of the solute B, m_B, is given by

$$m_B \propto r_{B,S}^{-2},$$

and the selectivity β_B is given by

$$\beta_B \propto \left(\frac{r_{A,S}}{r_{B,S}} \right)^2,$$

where $r_{i,j}$ is the distance between points i and j. The optimal solvent for extraction thus lies on a line passing through the solute and parent liquor close to the solute for large distribution but far from the parent liquor for high selectivity.

Figure 5 shows the design space for the problem of designing a solvent to extract acetic acid from water. The target area is the lower left-hand portion of the acetic acid–water line. Figure 5 shows the group vectors for acetone near the target region. Table XVIII shows several solvents that

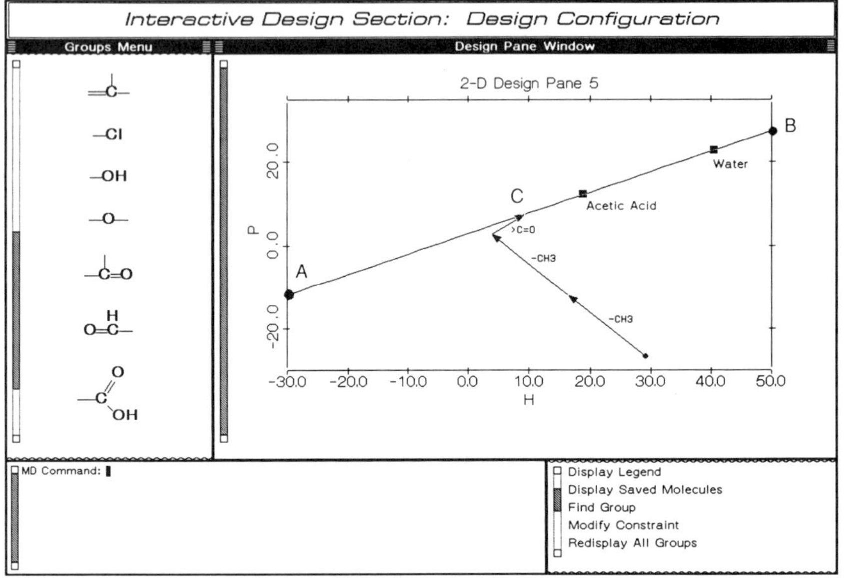

FIG. 5. Solvent design: a solution.

were interactively designed along with estimated values for their solubility parameters.

The $-CH_2-$ group has a δ_p contribution of $-.328$, and δ_H contribution of $-.512$. The slope of the $-CH_2-$ group vector in the interactive design space is .64. This is very close to the acetic acid–water target line slope of .49. Adding several $-CH_2-$ groups to any of the solvents shown in Table XVIII thus results in an acceptable new solvent.

TABLE XVIII

LIQUID–LIQUID EXTRACTION SOLVENTS

Solvent	δ_p	δ_H
$CH_3 - Cl$	7.2	6.2
$CH_2 = CH - CH_3$	4.7	4.0
$CH_2 = CH - Cl$	9.2	6.3
$CH_3 - O - CH_3$	3.8	5.7
$CH_3 - CO - CH_3$	7.5	9.5
$CH_2 = C(CH_3)_2$	4.6	4.1
$C(CH_3)_4$	0.5	1.2
$CCl_2(CH_3)_2$	9.5	5.8

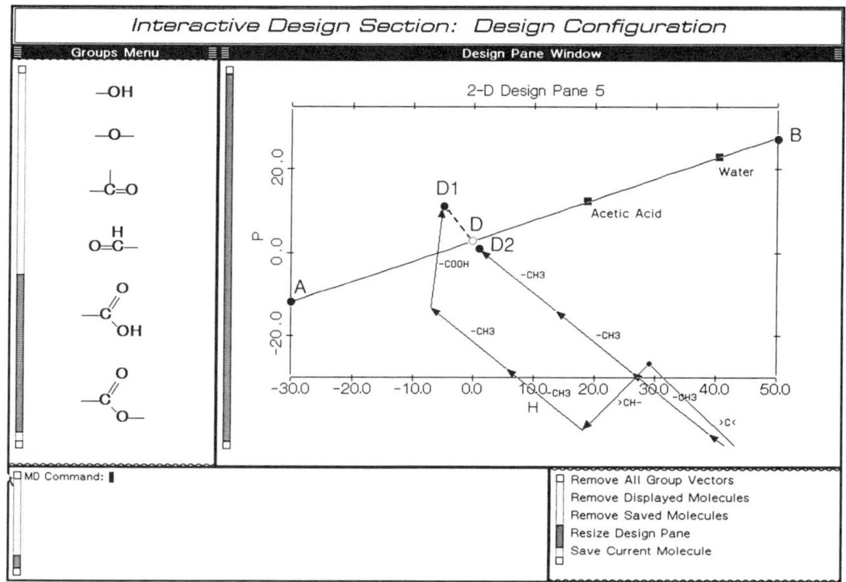

FIG. 6. Interactive design of solvent mixtures.

The direction of the $-CH_2-$ group vector is toward lower δ_p and δ_H values. This reduces the distribution coefficient with increasing number of $-CH_2-$ groups. This result is in accordance with experimental observation (Lo *et al.*, 1983).

Linear mixing rules for solubility parameters enable the design of solvent mixtures. In a δ_p–δ_H design space a binary mixture's solubility parameters lie on a straight line joining the components' solubility parameters. Figure 6 shows the group vectors for the $[CH_3-CH(COOH)-CH_3]-[C(CH_3)_4]$ solvent pair (point D resulting from the mixing of the two solvents represented by points D_1 and D_2). Such an approach can be used for any mixture property approximated by a linear mixing rule. Research is continuing on the use of nonlinear mixing rules for the interactive design of mixtures.

D. CASE STUDY: INTERACTIVE DESIGN OF A PHARMACEUTICAL

This case study presents an example taken from work done by Cramer (1980c; Cramer *et al.*, 1979). It demonstrates how the interactive design procedure assists in the second step of the quantitative structure activity relationship (QSAR) approach to drug design.

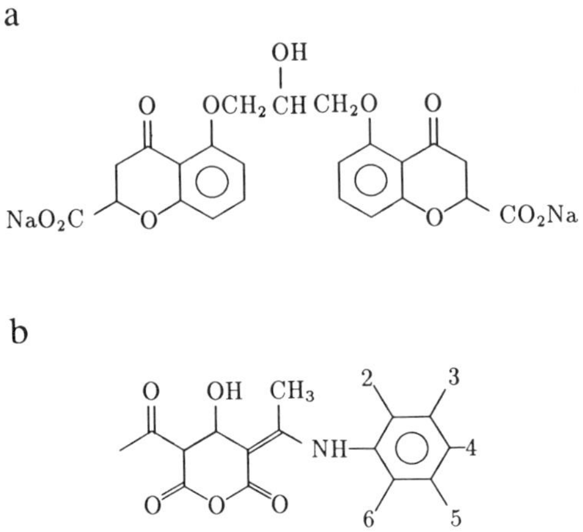

FIG. 7. Current and lead compounds for the drug design case study: (a) existing antiallergic, sodium chromoglycate; (b) a lead compound.

The QSAR approach to drug design takes a two-step approach to the correlation of biological activity with structure. The first step correlates a measure of biological activity, usually the reciprocal concentration having a 50% effect, with a number of physical properties. This correlation yields a model that is used to determine the optimal value of the physical properties giving the maximum potency. The second step of the approach is to identify group substitutions that lead to structures possessing these optimal physical property values.

Sodium chromoglycate (Fig. 7a) is effective at warding off asthmatic attacks. However, it must be administered by inhalation. Cramer *et al.* (1980c) performed a QSAR study to find a more potent pharmaceutical that could be administered orally. Searching through a database of approximately 1000 compounds they selected the pyranenamies (Fig. 7b) as their lead compound. The pyranenamines have biological properties sufficiently promising to merit a synthetic search for structurally related compounds having improved properties. The task was to develop highly active compounds by varying phenyl ring substituents.

Potency of the developed candidates was measured by the log of the concentration causing a 50% inhibition in asthmatic activity. This measure is represented as pI_{50}. To relate pI_{50} to molecular structure Cramer *et al.* (1980c) followed the two step QSAR approach. Nineteen compounds were synthesized with varying substituents. pI_{50} was measured for each analog. Values for the octanol–water partition logarithm, π, and Hammett's constant, σ, were computed for each of the substituent sets.

These data were regressed to develop a relationship between pI_{50} and the biochemical parameters π and σ. The model developed was

$$pI_{50} = -0.72 - 0.14\Sigma\pi - 1.35(\Sigma\sigma)^2. \tag{14}$$

Group-contribution estimation techniques were developed for π and σ. The contributions were obtained by regressing the substituent constants tabulated by Hansch and Leo (1979). Overall the estimation techniques have considerable error and cannot be used for quantitative estimation. However, they provide correct overall trends and thus inform the designer of promising substitutions.

Equation (14) shows that to perform an interactive design we would use a two-dimensional σ vs. π physical property design space. The contours represent solutions of Eq. (14) with pI_{50} equal to -1.0, -0.5, and 0.0. The symbols correspond to the values:

Symbol	pI_{50}
Square	-1.0
Triangle	-0.5
Circle	0.0

The direction of maximum activity is thus near σ equal to zero and π large and negative. Figure 8 shows a pair of metasubstituents in our target

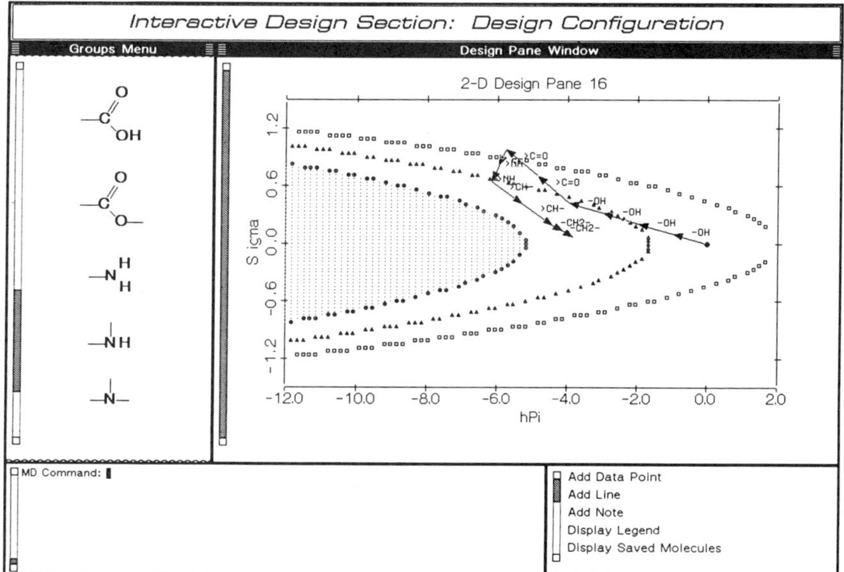

FIG. 8. Two high-activity metasubstituents during the interactive drug design.

TABLE XIX

COMPARISON OF EXPERIMENTAL AND MODEL ACTIVITIES

| | | | $p\mathrm{I}_{50}$ | |
Substituents	π	σ_M	Model	Experimental
1. 3 — $NHCO(CHOH)_2H$ 5 — $NHCO(CHOH)_2H$	-3.812	0.064	-0.19	3.0
2. 3 — $NHCOCH_2CH_3$ 5 — $NHCOCH_2CH_3$	-0.964	0.269	-0.68	2.5
3. 3 — $NHCOCH_3$ 5 — $NHCOCH_3$	-1.552	0.391	-0.71	1.9
4. 3 — $NHCOCH_3$ 5 — OH	-1.763	0.301	-0.60	1.7
5. 3 — $NHCOCOOCH_2CH_3$ 5 — $NHCOCOOCH_2CH_3$	-1.890	0.838	-1.40	1.7
6. 3 — $NHCOCH_2CH_2CH_3$ 5 — $NHCOCH_2CH_2CH_3$	-0.376	0.147	-0.70	1.3
7. 3 — $NHCO(CHOH)_2H$	-1.906	0.032	-0.45	1.3
8. 3 — $NHCOCH_3$ 5 — NH_2	-1.653	0.157	-0.49	1.0
9. 3 — $NHCOCH_2CH_3$	-0.482	0.134	-0.68	0.7
10. 3 — $NHCOCH_3$	-0.776	0.196	-0.66	0.7

direction. These metasubstituents correspond to entry 1 of Table XIX. Cramer found these substituents to have one thousand times more activity than the original unsubstituted compound.

IV. The *Molecule-Designer* Software System

A. GENERAL DESCRIPTION

The *Molecule-Designer* is the software system constructed to implement the interactive and automatic procedures for the design of molecules discussed in Sections II and III. It consists of approximately 20,000 lines of LISP code with an additional 17,000-line databank. It is implemented in Common LISP on a LISP Machine. The system is divided into eight sections each corresponding to a section of the overall methodology. The eight sections of the system are (1) login section configuration, (2) database section, (3) interactive design section, (4) molecule evaluation section, (5) problem formulation section, (6) target transformation section, (7) automatic design section, and (8) group-contribution section.

B. Interactive-Design-Relevant Sections

The following sections are specifically relevant to interactive design.

1. Problem Formulation Section

The problem formulation section provides an interface with which the designer can enter physical property constraints. The system displays the properties stored in its database (currently about 40 properties are present) and provides a simple constraint editor for entering and modifying physical property constraints.

2. Group-Contribution Section

Physical property estimation procedures are at the heart of the design procedures. The group-contribution section provides facilities for entering group contribution and equation oriented correlations. Models in the form of LISP code are entered for both types of estimation techniques. Additionally, groups and their contributions are specified for group contribution techniques.

The section is divided into two configurations: (1) an editing configuration that provides facilities for entering new groups to the system's database and (2) a model entry configuration that provides facilities for entering contributions for these groups, models for group-contribution techniques, and models for equation oriented estimation techniques. Figure 9 shows an example screen of the editing configuration with a new group being constructed.

3. Target Transformation Section

Once physical property constraints are entered in the problem formulation section, it is necessary to instruct the system how to estimate the properties contained in these constraints. This involves the creation of estimation procedures for each physical property used in the design constraints. The target transformation section provides facilities for collecting estimation techniques into estimation procedures. The chosen

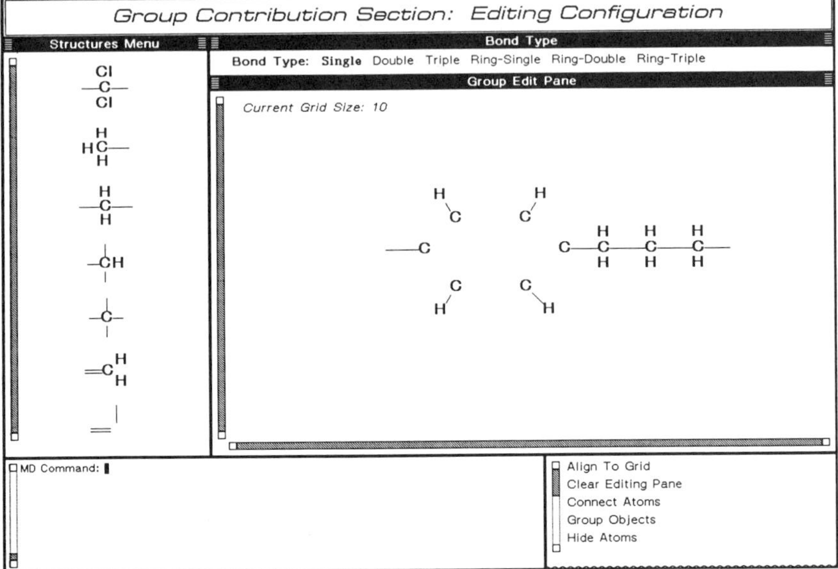

FIG. 9. Group contribution editing, configuration screen.

estimation techniques are used by the system to form an evaluation function to determine the feasibility of each point in a design space.

Figure 10 shows an example screen of the target transformation section. The constraints being transformed are shown in the window on the left. The estimation techniques that the designer can choose are shown in the window on the right. Currently the system has over 80 estimation techniques stored in its database.

4. Interactive Design Section

The interactive design section displays design spaces and group vectors. The configuration provides facilities to assist in selecting groups and in manipulating design constraints. Group vectors can be sorted by angle, magnitude, or closeness to another vector. The groups available can be restricted to those satisfying constraints on global valence, bond or atom types, and designer preferences. Each constraint can be interactively respecified enabling a designer to investigate the sensitivity of the feasible region. Example screens were previously shown when discussing the case studies in Sections III.B–D.

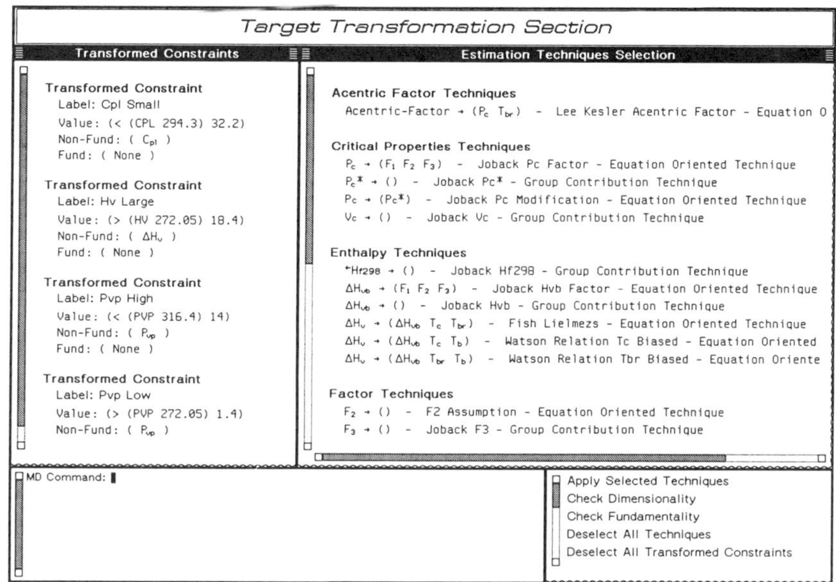

FIG. 10. Target transformation section, configuration screen.

5. *Molecule Evaluation Section*

The evaluation section provides facilities for estimating a molecules' physical properties. Estimating physical properties is especially useful when formulating the design target and evaluating designed molecules. During problem formulation we may need to estimate the properties of compounds currently in use. In final evaluation we would like to have the property profile of the designed compounds available for inspection.

One of the major objectives of the evaluation section is to provide estimation techniques of the highest accuracy. These estimation techniques may not be appropriate for use in the design procedures. They would thus serve as an additional check to verify the efficacy of any molecules designed.

V. Concluding Remarks

The identification of molecules with desired physical properties' values is becoming at an increasing frequency, a deliberate task with a dominant feedforward design philosophy and a decreasing number of feedback adjustments, coming from experimental results. This attitude will be reinforced as new theoretical and empirical physical property estimation techniques will provide improved estimates.

Using functional groups as the essential building blocks, one may compose any molecular structure and simultaneously employ group contribution techniques for the estimation of the molecule's properties. Clearly, a common alphabet for (1) the description of molecules and (2) the evaluation of their properties, is the most efficient but not necessarily the only alternative. The best representation for the design of molecules will be determined by the representation offering the best property estimation techniques, provided that it is at a higher-than-atomic level. Functional groups and group contribution techniques offer, at the present, the only available, consistent framework.

Once the representation of a molecular structure has been resolved, selecting the molecule(s), which satisfies a set of constraints, is simply a task of searching through a large space of discrete solutions. One could bring forth any technique for solving such problem, e.g., specific branch-and-bound algorithm to solve the underlying MINLP problem. Instead of *one, implicitly located solution*, we have opted for a search strategy that *explicitly locates all feasible solutions*. Such a preference has been motivated by the fact that during the initial design phase one is interested in all options rather than just one. Several additional considerations will be taken into account before a final choice is made; considerations that are not normally included during the initial phase of the design. To tame the combinatorial complexity of the design problem, we have used a hierarchical strategy with successive refinement of molecular representation. So, the clearly infeasible molecules can be easily screened out fairly early on. As more detail is added to the molecular representations, if the number of retained molecules is still very high, then, clearly, the initial design specs are fairly loose and should be tightened. Also, as the number of alternative molecules satisfying the set of property constraints decreases as a result of systematic search, more advanced techniques can be employed for the estimation of physical properties, and increased experimentation offers high returns.

As the limitations of the additive group-contribution techniques become more apparent, new representational models will be required to solve the product design problem. These models must maintain some simplicity in the *structure–property* relationships, which can be inverted in an efficient and explicit manner to yield the structure(s) of the feasible molecule(s). Such models have yet to be invented, but it is important to keep in mind the needs of the product design, as new theories and techniques are being written for the estimation of physical properties.

References

Alternburg, von K., Die Abhängigkeit der Siedetemperatur isomerer Kohlenwasserstoffe von der Form der Moleküle. *Brennst.-Chem.* **47**(11), 331–336 (1966).

American Society of Heating, Refrigerating and Air-Conditioning Engineers (ASHRAE), "Handbook of Fundamentals." ASHRAE, New York, 1972.

Barton, A. F. M., "CRC Handbook of Solubility Parameters and Other Cohesion Parameters." CRC Press, Boca Raton, FL, 1983.

Berg, L., Selecting the agent for distillation processes. *Chem. Eng. Prog.* **65**(9), 52–57 (1969).

Brignole, E. A., Bottini, S., and Gani, R., A strategy for the design and selection of solvents for separation processes. *Fluid Phase Equilib.* **29**, 125–132 (1986).

Constantinou, L., Gani, R., Fredenslund, Aa., Klein, J. A., and Wu, D. J., Computer-aided product design, problem formulation and application. *Proc. PSE'94*, Kyongju, Korea (1994).

Cramer, R. D., BC(DEF) Parameters. 1. The intrinsic dimensionality of intermolecular interactions in the liquid state. *J. Am. Chem. Soc.* **102**(6), 1837–1849 (1980a).

Cramer, R. D., BC(DEF) Parameters. 2. An empirical structure-based for the prediction of some physical properties. *J. Am. Chem. Soc.* **102**(6), 1849–1859 (1980b).

Cramer, R. D., A QSAR success story. *CHEMTECH*, December, pp. 744–747 (1980c).

Cramer, R. D., Snader, K. M., Willis, C. R., Chakrin, L. W., Thomas, J., and Sutton, B. M., Application of quantitative structure-activity relationships in the development of the antiallergic pyranenamines. *J. Med. Chem.* **22**(6), 714–725 (1979).

Derringer, G. C., and Markham, R. L., A computer-based methodology for matching polymer structures with required properties. *J. Appl. Polym. Sci.* **30**, 4609–4617 (1985).

Dillinger, T., "VLSI Engineering." Prentice-Hall, Englewood Cliffs, NJ, 1988.

Dossat, R. J., "Principles of Refrigeration." Wiley, New York, 1981.

Douglas, J. M., "Conceptual Design of Chemical Processes." McGraw-Hill, New York, 1988.

Fogiel, M., "Modern Microelectronics." Research and Education Association, New York, 1972.

Francis, A. W., Solvent selectivity for hydrocarbons. *Ind. Eng. Chem.* **36**(8), 764–771 (1944).

Franke, R., "Theoretical Drug Design Methods." Elsevier, Amsterdam, 1984.

Fredenslund, A., Gmehling, J., and Rasmussen, P., "Vapor-Liquid Equilibria using UNIFAC." Elsevier, Amsterdam, 1977.

Gani, R., and Brignole, E. A., Molecular design of solvents for liquid extraction based on UNIFAC. *Fluid Phase Equilib.* **13**, 331–340 (1983).

Gani, R., and Fredenslund, Aa., Computer-aided molecular and mixture design with specific property constraints. *Fluid Phase Equilib.* **82**, (1993).

Gani, R., Nielsen, B., and Fredenslund, Aa., A group contribution approach to computer-aided molecular design. *AIChE J.* **37**, 1318 (1991).

Godfrey, N. B., Solvent selection via miscibility number. *CHEMTECH*, June, pp. 359–363 (1972).

Goosey, M. T., Permeability of coatings and encapsulants for electronic and optoelectronic devices, *in* "Polymer Permeability" (J. Comyn, ed.), pp. 309–339. Elsevier, London, 1985.

Gordon, M., and Scantlebury, G. R., Non-random polycondensation, statistical theory of the substitution effect. *Trans. Faraday Soc.* **60**, 604–621 (1964).

Gray, N. A. B., "Computer-Assisted Structure Elucidation." Wiley, New York, 1986.

Hammett, L. P., Some relations between reaction rates and equilibrium constants. *Chem. Rev.* **17**(1), 125–136 (1935).

Hansch, C., and Leo, A., "Substituent Constants for Correlation Analysis in Chemistry and Biology." Wiley, New York, 1979.

Hansch, C., Muir, R. M., Fujita, T., Maloney, P. P., Geiger, F., and Streich, M., The correlation of biological activity of plant growth regulators and chloromycetin derivatives with Hammett constants and partition coefficients. *J. Am. Chem. Soc.* **85**, 2817–2824 (1963).

Hayes-Roth, F., Waterman, D. A., and Lenat, D. B., "Building Expert Systems." Addison-Wesley, Reading, MA, 1983.

Horvath, A. L., "Molecular Design–Chemical Structure Generation from the Properties of Pure Organic Compounds." Elsevier, Amsterdam, 1992.

Hosoya, H., and Murakami, M., Topological index as applied to p-electronic systems. II. Topological bond order. *Bull. Chem. Soc. J.* **48**(12), 3512–3517 (1975).

Joback, K. G., A unified approach to physical property estimation using multivariate statistical techniques. Master's Thesis, Massachusetts Institute of Technology, Cambridge, MA (1984).

Joback, K. G., and Reid, R. C. Estimation of pure-component properties from group-contributions. *Chem. Eng. Commun.* **57**, 233–243 (1987).

Joback, K. G., and Stephanopoulos, G. Designing molecules possessing desired physical property values. "Foundations of Computer-Aided Process Design." Elsevier, Amsterdam, 1990.

Kier, L. B., and Hall, L. H., "Molecular Connectivity in Structure-Activity Analysis." Wiley, New York, 1986.

Klincewicz, K. M., Prediction of critical temperatures, pressures, and volumes of organic compounds from molecular structure. Master's Thesis, Massachusetts Institute of Technology, Cambridge, MA, (1982).

Langley, B. C., "Refrigeration and Air Conditioning." Prentice-Hall, Englewood Cliffs, NJ, 1986.

Lo, T. C., Baird, M. H. I., and Hanson, C., "Handbook of Solvent Extraction." Wiley, New York, 1983.

Lyman, W. J., Reehl, W. F., and Rosenblatt, D. H., "Handbook of Chemical Property Estimation Methods." McGraw-Hill, New York, 1982.

Macchietto, S., Odele, O., and Omatsone, O., Design of optimal solvents for liquid–liquid extraction and gas absorption processes. *Trans. Inst. Chem. Eng.* **68**, 429 (1990).

Martin, Y. C., "Quantitative Drug Design: A Critical Introduction." Dekker, Inc., New York, 1978.

Mih, W. C., Catalysts for epoxy molding compounds in microelectronic encapsulation. *ACS Symp. Ser.* **242**, 273–283 (1984).

Moore, R. E., "Methods and Applications of Interval Analysis." Society for Industrial and Applied Mathematics, Philadelphia, 1979.

Nielsen, B., Gani, R., and Fredenslund, Aa., A group contribution approach to computer aided molecular design. *AIChE J.* (1995) (in press).

Odele, O., and Macchietto, S., Computer-aided molecular design, a novel method for optimal solvent selection. *Fluid Phase Equilib.* **82**, 47 (1993).

Perry, R. H., and Chilton, C. H., "Chemical Engineers' Handbook." McGraw-Hill, New York, 1973.

Randić, M., On characterization of molecular branching. *J. Am. Chem. Soc.* **97**(23), 6609–6615 (1975).

Reid, R. C., Prausnitz, J. M., and Poling, B. E., "The Properties of Gases and Liquid." McGraw-Hill, New York, 1987.

Rouvray, D. H., Predicting chemistry from topology. *Sci. Am.* **255**(3), 40–47 (1986).

Salame, M., Prediction of gas barrier properties of high polymers. *Polym. Eng. Sci.* **26**(33), 1543–1546 (1986).

Stephanopoulos, G., and Townsend, D. W., Synthesis in process development. *Chem. Eng. Res. Dev.* **64**, 160–174 (1986).

Taft, R. W., Separation of polar, steric, and resonance effects in reactivity. *In* "Steric Effects in Organic Chemistry" (M. S. Newman, ed.). Wiley, New York, 1956.

Tortorello, A., and Kinsella, M. A., Solubility parameter concept in the design of polymers for high performance coatings. I. *J. Coat. Technol.* **55**(696), 99–38 (1983a).

Torterello, A., and Kinsella, M. A., Solubility parameter concept in the design of polymers for high performance coatings, II. *J. Coat. Technol.* **55**(697), 29–38 (1983b).

van Krevelen, D. W., "Properties of Polymers, Correlations with Chemical Structure." Elsevier, Amsterdam (1976).

Venkatasubramanian, V., Chan, K., and Carathers, J. M., Computer-aided molecular design using genetic algorithms. *Comput. Chem. Eng.* **18**, 883 (1994).

Verloop, A., The use of linear free energy parameters and other experimental constants in structure-activity studies. *In* "Drug Design" (E. J. Ariens, ed.) Academic Press, New York, 1972.

Watson, K. M., Thermodynamics of the liquid state. *Ind. Eng. Chem.* **35**, 398–405 (1943).

Wiener, H., Correlation of heats of isomerization, and differences in heats of vaporization of isomers, among the paraffin hydrocarbons. *J. Am. Chem. Soc.* **69**(11), 2636–2638 (1947).

NONMONOTONIC REASONING: THE SYNTHESIS OF OPERATING PROCEDURES IN CHEMICAL PLANTS

Chonghun Han, Ramachandran Lakshmanan,[1]
Bhavik Bakshi,[2] and George Stephanopoulos

Laboratory for Intelligent Systems in Process Engineering
Department of Chemical Engineering
Massachusetts Institute of Technology
Cambridge, Massachusetts 02139

[1] Present address: Department of Chemical Engineering, University of Edinburgh, Scotland, UK.

[2] Present address: Department of Chemical Engineering, Ohio State University, Columbus, OH 43210, USA.

ADVANCES IN CHEMICAL ENGINEERING, VOL. 22

Planning a sequence of actions to take you from one point to another is usually a proposition whose inherent difficulty increases as more obstacles are placed in your way. The number of these obstacles (constraints), which you must skirt around, determines the complexity of the task, because any time you run into one of them you must *backtrack* and try an alternative step or path of steps. Such *serial* (or *linear* or *monotonic*) construction of a plan is fraught with pitfalls and repeated backtracking. The more the constraints, the more inefficient the monotonic planning. If, on the other hand, an action step (a Clobberer) leads to the violation of a constraint, then *do not backtrack*. Take another action step (a White Knight), which, when it precedes a Clobberer, negates the impact of the Clobberer, and you never need to backtrack. So, the more constraints, the more efficient your planning process. Such *nonserial* (or *nonlinear* or *nonmonotonic*) reasoning has become the essence of all modern and efficient planners, whether they are *logic-based* and *explicit*, or *implicit* enumerators of alternative plans. The purpose of this choice is twofold: (1) to introduce the ideas of *nonmonotonic reasoning* in the planning of process operations and (2) to demonstrate how nonmonotonic planning can be used to synthesize operating procedures for chemical processes, either off-line for standard tasks (e.g., routine startup or shutdown), or on-line for real-time response to large departures from desired conditions. It is shown that hierarchical modeling of process operations and operators is essential for the efficient deployment of nonmonotonic planning, and that the tractability of the resulting algorithms is strictly dependent on the form of the operators. In this regard, the ideas on modeling in this chapter draw heavily from the material of the first chapter in Volume 21 of this series. Nonmonotonic planners handle with superb efficiency constraints on (1) the temporal ordering of operations, (2) avoidable mixtures of chemical species, and (3) bounding quantitative conditions on the state of a process. Consequently, they could be used to generate explicitly all feasible operating procedures, leaving a far smaller search space for the selection of the optimum procedure by a numerical optimizer.

I. Introduction

The planning and scheduling of process operations, beyond the confines of single processing units, are fairly complex tasks. Thus, the synthesis of operating procedures for the routine startup or shutdown of a plant, equipment changeover, safe fallback from hazardous situations, changeover of the process to alternate products, and others, involves (1) a long list of noncommensurable objectives (operating economics, safety, environmental impact), (2) models describing the behavior of all units in a plant, (3) human supervision and intervention, and (4) the degree of the required automation. While mathematical programming techniques, such as branch and bound algorithms for mixed-integer linear or nonlinear (MILP or MINLP) formulations, have been used (Crooks and Macchietto, 1992) to locate the "optimal" plan, computer-based process control systems (Pavlik, 1984) "do not know" how to plan, schedule, and implement complex, nonserial schedules of process operations. The latter is considered to be largely a supervisory task left to the human operator. Nevertheless, there exists a natural tendency for increased automation in the control room (Garrison *et al.*, 1986), which takes one or both of the following forms: (1) a priori synthesis of operating procedures and subsequent implementation through programmed-logic control systems or (2) online, automatic planning and scheduling of operating procedures with real-time implementation through distributed control systems. The former approach is favored for the routine tasks such as startup, shutdown, and changeover, whereas the latter is essential for the optimization of operating performance, the intelligent response to large departures from operating norms, and the planning of emerging fallback operations in the presence of unsafe conditions.

The synthesis of operating procedures involves the specification of an ordered sequence of primitive operations, such as, opening or closing manual valves, turning on or off motors, and changing the values of setpoints, which when applied to a process will change the state of a plant from some given initial state and eventually bring it to a desired *goal* state (Lakshmanan and Stephanopoulos, 1988a). Typically, the transformation from the initial to the goal state is attained through a series of intermediate states (Fig. 1), each of which must satisfy a set of constraints, such as

1. *Physical constraints*, imposed by mass, energy, and momentum balances, chemical and phase equilibrium conditions, and rate phenomena.
2. *Temporal constraints*, imposed by various considerations; e.g., you cannot reach the state of a full tank if the state of the feeding pump has not been previously turned on.
3. *Logical, mixing constraints*, to avoid explosive mixtures, poisoning of catalysts, generation of toxic materials, deterioration of product quality, and other consequences resulting from the unintentional mixing of various chemicals.
4. *Quantitative inequality constraints*, requiring that the values of temperatures, pressures, flowrates, concentrations, and material or energy accumulations remain within a range defined by lower and upper bounds, in order to maintain the safety of personnel and

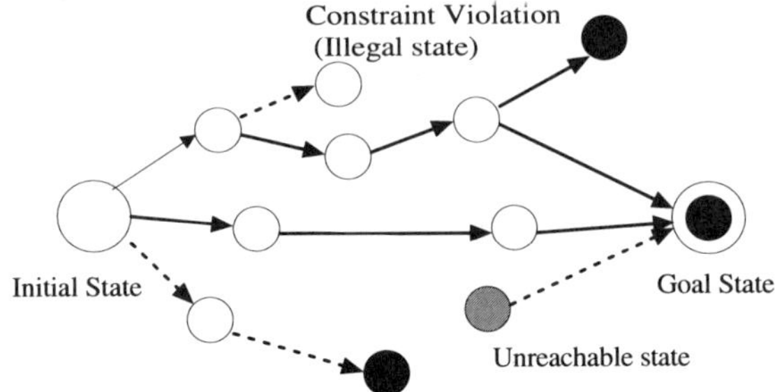

FIG. 1. Schematic of a typical planning problem (solid line signifies a feasible path and dotted line an infeasible path).

equipment, environmental regulations, and production specifications.

5. *Performance constraints*, in terms of operating cost, product turnaround time, and others.

A. PREVIOUS APPROACHES TO THE SYNTHESIS OF OPERATING PROCEDURES

Domain-independent theories of planning have attempted to generate formulations and solution methodologies that do not depend on the characteristics of the particular area of application, whereas *domain-dependent* approaches have attempted to capitalize on the specific characteristics of a given problem domain, in order to construct special-purpose planning programs. Although a domain-independent planning theory would be preferable, there exists sufficient theoretical evidence to suggest that building a probably correct, complete, domain-independent planner that is versatile enough to solve real-world planning problems, is impossible (Chapman, 1985).

Rivas and Rudd (1974) should be credited with the pioneering effort of synthesizing operating sequences that achieve certain operational goals, while ensuring that safety constraints were not violated. Chemical processes were modeled as networks of *valves* and *connectors* (e.g., pipes, vessels, unit operations), and the safety constraints were expressed as statements about the species that should never be present simultaneously in any location of the plant. The sequential logic algorithm that they proposed is equivalent to a restricted form of constraint propagation. Thus, high-level, abstractly expressed goals (input by the user) such as "start-a-heater" werc propagated to lower-level goals of higher detail and were converted into a series of valve operations. Unfortunately, keeping track of the dependencies in order to generate *explainable* plans (a highly desirable feature) was not possible, as it was impossible to account for additional classes of constraints.

Kinoshita *et al.* (1982) divided a plant into sections around "key" processing equipment (including peripherals and attached pipes and valves) and identified a list of individual operations for each section. During the first phase of their methodology, they generate a tree of all possible sequences of operations for each section. Each tree described the *transition relationships among the states* of each section, and was endowed with the following information: (1) minimum and maximum time required for each operation, (2) the cost associated with each operation, and (3) constraints on the states of adjacent units. During the second phase of their work, Kinoshita *et al.* attempted to coordinate the sequences of the independent trees in an effort to achieve consistency among the operations of interacting units, by adjusting the timing and thus the temporal order of the various primitive operations.

In a similar spirit, Ivanov *et al.* (1980) had previously constructed *state transition networks* for the representation of operating states taken on by a plant as it goes from the initial state to the goal state. While the nodes of the network represent the states of a plant, the edges (arcs) signify the primitive operations that carry the processing units from one state to the next. To account for quantitative performance measures, Ivanov *et al.* assigned to each arc a weighting factor, which depended on the type of the performance measure. Within such framework, the synthesis of the "optimal" operating procedure becomes a search problem through a set of discrete alternatives. Ivanov *et al.* applied this formulation to the synthesis of optimal startup sequences of operations.

Tomita, Nagata, and O'Shima (1986), on the other hand, represented the topology of a plant through a directed graph with the nodes signifying the constituent equipment and the directed edges representing the pipelines. The valves were assigned to the appropriate edges of the digraph. Propositional logic was used to express the topology of the digraph and the temporal ordering of the various operations, thus leading to the generation of logically feasible operating plans. Subsequent work (Tomita *et al.*, 1989a, b) focused on the use of logic-based techniques and other ideas from artificial intelligence toward the development of an automatic synthesizer of operating procedures.

The work of Fusillo and Powers (1987) has attempted to define a formal theory for the synthesis of operating procedures. First, it introduced more expressive descriptions of the plants and their behavior, going beyond the

Boolean formulations favored by the pioneers in the field. Second, the planning methodology allowed the incorporation of global and local constraints, all of which were checked at each intermediate state to ensure that the evolving operating procedure was feasible. Third, they introduced the concept of a "stationary state," i.e., a state at which the plant itself does not change significantly with time (e.g., can wait until next action is taken), and where the operating goals are partially met. The "stationary state" was a breakthrough concept. It allows a natural breakoff point from a large schedule of operations, thus leading to an automatic generation of *intermediate operating goals*, which simplify the search for feasible operating procedures by decomposing the overall planning problem to smaller subproblems. In addition, a "stationary state" represents a convenient stopping-off point for recovery from operational errors, thus allowing the process to retreat back to an intermediate state thereby avoiding a complete shutdown. In subsequent publications (Fusillo and Powers, 1988a, b) the same authors introduced local quantitative models, and they also carried out detailed studies on the synthesis of purge operations.

Foulkes *et al.* (1988) have approached the synthesis of operating procedures from a more empirical angle. They have extended the work of Rivas and Rudd (1974) for the synthesis of complex pump and valve sequencing operations, relying on the use of logical propositions (implemented as rule-based expert systems), which capture the various types of constraints imposed on the states of a processing system.

B. The Components of a Planning Methodology

From Fig. 1 it is clear that the synthesis of an operating procedure could start from (1) the initial state and proceed forward to the goal state, (2) the goal state and proceed backward to the initial state, or (3) any known intermediate state(s) and proceed in both directions to bridge the gap with the specified initial and goal states. Furthermore, it should also be clear that there exist many paths connecting the initial to the goal state, which can be evaluated in terms of (1) their *feasibility*, i.e., whether their corresponding constituent states violate the specified constraints or not, and (2) a *performance index*, i.e., associated cost, total operational time, or other. To check the feasibility of a path we need an explicit *set of constraints* determining the allowable states of the process, and to evaluate performance we need a quantitative description of the desired performance index (or a vector of performance indices). Finally, in order to be able to evaluate the state of the process, we need *models* that describe the underlying physicochemical phenomena and the variety of primitive operations.

No planning methodology can be efficient or credible for the synthesis of process operating procedures, if it focuses only on the search for the "best" plan, delegating the modeling needs and the articulation of the requisite constraints to inferior tasks. The reverse is also true. Consequently, a concise methodology for the synthesis of operating procedures should provide a unified treatment of all three pivotal components, namely

1. Modeling of processing systems and primitive operations.
2. Explicit articulation of all constraints that an operating procedure should satisfy.
3. Search over the set of discrete plans and the identification of the feasible, and possibly of the "optimal" among them.

1. Formal Statement of the Operations Planning Problem

Consider a processing system composed of N subsystems, whose topological interconnections are determined by a set of streams, $S = \{s_{i,j}^{(p,q)}\}$; $i, j = 1, 2, \ldots, N$, with $s_{i,j}^{(p,q)} = 1$ if the pth output of the ith subsystem is the qth input to the jth subsystem, and zero otherwise. What is denoted as a subsystem could vary with the level of detail considered in the description of the processing system. At a high level of detail, a subsystem could be a processing equipment, an actuator, a safety device, or a controller. At a low level of detail, a subsystem could be a processing section, composed of several interconnected processing equipment with their actuators and controllers.

Let $\mathbf{x}_i$ denote the vector of state variables describing the behavior of the ith subsystem, and $\mathbf{u}_i^{(q)}, \mathbf{y}_i^{(p)}$ the vectors describing its qth input and pth output streams. The modeling relationships around each subsystem yield

$$\frac{d\mathbf{x}_i(t)}{dt} = \mathbf{f}_i\big[\mathbf{x}_i(t), \mathbf{u}_i^{(q)}(t); \quad q = 1, 2, \ldots, k_i\big]; \quad i = 1, 2, \ldots, N, \quad (1)$$

$$\mathbf{y}_i^{(p)}(t) = \mathbf{g}_i^{(p)}\big[\mathbf{x}_i(t), \mathbf{u}_i^{(q)}; \quad q = 1, 2, \ldots, k_i\big];$$

$$i = 1, 2, \ldots, N, \quad p = 1, 2, \ldots, m_i. \quad (2)$$

If the surrounding world is denoted as the zeroth subsystem, then the variables, $\mathbf{y}_0^{(p)}$; $p = 1, 2, \ldots, m_0$ represent the external inputs to the processing system, and in general they can be divided into two classes; the *real-valued*, $\mathbf{y}_{0,\mathrm{RV}}^{(p)}$ and the *integer-valued*, $\mathbf{y}_{0,\mathrm{IV}}^{(p)}$. The $\mathbf{y}_0^{(p)}$ variables can be used to represent the actions of human operators of supervisory control systems.

The values of the inputs, $\mathbf{u}_i^{(q)}$, $q = 1, 2, \ldots, k_i$; $i = 1, 2, \ldots, N$, are determined by a set of operations, such as opening or closing a manual valve, turning on or off a motor, or changing the value of a controller's setpoint. Each of these operations can be associated with a particular input and describes how the value of input, $\mathbf{u}_j^{(q)}$, is determined from the value of outputs, $\mathbf{y}_i^{(p)}$;

$$\mathbf{u}_j^{(q)}(t) = Op_j^{(q)}\left[\mathbf{y}_i^{(p)}(t); \qquad p = 0, 1, 2, \ldots, m_i \quad \text{and}\right.$$

$$\left. \forall i: S_{ij}^{(p,q)} = 1; \quad \alpha_j^{(q)}(t)\right] \qquad (3)$$

where $j = 1, 2, \ldots, N$.

Operations $Op_j^{(q)}$, can be seen as *operators* that can take on a *logical integer*, *analytic* (static or dynamic), or *hybrid* form. With each operation we have associated an integer-valued variable, $\alpha_j^{(q)}(t)$, which determines the discrete state that the operator $Op_j^{(q)}$, takes on with time.

Given an initial state of a processing system and the outputs of the surrounding world, one may solve the set of Eqs. (1), (2), and (3) and thus find the temporal evolution of the state describing the process behavior. As the various operators $Op_j^{(q)}$ take on different discrete states over time, the trajectory of the process' state goes through changes at discrete time points. Clearly, the order in which the various operations change state and the time points at which they change value, affect the temporal trajectory of the process evolution and determine the value of the performance index. Therefore, we are led to a *mixed-integer*, *nonlinear programming* (MINLP) problem with the following decision variables:

1. *Integer decisions*
   (a) $\alpha_j^{(q)}(t)$, i.e., the discrete state of each operator, $Op_j^{(q)}$.
   (b) The temporal ordering of the discrete-state changes of the various operators, e.g., $t_1 < t_2 < t_3$, where $\alpha_2^{(1)}(t_1) = 1$ (from 0), $\alpha_4^{(1)}(t_2) = 0$ (from 1), $\alpha_3^{(1)}(t_3) = 2$ (from 1).
   (c) $y_{0,\text{IV}}^{(p)}(t)$, i.e., integer-valued variables, which are inputs from the surrounding world.
   (d) The temporal ordering of the discrete-value changes of the integer-valued inputs $y_{0,\text{IV}}^{(p)}(t)$, from the surrounding world.
2. *Continuous decisions*. The time-dependent trajectories of the real-valued variables that are inputs from the surrounding world.

In addition to these above, the values that the states, $\mathbf{x}_i$, $i = 1, 2, \ldots, N$; the outputs, $y_i^{(p)}$; $p = 1, 2, \ldots, m_i$; $i = 1, 2, \ldots, N$; and the inputs, $\mathbf{u}_i^{(q)}$; $q = 1, 2, \ldots, k_i$; $i = 1, 2, \ldots, N$ can take are limited by a set of engineering constraints, which are dictated by safety and operability considerations, physical limitations, or/ and production specifications. Finally, the selection of the "best" plan of operations requires the formulation of an objective function, or a vector of objective performance indices. Thus, for the general multiobjective synthesis of operating procedures we can state the following mixed-integer, dynamic optimization problem (termed *Problem* 1):

$$\underset{\mathbf{A};\, \mathbf{y}_0^{(p)}(t)}{\text{Minimize}}\; P = [P_1, P_2, \ldots, P_l]$$

subject to

$$\frac{d\mathbf{x}_i(t)}{dt} = \mathbf{f}_i[\mathbf{x}_i(t), \mathbf{u}_i^{(q)}(t); \qquad q = 1, 2, \ldots, k_i]; \quad i = 1, 2, \ldots, N, \tag{1}$$

$$\mathbf{y}_i^{(p)}(t) = \mathbf{g}_i^{(p)}[\mathbf{x}_i(t), \mathbf{u}_i^{(q)}; \qquad q = 1, 2, \ldots, k_i];$$

$$p = 1, 2, \ldots, m_i, \quad i = 1, 2, \ldots, N. \tag{2}$$

$$\mathbf{u}_j^{(q)}(t) = Op_j^{(q)}[\mathbf{y}_i^{(p)}(t), \qquad p = 1, 2, \ldots, m_i, \forall i: s_{ij}^{(p,q)} = 1; \quad \alpha_j^{(q)}(t)];$$

$$q = 1, 2, \ldots, k_j \quad \text{and} \quad j = 1, 2, \ldots, N, \tag{3}$$

$$\mathbf{h}_i(\mathbf{x}_i; \mathbf{u}_i^{(q)}; \qquad q = 1, 2, \ldots, k_i; \quad \mathbf{y}_i^{(p)};$$

$$p = 1, 2, \ldots, m_i) \le 0; \quad i = 1, 2, \ldots, N, \tag{4}$$

where $\mathbf{A}$ is the vector of all $\alpha_j^{(q)}(t)$, for $q = 1, 2, \ldots, k_j$ and $j = 1, 2, \ldots, N$.

Problem-1 is a formidable challenge for mathematical programming. It is an NP-hard problem, and consequently all computational attempts to solve it cannot be guaranteed to provide a solution in polynomial time. It is not surprising then that all previous efforts have dealt with simplified versions of Problem-1. These simplifications have led to a variety of

(a) Representation models for the behavior of the subsystems, i.e., the form of the f_i and $\mathbf{g}_i^{(p)}$ functions.
(b) Descriptions for the operations, i.e., the form of the $Op_j^{(q)}$ operators.
(c) Search techniques for feasible or optimal plans.

2. The Character of Constraints and the Modeling Requirements

The character of the constraining functions, $\mathbf{h}_i$, given by Eq. (4), determines to a large extent the form of the required modeling functions, $\mathbf{f}_i$, $\mathbf{g}_i^{(p)}$, and $Op_i^{(p)}$. For example, Rivas and Rudd (1974) stipulated that the only constraints they were concerned were related to the avoidance of certain mixtures of chemicals, anywhere in the process. This being the case, all states, input and output variables can be characterized by Boolean values, and the required models are simple expressions of propositional calculus. Subsequently, the work of Fusillo and Powers (1987, 1988a, b) introduced qualitative and static quantitative constraints, which required more detailed $\mathbf{f}_i$ and $\mathbf{g}_i^{(p)}$ descriptions.

Another important consideration regarding the character of constraints, $\mathbf{h}_i$, is their *hierarchical articulation at multiple levels of detail*. For example, suppose that one is planning the routine startup procedure for a chemical plant. Safety considerations impose the following constraint on the temporal ordering of operations:

Start the reaction section last,

which is expressed at the abstract level of processing section. On the other hand, the constraint

in starting a heat exchanger, start the cold side first

is expressed at the more detailed level of processing equipment. Since both types of constraints must be accounted for during the planning of an operating procedure, it is clear that the computer-aided modeling of a processing system should allow for hierarchical representations at various levels of detail.

The final comment on the character of constraints is related to the available technology for equation solving. The general form of constraints (4) involves restrictions on the time-dependent trajectories of states, inputs, and outputs. Since the set of decisions of Problem-1 involves both integer and continuous variables, we need numerical methodologies that can solve large sets of differential algebraic equations [i.e., Eqs. (1), (2), and (3)] with continuous and discrete variables. The recent works in dynamic simulation (Pantelides and Barton, 1993; Barton, 1992; Marquardt, 1991; Holl *et al.*, 1988) are spearheading the developments in this area, but until these efforts produce efficient and robust solution algorithms, the formulations of Problem-1 will be forced to accept simple statements of constraints [Eq. (4)].

3. Search Procedures

The solution of Problem-1 requires extensive search over the set of potential sequences of operations. Prior work has tried either to identify all feasible operating sequences through *explicit* search techniques, or locate the "optimum" sequence (for single-objective problems) through the *implicit* enumeration of plans. The former have been used primarily to solve planning problems with Boolean or integer variables, whereas the latter have applied to problems with integer and continuous decisions.

Explicit search techniques for the identification of feasible operating procedures originated in the area of artificial intelligence. Typical examples involve the pioneering "means–end analysis" paradigm of Newell *et al.* (1960), and the various flavors of *monotonic planning* (e.g., Fikes and Nilsson, 1971), or *nonmonotonic planning* (e.g., Sacerdoti, 1975; Chapman, 1985). With the exception of the work of Lakshmanan and Stephanopoulos (1988a, b, 1990), who have used nonmonotonic planning, all other works on the synthesis of operating procedures for chemical processes have employed ideas of monotonic planning.

Starting either from the desired goal or the given initial state of a processing system, monotonic planning constructs an operating procedure step by step, moving "closer" all the time to the other end (i.e., initial or goal states, respectively). After each operating step the values for all state, input, and output variables are known by solving Eqs. (1), (2), and (3), respectively, and the satisfaction of constraints [Eq. (4)] is tested. The last operating step is revoked (or modified) if any of the constraints (4) is violated. Nonmonotonic planning, on the other hand, works by constructing and refining partial plans. A *partial plan* is a set of operating steps that leaves a certain amount of information unspecified. Perhaps the temporal order in which two operating steps are executed is not specified, or a particular step of the operating procedure may be assigned a *set* of potential feasible operations, rather than a *single* operation. Thus, at the end, *a single partial plan actually describes a large number of "completions" or total operating procedures.* This feature is extremely important for two reasons:

1. *Efficiency.* By conducting planning in terms of partial plans, non-monotonic reasoning allows a single planning decision to represent a large number of plans, resulting in possibly exponential savings in efficiency (Chapman, 1985; Lakshmanan and Stephanopoulos, 1989).
2. *Integration.* The fact that nonmonotonic planning can capture large sets of feasible plans through single decisions implies that it can be integrated very naturally with the mixed-integer mathematical programming techniques if the identification of the "optimum" plan is required.

Implicit enumeration techniques for the identification of the optimum operating procedure originated in the area of operations research and are based largely on variants of the *branch-and-bound* paradigm. Through recent developments, they have become fairly efficient in locating the "optimum" solution, but they are still awkward in defining the set of feasible plans. Continued progress in the development of more efficient formulations of the planning problem and more efficient branch-and-bound strategies, to handle decision variables with many integer values, will provide a *practical tractability* for the solution of larger and richer versions of Problem-1.

C. OVERVIEW OF THE CHAPTER'S STRUCTURE

Section II addresses the most important facet of an efficient planning methodology, namely, "how to represent the behavior of processes and process operating steps." Three basic operators—simple propositional, conditional, and functional—are explored for their representational power to model process operations. In addition, the representation of process operational constraints dictates the need for a hierarchical modeling of processes and their subsystems, thus leading a modeling framework very similar to that of MODEL.LA. (see first chapter in Volume 21). In fact, we have used the formalism of MODEL.LA. to articulate all modeling needs, related to nonmonotonic planning of process operations.

Section III introduces the concept of nonmonotonic planning and outlines its basic features. It is shown that the tractability of nonmonotonic planning is directly related to the form of the operators employed; simple propositional operators lead to polynomial-time algorithms, whereas conditional and functional operators lead to NP-hard formulations. In addition, three specific subsections establish the theoretical foundation for the conversion of operational constraints on the plans into temporal orderings of primitive operations. The three classes of constraints considered are (1) temporal ordering of abstract operations, (2) avoidable mixtures of chemical species, and (3) quantitative bounding constraints on the state of processing systems.

In Section IV we provide illustrations of the modeling concepts presented in Section II and how the strategy of nonmonotonic planning has been used to synthesize the switchover operating strategy for a chemical process.

Section V extends the above ideas into a different problem, i.e., how to revamp the design of a process so that it can accommodate feasible operating procedures.

II. Hierarchical Modeling of Processes and Operations

In this section we will discuss a systematic approach for the representation of processes and operating steps. These representations, or *models*, conform to the general requirements presented in Section I,B; they can

(1) capture Boolean, integer (i.e., categorical, or qualitative), or real-valued variables of a process; and (2) emulate descriptions of the process or of the operations at multiple levels of detail with internal consistency.

A. MODELING OF OPERATIONS

The characterization of operating steps, i.e., the characterization of the operators, $Op_j^{(q)}$, depends on (1) the type of operation (i.e., Boolean, integer, analytic real-valued) and (2) the level of the desired detail (abstract operation over a section of a plant, primitive operations on valves, pumps, etc.). Let us see the classes of possible models, all of which could be available and used even within the same planning problem.

The STRIPS-Operator

Figure 2a shows the simplest possible model of an operating step, introduced by Fikes and Nilsson (1971) and known as the STRIPS operator. The *preconditions* are statements that must be true of the system and its surrounding world before the action can be taken. Similarly, the *postconditions* are statements that are guaranteed to be true after the actions corresponding to the specific operation have been carried out. The allowable forms of pre-, and postconditions should satisfy the following requirements:

(a) Pre- and postconditions must be expressed as propositions that have content, which is a tuple of elements, and can be negated.
(b) The elements of the preceding tuples can be variables or constraints and could be infinitely many of them.
(c) Functions, propositional operators, and quantifiers are not allowed.

The STRIPS operator is very simple and intuitively appealing. In attempting to use it for the synthesis of operating procedures for chemical processes, one finds immediately that it suffers from severe limitations, such as the following:

1. The temporal preconditions and postconditions in the STRIPS model vary from process to process, thus making impossible the creation of a generic set of operators a priori.
2. Given a set of preconditions [i.e., the values of $\mathbf{y}_i^{(p)}$ and $\alpha_j^{(q)}$; see Eq. (4)], the actions of process operation lead to new values for $\mathbf{u}_j^{(q)}$, which when propagated through Eq. (1) generate the new state, i.e., the action's postconditions. But Eq. (1) does include functional relationships, $\mathbf{f}_i$, thus leading to an overall functional operator, which is disallowed in a strict STRIPS model.

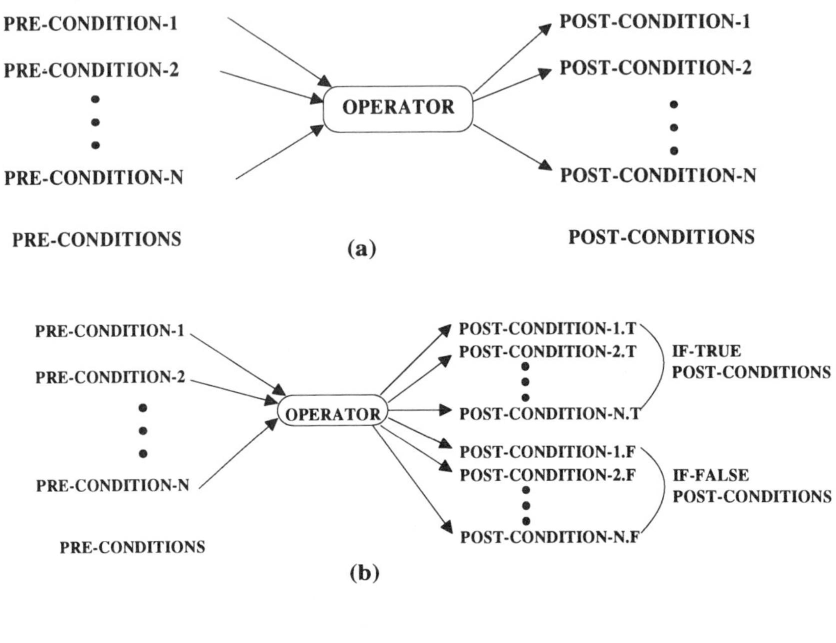

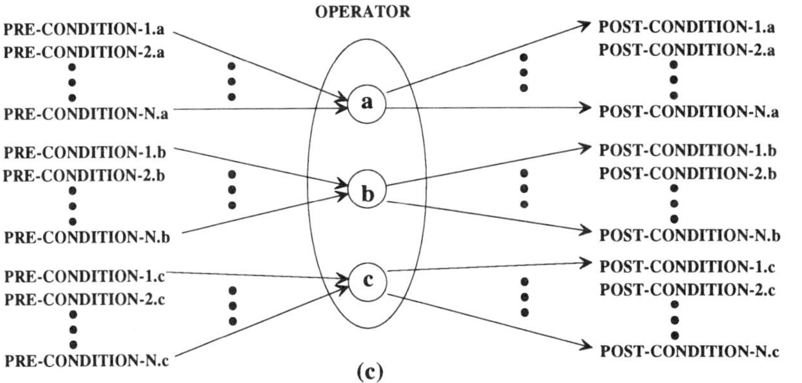

FIG. 2. The three planning operators: (a) STRIPS, (b) conditional, (c) functional.

Nevertheless, early work in the synthesis of operating procedures (Rivas and Rudd, 1974; Ivanov *et al.*, 1980; Kinoshita *et al.*, 1982; Fusillo and Powers, 1987) did employ various variants of the STRIPS operator on limited-scope problems with very useful results. Lakshmanan and Stephanopoulos (1987) also demonstrated the value of the STRIPS operators in ordering sequences of feasible operations during routine startup.

2. Conditional Operators

To overcome the limitations of the STRIPS operator, Chapman (1985) suggested the use of *conditional operators*, which produce two sets of postconditions (Fig. 2b), depending on whether all preconditions are true or any one of them is false. Conditional operators can describe a broader class of operations, but they do possess similar drawbacks when they are considered for the synthesis of operating procedures. In this regard, it is important to realize that the success of the operator-based approaches depend on the ability of the user to define, a priori, operator models for individual processing units, since such operators represent a fusion (i.e., simultaneous solution) of Eqs. (1)–(3) so that they can relate preconditions to the resulting postconditions. It is clear that their use is severely limited and cannot cover planning of operating procedures for (1) equipment/ product changeover, (2) optimizing control, and (3) certain types of safety fallback strategies.

3. Functional Operators

Instead of a conjunction of preconditions, as used by the STRIPS and conditional operators, the *functional* operator has a *set* of conjunctions of preconditions (Fig. 2c). Each element in the set describes some possible situation that might exist before the operator is applied. For each element of the set of preconditions, there is a corresponding element in the set of postconditions. The functional operator is a more flexible model than the STRIPS or conditional operators. It comes closer to the modeling needs for the synthesis of operating procedures for chemical processes, but as we will see in the next section, we need to introduce additional aspects in order to capture the network-like structure of chemical processes.

4. An Illustration

Consider the simple pipes-and-valves system shown in Fig. 3. Initially, oxygen is flowing into the system through inlet 1, and out through the outlet. We would like to develop an operating procedure to solve the following operational problem: "Route the flow of methane gas from the inlet 2 to the outlet without running the risk of explosion (i.e., avoid mixing oxygen and the hydrocarbon)."

Since the only conditions of interest relate to the presence or absence of materials in the various piping segments, the required models for the operations are very simple and can be given by the following three

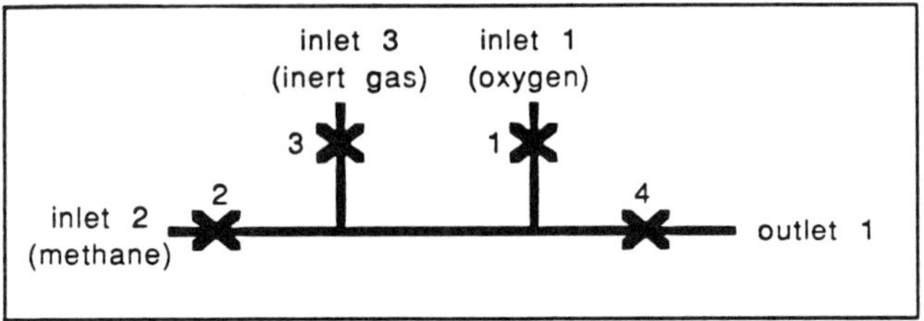

INITIAL-STATE	GOAL-STATE
(flowing oxygen)	(flowing methane)
not(explosion)	not(explosion)

OPERATORS

(stop-flow x)
pre-conditions: (flowing x)
post-conditions: not(flowing x)

(establish-flow x)
pre-conditions: ()
post-conditions: (flowing x)

(purge x)   pre-conditions:
not(flowing x);   not(equal inert-gas x)
post-conditions: not(present x); (flowing inert-gas)

FRAME AXIOMS

(present methane) & (present oxygen) => (explosion!)

(flowing x) => (present x)

FIG. 3. Example operations planning problem. (Reprinted from *Comp. Chem. Eng.*, **12**, Lakshmanan, R. and Stephanopoulos, G., Synthesis of operating procedures for complete chemical plants, Parts I, II, p. 985, 1003, Copyright 1988, with kind permission from Elsevier Science Ltd., The Boulevard, Langford Lane, Kidlington OX5 1GB, UK.)

STRIPS-like operators:

1. Operator-1: (STOP_FLOW x)
       Preconditions: (flowing x)
       Postconditions: not(flowing x)

2. Operator-2: (ESTABLISH_FLOW x)
       Preconditions: ( )
       Postconditions: (flowing x)

3. Operator-3: (PURGE x)
       Preconditions: not(flowing x); not(equal inert-gas x)
       Postconditions: not(present x); (flowing inert-gas)

Using the initial and goal states as indicated in Fig. 3, it is easy to construct the following sequence of operations that satisfy the desired objectives:

(STOP_FLOW Oxygen),
   (PURGE Oxygen), (ESTABLISH_FLOW Methane)

B. Modeling of Process Behavior

A processing facility is a network of unit operations, connected through material and energy flows. Through these connections, the effect of an operation is not confined to the operational behavior of the processing unit it directly affects and its immediate vicinity, but it may propagate and effect the state of all units in the plant. Thus, in addition to the complexity introduced by the functional dependence of process behavior on initial preconditions (see Section II,A), the conditions describing the state of a process resulting after the application of an operation must satisfy the constraints imposed by the network-like structure of the plant. Since the actual structure of the plant is not known at the time when we formulate the operator-models, we must rely on some other means for modeling the constraints characterizing the interactions among different units.

In the first chapter of Volume 2 (hereinafter referred to as **21**:1) we presented the general framework of MODEL.LA., a modeling language that can capture the hierarchical and distributed character of processing systems. We will employ all aspects of MODEL.LA. in order to develop a complete and consistent description of plants that will satisfy the modeling needs for the synthesis of operating procedures.

1. The Hierarchical Description of Process Topology

MODEL.LA.'s primitive modeling elements (see **21**:1), *Generic-Unit*, *Port*, and *Stream*, are capable of depicting any topological abstraction of a chemical process. In particular, the specific subclasses emanating from the *Generic-Unit* allow us to describe a process as (see **21**:1): (1) an overall *Plant-Sections*, where each section represents a grouping of units with a common operating framework, e.g., train of distillation columns; (2) a network of *Augmented-Units*, which encapsulate the structure of processing units along with their ancillary equipment, e.g., a distillation column with its feed preheater, condenser, and reboiler; and (3) a network of *Units*. Figure 4 shows a schematic of this topological hierarchy.

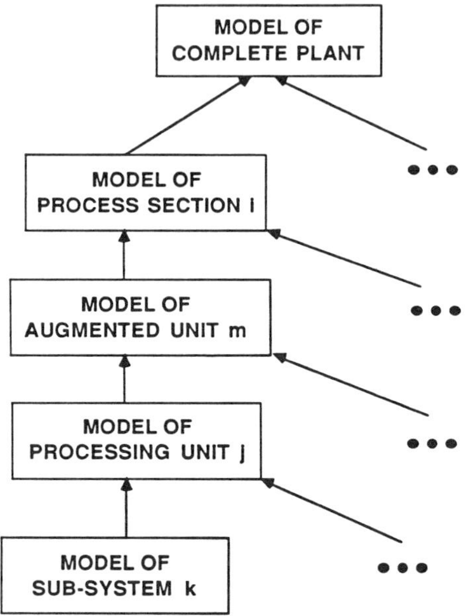

FIG. 4. Hierarchical description of process flowsheets.

Figure 5 shows the sections of a plant producing gasoline from the dimerization of olefins. Using the explicit hierarchical descriptions given above, we can formulate the following series of abstract operators, which (as we will see in Section III) can provide significant help in reducing the number of alternative plans:

Plant level. (START_UP_PLANT);

Section level. (START_UP_FEED_PREPARATION), (START_UP_REACTION), (START_UP_RECOVERY), (START_UP_RE-FINING);

Augmented-Unit level. (START_UP_DEPROPANIZER), (START_UP_DEBUTANIZER), etc.

Any of these operators can be represented as a STRIPS, conditional, or functional operator, depending on the character of the constraints imposed on the particular operating procedure to be synthesized.

2. The Hierarchical Description of the Operating State

Whether we are dealing with an overall plant and its sections, augmented units, or individual units, we must view each system as a thermody-

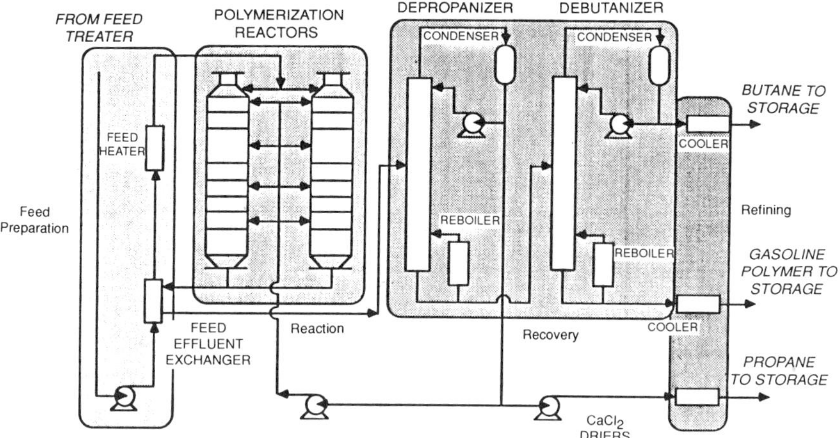

Fig. 5. Decomposition of gasoline polymerization plant.

namically "simple" system (Modell and Reid, 1983), which is characterized by a "state." From an operations planning point of view, this "state" can be viewed at different levels of detail, thus giving rise to a hierarchy of descriptions. MODEL.LA. provides the requisite modeling elements, which can be used to provide the necessary hierarchical descriptions of operating states:

a. Operational State of Variables. The modeling element, Generic-Variable (see **21**:1), allows the representation of the structured information that describes the behavior of "temperature," "pressure," "concentration," "flow," etc. Each of these instances of the *Generic-Variable* is an object with a series of attributes such as "current-value," "current-trend," "range-of-values," "record-over-x-minutes," and "average-over-x-minutes." Through these attributes we can capture all possible preconditions and postconditions that may be needed by the individual operators.

b. Operational State of "Terms". MODEL.LA.'s *Generic-Variable* possesses a series of subclasses (see **21**:1), whose sole purpose is to provide declarative description of certain compound variables, called *Terms*, such as "flow-of-component-x," "enthalpy-flow," "heat-flow," "diffusive-mass-flow," and "reaction rate." The declarative, rather than procedural, representation of compound variables enables the computer-aided planner to have an explicit description of the quantities involved in the preconditions and postconditions of an operation, thus replacing a numerical procedure by a series of explicit numerical or logical inferences, as the planning needs may be.

c. Operational State of Constraining Relationships. The operational state of "terms" is constrained by the laws of physics and chemistry and relationships that arise from engineering considerations. Typical examples are (1) the conservation principle (applied on mass, energy, and momentum); (2) phase or reaction equilibria; (3) reaction rates or transport rates for mass, energy, and momentum; (4) limits on pressure drops, heat flows, or work input, dictated by the capacity of processing units; and (5) experimental correlations between input and output "terms," etc. MODEL.LA.'s modeling element, *Constraint*, and its subclasses, *Equation, Inequality, Order-of-Magnitude, Qualitative*, and *Boolean* (see **21**:1) offer the necessary data structures to capture the information associated with any type of constraining relationships.

d. Operational State of a Process System. MODEL.LA's modeling element, *Modeling-Scope*, captures the consistent set of modeling constraints (described in the previous paragraph), which apply on a given process system. As the operational state of a system changes, the content of the *Modeling-Scope* may change; e.g., transition from a single to a two-phase content, transition from laminar to turbulent flow.

e. Example. The four-level hierarchy for the description of process operational states as shown in Fig. 6a offers a specific example. It offers a very rich and flexible modeling framework to capture any conditions (pre- or post-) associated with the definition of operators.

3. Maintaining Consistency among Hierarchical Descriptions of Process Topology or State

Operators, describing process operations, can be declared at any level of abstraction, but they should maintain consistent relations with each other, since they refer to the same process. For example, the top-level operator, (START_UP_PLANT), could be refined to the following sequence of operators (see also Fig. 5): (START_UP_RECOVERY), (START_UP_REFINING), (START_UP_FEED_PREPARATION), and (START_UP_REACTION). Clearly, the preconditions of (START_UP_PLANT) are distributed and represent a subset of the preconditions for all four more detailed operations. Similarly, the postconditions, derived from the startup operation of the four sections, should be

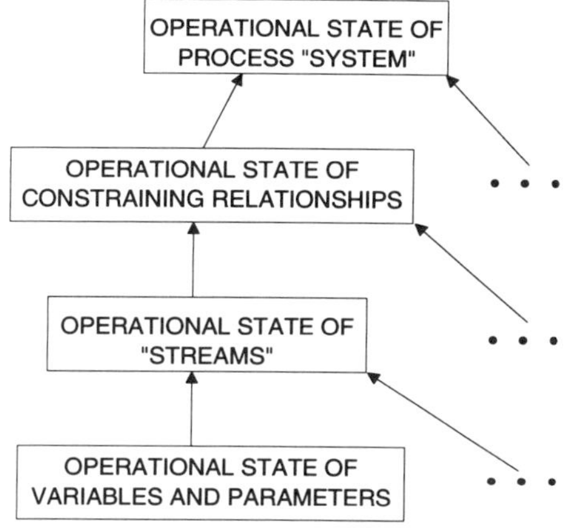

(a)

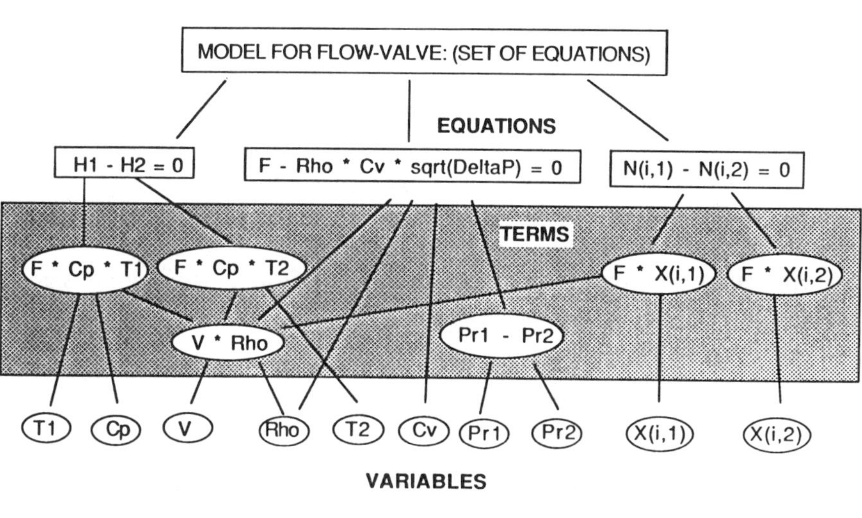

(b)

Fig. 6. Hierarchical description of (a) operational states and (b) operational relationships. (Reprinted from *Comp. Chem. Eng.*, **12**, Lakshmanan, R. and Stephanopoulos, G., Synthesis of operating procedures for complete chemical plants. Parts I, II, p. 985, 1003, Copyright 1988, with kind permission from Elsevier Science Ltd., The Boulevard, Langford Lane, Kidlington 0X5 1GB, UK.)

consistent with the postulated postconditions of the overall (START_UP_PLANT).

To achieve these consistencies, MODEL.LA. provides a series of semantic relationships among its modeling elements, which are defined at different levels of abstraction. For example, the semantic relationship (see **21**:1), **is-disaggregated-in**, triggers the generation of a series of relationships between the abstract entity (e.g., overall plant) and the entities (e.g., process sections) that it was decomposed to. The relationships establish the requisite consistency in the (1) topological structure and (2) the state (variables, terms, constraints) of the systems. For more detailed discussion on how MODEL.LA. maintains consistency among the various hierarchical descriptions of a plant, the reader should consult **21**:1.

III. Nonmonotonic Planning

Let us return to the simple operations planning problem of Fig. 3 (see also Section II,A,4). A non-monotonic planner faced with this problem proceeds as follows:

Step 1. Choose one of the desired goals (e.g., flowing methane).

Step 2. Select an operator that achieves the selected goal, e.g., (ESTABLISH_FLOW METHANE), which produces the desired goal as its postcondition.

Step 3. Propagate derived postconditions through the frame axioms (see Fig. 3) and take:

$$(\text{flowing methane}) \Rightarrow (\text{present methane})$$
$$(\text{flowing oxygen}) \Rightarrow (\text{present oxygen})$$

Step 4. Check for violation of constraints. In this case the constraint is also a frame axiom:

$$(\text{present methane}) \text{ AND } (\text{present oxygen}) \Rightarrow (\text{explosion})$$

and is confirmed.

Step 5. Since the constraint was violated, the non-monotonic planner attempts to identify the proper operator among the set of available operators, which can alleviate the constraint violation, if this operator were "forced" to be applied before the constraint violation. In this example such an operator is (PURGE X). Thus, (PURGE X) must be applied before (ESTABLISH-FLOW METHANE). The variable X remains temporarily unbound to any particular value.

Step 6. Operation (PURGE X) stipulates the a priori satisfaction of the following two preconditions:

not(flowing X) AND not(equal inert-gas X)

Since the final goal state requires that (flowing methane) is true, then the value of X can be bound to oxygen. Then the following postconditions of (PURGE OXYGEN) are true:

not(present oxygen) (flowing inert-gas)

Step 7. The first precondition of (PURGE OXYGEN) becomes the new goal and requires that the following operation must take place before (PURGE OXYGEN):

(STOP_FLOW OXYGEN) producing not(flowing oxygen)

So the solution to the planning problem is now complete, and is given by the following sequence of elementary operations:

(STOP_FLOW OXYGEN) > (PURGE OXYGEN)
> (ESTABLISH_FLOW METHANE)

where the symbol > signifies the relative temporal ordering of two operations.

In summary, nonmonotonic planning of process operations has the following distinguishing features:

1. *Property 1: goal-driven approach.* This always starts from a desired goal state and tries to identify the requisite operations that will achieve it.

2. *Property 2: constraint-posting without backtracking.* Instead of backtracking, when a constraint violation is detected, nonmonotonic planning attempts to find an operator that would negate the preconditions leading to the constraint violation and forces this operator to be applied *before* the operation that causes the constraint violation. It leads to the posting of a new constraint determining the temporal ordering of operations. The operator that causes the violation of a constraint, negates, prevents, or undoes one or more of the planning goals, is called *Clobberer*. On the other hand, an operator that rectifies, or prevents any of the damage caused by a Clobberer will be called *nonmonotonic White Knight* (Chapman, 1985). Clearly, the success of any nonmonotonic planning methodology lies in its ability to identify these Clobberers and White Knights without having to resort to an exhaustive *generate-and-test* approach. Once these operators have been identified, the planner must

ensure that whenever a Clobberer is used in the plan, an accompanying White Knight is present.

3. *Property 3: generator of partial plans*. Nonmonotonic planning generates partially specified plans where a certain amount of information is not bound to specific values. A typical example was the value of X in the operation (PURGE X), that we saw in the illustration at the beginning of this section. Partial plans are abstractions, which can represent large sets of plans, and can lead to every efficient search strategies.

A. Operator Models and Complexity of Nonmonotonic Planning

The correctness of complete plans is encapsulated by the following theorem (Chapman, 1985).

Theorem 1 (Truth Criterion of Complete Plans).   *In a complete plan, a proposition, **p**, is necessarily true in a situation, **s**, if and only if there exists a situation, **t**, previous or equal to **s** in which **p** is asserted, and there is no step between **t** and **s** that denies **p**.*

In this theorem, any proposition **p** can represent an operator (i.e., an operation step), whereas the situations **t** and **s** represent any intermediate state of the process. Although the validity of the theorem is general, its practical utility is confined to monotonic planning with STRIPS-like operators. For example, in nonmonotonic planning the plans are at any point when partially specified and a new mechanism is needed to guarantee that when the partial plan is completed, a given proposition (i.e., a given operation) is still true (i.e., consistent).

Chapman's work produced the following theorem, which provides the necessary and sufficient conditions for guaranteeing the truth of any given statement in a partial plan, if all operations are modeled by STRIPS-like operators.

Theorem 2 (Modal Truth Criterion).   *A proposition **p** is necessarily true in a situation **s** if and only if the following two conditions hold:*

(a) *There is a situation **t** equal or necessarily previous to **s** in which **p** is necessarily asserted.*

(b) *For every step **C**, possibly before **s**, and every proposition, **q**, possibly codesignating with **p**, which **C** denies, there is a step **W** necessarily between **C** and **s** that asserts **r**, a proposition such that **r** and **p** codesignate whenever **p** and **q** codesignate.*

So, if **C** is the step in an operating procedure that introduces the

Clobberer, **q**, denying the validity of the postconditions of operation **p**, then **W** is the operating step (necessarily after **C**) at which the White Knight, **r**, comes to rescue and negates the effects of **q**. Theorem 2 also indicates that for each operation **p**, any preceding Clobberer must have the corresponding White Knight, if the nonmonotonic plan is to be correct.

For STRIPS-like operators, Chapman (1985) developed a polynomial-time algorithm, called TWEAK, around five actions that are necessary and sufficient for constructing a correct and complete plan. As soon as we try to extend these ideas to nonmonotonic planning with conditional opera-tors, we realize that no polynomial-time algorithm can be constructed, as the following theorem explicitly prohibits (Chapman, 1985):

Theorem 3 (First Intractability Theorem). *The problem of determining whether a proposition is necessarily true in a nonmonotonic plan whose action representation is sufficiently strong to represent conditional actions is NP-hard.*

It is not surprising, then that, as we require functional operators to approximate more closely the majority of operations during the synthesis of process operating procedures, we must give up any expectation for a tractable algorithm (Lakshmanan, 1989):

Theorem 4 (Second Intractability Theorem). *The problem of determining whether a proposition is necessarily true in a nonmonotonic plan whose action representation employs functional operators is NP-hard.*

B. Handling Constraints on the Temporal Ordering of Operational Goals

This section and the following two will cover the treatment of three distinct classes of constraints within the general framework of nonmono-tonic planning. Let us first consider constraints on the temporal ordering of operational goals. They are of the form "GOAL-A must be achieved before achieving GOAL-B" and appear at a certain level of process abstraction. Examples are

(a) (START_UP_FEED_TREATMENT_SECTION) before (START_UP_REACTOR).
(b) (START_FLOW_IN_COLD_SIDE) before (START_FLOW_IN_HOT_SIDE) of a heat exchanger.
(c) (OPEN_VALVE_1) before (OPEN_VALVE_2).

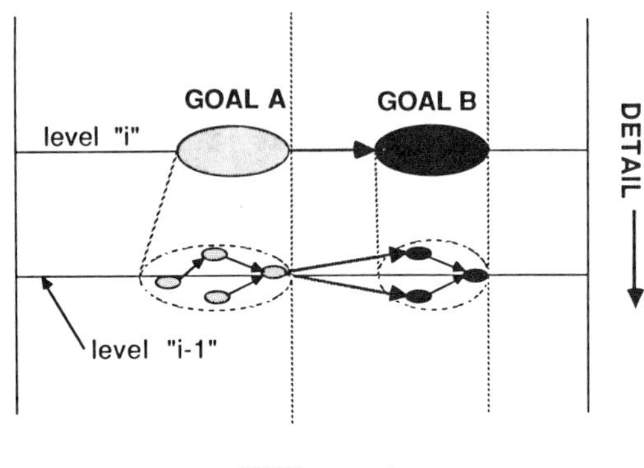

FIG. 7. Downward transformation of temporal constraints. (Reprinted from *Comp. Chem. Eng.*, **12**, Lakshmanan, R. and Stephanopoulos, G., Synthesis of operating procedures for complete chemical plants. Parts I, II, p. 985, 1003, Copyright 1988, with kind permission from Elsevier Science Ltd., The Boulevard, Langford Lane, Kidlington 0X5 1GB, UK.)

A temporal constraint at a particular level of the abstraction hierarchy specifies a partial ordering on the goals in the level just below (more detailed), *causing the posting of new temporal constraints* on the goals of this level. Each of these new constraints specifies in turn a set of new temporal constraints at the level below it. In this manner, constraints at a high-level of abstraction are successively transmitted from level to level until they reach the most detailed level of individual processing units. All the newly generated constraints must be taken into account and satisfied by the non-monotonic planner. Figure 7 shows a schematic of the propagation of temporal constraints through two levels of abstraction. The directed path indicates the temporal ordering of goals.

Lakshmanan and Stephanopoulos (1988b) developed an algorithm, which automates the downward propagation of temporal constraints all the way to the level of processing units, where these constraints can be incorporated in planning of primitive operators (e.g., open or close valve, turnon or shutdown motor). This algorithm is provably correct having the following property.

1. Property 1: Correctness of the Algorithm for the Downward Propagation of Temporal Constraints

Let C_i be a temporal constraint, stated between two goals, A_i and B_i, at the ith level of the modeling hierarchy (see Section II,B), which indicates that A_i must be achieved before B_i. Let $A_{(i-k)}$ and $B_{(i-k)}$, with $k > 0$, be subgoals at the $(i-k)$th level of A_i and B_i, respectively. Then, the algorithm for the downward propagation of temporal constraints guarantees, on termination, that, for all $[A_{(i-k)}, B_{(i-k)}]$, a directed path from $A_{(i-k)}$ to $B_{(i-k)}$ exists in the constraint network $N_{(i-k)}$ and no corresponding path exists from $B_{(i-k)}$ to $A_{(i-k)}$.

The successive introduction of new constraints in the temporal ordering of operational steps at lower levels of the goal hierarchy, may create internal conflicts. For example, the propagation of constraint C_i may result in ordering A_i before B_i, while the propagation of constraint C_j may lead to ordering B_i before A_i. These conflicts manifest themselves in the form of directed circuits (cycles) at one or more levels in the goal hierarchy. A polynomial-time algorithm has been developed to detect the generation of directed cycles as new constraints are introduced (Lakshmanan and Stephanopoulos, 1988b). The user is informed and the offending constraint may be detracted.

a. Transformation of Integer–Goal Ordering Constraints onto Orderings of Primitive Operators. The temporal ordering of goals at higher levels of the goal hierarchy is successively transformed to temporal orderings of goals at lower levels. It now remains to be seen how the temporal ordering of goals at the lowest level of the hierarchy can be transformed into constraints on the ordering of primitive operators. Lakshmanan and Stephanopoulos (1988b) have developed such an algorithm which possesses the following property.

2. *Property 2: Correctness of the Constraint Transformation Algorithm*

The constraint transformation algorithm accepts a network of goals partially ordered by constraints, and generates a constraint network of primitive actions, such that, if there exists a directed path from goal A to goal B (i.e., A must be achieved before B) in the first network, and if OP-A is the primitive action that achieves goal A, and OP-B the action that accomplishes B, then OP-A and OP-B are labels on nodes in the generated network, and there exists a directed path from the node labeled with OP-A to the node labeled with OP-B.

This property guarantees that constraints on the temporal ordering of goals are correctly transformed into constraints on the temporal ordering of primitive operations that achieve these goals.

C. HANDLING CONSTRAINTS ON THE MIXING OF CHEMICALS

Safety, environmental, health, and performance considerations dictate that certain chemicals should not coexist anywhere in the plant. Rivas and Rudd (1974), O'Shima (1982), and Fusillo and Powers (1988a) have dealt with this important class of constraints, and have proposed approaches all of which are monotonic in character. In its most general form, a constraint on the mixing of chemicals involves the specification of a set of chemical species, and composition ranges for those species that ought to be avoided anywhere in the plant. Very often, though, the high cost of the consequences of an unsafe operation (e.g., an explosion) forces the operations planners to adopt a more conservative posture and require that the coexistence of a set of chemicals be avoided altogether. This class of constraints is expressed as an unordered set of chemical species $(x_1, x_2, \ldots, x_n)$ that should never be present at the same time anywhere in the plant. In this section we will examine how nonmonotonic planning handles such constraints and converts them into constraints on the temporal ordering of primitive operations through the following sequence of steps:

Step 1. *Construction of the influence graphs*, which represent the extent to which a given chemical species is present within the process in a given operational state.

Step 2. *Identification of potential constraint violations*, resulting from selected operations during the construction of partial plans.

Step 3. *Generation of temporal constraints on primitive operations*, to negate the violation of mixing constraints.

In the following paragraphs we will discuss the technical details involved in each of the preceding steps.

1. Construction of the Influence Graphs

Consider the directed graph representing the topology of a chemical process at a given operating state, i.e., the nodes of the graph represent process equipment and the directed edges of the graph the material flows. If the direction of the flow has not been computed or is uncertain, the corresponding edges are bidirectional. Each edge conveys information related to the type of chemical species flowing and the directionality of flow. All other state information (e.g., temperature, concentration) has been suppressed. Then, using the directed graph that represents the topology of the chemical process, the construction of the influence graphs (IGs) proceeds as follows (Lakshmanan and Stephanopoulos, 1990):

1. Locate all sources, s_i, of the chemical species, x_i.
2. Remove all edges that are immediately adjacent to closed valves.
3. Any node, v, which is on a directed path from any source, s_i, is labeled with x_i. Let N_i be the set of nodes labeled with x_i and let E_i be the set of associated edges. Nodes labeled with x_i and the attached edges contain the chemical species.

At the end of this three-step procedure we have generated a graph (i.e., a subgraph of the process graph) whose edges indicate the flow of species x_i and whose nodes contain the chemical x_i, at the given operating state.

2. Identification of Constraint Violations

Violations of the mixing constraints may occur at the *initial*, *goal*, or any *intermediate* state in the course of an operating procedure. Since the initial state is presumed to be a known, feasible state, it is generally not the case that it will contain potential mixing constraint violations. It is possible though that the human user or the planning program may specify a goal operating state that violates the mixing constraints.

The detection of potential mixing constraints violations at an intermediate operating state is based on a "worst-case scenario" with the maximum potential for the species to coexist, and which corresponds to the following situation:

(a) Keep open all the valves which are OPEN at the initial state.
(b) Keep open all the valves which must be OPEN at the goal state.

Construct the IGs for all species present in the various mixing constraints. Then:

If a node, v, carries both labels, x_i and x_j, where (x_i, x_j) is an unacceptable pair of coexisting species, then the potential for a mixing constraint violation has been detected.

Clearly, if under the worst-case scenario no node can be found in the IGs that is labeled with all species present in a mixing constraint, then no potential violation of a mixing constraint exists. If, on the other hand, a potential violation has been detected, then we need to *generate additional constraints on the temporal ordering of primitive operations so that we can prevent or negate the preconditions of a mixing constraint*.

3. Generation of Constraints on the Temporal Ordering of Primitive Operations

The potential violation of a mixing constraint is equivalent to the presence of a potential *Clobberer*. Therefore, according to Theorem 2, we need to identify a *White Knight* and force him to proceed the action of the Clobberer. In terms of the synthesis of operating procedures, the "Clobberers" are the valves that are open under the worst-case scenarios, described in the previous paragraph. Obviously, the "White Knights" represent the set of valves that, when closed, will prevent the chemicals from mixing. We will call this set the *minimal separation valve set* (MSVS). Once the valves that should be closed have been identified, temporal ordering constraints are imposed, which specify that these valves be closed, *before* the valves that lead to a mixing constraint violation were opened. Lakshmanan and Stephanopoulos (1990) have developed a graph-based algorithm with polynomial-time complexity, which identifies the MSVS. However, the MSVS may not be an *acceptable separation valve set* (ASVS) for the given planning problem, since (1) some valve in the MSVS is required to be open in the goal state (in order to achieve a certain operating goal), or (2) the user may have some subjective preferences regarding the state of some valves. A concrete algorithm exists (Lakshmanan and Stephanopoulos, 1990) to convert the initial MSVS to an ASVS.

Once an ASVS has been found, the constraints on the temporal ordering of the valve operations in the ASVS must be established. This is systematically accomplished through an algorithm that undergoes through the following steps.

1. Let P be the set of valves that must be open under the worst-case scenario.
2. If p is an element of P and q is an element of ASVS, then, to negate the violation of a mixing constraint, the following temporal constraint must be generated:

$$\text{CLOSE}(q) \text{ before } \text{OPEN}(p)$$

4. Purging a Plant from Offending Chemicals

Once the ASVS has been identified and the temporal ordering of valve operations has been determined, no mixing constraints will be violated in the steady state. However, it is possible to violate the mixing constraints during the transient from one steady state to another. For example, during the changeover from oxygen to methane in the network of Fig. 3, it is possible that oxygen and methane will come into contact with each other even if the upstream oxygen inlet valve is closed, before the methane is allowed to enter. To avoid the contact between the offending chemicals, the first (e.g., oxygen) must be purged, before the second (e.g., methane) is allowed to flow in.

Lakshmanan (1989) has indicated that *"purging or evacuation operations are necessary, whenever the intersection of the IG of one component in the initial state with the IG of another (offending) component in the goal state is non-zero."* In such case, nonmonotonic planning generates a new intermediate goal, i.e., purge the plant of chemical x_i, and tries to achieve this goal first before admitting the offending chemical x_j in the process.

Consequently, the overall planning is broken into two phases and proceeds as follows (Lakshmanan and Stephanopoulos, 1990):

a. Phase 1. Find the path and destination of purge operations.

1. Identify all locations of a plant where a certain species must be purged. To accomplish this it is sufficient to scan the labels of the nodes in the IGs and identify those nodes that are labeled, e.g. x_i and x_j in two successive operating states, where (x_i, x_j) is a prohibited mixture of chemicals.
2. For each node v, where the offending chemicals (x_i, x_j) are present in the initial and goal states, respectively, find a directed path that leads the purgative material from its source to the sink through node v.

b. Phase 2. Generation of the purge planning island, i.e., intermediate planning goals.

1. Create an Intermediate operating State, i.e., introduce the intermediate goals requiring that *all valves along the purge path be open.*
2. Place the intermediate operating state between the initial and goal states and constrain the planner to achieve the goals in the intermediate state before opening the valves in the initial state, which were earlier identified as being open in the goal state (viz., worst-case scenario).

D. Handling Quantitative Constraints

During the synthesis of operating procedures, the operating state may be required to satisfy certain quantitative constraints. For example, during the operation of a reactor, the temperature may not be allowed to exceed a certain maximum, or the distillate from a distillation column may be required to have a minimum purity. Two chemicals may not be allowed to mix if the temperature is below a minimum. These constraints involve numerical values and may be stated as follows:

$$\{\text{HCN should not mix with HCHO at } T \geq T_{\text{max}}\}$$
$$\{[\text{HCN}] \leq 0.1 \text{ mol}/\text{L}\}, \text{ in the reactor at all times.}$$

Such constraints commonly arise from safety requirements, product specifications, environmental regulations, etc.

In this section we will present a formalized methodology that allows the transformation of quantitative bounding constraints into constraints on the temporal ordering of operators within the spirit of nonmonotonic planning.

1. Truth Criterion for Quantitative Constraints

The efficiency of nonmonotonic planning depends on its ability to identify Clobberers and White Knights efficiently. Theorems 3 and 4 (see Theorems 1–4 in Section III,A) have established the intractability of nonmonotonic planning for conditional and functional operators, which are rich enough to ensure adequate representation of process operations. Consequently, practical solutions can be derived and intractability can be avoided only through the use of domain-specific knowledge. A truth criterion, valid for the specific domain, should be identified and proved. Also, efficient ways of evaluating the domain-specific truth criterion should be given. Such a truth criterion has been identified and proved for synthesizing plans that are guaranteed to satisfy quantitative constraints in chemical plants. For posting quantitative constraints of the form $x.*.k^2$ the truth criterion is stated as follows:

Theorem 5: (Modal Truth Criterion for Quantitative Constraints). *A constraint $x.*.k$ is necessarily satisfied in a situation s iff there is a situation t equal to or necessarily previous to s in which $x.*.k$ is satisfied and for every step C possibly before s that possibly violates $x.*.k$, there is a step W necessarily before C that ensures $x.*.k$ in s.*

The statement and proof of the truth criterion (see Lakshmanan, 1989) for quantitative constraints is along the lines of Chapman's truth criterion for domain independent nonmonotonic planning. From this criterion, the plan modification operations that would ensure satisfaction of constraints are

1. If C is constrained to lie after s, its clobbering effect will not be felt at situation *s*. This process of constraining C to lie after s is called *demotion*.
2. A White Knight, W, may be selected and constrained to lie before C and therefore s, so as to ensure that the clobbering effect of C will not be felt.

The time required to evaluate the truth criterion depends on the complexity of the procedures for determining clobberers and white knights. The

---

[3] The notation '.*.' indicates relationships of the type $\langle,\rangle, =$, etc. between X and k.

algorithms for these operations are presented below. Clobbering of a constraint may often be caused by a set of valve operations that achieve an effect together. Such a set of valves is identified as an abstracted operator.

2. Identification of Clobberers of Quantitative Constraints

A goal is often achieved by multiple primitive (e.g., valve) operations. All the primitive operations that will together result in violation of a constraint will be Clobberers of that constraint. Consequently, the first step in the quantitative constraint-posting methodology is to identify abstracted operators as the sets of primitive operations. This is accomplished through the following procedure:

> **procedure** ABSTRACT_OPERATOR
>
> **Input**: Initial state, I; Final state, F; List of goals to be achieved,
> G and list of primitive valve operations required
> to achieve the goal state, V.
>
> **Output**: Abstract operators, A consisting of the set of valve operations
> necessary to start or stop flow from each source to point
> where goal is specified
>
> **begin**
>
> **for** each $g \in$ G do
>
>   **if** RHS(g) = 0 **then** STATE := I
>
>   {**comment**: if goal is to start/ stop flow, consider}
>
>   else STATE := F   { final/ initial state as 'state' }
>
>   remove closed valves from STATE
>
>   j = 0
>
>   **for** each point of goal specification, e_i **do**
>
>   **begin**
>
>   j = j + 1
>
>   **for** each source, s_j **do**
>
>   FIND_PATH(e_i, s_j)
>
>   {**comment**: Search for paths from point where the goal
>
>   is specified to sources of species and mark valves along it}
>
>   AO-j := Intersection (open/closed valves in V,
>
>   marked valves), for start/stop flow
>
>   **end**
>
> **end**

This algorithm provides a complete identification of all requisite operators, simple or abstract (i.e., sets of operators), as the following theorem guarantees (Lakshmanan, 1989).

Theorem 6: *Given the list of primitive operations to achieve a goal state, the algorithm* ABSTRACT_OPERATOR *will detect every simple or abstracted operator corresponding to the set of operations required for starting or stopping flow of materials from each source, s_j, to the point of constraint specification, e_i, necessary to achieve the goal state.*

The algorithm ABSTRACT_OPERATOR runs in time $O[g(V + s(V + E + V \ln V))]$, where g is the number of goals, s is the number of sources, V is the number of operators (e.g., valves) and E is the number of directed edges (i.e., connections) in the graph representing the topology of a process flowsheet. Any set of primitive operators that results in possible violation of a quantitative constraint will be a Clobberer of that constraint. For satisfaction of the truth criterion (Theorem 5), we need to identify those operators that could result in constraint violation. Clobberers of quantitative constraints may be identified efficiently only when the effect of just one operator is considered at a time. When multiple operators are considered simultaneously, the quantitative effect of each operator will depend on the magnitude of the others. Numerical simulation based on assumptions of the values of the variables may then be required to detect Clobberers. Since each variables could take any value within a range, such simulation would be intractable.

For the detection of Clobberers, it is necessary to propagate the quantitative effects of each abstracted operator through the plant to the point where the constraint is applicable. The algorithm for performing this task is given below.

```
        procedure CLOBBERER_DETECTION
Input: Abstracted operators, A; final state, F;
        quantitative constraints, Q
Output: Abstracted operators that are Clobberers, C
begin
for each q ∈ Q do
      for each a ∈ A do
      begin
            QUANTITATIVE_PROPAGATE(q, eᵢ, F)
            {comment: propagate the quantitative values
             of the variables through the
            model equations}
            if constraint, q is violated then a ∈ C
      end
end
```

The following theorem guarantees the completeness of the above algorithm:

Theorem 7. *Every abstract operator, that will result in violation of a quantitative constraint, if implemented first and alone, will be detected as a Clobberer by the procedure* CLOBBERER_DETECTION.

3. Selection of Plan Modification Operations: The White Knight for Quantitative Constraints

Once the Clobberers that lead to violations of quantitative constraints have been identified, we must search for modifications of the operating procedure, which will negate the effects of the Clobberers on the procedure and thus restore its feasibility. There exist two general mechanisms for achieving this objective, and in the following paragraphs we will discuss each one of them.

a. Mechanism 1: Demotion of Clobberers. Demotion of the Clobberer involves constraining the clobberer to lie after the situation at which constraint satisfaction is being considered. Thus, if $\{AO - 1, AO - 2, AO - 3\}$ are three unordered operators, and $AO - 2$ is a Clobberer of a quantitative constraint, the plan may be modified by demotion of $AO - 2$ to give $\{AO - 1, AO - 3\} > \{AO - 2\}$, i.e., $AO - 1$ and $AO - 3$ may be applied at situation s, but $AO - 2$ is constrained to lie after s. Thus, posting quantitative constraints provide greater temporal order to the operators in the plan.

b. Mechanism 2: Identification of White Knights. If all operators for obtaining a feasible plan (i.e., satisfies both goals and constraints) have been correctly identified, then the quantitative constraints can be transformed into constraints on the temporal ordering of operations and demotion will always work. Nevertheless, if every operator identified for a single equipment is a potential Clobberer, then quantitative constraint violation cannot be avoided by the current set of operators and a White Knight has to be identified. The procedure for identifying White Knights is composed of two steps:

Step 1. It is necessary to identify the constraint being possibly violated and the qualitative change required in the variable to avoid the violation. For example, if a "$T < T_{\text{max}}$" constraint is violated, then the qualitative change required in T to avoid the violation is "decrease T." Having identified the property to be affected and the direction of desired change, we now need to identify manipulations that may help us meet our objective.

Step 2. The qualitative value of the desired change is propagated through the steady-state model equations of the plant equipment, following the constraint propagation procedure of Steele (1980). Manipulations that cause the desired change and that are feasible are identified as White Knights and are constrained to lie before the situation of interest s, in accordance with the truth criterion.

The complete algorithm for the identification of White Knights is as follows:

procedure WHITE_KNIGHT
Input: Plant flowsheet in final state, F; constraint being clobbered, q.
Output: White Knight, W for the constraint q.
begin
for q **do**
    begin
    if q $\equiv$ 'X$(<, \leq)$k' **then** L := 'decrease X' {**comment**: L − load}
    if q $\equiv$ 'X$(>, \geq)$k' **then** L := 'increase X'
    if q $\equiv$ 'X = k' and X $<$ k **then** L := 'increase X'
    if q $\equiv$ 'X = k' then X $>$ k **then** L := 'decrease X'
    { **comment**: Such rules may be written for each type
     of constraint encountered }
    end
while v $\neq$ manipulatable valve **do**
    QUALITATIVE − PROPAGATE(L, F)
return v and required qualitative change
end

The following theorem (Lakshmanan, 1989) guarantees the completeness of the preceding algorithm.

Theorem 8. *The algorithm* WHITE-KNIGHT *will detect every manipulation, that will possibly avoid a quantitative constraint violation, as a White Knight.*

E. Summary of Approach for Synthesis of Operating Procedures

The synthesis of operating procedures for a chemical plant, using nonmonotonic planning ideas, consists of two distinct phases: (1) formulation of the planning problem and (2) synthesis of operating plans. In the following two subsections we will examine the features of each phase and we will discuss the technical details for their implementation.

1. Phase 1: Formulation of the Planning Problem

The complete formulation of the planning problem implies the complete specification of the following four items.

a. Description of the Initial State. We must make sure that the specified starting operational state of a process is complete and satisfies the balances (mass, energy, and momentum), equilibrium, and rate phenomena in the process. This task is composed of the following steps:

1. Input of the initial operational state of the plant, using the linguistic primitives of MODEL.LA. (see **21**:1).
2. The specification of the initial state may be partial and lead to incomplete descriptions of various segments of the plant. The modeling facilities of MODEL.LA. contain a complete set of the balance equations, phase and chemical equilibrium, and rate relationships. These relationships are used to propagate the user-supplied specifications and thus complete the description of the initial state throughout the plant.
3. During the specification of the initial state, it is conceivable that the user-supplied information is not consistent with the physical constraining relationships. The planning program should possess procedures for detecting such conflicts and provide mechanisms for their resolution.

b. Description of the Final, Goal State. This proceeds in a manner similar to that for the description of the initial state.

1. Input the specifications of the desired goal state.
2. Check for potential conflicts in the specification of the goal state, and provide resolution for its consistent definition.
3. Generate concurrent goals, which usually come from the interaction among the various subsystems of the plant. This is accomplished through constraint propagation mechanisms, using the set of modeling relationships describing the process, analogous to the mechanism that ensures the completeness in the description of the initial state.

c. Specification and Identification of Operational Constraints in the Plan. In addition to the physical modeling constraints that govern the operation of the chemical process, we require that several "operational" constraints be met by an operating procedure, such as (1) temporal ordering in the execution of process operations, (2) avoiding the creation of undesirable mixtures of chemical species, and (3) maintaining the state variables within specific ranges of numerical values, e.g., bounding quantitative constraints.

d. Specification, Identification of Planning Islands. When an operating procedure is expected to be long and to involve a large number of intermediate states, it may be advantageous to explicitly identify intermediate goals, known as *planning islands* (Chakrabarti *et al.*, 1986), through which the operating procedure must pass, as it goes from the initial to the goal state. These planning islands are partial descriptions of intermediate states, which are known to lie on the paths of the most efficient, or feasible, plans, thus decomposing the overall operating plan into a series of subplans, which can be synthesized separately in a sequence. Lakshmanan (1989) has established specific procedures for the identification of planning islands.

2. Phase II: Synthesis of Operating Plans

Having defined the planning problem in a form that is "understood" by the computer, we can proceed to the synthesis of operating procedures through the following sequence of steps.

a. Identification of the Primitive Operators. The "means–ends analysis" of Newell *et al.* (1960) is employed to (1) identify the differences between the goal and initial states and (2) select the operators (i.e., operating actions) that would eliminate these differences.

b. Construction of Partial Plans. This consists of deriving a partial temporal ordering in the application of primitive operators. The partial ordering is driven by the constraints placed on the states of the desired operating plan, and proceeds as follows:

1. User-specified, temporal ordering of operational goals at higher levels of abstraction is propagated downwards in the hierarchy goals and is ultimately expressed as temporal ordering of primitive operations (see Section III,B).
2. Constraints on the disallowed mixtures of chemical species (a) are transformed into temporal orderings of primitive operations, or/ and

(b) introduce additional intermediate goals, e.g., planning islands for purge or/ and evacuation operations (see Section III,C).

3. Quantitative constrains (a) are transformed into temporal orderings of primitive operations (e.g., demotion of Clobberers) or (b) dictate the introduction of new operators (e.g., White Knights) to recover the feasibility of plans.

c. Synthesis of Complete Plans. At the present, nonmonotonic planning covers the three classes of constraints discussed above. For other classes of constraints, not amenable to direct transformation into temporal constraints between goals, the generate-and-test strategy of monotonic planning is employed. Thus, the partial plans generated by the constraint propagation of nonmonotonic planning are put together to develop feasible plans to solve the problem.

IV. Illustrations of Modeling and Nonmonotonic Operations Planning

In this section we will offer several illustrations of the various aspects of nonmonotonic operations planning (discussed in earlier sections) including the following: (1) development of hierarchical models for the process and its operations, (2) conversion of constraints to temporal orderings of primitive operations, and (3) synthesis of complete plans.

A. CONSTRUCTION OF HIERARCHICAL MODELS AND DEFINITION OF OPERATING STATES

Consider the flowsheet shown in Fig. 8. It represents the power forming section of a refinery, along the ancillary storage tanks and piping network required for the catalyst regeneration. During the operation of the plant, the activity of the catalyst (chloroplatinic acid on alumina base) in the operating reactor, e.g., REACTOR-1, decays as a result of (1) coke formation, (2) reduction in the chloride content, and (3) size growth of the platinum crystals. We want to synthesize an operating procedure that will discontinue the operation of REACTOR-1, start the regeneration of its catalyst (i.e., remove coke, add chloride, redisperse the platinum crystals), and initiate the operation of the standby REACTOR-2. For a fairly detailed description of the plant, its modeling, and the synthesis of

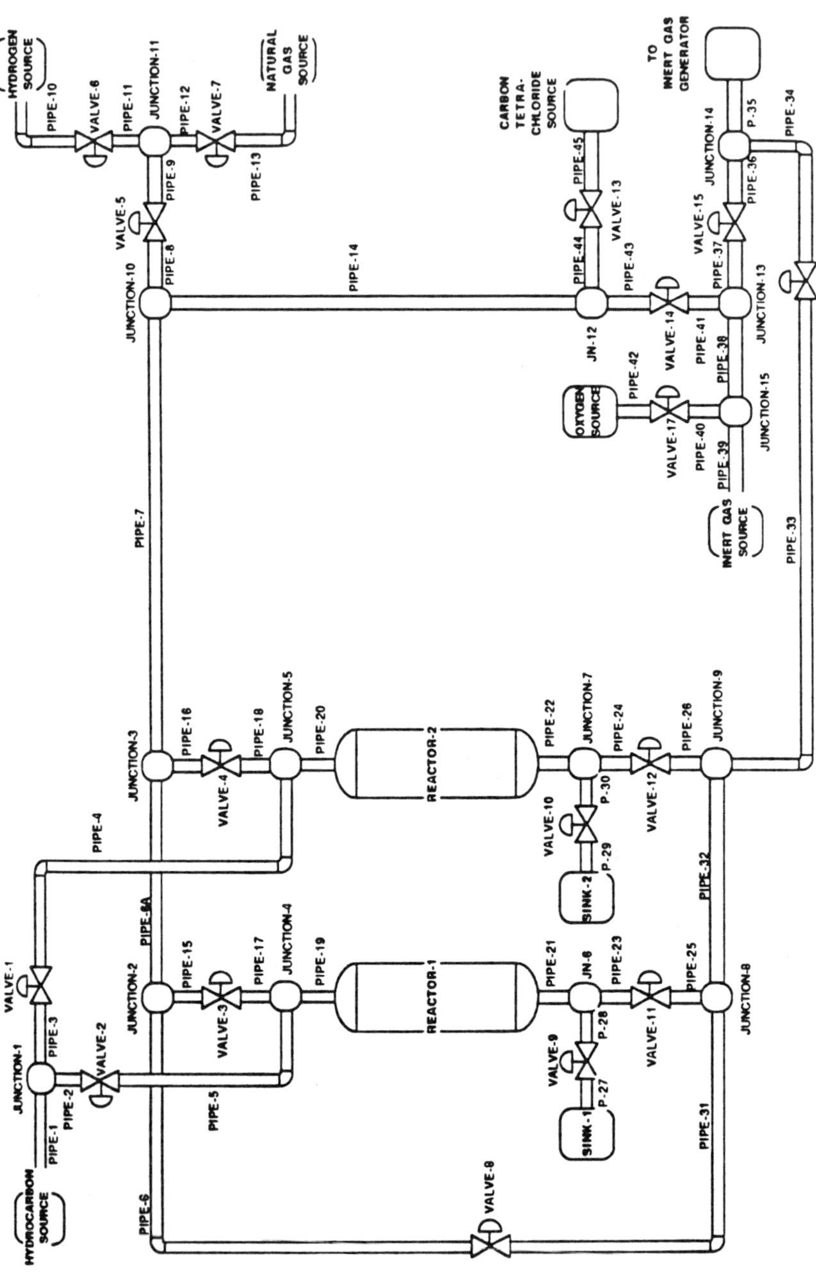

Fig. 8. Structure of the catalyst regeneration system. (Reprinted from *Comp. Chem. Eng.*, **12**, Lakshmanan, R. and Stephanopoulos, G., Synthesis of operating procedures for complete chemical plants, Parts I, II, p. 985, 1003, Copyright 1988, with kind permission from Elsevier Science Ltd., The Boulevard, Langford Lane, Kidlington 0X5 1GB, UK.)

changeover operating procedures, the reader is referred to Lakshmanan (1989).

The flowsheet editing facilities of *Design-Kit* (Stephanopoulos *et al.*, 1987) allow the user to construct an iconic representation of the flowsheet shown in Fig. 8, while the system at the same time creates the requisite data structures to represent the topology of the process (i.e., units and their interconnections). These data structures represent instances of the basic modeling elements of MODEL.LA. (see **21**:1). For example, Fig. 9a shows the relevant attributes in the declarative description of "Power-Forming-Plant," which is an instance of the modeling element (see Section II,B), *Plant*. The latter is, in turn, a subclass of the generic modeling element, *Generic-Unit*. Every device in the plant of Fig. 9a is an instance of the modeling element, *Unit*. Figure 9b shows the relevant declarative description of the unit REACTOR-1. Since abstraction in the representation of a plant is essential for the synthesis of operating procedures, the user can represent the "Power-Forming-Plant" as composed of two instances of *Plant-Section* called "Power-Forming-Reaction-Section" and "Power-Forming-Catalyst-Regeneration-Section" (see Fig. 9c). Figure 9d shows the relevant declarative elements of the "Power-Forming-Reaction-Section," whose boundaries (i.e., component units and streams) have been defined by the user. MODEL.LA. allows the creation of mathematical models around any abstraction of the processing units, while maintaining internal consistency between the model of an abstract entity (e.g., Plant-Section) and those of its constituents. Figure 10 shows the classes modeling a flow-valve and their interrelationships. For more details on the generation of the mathematical models, see **21**:1.

1. Defining an Operating State and Ensuring Its Completeness and Consistency

Any operations planning problem requires the *complete* and *consistent* definition of the initial and the goal states. In addition, during the evolution of an operating procedure any intermediate state must be fully specified by the operating actions and the modeling relations. Let us focus our attention on one segment of the desired operating procedure, e.g., "take REACTOR-1 off-line and bring REACTOR-2 on-line."

In the *initial state*, REACTOR-1 is operating while REACTOR-2 is on standby. Consequently,

> Condition-1: Hydrocarbon-Flow-in-REACTOR-1 > 0
> Condition-2: Product-Flow-from-REACTOR-1 > 0
> Condition-3: Hydrocarbon-Flow-in-REACTOR-2 $= 0$
> Condition-4: Product-Flow-from-REACTOR-2 $= 0$

Instance: POWER-FORMING-PLANT
Is-a-member-of: PLANT
Is-composed-of: [REACTOR-1, REACTOR-2, VALVE-1, VALVE-2, ..., VALVE-17,
        JUNCTION-1, ..., JUNCTION-15, PIPE-1, ..., PIPE-42,
        HYDROCARBON-SOURCE, SINK-1, SINK-2, HYDROGEN-SOURCE,
        NATURAL-GAS-SOURCE, CARBON-TETRACHLORIDE-SOURCE,
        OXYGEN-SOURCE, INERT-GAS-SOURCE, TO-REGENERATOR]

Note: "X" is-a-member-of "Unit"
    where "X" is any device in the above list.

(a)

Instance: REACTOR-1
Is-a-member-of: UNIT
Is-composed-of: []
Ports: [PORT-1, PORT-2]

Note: "PORT-1" is-a-member-of "PORT"
    "PORT-2" is-a-member-of "PORT"

(b)

Instance: POWER-FORMING-PLANT
Is-a-member-of: PLANT
Is-composed-of: [POWER-FORMING-REACTION-SECTION,
        POWER-FORMING-CATALYST-]-REGENERATION-SECTION]

(c)

Instance: POWER-FORMING-REACTION-SECTION
Is-a-member-of: PLANT-SECTION
Is-composed-of: [REACTOR-1, REACTOR-2, VALVE-1, ..., VALVE-4, VALVE-9, VALVE-12,
        JUNCTION-1, ..., JUNCTION-9, PIPE-1, ..., PIPE-7, PIPE-15, ..., PIPE-26,
        PIPE-31, ..., PIPE-33,   HYDROCARBON-SOURCE, SINK-1, SINK-2]

(d)

FIG. 9. Object-oriented descriptions of the Power-Forming-Plant at two levels of abstraction: (a) in terms of processing units, and (c) in terms of processing sections. The descriptions of a processing unit (b) and a processing section (c).

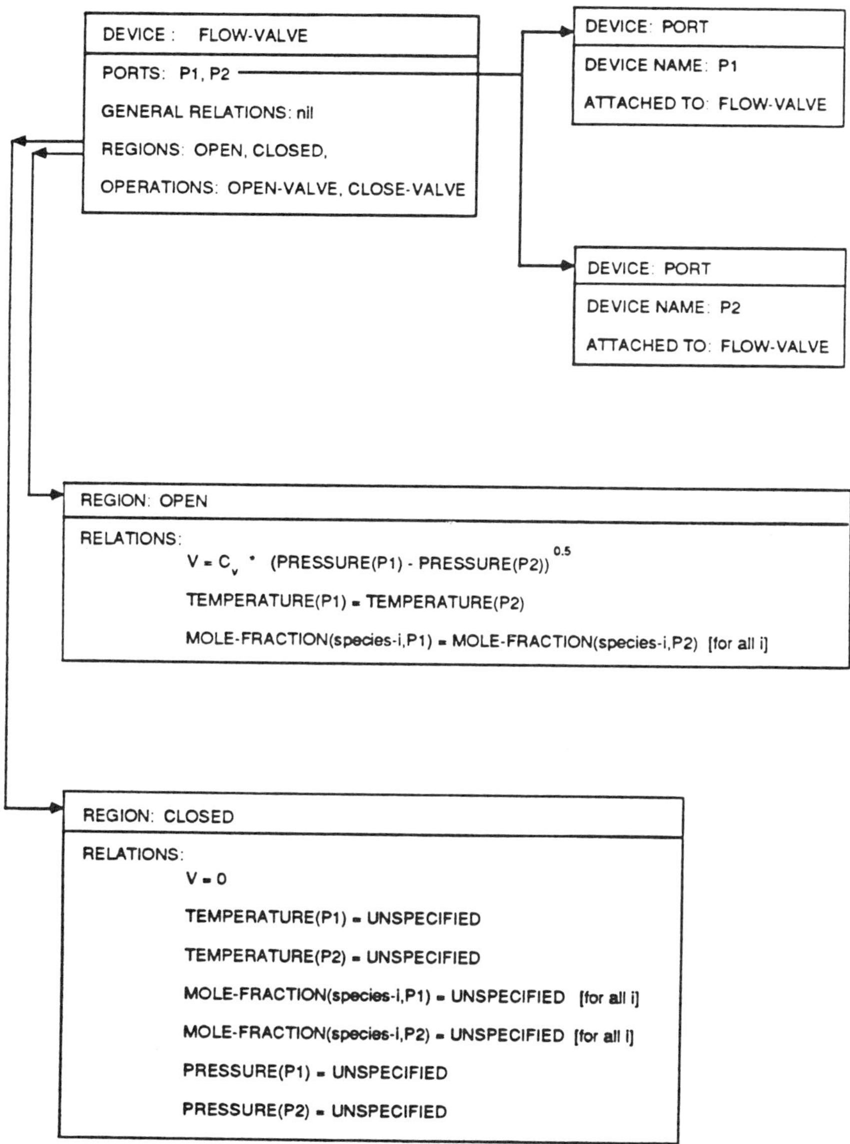

FIG. 10. Data structure modeling a flow-valve. (Reprinted from *Comp. Chem. Eng.*, **12**, Lakshmanan, R. and Stephanopoulos, G., Synthesis of operating procedures for complete chemical plants, Parts I, II, p. 985, 1003, Copyright 1988, with kind permission from Elsevier Science Ltd., The Boulevard, Langford Lane, Kidlington 0X5 1GB, UK.)

Condition-1, when propagated through the modeling relationships, yields the following; VALVE-2 is OPEN and VALVE-3 is CLOSED. These values are then stored in the instances of the corresponding valves. Similarly, we take the following associations:

Condition-2 $\Rightarrow$ [VALVE-9 is OPEN; VALVE-11 is CLOSED];
Condition-3 $\Rightarrow$ [VALVE-1 is CLOSED; VALVE-4 is CLOSED];
Condition-4 $\Rightarrow$ [VALVE-10 is CLOSED; VALVE-12 is CLOSED]

The propagation of these conditions through the modeling relationships is fairly straightforward, given the declarative richness of the data structures (e.g., see Fig. 10) representing the units of the plant, their interconnections and their physical behavior. Figure 11 shows the information flow through the modeling dependencies of variables, as Conditions 1–4 are propagated and produce the states of the associated operators (valves).

2. Ensuring the Completeness of the Initial State

The propagation of initial conditions through the modeling relationships is also used to ensure completeness in the definition of an initial state. For example, Condition-1 implies that the state of PIPE-1, PIPE-2, and PIPE-3 is characterized by the state of the flowing (or, stagnant, in PIPE-3) hydrocarbon. Similarly, Condition-2, when propagated backward, implies that the state of PIPE-27, PIPE-28, PIPE-21, and PIPE-23 is determined by the state of the flowing or stagnant product. The available initial conditions may not be able to define all facets of the initial state of a plant. For example, Conditions-1–3 cannot produce unambiguously the value of the state for VALVE-8, PIPE-6, and PIPE-31 (see Fig. 8). This is due to the fact that there is no directed path in the graph representing all modeling relationships, between the variables of the four conditions and the variables describing the state of VALVE-8, PIPE-6, and PIPE-31. In such cases the user must intervene and complete the specification for the initial state of the whole plant.

3. Checking the Consistency of the Initial State

Whereas the propagation of conditions through the modeling relationships always produces consistent completions in the definition of the initial state, this may not be true with the user-driven specifications. It is possible that a valve specified by the user to be OPEN, is "found" by the propagation of other conditions to be CLOSED. To detect such potential inconsistencies, we have developed a *dependency network* (Steele, 1980) to keep track of the flow of computations during constraint propagation. The dependency network is built on the undirected graph that represents the

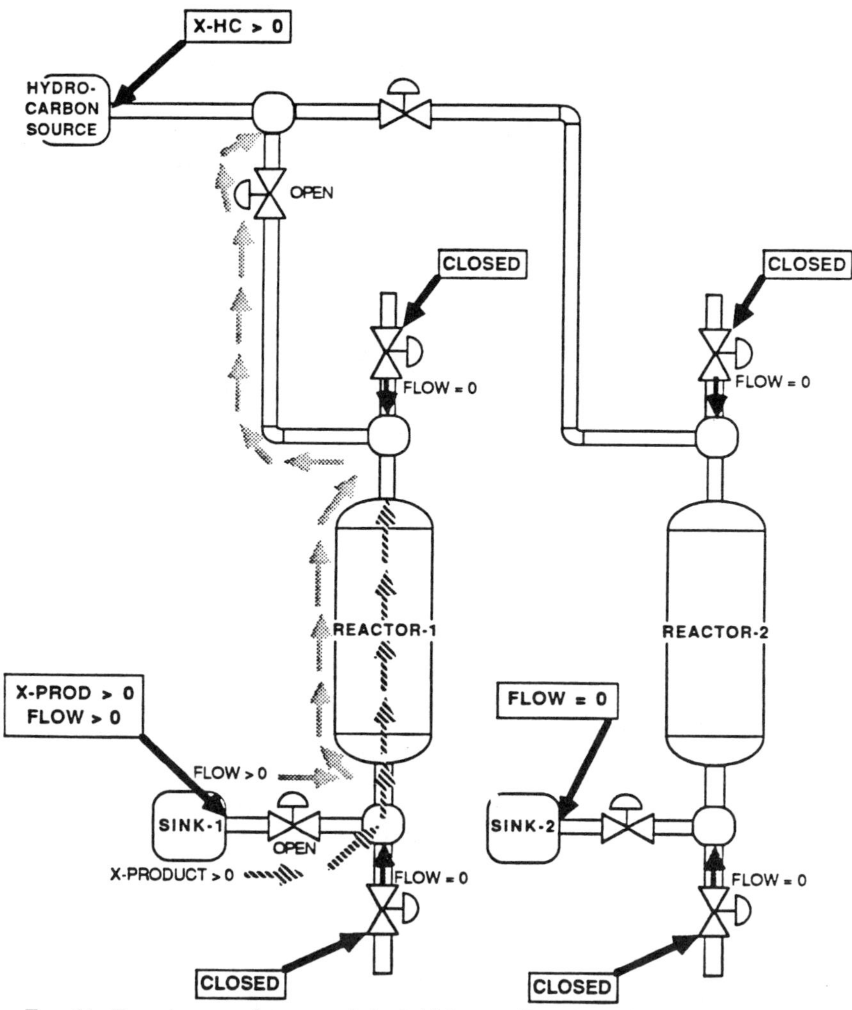

FIG. 11. Ensuring completeness of the initial state. (Reprinted from *Comp. Chem. Eng.*, **12**, Lakshmanan, R. and Stephanopoulos, G., Synthesis of operating procedures for complete chemical plants, Parts I, II, p. 985, 1003, Copyright 1988, with kind permission from Elsevier Science Ltd., The Boulevard, Langford Lane, Kidlington 0X5 1GB, UK.)

set of modeling relationships. With the given initial states as input variables, the undirected graph of modeling relationships is converted to a directed graph (dependency network), indicating the directionality of dependencies during the computation of initial states. If a conflict is encountered, backtracking through the dependency network reveals the source of conflict.

4. Complete and Consistent Definition of the Goal State

The mechanics for the definition of the goal state are fairly similar to those used for the definition of the initial state. Since the goal state is far more vague in the planners' mind than the initial state, the potential for conflicts is higher. Figure 12 shows the partial specification of the goal state, as well as the propagation of conditions to ensure completeness in the definition of the goal state.

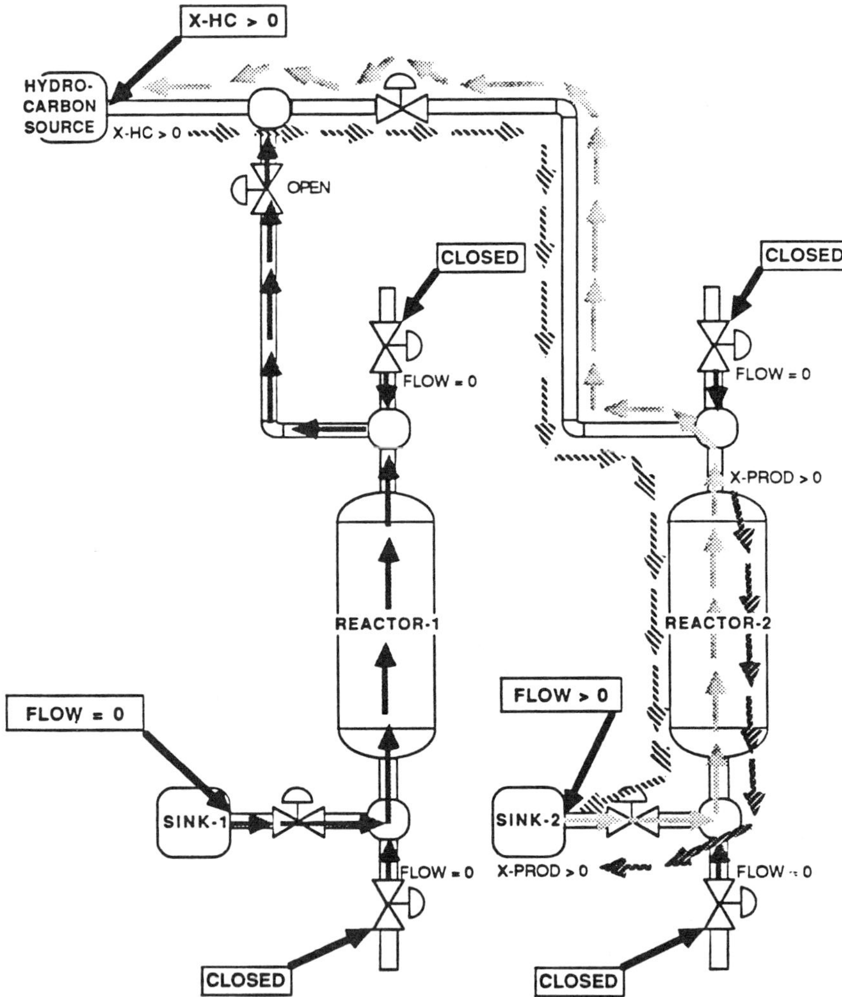

Fig. 12. Input of goal state and generation of concurrent goals. (Reprinted from *Comp. Chem. Eng.*, **12**, Lakshmanan, R. and Stephanopoulos, G., Synthesis of operating procedures for complete chemical plants, Parts I, II, p. 985, 1003, Copyright 1988, with kind permission from Elsevier Science Ltd., The Boulevard, Langford Lane, Kidlington 0X5 1GB, UK.)

B. Nonmonotonic Synthesis of a Switchover Procedure

Ethylenediaminetetraacetic acid (EDTA) is manufactured by the cyanomethylation process, whose main reactions are

$$
HCHO \;+\; HCN \;\xrightarrow{\;H_2SO_4\;}\; H\text{-}\underset{CN}{\overset{H}{\underset{|}{\overset{|}{C}}}}\text{-}OH \qquad \text{(Cyanohydrin)}
$$

$$
4\; H\text{-}\underset{CN}{\overset{H}{\underset{|}{\overset{|}{C}}}}\text{-}OH \;+\; H_2N\text{-}C_2H_4\text{-}NH_2 \;\xrightarrow{\;NAOH\;}\;
\begin{matrix} HOOCH_2C \\ HOOCH_2C \end{matrix}\!\!\Big\rangle N\text{-}C_2H_4\text{-}N\Big\langle\!\!\begin{matrix} CH_2COOH \\ CH_2COOH \end{matrix}
$$

$$
\text{(EDA)} \qquad\qquad\qquad\qquad\qquad\qquad\qquad \text{(EDTA)}
$$

$$
+\; 4NH_3
$$

Sulfuric acid, formaldehyde, and hydrogen cyanide are pumped into a glass-lined mixer (mixer 1, M1, of Fig. 13). Particular care is exercised so that the three charge operations are carried out in the order indicated above, to ensure the stability of the mixture at all times. In a separate segment of the plant, ethylenediamine (EDA) and dilute sodium hydroxide are charged and mixed in mixer 3 (M3 in Fig. 13). The solutions from mixer 1 and mixer 3 are pumped to the reactor (REACTOR, R1, in Fig. 13). When the reaction is complete, the reaction mixture is tested for traces of hydrogen cyanide. Dilute solution of formaldehyde is prepared in mixture 2 and is added to the reaction mixture, if there is any HCN present.

1. The Operating Situation Requiring a Switchover Procedure

The specific situation for which we want to synthesize an operating procedure is as follows. Hydrogen cyanide is handled by the top most line of piping, pumps, and valves (Fig. 13). Sulfuric acid and formaldehyde are handled by the next two flowlines in the flowsheet. The fourth flowline, which passes through pump PO4, has been shut down for routine maintenance. At some point during the operation of the plant, a leak is detected in the cooling jacket of mixer 1 and it has to be shut down in order to be repaired. Mixer 2 can be used in emergencies for limited periods of time only, because it is not a glass-lined vessel. As a result of this situation, we need to develop an operating procedure that will (1) shut mixer 1 down and disconnect it from the rest of the plant and (2) divert the flow of these materials, i.e., hydrogen cyanide, sulfuric acid, and formaldehyde, to mixer 2.

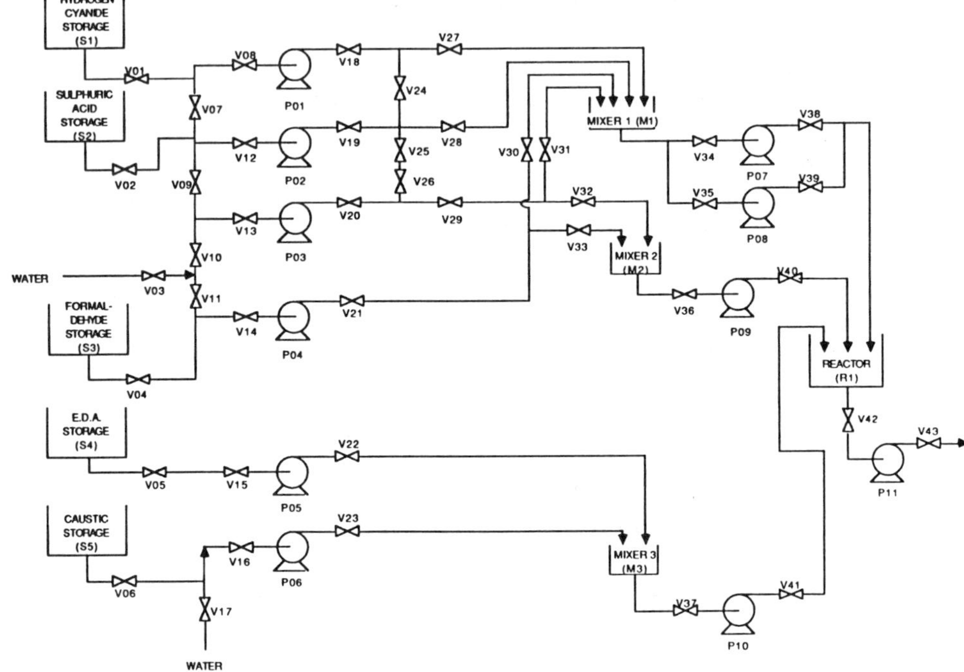

Fig. 13. Flowsheet of the ethylenediaminetetraacetic acid (EDTA) plant.

2. Constraints on the Operating Procedure

As we switch the operation from mixer 1 to mixer 2, certain constraints must be kept satisfied at all times. These constraints are as follows:

Constraint-1. Nowhere in the plant is HCN allowed to mix with formaldehyde.

Constraint-2, 3. The concentrations of HCN and H_2SO_4 in the mixer should satisfy the following constraints at all times:

$$[HCN] \leq 0.1 \, mol/L$$

$$[H_2SO_4] \geq 1 \, mol/L$$

Constraint-4. The temperature in the mixer should not exceed 30°C:

$$T_{M2} \leq 30°C$$

3. Initial and Goal States

The starting situation is partially defined by the state of the following valves:

$$\{\text{CLOSED-VALVES}\} = \{V03, V07, V09, V14, V17, V24, V25,$$
$$V26, V30, V32\}$$
$$\{\text{OPEN-VALVES}\} = \{V22, V23, V27, V28, V29\}$$

Propagation of the above states through the modeling relationships completes the definition of the initial state of the plant. The results are shown in Fig. 14a. (Note that valves V34 through V43, which are downstream of the mixers, are considered closed, and they do not affect the synthesis of the switchover procedure.) The desired goal state is operationally defined by the following conditions:

$$\text{Mixer-1-input-flow-i} = 0 \qquad i = HCN, H_2SO_4, HCHO$$
$$\text{Mixer-1-output-flow} = 0$$
$$\text{Mixer-2-input-flow-i} = \text{positive} \qquad i = HCN, H_2SO_4, HCHO$$

Propagating these values through the modeling equations, we can complete the definition of the goal state (see Fig. 14b). Comparing initial and

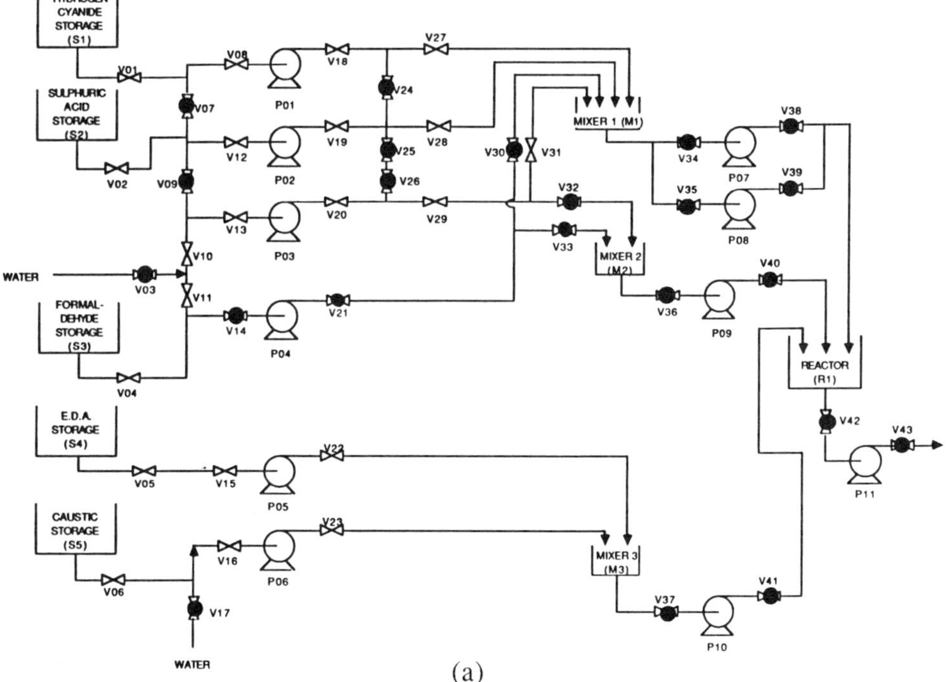

Fig. 14. The (a) initial and (b) goal states for the EDTA planning problem.

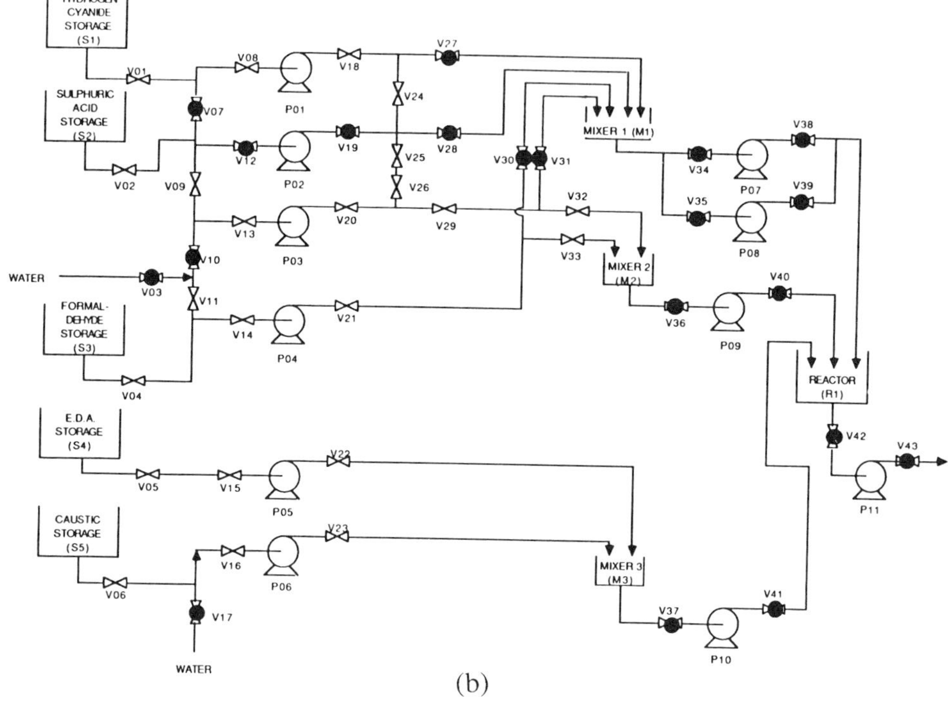

(b)

Fig. 14. *Continued.*

goal states, we conclude that the means–ends analysis requires the availability of the following operations to achieve the desired goal state:

V = {CLOSE V27, CLOSE V28, CLOSE V31, CLOSE V10, OPEN V9, CLOSE V12, CLOSE V19, OPEN V24, OPEN V25, OPEN V26, OPEN V32, OPEN V14, OPEN V21, OPEN V33}.

4. Abstraction of Operators

In Section III,D we presented an algorithm for the aggregation of primitive valve actions into abstract operators. Applying this algorithm over the set V of actions to achieve the goal state, we take:

AO-1: {CLOSE V27} $\Rightarrow$ Stop-Flow-HCN-M1
AO-2: {CLOSE V12, CLOSE V19, CLOSE V28} $\Rightarrow$ Stop-Flow-H$_2$SO$_4$-M1
AO-3: {CLOSE V10, CLOSE V31} $\Rightarrow$ Stop-Flow-HCHO-M1
AO-4: {OPEN V24, OPEN V25, OPEN V26, OPEN V32}
      $\Rightarrow$ Start-Flow-HCN-M2
AO-5: {OPEN V09, OPEN V32} $\Rightarrow$ Start-Flow-H$_2$SO$_4$-M2
AO-6: {OPEN V14, OPEN V21, OPEN V33} $\Rightarrow$ Start-Flow-HCHO-M2.

Then, the set V takes the following form:

$$V = \{AO\text{-}1, AO\text{-}2, AO\text{-}3, AO\text{-}4, AO\text{-}5, AO\text{-}6\}.$$

5. Transformation of Constraint-1 (Mixing Constraint)

In Section III,C we argued that one should identify the worst-case scenario for the violation of Constraint-1 and take action to ensure that a White Knight operation is carried out before any offending operation (i.e., Clobberer), in order to negate its impact (i.e., violation of mixing constraint). The worst-case scenario occurs when the plant has the following two sets of valves OPEN at the same time:

{Valves that are OPEN at the initial state}.

{Valves that are characterized as OPEN at the goal state}.

Then, we construct the influence graphs (IGs) of HCN and HCHO for the initial and goal states (Fig. 15), and from these we can identify the

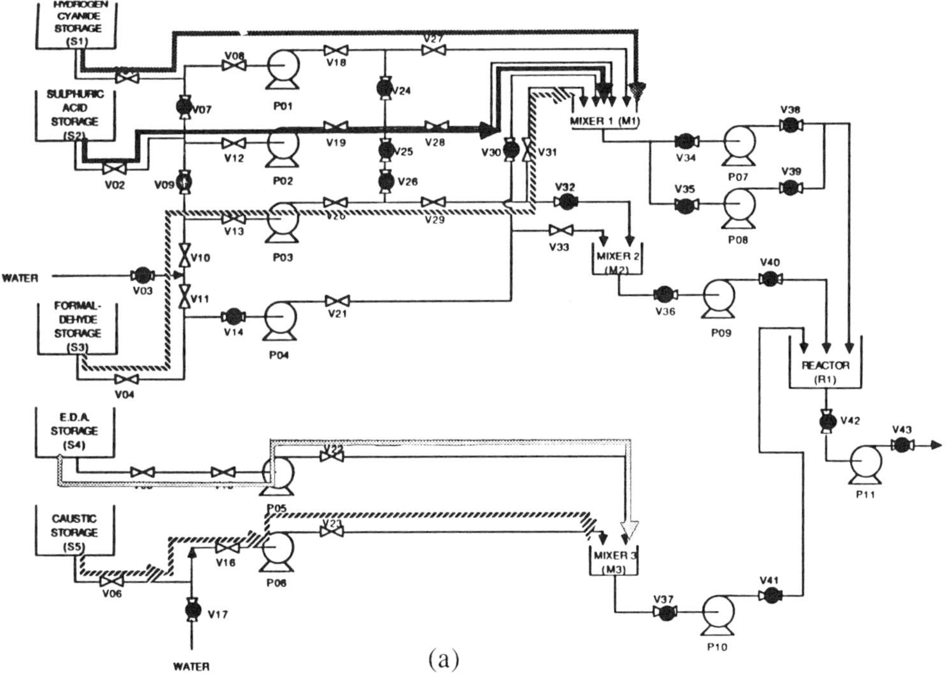

FIG. 15. Influence graphs in the (a) initial and (b) goal states. (Reprinted from *Comp. Chem. Eng.*, **14**, Lakshmanan, R. and Stephanopoulos, G., Synthesis of operating procedures for complete chemical plants, Parts III, p. 301, Copyright 1990, with kind permission from Elsevier Science Ltd., The Boulevard, Langford Lane, Kidlington 0X5 1GB, UK.)

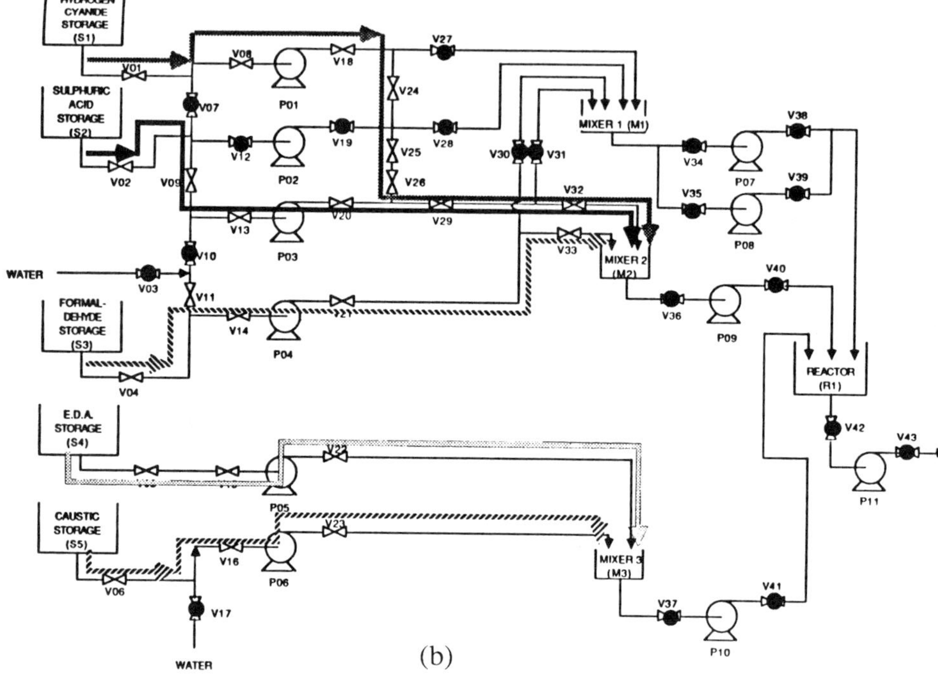

Fig. 15. *Continued.*

region of potential mixing between HCN and HCHO (intersection of IGs for HCN and HCHO) at the worst-case scenario (Fig. 16). From Fig. 16 we see that the IGs intersect at the valves V29 and V32. It is also clear that {V10} is the minimal separation valve set (MSVS; see Section III,C), i.e., the set of valves that, if closed, would prevent the mixing of HCN and HCHO. Thus, V10 must be closed before any valve in the initial state is opened. Looking at the elements of the set V in the previous paragraph, we derive the following temporal ordering of valve operations:

(CLOSE V10) before {OPEN V9, OPEN V24, OPEN V25, OPEN V26, OPEN V32, OPEN V14, OPEN V21, OPEN V33]   (Ordering-1)

So the mixing constraint has been transformed into the above constraint on the temporal ordering of primitive operations.

6. Generation of a Purge / Evacuation Procedure

Valves V29 and V32 and the associated pipes contain leftovers of HCHO and must be purged with H_2SO_4 before the other chemical,

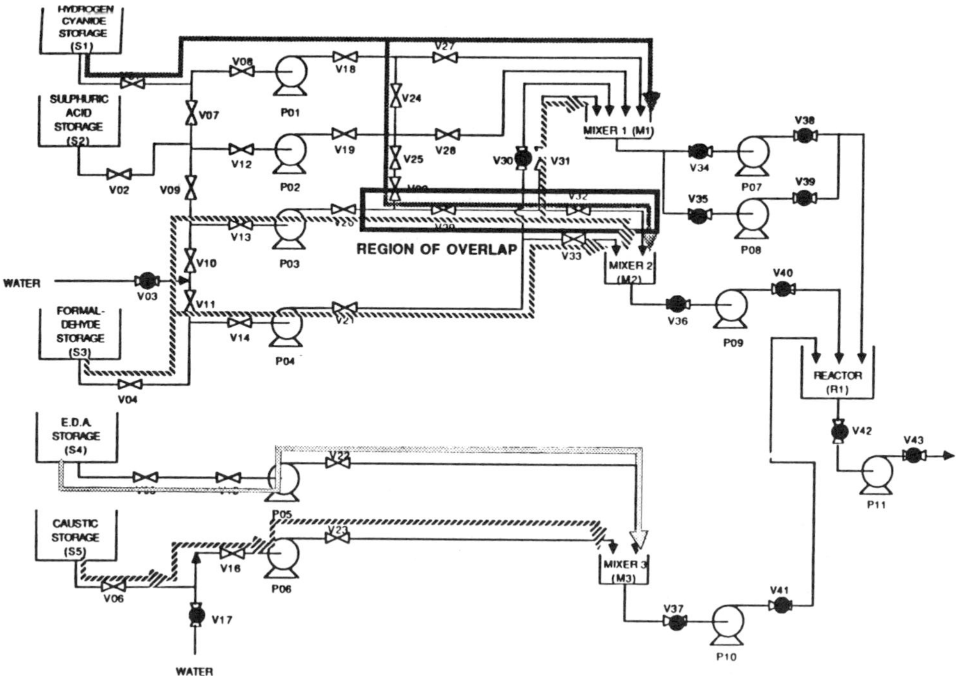

FIG. 16. Influence graphs in the worst-case state, showing the region of overlap. (Reprinted from *Comp. Chem. Eng.*, **14**, Lakshmanan, R. and Stephanopoulos, G., Synthesis of operating procedures for complete chemical plants, Parts III, p. 301, Copyright 1990, with kind permission from Elsevier Science Ltd., The Boulevard, Langford Lane, Kidlington OX5 1GB, UK.)

i.e., HCN, is allowed to flow through. (Remember the required order of addition: H_2SO_4 before HCHO before HCN.) The path-finding algorithm (see Section III,C) identifies the following path from the storage of the purgative material (i.e., H_2SO_4) to mixer 2:

Purge Path = {PUMP PO3; VALVE V20; VALVE V29; VALVE V32}

This path represents a "planning island," i.e., the goals of an operating subprocedure, which must be completed after the MSVS has been closed, and before the operations that allow HCN in the line. Therefore, we have the following temporal constraints:

(CLOSE V10) before (PURGE OPERATIONS) before (OPEN V24)

(Ordering-2)

where (PURGE OPERATIONS) is the sequence of operating steps required for the purging of HCHO and stands on its own as an independent operating procedure. The synthesis of primitive steps leads to

$$\text{(PURGE OPERATIONS)} = \{\text{OPEN V09, START PO3, OPEN V32,}$$
$$\text{CLOSE V32, STOP PO3, CLOSE V09}\}$$

7. Transformation of Constraints-2 and -3

Suppose that we were to load mixer M2 with the following sequence of abstract operators (see earlier paragraph in this section for definition of abstract operators):

$$\{(\text{Start-Flow-HCN-M2}), (\text{Start-Flow-H}_2\text{SO}_4\text{-M2}),$$
$$(\text{Start-Flow-HCHO-M2})\}.$$

If the initial concentrations of the raw material were (in moles per liter), $[\text{HCN}]_0 = 1$, $[\text{H}_2\text{SO}_4]_0 = 1.5$, and $[\text{HCHO}]_0 = 2$, then the preceding sequence of operations would immediately have violated Constraint-2, since $[\text{HCN}] = 1 > 0.1$. Similarly, Constraint-3 is violated by the preceding sequence, since $[\text{HSO}_4] = 0 < 1.5$. In Section III,D, we discuss the notion of "demotion" of Clobberers. Let us see how it works here and leads to temporal ordering of operations:

(a) (Start-Flow-HCN-M2) is a Clobberer of $[\text{HCN}] \leq 0.1$. Demoting this Clobberer, we require that

$$\text{(Start-Flow-HCN-M2) after (Start-Flow-H}_2\text{SO}_4\text{-M2; Start-Flow-HCHO-M2}$$

(b) (Start-Flow-HCN-M2) and (Start-Flow-HCHO-M2) are Clobberers of $[\text{H}_2\text{SO}_4] \geq 1$. Demoting both Clobberers, we establish the following temporal ordering constraints:

$$\text{(Start-Flow-H}_2\text{SO}_4\text{-M2) before}$$
$$\text{(Start-Flow-HCN-M2; Start-Flow-HCHO-M2)}$$

Taking (a) and (b) together, we see that Constraints-2 and -3 have been transformed to the following temporal ordering:

$$\text{(Start-Flow-H}_2\text{SO}_4\text{-M2) before (Start-Flow-HCHO-M2)}$$
$$\text{before (Start-Flow-HCN-M2)}$$
$$\text{(Ordering-3)}$$

8. Transformation of Constraint-4

Let the temperature of the three materials in the storage tanks be 35°C. Then, all three operations, (Start-Flow-HCN-M2), (Start-Flow-H$_2$SO$_4$-M2) and (Start-Flow-HCHO-M2), are Clobberers of the Constraint-4, since $T_{M2} = 35°C > 30°C$. No demotion of any of the three operators would restore feasibility, and we are led to the need for a White Knight. Searching the plant for relevant operations we identify the following White Knight:

$$(OPEN \; VALVE\text{-}CW) \Rightarrow Start\text{-}Cooling\text{-}Water\text{-}through\text{-}Jacket$$

and we constrain the White Knight to be carried out before any of three Clobberers, i.e.,

(OPEN VALVE-CW) before
[(Start-Flow-HCN-M2), (Start-Flow-H$_2$SO$_4$-M2), (Start-Flow-HCHO-M2)]
(Ordering-4)

9. Synthesis of the Complete Switchover Operating Procedure

In the previous paragraphs we saw how the ideas of nonmonotonic planning have transformed the required operational constraints into constraints ordering the temporal sequencing of operations. These temporal constraints define partial segments of the overall operating procedure. Now, let us see how they can be merged into a complete operating procedure. Let us recall the set V of operations, which was defined earlier through the means–ends analysis. Any complete plan must carry out the abstract operators AO-1 through AO-6. From each temporal ordering we take the following implications:

Ordering-1 $\Rightarrow$ (AO-3: Stop-Flow-HCHO-M1) before (AO-4; AO-5; AO-6).

Ordering-2 $\Rightarrow$ (AO-3) before (PURGE OPERATIONS) before (AO-4).

Ordering-3 $\Rightarrow$ (AO-5) before (AO-6) before (AO-4).

Ordering-4 $\Rightarrow$ (OPEN VALVE-CW) before (AO-4, AO-5, AO-6).

The following three sequences satisfy all the preceding constraints on the temporal ordering of operations:

Procedure-A. (AO-3) > (PURGE OPERATIONS) > (OPEN VALVE-CW) > (AO-5) > (AO-6) > (AO-4).

Procedure-B. (AO-3) > (OPEN VALVE-CW) > (PURGE OPERATIONS) > (AO-5) > (AO-6) > (AO-4).

Procedure-C. (OPEN VALVE-CW) > (AO-3) > (PURGE OPERATIONS) > (AO-5) > (AO-6) > (AO-4).

Each of the abstract operators, AO-3, AO-4, AO-5, AO-6, is composed of more than one operation. No constraints have been stated or generated to determine their relative order. For example, (AO-3) could be implemented as (CLOSE V10) > (CLOSE V31) or (CLOSE V31) > (CLOSE 10).

It is also important to note that abstract operators (AO-1) and (AO-2) are not participating in any temporal ordering constraint. Consequently, they could be placed anywhere in the chain of operations of a procedure. For example, *Procedure-A* can be completed with (AO-1) and (AO-2) in any of the following ways:

Procedure A-1. (AO-1) > (AO-2) > (AO-3) > (OPEN VALVE-CW) > . . . .

Procedure A-2. (AO-1) > (AO-3) > (AO-2) > (OPEN VALVE-CW) > . . . .

Procedure A-3. (AO-2) > (AO-1) > (AO-3) > (OPEN VALVE-CW) > . . . .

Procedure A-4. (AO-2) > (AO-3) > (AO-1) > (OPEN VALVE-CW) > . . . .

Procedure A-5. (AO-3) > $\cdots$ > (AO-5) > (AO-6) > (AO-1) > (AO-2).

Procedure A-6. (AO-3) > $\cdots$ > (AO-5) > (AO-6) > (AO-2) > (AO-1) etc.

V. Revamping Process Designs to Ensure Feasibility of Operating Procedures

During the planning of operating procedures for complete chemical plants, a very important stage is the one in which the computer tries to identify whether there is a possibility that one or more of the operating constraints will be violated. This is analogous to the detection of Clobberers in Chapman's program TWEAK. As is the case in TWEAK, when Clobberers are detected, they must be countered by a suitable White Knight unless they are subdued by demotion or promotion. In the planning methodology presented in this chapter, the worst-case scenario lies between the initial and goal states, and promotion or demotion after or before these two states is clearly not possible. Thus the only alternative is for the program to locate a White Knight to counter the Clobberer.

It should be obvious that there is no guarantee that a suitable White Knight can be found to counter at least one of the Clobberers. In these instances a modification must be made to the process flowsheets so as to introduce structures that will serve as White Knights. In this section we describe algorithms that allow a computer-based methodology to automatically generate the necessary process modifications.

Let us look more closely at the situation where a Clobberer has been detected. In the case of qualitative mixing constraints, this involves an overlap of influence graphs in the worst-case scenario. The White Knights that are needed to counter this Clobberer are (a) an acceptable separation valve set (ASVS), and (b) a purgative and a purge route. In this analysis, we shall assume that the absence of White Knights is due to one of these two situations. Thus the modifications to the flowsheet must provide either an ASVS or a purgative and purge route (or both) depending on the needs of the situation at hand. The remainder of this section provides a methodology for accomplishing both these tasks.

A. Algorithms for Generating Design Modifications

In this subsection, we present the algorithms that are to be used by a computer program that attempts to make design modifications to a flowsheet structure so as to allow the system to generate feasible operating plans where none existed before. The algorithms are classified into two types: those that are used to ensure the existence of an ASVS, and those that are used for generating a purge source and a purge route.

1. Flowsheet Modifications for Generating an ASVS

The methodology for proposing flowsheet modifications to ensure the existence of an ASVS is described in this section.

a. Stage 1. Determine a Minimal Separation Pipe Set

1. Construct the reduced pipe network corresponding to the topological graph of the worst-case state.
   (a) If the current node represents a PIPE or a "supersource," do nothing.
   (b) If the current node represents any other type of processing equipment, do the following:
       (i) Delete the current node from the topological graph.
       (ii) Let U be the set of nodes incident to the in-edges of the current node.
       (iii) Let V be the set of nodes incident to the out-edges of the current node.
       (iv) For every pair (u, v), where u is in U and v is in V, create a directed edge (u, v) and add it to the topological graph.

In the transformed graph, the only nodes are pipes and "supersources" and an edge leads from one node to another iff there is a path connecting the corresponding nodes in the topological graph, and no other pipe or source lies on this path.

2. Now, replace all directed edges with undirected edges. The problem of finding a minimal separation pipe set (MSPS) now reduces to the problem of determining a *vertex separator* in the reduced pipe network for the pair of supersources for the dangerous species under consideration. A polynomial–time algorithm exists for determining a vertex separator (see Even, 1979).

The MSPS may not be an acceptable separation pipe set (ASPS) for the given planning problem. This is because it is possible that some pipe in the MSPS is required to have flow through it in the goal state, or the user may simply have some preference regarding the state of flow through some of the pipes. However, from a heuristic standpoint, the minimal set is a good starting point for evolving into an ASPS, since disturbing the flow in these pipes upsets the overall flow only minimally. The procedure for evolving from an MSPS to an ASPS is described in stage 2.

b. Stage 2. Determine an Acceptable Separation Pipe Set

1. Let the current separation pipe set (SPS) of edges be the MSPS already determined.
2. Identify the pipes (if any) in the current SPS that are constrained to have positive or negative flows in the goal state. If there are no such pipes, then the current SPS is feasible. Go to step 3. If, on the other hand, there are some such pipes, mark them as unsatisfactory and proceed to step 4.
3. Present the MSPS graphically to the user (by highlighting the pipes on the flowsheet, for example). Ask the user to specify any pipes that he or she prefers to keep open. Mark all such pipes as "unsatisfactory." If there are no unsatisfactory pipes, return the current SPS as the ASPS.
4. Remove all "unsatisfactory" pipes from the reduced pipe network. Go to stage 1, step 2.

Once an ASPS has been found, the final stage of the methodology can be started.

c. Stage 3. Generation of an ASVS. For each pipe in the ASVS, locate a corresponding pipe in the topological graph. For each pipe located in the topological graph, place a valve on the pipe. The set of valves that are added to the flowsheet in this step will form an ASVS for the planning problem under consideration.

Thus, if the planner fails to come up with a feasible plan, and the reason for failure is that no ASVS was found, the preceding algorithm may be used to propose flowsheet modifications (in the form of extra valves) to create an appropriate ASVS.

2. Flowsheet Modifications for Purge Sources and Routes

Another reason why the planner may fail to find a feasible plan is that no suitable purgative or purge route can be found. The modification of the flowsheet structure to allow for this purging is described in this section. The algorithms provided here will, however, address only the structural modifications that need to be performed, and will not address the chemistry-related issues of deciding on a suitable chemical species to introduce as a purgative. This aspect of the problem can be addressed by encoding chemistry knowledge in the form of a rule-based system or by asking the user to choose an inert component. We also make the assumption that all of the flows in the worst-case scenario are defined. This may not necessarily be the case, but as yet no algorithm is known that will propose structural modifications to the flowsheet in the absence of such flow information.

ALGORITHM **CREATE-PURGE-ROUTE**

1. Construct a copy of the initial state topological graph. In this graph close all the valves that are members of the ASVS. This will partition the topological graph into two subgraphs G_A and G_B, each of which will contain the source of only one of the two dangerous components A and B, respectively.
2. Locate all the nodes in G_A where B was present in the initial state. Let the subgraph that consists of these nodes be called G'. Similarly, locate all the nodes in G_B where A was present in the initial state and create a subgraph G'' that consists of these nodes and their incident edges. For each of the graphs G' and G'', do the following:
   (a) Find all the minimal and maximal nodes in G' (or G'') (see Lakshmanan, 1989).
   (b) Create a supersource of the chemical species that was identified (by either the user or a rule-based system) as a suitable purgative. For each minimal node in G' introduce a pipe node and appropriate connecting edges leading from the supersource to the minimal node.
   (c) Create a *sink* for the chemical species used as a purgative. For each maximal node in G' introduce a pipe node and the appropriate connecting edges leading from the maximal nodes to the sink.

At the end of this construction process, the flowsheet will have been suitably modified so as to allow the purging of one dangerous species before admitting the second.

3. Case Studies

In this subsection we examine two modifications of the EDTA plant discussed earlier, in Section IV. The purpose of the illustration is to demonstrate the working of the algorithms that generate flowsheet modifications when (1) no MSVS or ASVS can be found and (2) no purge source or purge route is available.

In this example the flowsheet of the EDTA process has been modified by removing the following valves: v10, v11, v13, and v20. The result of the modifications is shown in Fig. 17.

When the planning algorithms are run on this plant no MSVS is found since, in the worst-case state, there are no valves that will prevent hydrogen cyanide and formaldehyde from mixing in the region of overlap. Thus the computer will try to use the methodology described in the

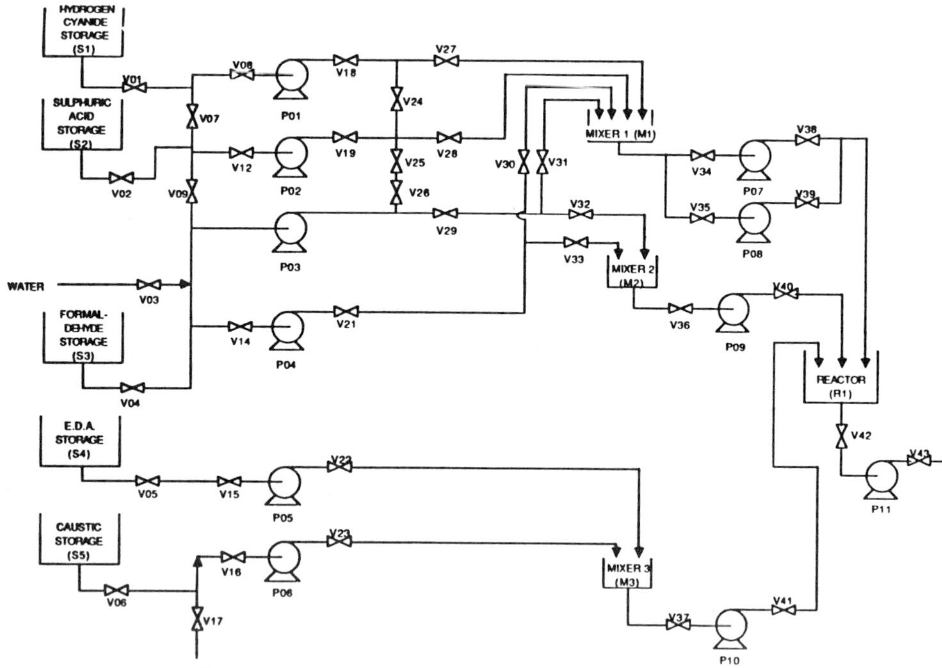

FIG. 17. Flowsheet of the modified EDTA plant.

previous paragraphs to come up with flowsheet modifications that will allow the computer to find an MSVS.

The algorithms described earlier are run on the modified plant, and the computer suggests that a valve be placed between the water inlet and the junction just above it. This modification is shown in Fig. 18, and this valve serves as the MSVS in the planning problem.

The second example that is used to illustrate the design methodologies is a modification to the EDTA problem as follows. The structure of the flowsheet is exactly the same as the one presented in Section IV. The only change to the problem is the statement that sulfuric acid and formaldehyde should not be allowed to come into contact with each other.

It may be recalled that in the initial analysis sulfuric acid was used as a purge species. It is obvious that this is not possible anymore since one of the dangerous components, namely formaldehyde, cannot be brought into contact with sulfuric acid. Thus the computer must generate new piping and source and sink structure to allow the computer to find a purge route.

The algorithm CREATE-PURGE-ROUTE described earlier is run under the worst-case scenario of the original EDTA plant with the added

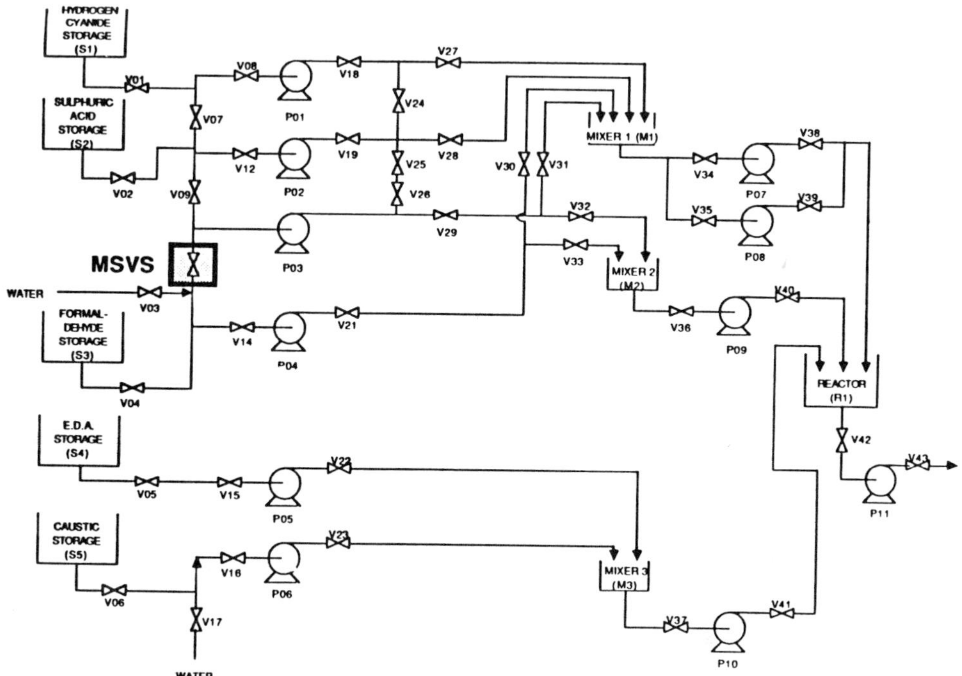

FIG. 18. Flowsheet of the EDTA plant with MSVS.

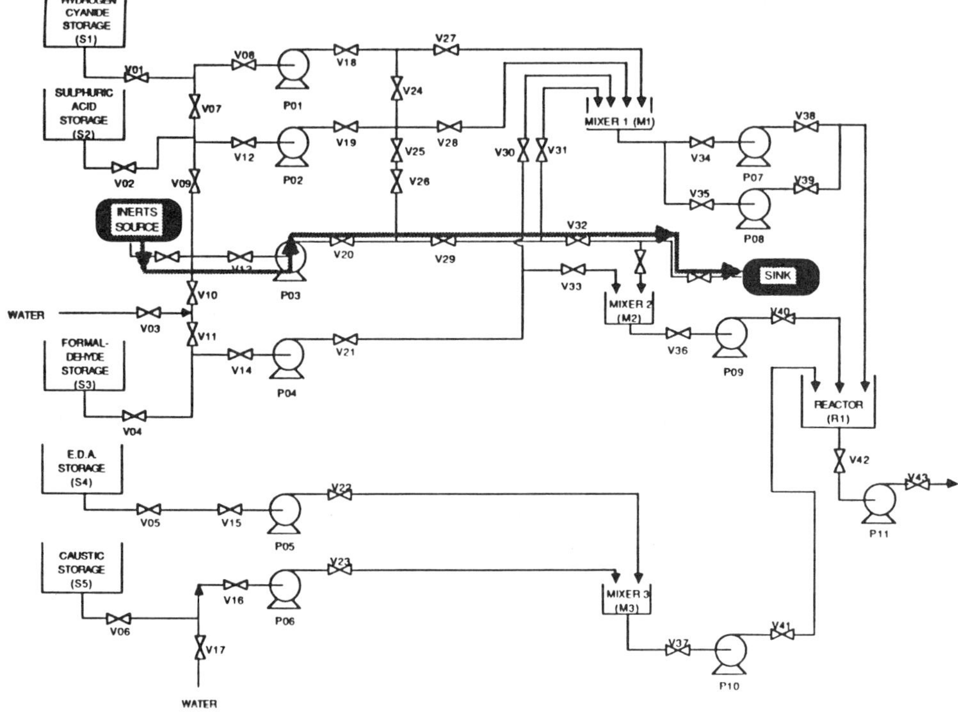

FIG. 19. Flowsheet of the EDTA plant showing purge route.

mixing constraint, and piping modifications are generated in the flowsheet. The final flowsheet with the purge route is shown in Fig. 19.

Note that, as mentioned earlier, the program does not attempt to specify what component is actually used as the inert purgative. A rule-based system with extensive knowledge of explosive mixtures is needed to accomplish this task. The construction of such a knowledge base, while useful, is assumed to be outside the scope of the planning problem. The program is designed only to propose the piping and valving modifications to the flowsheet.

VI. Summary and Conclusions

The synthesis of operating procedures for continuous chemical plants can be represented as a mixed-integer nonlinear programming problem, and it has been addressed as such by other researchers. In this chapter we have attempted to present a unifying theoretical framework, which addressed the logical (or integer) components of the problem, while allowing a smooth integration of optimization algorithms for the optimal selection of the values of the continuous variables. Nonmonotonic reasoning has been found to be superior to the traditional monotonic reasoning approaches, and capable to account for all integer decision variables. Critical in its success is the modeling approach, which is used to represent process operational tasks at various levels of abstraction.

Nevertheless, a series of interesting issues arise, which must be resolved before these ideas find their way to full industrial implementation:

1. *Dynamic simulation with discrete-time events and constraints.* In an effort to go beyond the integer (logical) states of process variables and include quantitative descriptions of temporal profiles of process variables one must develop robust numerical algorithms for the simulation of dynamic systems in the presence of discrete-time events. Research in this area is presently in full bloom and the results would significantly expand the capabilities of the approaches, discussed in this chapter.

2. *Design modifications for the accommodation of feasible operating procedures.* Section V introduced some early ideas on how the ideas of planning operating procedures could be used to identify modifications to a process flowsheet, which are necessary to render feasible operating procedures. More work is needed in this direction. Clearly, any advances in dynamic simulation with discrete-time events would have beneficial effects on this problem.

3. *Real-time synthesis of operating procedures.* Most of the ideas and methodologies, presented in this chapter, are applicable to the a priori, off-line, synthesis of operating procedures. There is a need though to address similar problems during the operation of a chemical plant. Typical examples are the synthesis of operational response (i.e., operating procedure) to process upsets, real-time recovery from a fallback position, and supervisory control for constrained optimum operation.

References

Barton, P. I., The modeling and simulation of combined discrete/continuous processes. Ph.D. Thesis, University of London (1992).

Chakrabarti, P. P., Ghose, S., and DeSarkar, S. C., Heuristic search through islands. *Artif. Intell.* **29**, 339 (1986).

Chapman, D., "Planning for Conjunctive Goals," MIT AI Lab. Tech. Rep. AI-TR-802. Massachusetts Institute of Technology, Cambridge, MA, 1985.

Crooks, C. A., and Macchietto, S., A combined MILP and logic-based approach to the synthesis of operating procedures for batch plants. *Chem. Eng. Commun.* **114**, 117 (1992).

Even, S., "Graph Algorithms." Computer Science Press, Rockville, MD, 1979.

Fikes, R. E., and Nilsson, N. J., STRIPS: A new approach to the application of theorem proving to problem solving. *Artif. Intell.* **2**, 198 (1971).

Foulkes, N. R., Walton, M. J., Andow, P. K., and Galluzzo, M., Computer-aided synthesis of complex pump and valve operations. *Comput. Chem. Eng.* **12**, 1035 (1988).

Fusillo, R. H., and Powers, G. J., A synthesis method for chemical plant operating procedures. *Comput. Chem. Eng.* **11**, 369 (1987).

Fusillo, R. H., and Powers, G. J., Operating procedure synthesis using local models and distributed goals. *Comput. Chem. Eng.* **12**, 1023 (1988a).

Fusillo, R. H., and Powers, G. J., Computer aided planning of purge operations. *AIChE J.* **34**, 558 (1988b).

Garrison, D. B., Prett, M., and Steacy, P. E., Expert systems in process control. A perspective. *Proc. AAAI Workshop Control*, Philadelphia (1986).

Holl, P., Marquardt, W., and Gilles, E. D., Diva—a powerful tool for dynamic process simulation. *Comput. Chem. Eng.* **12**, 421 (1988).

Ivanov, V. A., Kafarov, V. V., Kafarov, V. L., and Reznichenko, A. A., On algorithmization of the startup of chemical productions. *Eng. Cybern. (Engl. Transl.)* **18**, 104 (1980).

Kinoshita, A., Umeda, T., and O'Shima, E., An approach for determination of operational procedure of chemical plants. *Proc. PSE'82, Kyoto* 114 (1982).

Lakshmanan, R., and Stephanopoulos, G., "Planning and Scheduling Plant-Wide Startup Operations," LISPE Tech. Rep. (1987).

Lakshmanan, R., and Stephanopoulos, G., Synthesis of operating procedures for complete chemical plants. I. Hierarchical, structured modeling for nonlinear planning. *Comput. Chem. Eng.* **12**, 985 (1988a).

Lakshmanan, R., and Stephanopoulos, G., Synthesis of operating procedures for complete chemical plants. II. A nonlinear planning methodology. *Comput. Chem. Eng.* **12**, 1003 (1988b).

Lakshmanan, R., Synthesis of operating procedures for complete chemical plants. Ph.D. Thesis. Massachusetts Institute of Technology, Dept. Chem. Eng., Cambridge, MA (1989).

Lakshmanan, R., and Stephanopoulos, G., Synthesis of operating procedures for complete chemical plants. III. Planning in the presence of qualitative, mixing constraints. *Comput. Chem. Eng.* **14**, 301 (1990).

Marquardt, W., Dynamic process: Simulation-recent progress and future challenges. *In* "Chemical Process Control, CPC-IV" (Y. Arkun, and W. H. Ray, eds.) CACHE, AIChE Publishers, New York, 1991.

Modell, M., and Reid, R. C., "Thermodynamics and its Applications," 2nd ed., Prentice-Hall, Englewood Cliffs, NJ, 1983.

Newell, A., Shaw, J. C., and Simon, H. A., Report on a general problem solving program. *Proc. ICIP*, 256 (1960).

Pantelides, C. C., and Barton, P. I., Equation-oriented dynamic simulation. Current status and future perspectives. *Comput. Chem. Eng.* **17**, S263 (1993).

Pavlik, E., Structures and criteria of distributed process automation systems. *Comput. Chem. Eng.* **8**, 295 (1984).

Rivas, J. R., and Rudd, D. F., Synthesis of failure-safe operations. *AIChE J.* **20**, 311 (1974).

Sacerdoti, E. D., The non-linear nature of plans. *Adv. Pap. IJCAI, 4th*, 206, (1975).

Steele, G. L., "The Definition and Implementation of a Computer Programming Language Based On Constraints," AI Lab. Tech. Rep. AI-TR-595. Massachusetts Institute of Technology, Cambridge, MA, 1980.

Tomita, S., Hwang, K., and O'Shima, E., On the development of an AI-based system for synthesizing plant operating procedures. *Inst. Chem. Eng. Symp. Ser.*, 114 (1989a).

Tomita, S., Hwang, K., O'Shima, E., and McGreavy, C., Automatic synthesizer of operating procedures for chemical plants by use of fragmentary knowledge. *J. Chem. Eng. Jpn.* **22**, 364 (1989b).

Tomita, S., Nagata, M., and O'Shima, E., Preprints IFAC Workshop, Kyoto, 66 (1986).

INDUCTIVE AND ANALOGICAL LEARNING: DATA-DRIVEN IMPROVEMENT OF PROCESS OPERATIONS[1]

Pedro M. Saraiva

Department of Chemical Engineering
University of Coimbra
3000 Coimbra, Portugal

[1] The work reported in this chapter was performed while the author was on leave as a Ph.D. student in the Department of Chemical Engineering, Massachusetts Institute of Technology, Cambridge, MA, 02139.

Informed and systematic observation of data, naturally generated, can lead to the formulation of interesting and effective generalizations. Whereas some statisticians believe that experimentation is the only safe and reliable way to "learn" and achieve operational improvements in a manufacturing system, other statisticians and all the empirical machine learning researchers contend that by looking at past records and sets of examples, it is possible to extract and generate important new knowledge. This chapter draws from *inductive and analogical learning* ideas in an effort to develop systematic methodologies for the extraction of structured new knowledge from operational data of manufacturing systems. These methodologies do not require any a priori decisions and assumptions either on the character of the operating data (e.g., probability density distributions) or the behavior of the manufacturing operations (e.g., linear or nonlinear structured quantitative models), and make use of *instance-based learning* and *inductive symbolic learning* techniques, developed by artificial intelligence (AI). They are aimed to be complementary to the usual set of statistical tools that have been employed to solve analogous problems. Thus, one can see the material of this chapter as an attempt to fuse statistics and machine learning in solving specific engineering problems. The framework developed in this chapter is quite generic and, as the subsequent sections illustrate, it can be used to generate operational improvement opportunities for manufacturing systems (1) that are simple or complex (with internal structure), (2) whose performance is characterized by one or multiple objectives, and (3) whose performance metrics are categorical (qualitative) or continuous (real numbers). A series of industrial case studies illustrates the learning ideas and methodologies.

I. Introduction

With rapid advances in hardware, database management and information processing systems, efficient and competitive manufacturing has become an information-intensive activity. The amounts of data presently collected in the field on a routine basis are staggering, and it is not unusual to find plants where as many as 20,000 variables are continuously monitored and stored (Taylor, 1989).

It has also been recognized that the ability of manufacturing companies to become *learning* organizations (Senge, 1990; Shiba *et al.*, 1993; Hayes *et al.*, 1988), achieving high rates of continuous and never-ending improvements (with a special focus on *total manufacturing quality*), is a critical factor determining the survival and growth of the company in an increasingly competitive global economy (Deming, 1986; Juran, 1964).

As a result, there has been a significant and increasing awareness and interest on how the accumulation of data can lead to a better management of operational quality, which involves two complementary steps (Saraiva and Stephanopoulos, 1992a): (a) *control within prespecified limits* and (b) *continuous improvement of process performance*. The aim of the first is to detect "abnormal" situations, identify and eliminate the *special causes* that produced them. But, bringing the process under bounded, statistical control is not enough. The final level of variability thus achieved is the result of *common and sustained causes*, present within the process itself, and must not be considered as unavoidable. Therefore, the second step dictates that one should move from process control to process improvement, i.e., continuously search for common causes, ways of reducing their impact and challenge the current levels of performance (Juran, 1964).

The bulk chemical commodity producing companies (e.g., refineries, petrochemicals) have been practicing this philosophy for some time, using dynamic models to contain operational variability through feedback controllers, and employing static models to determine the optimal levels of operating conditions (Lasdon and Baker, 1986; Garcia and Prett, 1986).

On the other hand, plants involving solids processing (e.g., pulp and paper) with poorly understood physicochemical phenomena, not lending themselves to description through first-principles models, have not been using effectively the mountains of accumulated operational data. This lack of exploration and analysis of data is particularly severe for records of data collected while the processes were kept under "normal" operating contexts. Thus, given that less than 5% of the points contained in a conventional 6σ control chart are likely to attract any attention, the bulk of the acquired operating records is simply "stored," i.e., wasted. For such manufacturing plants there is a strong need and incentives to develop theoretical frameworks and practical tools that are able to extract useful and operational knowledge from existing records of data, leading to continuous improvement of process operational performance. This chapter represents a specific effort that attempts to fulfill that need.

Several statistical, quality management, and optimization data analysis tools, aimed at exploring records of measurements and uncover useful information from them, have been available for some time. However, all of them require from the user a significant number of assumptions and a priori decisions, which determine in a very strict manner the validity of the final results obtained. Furthermore, these classical tools are guided essentially by an academic attitude that emphasizes numerical accuracy over the extraction of substantive knowledge, whereas the formats chosen to express the solutions are often difficult to be interpreted and understood by human operators or to be implemented by them in manufacturing operations.

Rather than trying to replace any of the above traditional techniques, this chapter presents the development of complementary frameworks and methodologies, supported by symbolic empirical machine learning algorithms (Kodratoff and Michalski, 1990; Shavlik and Dietterich, 1990; Shapiro and Frawley, 1991). These ideas from machine learning try to overcome some of the weaknesses of the traditional techniques in terms of both (1) the number and type of a priori decisions and assumptions that they require and (2) the knowledge representation formats they choose to express final solutions.

One can thus state the primary goal of the approaches summarized in this chapter as follows:

$$
\begin{aligned}
&\textit{To develop and apply assumption-free learning frameworks and methodologies, aimed} \\
&\textit{at uncovering and expressing in adequate solution formats performance improvement} \\
&\textit{opportunities, extracted from existing data which were acquired from plants that} \\
&\textit{cannot be described effectively through first-principles quantitative models.}
\end{aligned}
\tag{1}
$$

The remaining material of this chapter has the following structure: Section II provides a more specific definition of the problems that are addressed, and it identifies the scope of applications and the type of manufacturing systems that the presented methodologies aim to cover. Then, Section III introduces a generic framework for the development and description of the learning algorithms that will allow us to present the several methodologies on a common basis. In the same section we will also enumerate the most critical characteristics and features that differentiate the approaches of this chapter from previous ones. Sections IV and V illustrate the problem statements and solution methodologies employed when the manufacturing system's performance is measured by a *categorical* (qualitative) or *continuous* (real-valued) variable. Section VI extends the ideas presented in Sections IV and V to manufacturing systems whose performance is characterized by multiple, possibly noncommensurable objectives, and Section VII makes extensions to complex manufacturing systems with internal structure. Concrete applications are presented at the end of these sections. Some final conclusions together with a critical summary of the main results are presented in Section VIII.

II. General Problem Statement and Scope of the Learning Task

In its most general form, the problem that we address in this chapter can be stated as follows:

> **Given** a set of existing $(\mathbf{x}, y)$ data records,
> where $\mathbf{x}$ is a vector of operating or decision variables, which
>   are believed to influence the values taken on by y;
> y is a performance metric, usually assumed to be a quality
>   characteristic of the product or process under analysis;
> **Learn** how to improve the system performance above its current levels   (2)

The work described in this chapter revisits this old problem by adopting a new perspective, exploring alternative formats for the presentation of solutions and problem-solving strategies. Such alternative problem-solving strategies are particularly useful in addressing systems where the weaknesses of traditional approaches become particularly severe. It is thus important to clarify the scope of application and the type of situations within the broad area covered by the problem statement (2) that the methodologies of this chapter aim to cover. That's what we will try to do in the following paragraphs.

1. *Supervisory control layer of decisionmaking*. Although learning capabilities and data analysis activities should be present across all levels of decisionmaking (Fig. 1), countless studies and authors (National Research Council, 1991; Latour, 1976, 1979; Launks *et al.*, 1992; Klein, 1990; Sargent, 1984; Ellingsen, 1976; Moore, 1990) have shown that the greatest

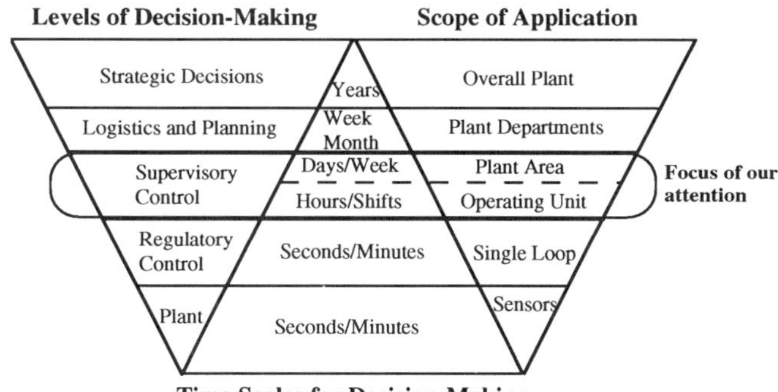

FIG. 1. Levels, time scales, and application scopes of decisionmaking activities.

opportunities and benefits are associated with improvements at the steady-state optimization or supervisory control layer. Shinnar (1986) makes clear this point by stating that in the chemical industry 80% of all practical control problems and 90% of the financial gains must be found and addressed at this level. Accordingly, we will be performing a steady-state analysis of data that has been collected and defined according to the supervisory control layer of decisionmaking.

2. *Plants lacking credible first-principles models.* The learning methodologies of this chapter require very few assumptions and decisions to be made a priori by the user, and rely on an exploratory analysis of data (Tukey, 1977) to uncover operational knowledge. Thus, they are particularly useful to analyze systems poorly understood and for which no first-principles models with acceptable accuracy are available. In such plants supervisory control decisions are quite often taken by human operators, rather than by computers. Reflecting on this fact, across the scale of automation in the spectrum of human–machine interaction (Sheridan, 1985) the approaches of this chapter fall within the scope of decision support systems (Turban, 1988); i.e., they do not intend to replace human intervention, but instead to interact with the users, and provide them as much useful information as possible. Yet, it is still the human operator who is responsible for the choice of a particular final solution, among a number of promising alternatives presented to him or her, and for selecting the particular course of action to follow. That being the case, it is critical for the knowledge extracted from the data to be represented in operational, explicit, and easy-to-understand formats, so that better use of that knowledge can be made by the human operators, who are not necessarily experts in pattern recognition, optimization, or statistics. Thus, the learning methodologies must reflect these concerns. Indeed, a strong emphasis is placed on the language used by the techniques of this chapter to express final solutions.

3. *Considerable amounts of data available on routine basis.* In learning activities, there is a fundamental tradeoff between the a priori existing knowledge and data intensity (Gaines, 1991). The more one already knows, the less data are needed to reach a given conclusion. As was said earlier, we are especially interested in studying complex systems for which no accurate quantitative models of behavior are available. Therefore, the learning methodologies must not rely or depend on the validity of any strong assumptions made about the system under study. A price has to be paid, as a result of this strategic choice; to reach the same level of accuracy and resolution in the final solutions found, more data are needed than if stronger assumptions could be made a priori and they happened to

be valid ones. This data-intensive nature of the approaches is not however likely to be problematic, since today in most manufacturing organizations there is no real shortage of data, but rather large amounts of unexplored measurement records. Furthermore, the case studies that were conducted show that even with moderate amounts of data (always less than 1000 records), it is possible to find solutions that result in quite significant improvements over the current levels of performance.

4. *Flexibility in problem definition.* Within the scope of application problems, outlined in the previous paragraphs, the learning methodologies are quite flexible and can handle a number of different situations and problem formulations. Both types of systems, where the performance metric used to evaluate performance, y, is continuous and categorical (i.e., can assume only one among a discrete number of possible values, such as "good," "bad," and "excellent"), can be analyzed. In systems of nontrivial size several performance objectives must often be taken into account. Extensions of the basic learning methodologies to handle such multiobjective problem formulations have been developed and tested. Finally, besides simple systems without internal structure, assumed to be isolated from the remaining world and self-sufficient for decisionmaking purposes, one has frequently to consider more complex systems, such as a complete plant, composed of a number of interconnected subsystems. A learning architecture for these complex systems was conceived and applied to a specific case study. It is based on the same procedures applied for simple systems, on top of which we have added articulation, coordination and propagation procedures. Thus, the learning methodologies of this chapter cover a broad range of possible industrial applications, and are able to handle simple or complex systems, with single or multiple objectives, and categorical or continuous performance evaluation variables.

5. *Prototypical application examples.* To provide a more concrete notion of the type of systems where our approaches are expected to be particularly helpful and useful, we conclude this section with a sample of prototypical examples of what the performance metric, y, in the problem statement (2) may represent, together with a definition of the corresponding systems:

Example 1: kappa (κ) index of the unbleached pulp produced by a Kraft digester

Example 2: activity, yield, and total production cost of proteins from fermentation units

Example 3: composition, molecular-weight distribution, and structure of complex copolymers

Example 4: selectivity and uniformity of etching rates at a chemical vapor deposition unit that produces silicon wafers

Example 5: analysis of the operation of an activated sludge wastewater treatment unit or of an entire pulp plant

As can be inferred from this sample of examples, industrial sectors where the learning methodologies of this chapter look particularly promising are pulp and paper, biotechnology, cement and ceramics, microelectronics, and discrete-parts manufacturing, i.e., complex plants or pieces of equipment that include solids processing, whose behavior is poorly understood and for which no accurate first-principles models exist. It should not be construed, though, that the methodologies of this chapter cannot also be applied to the analysis of other well-understood systems. However, as we move to such well-known systems, for which reliable quantitative models are already available, the suggested problem-solving approaches gradually loose their competitive advantages over traditional ones.

III. A Generic Framework to Describe Learning Procedures

In this section we present a formal framework to describe learning procedures that is sufficiently rich and general to allow us to concisely express (1) all the previous approaches that have been proposed for the formulation and solution of the problems stated in statement (2), and (2) all the different alternative definitions and methodologies that will be discussed in later sections of the chapter. This description language provides a common basis to introduce our different application contexts (categorical or continuous y, single or multiple objectives, simple or complex systems) and corresponding algorithms. It also allows us to make a clear statement of their most important departures from previous conventional approaches, thus facilitating a comparative analysis.

Subsequently, we use the common description framework in order to identify and analyze the following three most important distinguishing features that differentiate the learning frameworks of this chapter from traditional optimization and statistical techniques:

(a) Formats chosen to express the solutions
(b) Criteria used to evaluate the solutions
(c) Procedures used to estimate the solutions' evaluation criteria

A. A Generic Formalism

Any learning procedure, aimed to address and solve problems given by the problem statement (2) at the supervisory control level of decisionmaking, can be expressed by the following quartuple:

$$L = (\xi, \psi, f, S), \tag{3}$$

where

(a) $\xi \in \Xi$ represents a generic solution, ξ, defined in the solution space, Ξ.

(b) $\psi \in \Psi$ is the performance criterion, ψ, defined in the performance space, Ψ, that one chooses to evaluate the merit of a generic solution ξ.

(c) f is the model or procedure that maps the solution into the performance space, i.e., it allows one to compute an estimate of the performance criterion, ψ^{est}, for any potential solution ξ, $\psi^{\text{est}} = f(\xi)$.

(d) S is a search procedure that explores f in order to identify specific solutions, $\xi^* \in \Xi$, that look particularly promising according to their estimated performances, $f(\xi^*)$. This final fourth element is optional and it is absent from conventional approaches whose goal is strict estimation (prediction of ψ for a given particular choice of ξ).

B. Major Departures from Previous Approaches

An examination of previous classical learning procedures reveals that they differ from each other only with respect to the choices of ψ, f, and S. All of them share the same basic format for ξ and the corresponding solution space, Ξ. Let's assume that each $(\mathbf{x}, y)$ pair in the problem statement (2) contains a total of M decision variables:

$$\mathbf{x} \equiv [x_1, \ldots, x_m, \ldots, x_M]^T \in \Xi_{\text{decision}}, \tag{4}$$

where Ξ_{decision} stands for the decision space, composed of all feasible $\mathbf{x}$ vectors, a subspace of $\mathscr{R}^M$.

Traditional approaches adopt a solution space Ξ that coincides with Ξ_{decision}, and thus any final solution (ξ^* or $\mathbf{x}^*$) has the same format as $\mathbf{x}$, consisting of a real vector that defines a single point in the decision space.

By considering such a language to express solutions, one ignores the fact that in the type of problems we want to study decision variables behave as random variables, and there is always some variability associated with them. No matter how good control systems happen to be, in reality we will always have to live with ranges of values for the decision variables (concentrations, pressures, flows, etc.), eventually bounded within a narrow, but not null, operation window. As a consequence of not taking into account that one has to operate within a given zone of the decision space, rather than at a single point, the final solutions found by conventional approaches may be suboptimal even when perfect $f(\mathbf{x})$ models are available, since their evaluation criterion, ψ, reflects only the performance achieved at a particular point in the decision space, and completely ignores the system behavior around that point. The zone of the decision space that surrounds the $\mathbf{x}^*$ solution with the best $\psi(\mathbf{x}^*)$ does not correspond in general to the zone of the decision space, $\mathbf{Z}^*$, where best average performance, $\psi(\mathbf{Z}^*)$, can be achieved.

1. Hyperrectangles as a Convenient Solution Format

As a result of the above observations, once one chooses to adopt solution formats where each ξ represents a region rather than a point in the decision space, it has yet to be decided what type of zones and shapes should be considered. Since our goal is not to achieve full automation, but to provide support to human operators, it is critical for the new language, used to express solutions, to be understood by them and lead to results that are easy to implement. In that regard, after observing how people tend to articulate their reasoning activities in the control rooms we concluded that they essentially follow an "orthogonal thinking" paradigm: *human operators express themselves by means of conjunctions of statements about individual decision variables* (e.g., the concentration is high and the pressure low), x_m, *and not through some more or less intricate linear or nonlinear combination of them* (as is the case with multivariate statistical techniques such as principal-components analysis, partial least squares, factor analysis, or neural networks). Combining the need to identify zones in the decision space, rather than points, with the need to preserve the individuality of each decision variable, rather than losing it in linear or nonlinear combinations, hyperrectangles (conjunctions of ranges of x_m values) in the decision space appear as a very convenient and the natural format to express solutions.

Thus, a critical departure from previous approaches, common to all our learning methodologies, is the adoption of a solution format that consists of hyperrectangles (not points) defined in the decision space.

2. Interval Analysis Nomenclature

Thus, interval analysis (Moore, 1979; Alefeld and Herzberger, 1983) provides the adequate support and notation formalism to express solutions. To distinguish real numbers from intervals, we will use capital letters for intervals. Also, bold typing is employed to represent both real variable and real interval vectors. A real interval X is a subset of $\mathscr{R}$ of the form

$$X \in \mathbf{I} \equiv \{x \in \mathscr{R} \mid i(X) \le x \le s(X)\}, \tag{5}$$

where

(a) $\mathbf{I}$ is the space of all closed real intervals.
(b) $i(X)$ is the lower bound of X.
(c) $s(X)$ is the upper bound of X.
(d) $w(X) = s(X) - i(X)$ is the width of X.
(e) $m(X) = [i(X) + s(X)]/2$ is the midpoint of X.

Extending the notation to hyperrectangles in $\mathscr{R}^M$, an M-dimensional interval vector, $\mathbf{X}$, has as its components real intervals, X_m, defined by ranges of x_m:

$$\mathbf{X} \equiv [X_1, \ldots, X_m, \ldots, X_M]^T \in \mathbf{I}^M;$$

$$\mathbf{m}(\mathbf{X}) = [m(X_1), \ldots, m(X_m), \ldots, m(X_M)]^T \in \mathscr{R}^M. \tag{6}$$

Since a real vector is a degenerate interval vector whose components are null width intervals, previous conventional pointwise solution formats can be considered a particular case of the suggested alternative and more general solution space, obtained when the minimum allowed region size is reduced to zero, thus converting hyperrectangles into single points.

3. Major Differences

The notation introduced above allows us to make now a more explicit and condensed enumeration of the major characteristics and differences, with respect to the (ξ, ψ, f, S) key components, that separate our learning methodologies from other approaches.

Solution format, ξ. The solution space consists of hyperrectangles ($\xi = \mathbf{X} \in \mathbf{I}^M$), instead of points ($\xi = \mathbf{x} \in \mathscr{R}^M$), defined in the decision space.

Performance criterion, ψ. The quality of any potential solution, $\mathbf{X}$, is determined by the average system performance achieved within the zone of the decision space identified by $\mathbf{X}$, $\psi(\mathbf{X})$, not by the individual performance obtained at any particular point $\mathbf{x}$, $\psi(\mathbf{x})$.

Mapping procedure, f. The models that perform the mapping from the solution to the performance space have as argument a given hyperrectangle, $\mathbf{X}$, rather than a point, $\mathbf{x}$. They compute an estimate of the average performance that is expected within $\mathbf{X}$, $\psi^{\text{est}}(\mathbf{X}) = f(\mathbf{X})$, not a single pointwise prediction, $\psi^{\text{est}}(\mathbf{x}) = f(\mathbf{x})$.

Search procedure, S. The search procedures explore the modified mapping models, $\psi^{\text{est}}(\mathbf{X}) = f(\mathbf{X})$, in order to generate and identify a set of final solutions, $\mathbf{X}^*$, that look particularly promising according to the corresponding estimated performance scores, $\psi^{\text{est}}(\mathbf{X}^*)$.

As we will see in subsequent sections, the mapping procedures, f, adopted in our learning methodologies, are based on direct sampling approaches:

For any given $\mathbf{X}$ and ψ, we find those $(\mathbf{x}, y)$ pairs for which $\mathbf{x} \in \mathbf{X}$, and use this random sample to compute $\psi^{\text{est}}(\mathbf{X})$ and build confidence intervals for $\psi(\mathbf{X})$.

These direct sampling strategies provide $\psi^{\text{est}}(\mathbf{X})$ estimators that are consistent, unbiased, and do not require a priori assumptions about the system behavior, thus remaining consistent and unbiased regardless of the validity of any assumptions. In addition to a point estimate of performance, they also provide a probabilistic bound on the uncertainty associated with it, through the construction of confidence intervals for $\psi(\mathbf{X})$. Furthermore, the accuracy of the estimates obtained is limited only by the amounts of data that are available, not by any of the structural or functional form choices that have to be made with other mapping models. Finally, they are also computationally efficient, since all the effort involved consists of a search for those pairs of data falling inside $\mathbf{X}$, followed by a computation of the average and standard deviation among these records.

The preceding set of characteristics and properties of the $\psi^{\text{est}}(\mathbf{X})$ estimators makes our type of mapping procedures, f, particularly appealing for the kinds of systems that we are especially interested to study, i.e., manufacturing systems where considerable amounts of data records are available, with poorly understood behavior, and for which neither accurate first-principles quantitative models exist nor adequate functional form choices for empirical models can be made a priori. In other situations and application contexts that are substantially different from the above, while much can still be gained by adopting the same problem statements, solution formats and performance criteria, other mapping and search procedures (statistical, optimization theory) may be more efficient.

A solution space, Ξ, consisting of hyperrectangles defined in the decision space, $\mathbf{X}$, is a basic characteristic common to all the learning methodologies that will be described in subsequent sections. The same does not happen with the specific performance criteria ψ, mapping models f, and search procedures S, which obviously depend on the particular nature of the systems under analysis, and the type of the corresponding performance metric, y.

IV. Learning with Categorical Performance Metrics

The nature of the performance metric, y, is determined by the characteristics of the specific process under analysis. Since we are particularly interested in analyzing situations where y is related to product or process quality, it is quite common to find systems where a categorical variable y is chosen to classify and evaluate their performance. This may happen due to the intrinsic nature of y (e.g., it can only be measured and assume qualitative values, such as "good," "high," and "low"), or because y is derived from a quantization of the values of a surrogate continuous measure of performance (e.g., $y =$ "good" if some characteristic z of the product has value within the range of its specifications, and $y =$ "bad," otherwise).

In this section we will introduce the problem statements adopted for this type of performance metric, briefly describe the learning methodology employed to address it [for a more complete presentation, see Saraiva and Stephanopoulos (1992a)], and show a specific application case study.

A. Problem Statement

When y is a categorical variable, which can assume only one among a total of K discrete possible values, pattern recognition (Duda and Hart, 1973; James, 1985) provides an adequate context for the introduction of the learning methodologies. The most important features that separate these learning methodologies from existing classification techniques (such as linear discriminant functions, nearest-neighbor and other nonparametric classifiers, neural networks, etc.), are summarized in Table I, and briefly discussed below.

TABLE I

CONVENTIONAL PATTERN RECOGNITION AND SUGGESTED ALTERNATIVE

	Conventional classification	Suggested alternative
ξ	$\mathbf{x} \in \mathscr{R}^M$	$\mathbf{X} \in \mathbf{I}^M$
ψ	$y(\mathbf{x})$ or $p(y = j \mid \mathbf{x})$	$y(\mathbf{X})$ or $p(y = j \mid \mathbf{X})$
f	Technique-dependent	$\dfrac{n_j(\mathbf{X})}{n(\mathbf{X})}$
S	Nonexistent	Induction of decision trees

1. Conventional Classification Techniques

All conventional classification procedures are aimed at answering one of the following questions:

$$\text{Given a generic vector of values } \mathbf{x} \equiv [x_1, \ldots, x_m, \ldots, x_M]^T \in \mathscr{R}^M, \quad \text{(7a)}$$
what is the corresponding y value?

and/ or

$$\text{Given a generic vector of values } \mathbf{x} \equiv [x_1, \ldots, x_m, \ldots, x_M]^T \in \mathscr{R}^M,$$
what are reasonable estimates for the conditional probabilities (7b)
$p(y = j|\mathbf{x}), \; j = 1, \ldots, K?$

Thus, they share exactly the same solution (Ξ) and performance criteria (Ψ) spaces. Furthermore, since their role is simply to estimate y for a given $\mathbf{x}$, no search procedures S are attached to classical pattern recognition techniques. Consequently, the only element that differs from one classification procedure to another is the particular mapping procedure f that is used to estimate $y(\mathbf{x})$ and/ or $p(y = j|\mathbf{x})$. The available set of $(\mathbf{x}, y)$ data records is used to build f, either through the construction of approximations to the decision boundaries that separate zones in the decision space leading to different y values (Fig. 2a), or through the construction of approximations to the conditional probability functions, $p(y = j|\mathbf{x})$.

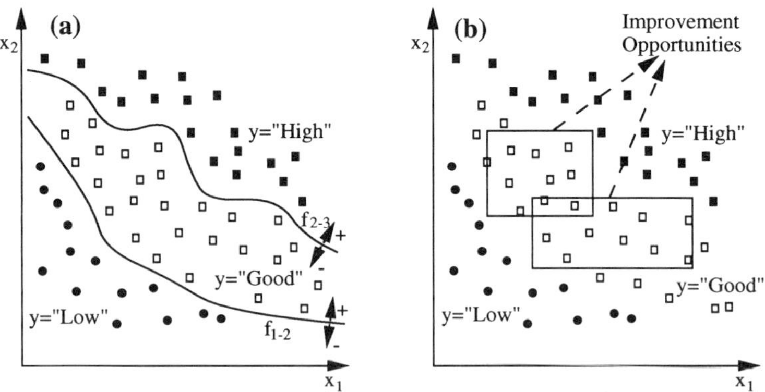

FIG. 2. (a) Conventional pattern recognition; (b) alternative problem statement and solution format.

2. The Learning Methodology

On the other hand, the question that we want to see answered is the following one (Fig. 2b):

What are hyperrectangles in the decision space, $\mathbf{X} \in \mathbf{I}^M$, inside which one gets only a desired y value, or at least a large fraction of that y value? (8)

The solution space thus consists of hyperrectangles in the decision space, $\mathbf{X} \in \mathbf{I}^M$, and the corresponding performance criteria are the conditional probabilities of getting any given y value inside $\mathbf{X}$, $p(y = j|\mathbf{X})$, $j = 1, \ldots, K$, or the single most likely y value within $\mathbf{X}$.

The mapping procedure, f, that allows us to compute $p(y = j|\mathbf{X})$ estimates, starts with a search performed over all the available $(\mathbf{x}, y)$ pairs that leads to the identification of the $n(\mathbf{X})$ cases for which $\mathbf{x} \in \mathbf{X}$. If we designate as $n_j(\mathbf{X})$ the number of such records with $y = j$, the desired estimates, $p^{\text{est}}(y = j|\mathbf{X})$, are given by

$$p^{\text{est}}(y = j|\mathbf{X}) = \frac{n_j(\mathbf{X})}{n(\mathbf{X})}, \qquad j = 1, \ldots, K. \tag{9}$$

Using the normal approximation to a binomial distribution, confidence intervals (CIs) for $p(y = j|\mathbf{X})$ can be established for a specific significance level, α:

$$\text{CI} = \left[p^{\text{est}}(y=j|\mathbf{X}) \pm t_{(\alpha/2, n(\mathbf{X})-1)} \cdot \sqrt{\frac{p^{\text{est}}(y=j|\mathbf{X}) \cdot (1 - p^{\text{est}}(y=j|\mathbf{X}))}{n(\mathbf{X})}} \right], \tag{10}$$

where t stands for the critical value of the Student's distribution.

The search procedure, S, used to uncover promising hyperrectangles in the decision space, $\mathbf{X}^*$, associated with a desired y value (e.g., $y =$ "good"), is based on *symbolic inductive learning algorithms*, and leads to the identification of a final number of promising solutions, $\mathbf{X}^*$, such as the ones in Fig. 2b. It is described in the following subsection.

B. Search Procedure, S

In order to introduce the search procedure, S, we start by showing how classification decision trees lead to the definition of a set of hyperrectangles, and how they can be constructed from a set of $(\mathbf{x}, y)$ data records.

Then we describe the conversion of the knowledge captured by the induced decision trees into a set of final solutions, $\mathbf{X}^*$.

1. Classification Decision Trees and Their Inductive Construction

A classification decision tree allows one to predict in a sequential way the y value (or corresponding conditional probabilities) that is associated with a particular $\mathbf{x}$ vector of values. At the top node of the tree (A in Fig. 3), a first test is performed, based on the value assumed by a particular decision variable (x_3). Depending on the outcome of this test, vector $\mathbf{x}$ is sent to one of the branches emanating from node A. A second test follows, being carried out at another node (B), and over the values of the same or a different decision variable (e.g., x_6). This procedure is

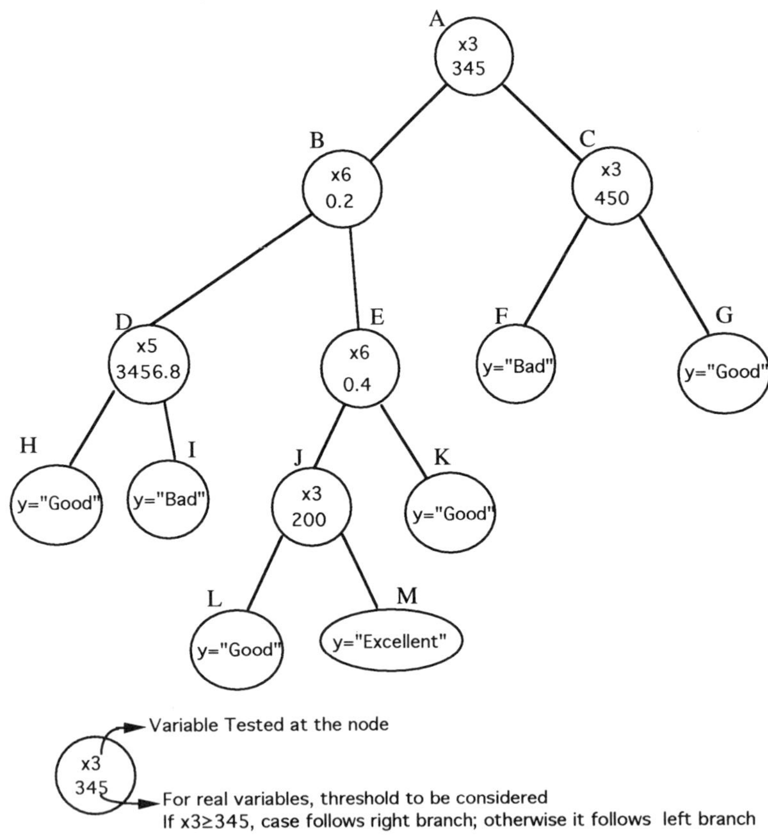

Fig. 3. A classification decision tree.

rcpeated until, after a last test, vector **x** follows a branch that leads to a terminal node (or leaf), labeled with a particular y value and/ or set of conditional probabilities, which provides the $y(\mathbf{x})$ and/ or $p(y = j|\mathbf{x})$ estimates.

Although decision trees contain a number of attractive features, including competitive accuracy, when considered strictly as classification devices (Saraiva and Stephanopoulos, 1992a), the most important point for our purposes is that each of the tree's terminal nodes identifies a particular hyperrectangle in the decision space, **X**, associated with a given y value. For example, node M defines a $y =$ "excellent" rectangle that corresponds to the following rule:

$$If\ 200 \le x_3 < 345\ and\ 0.2 < x_6 \le 0.4,\ then\ y\ =\ \text{“Excellent,”} \quad (11)$$

Once a decision tree such as this has been constructed from the available set of $(\mathbf{x}, y)$ pairs, by picking up those leaves labeled with the desired y value, one gets an initial collection of promising hyperrectangles. Some additional transformations, summarized in the next paragraph, convert this initial collection into a final set of solutions, $\mathbf{X}^*$.

The algorithm that we employ to build a classification decision tree from $(\mathbf{x}, y)$ data records belongs to a group of techniques known as *top–down induction of decision trees* (TDIDT) (Sonquist *et al.*, 1971; Fu, 1968; Hunt, 1962; Quinlan, 1986, 1987, 1993; Breiman *et al.*, 1984).

The construction starts at the root node of the tree, where all the available $(\mathbf{x}, y)$ pairs are initially placed. One identifies the particular split or test, s, that maximizes a given measure of information gain (Shannon and Weaver, 1964), $\Phi(s)$. The definition of a split, s, involves both the choice of the decision variable and the threshold to be used. Then, the $(\mathbf{x}, y)$ root node pairs are divided according to the best split found, and assigned to one of the children nodes emanating from it. The information gain measure, $\Phi(s)$, for a particular parent node t, is

$$\Phi(s) \ = \ -\sum_{k=1}^{K} P_k(t)\log_2 P_k(t) \ + \ \sum_{c=1}^{R} \frac{N(t_c)}{N(t)} \sum_{k=1}^{K} P_k(t_c)\log_2 P_k(t_c), \quad (12)$$

where

(a) $N(t)$ is the number of $(\mathbf{x}, y)$ pairs included in t.

(b) R is the total number of children nodes, t_c, $c = 1, \ldots, R$, emanating from t.

(c) $P_k(t) = N_k(t)/ N(t)$ is an estimate of the $p(y = k|t)$ conditional probability.

(d) $N_k(t)$ is the number of $(\mathbf{x}, y)$ pairs assigned to node t for which $y = k$.

This splitting procedure is now applied recursively to each of the children nodes just created. The successive expansion process continues until terminal nodes or leaves, over which no further partitions are performed, can be identified.

The preceding strategy for the construction of decision trees provides an efficient way for inducing compact classification decision trees from a set of $(\mathbf{x}, y)$ pairs (Moret, 1982; Utgoff, 1988; Goodman and Smyth, 1990). Furthermore, tests based on the values of irrelevant x_m variables are not likely to be present in the final decision tree. Thus, the problem dimensionality is automatically reduced to a subset of decision variables that convey critical information and influence decisively the system performance.

2. From Decision Trees to Final Solutions, $\mathbf{X}^*$

Once an induced decision tree has been found, it is subjected to an additional pruning and simplification treatment, whose details can be found in Saraiva and Stephanopoulos (1992a). Pruning and simplification lead to the definition of a revised and reasonably sized final decision tree, with greater statistical reliability. Each of the terminal nodes of this final decision tree corresponds to a particular hyperrectangle in the decision space, $\mathbf{X}$, which is labeled either with the y value that is most likely to be obtained within that hyperrectangle, or with the corresponding conditional probabilities estimates, $p^{\text{est}}(y = j|\mathbf{X})$, $j = 1, \ldots, K$. These terminal nodes are further refined through an expansion process aimed at enlarging the zone of the decision space that they cover, followed by statistical significance tests that may introduce additional simplifications (Saraiva and Stephanopoulos, 1992a). By the end of this refinement stage, we get a group of improved, statistically significant, simplified, and partially overlapped hyperrectangles. Those that lead predominantly to the desired y value constitute the final group of solutions, $\mathbf{X}^*$, and are presented to the user together with a number of auxiliary evaluation scores (Saraiva and Stephanopoulos, 1992a). It is the user's responsibility to analyze this set of hyperrectangles, $\mathbf{X}^*$, make a selection among them, and define the course of action to follow.

C. CASE STUDY: OPERATING STRATEGIES FOR DESIRED OCTANE NUMBER

To illustrate the potential practical capabilities of the learning methodology, we will now present the results obtained through its application to records of operating data collected from a refinery unit (Daniel and Wood, 1980). Additional industrial case studies can be found in Saraiva and Stephanopoulos (1992a).

The y variable that we will consider derives from a quantization of the octane numbers of the gasoline product, z, assuming one out of three possible values:

(a) $y = 1$ ("very low") for $z \leq 91$
(b) $y = 2$ ("low") when $91 < z < 92$
(c) $y = 3$ ("good") for $z \geq 92$

In this particular problem, one wishes to achieve values of z as high as possible, and thus to identify zones in the decision space where one gets mostly $y = 3$.

There are four decision variables: three different measures of the feed composition (x_1, x_2, x_3) and the value of an unspecified operating condition (x_4).

In Fig. 4 we present the final induced decision tree, as well as the partition of the (x_1, x_4) plane defined by its leaves, together with a projection of all the available $(\mathbf{x}, y)$ pairs on the same plane. These two decision variables are clearly influencing the current performance of the refinery unit, and the decision tree leaves perform a reasonable partition of the plane. To achieve better performance, we must look for operating zones that will result in obtaining mostly $y = 3$ values. Terminal nodes 2

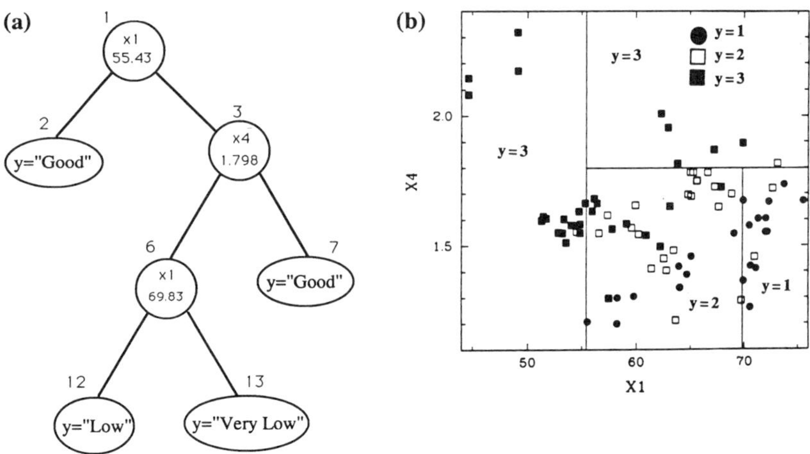

FIG. 4. (a) Induced decision tree; (b) partition of the plane defined by its leaves.

and 7 identify two such zones. The corresponding final solutions, $\mathbf{X}^*$, found after going through all the additional steps of refinement and validation, are

$$\mathbf{X}_1^* = \{x_1 \in [44.6, 55.4]\}, \text{ with } p^{\text{est}}(y = 3|\mathbf{X}_1^*) = 0.9,$$

$$\mathbf{X}_2^* = \{x_4 \in [1.8, 2.3]\}, \text{ with } p^{\text{est}}(y = 3|\mathbf{X}_2^*) = 1.0$$

It can be noticed from the conditional probability estimates that one should expect to get almost only "good" y values while operating inside these zones of the decision space, as opposed to the current operating conditions, which lead to just 40% of "good" y values.

V. Continuous Performance Metrics

In Section IV we considered a categorical performance metric y. Although that represents a common practice, especially when y defines the quality of a product or process operation, there are many instances where system performance is measured by a continuous variable. Even when y is quality-related, it is becoming increasingly clear that explicit continuous quality cost models should be adopted and replace evaluations of performance based on categorical variables.

This Section addresses cases with a continuous performance metric, y. We identify the corresponding problem statements and results, which are compared with conventional formulations and solutions. Then Taguchi loss functions are introduced as quality cost models that allow one to express a quality-related y on a continuous basis. Next we present the learning methodology used to solve the alternative problem statements and uncover a set of final solutions. The section ends with an application case study.

A. PROBLEM STATEMENT

A number of different techniques have been suggested and applied to address situations where y is a continuous variable. Table II summarizes the most important characteristics of our approach and major features that differentiate it from conventional procedures.

TABLE II

CONVENTIONAL APPROACHES AND SUGGESTED ALTERNATIVE

	Conventional approaches	Suggested alternative
ξ	$\mathbf{x} \in \mathscr{R}^M$	$\mathbf{X} \in \mathbf{I}^M$
ψ	$y(\mathbf{x})$ or $E(y\|\mathbf{x})$	$E(y\|\mathbf{X})$
f	Technique-dependent	$\dfrac{1}{n(\mathbf{X})} \displaystyle\sum_{j=1}^{n(\mathbf{X})} y_j$
S	Optimization	Exploratory search

1. Conventional Procedures

All conventional approaches (mathematical and stochastic programming, parametric and nonparametric regression analysis) adopt as a common solution format real vectors, $\mathbf{x} \in \mathscr{R}^M$, and as performance criterion, ψ, the expected value of y, $E(y|\mathbf{x})$, given $\mathbf{x}$, or the single y value that corresponds to a specific $\mathbf{x}$, $y(\mathbf{x})$, if one assumes a fully deterministic relationship between y and $\mathbf{x}$. Just as in Section IV, the element that essentially distinguishes the several techniques is the mapping procedure, f, used to compute $\psi^{\mathrm{est}} = f(\mathbf{x})$. In order to find a final solution $\mathbf{x}^*$, optimization algorithms form the search procedure, S, that leads to the identification of the particular point in the decision space that maximizes or minimizes $\psi^{\mathrm{est}} = f(\mathbf{x})$.

These conventional approaches usually follow a two-step sequential process:

Step 1. The $(\mathbf{x}, y)$ records of data are employed to build the f mapping.

Step 2. S does not make direct use of the original data, but rather employs f to find the final solution, $\mathbf{x}^*$, that optimizes $\psi^{\mathrm{est}} = f(\mathbf{x})$.

2. The Learning Methodology

For the reasons already discussed in Section III, our solution space consists of hyperrectangles in the decision space, $\mathbf{X} \in \mathbf{I}^M$, not single points, $\mathbf{x}$. The corresponding performance criterion used to evaluate solutions, ψ, is the expected y value within $\mathbf{X}$:

$$\psi(\mathbf{X}) = E(y|\mathbf{X}). \tag{13}$$

These conceptual changes in both solution formats and performance criteria are independent of the particular procedures chosen to estimate $\psi(\mathbf{X})$ and search for a set of final solutions, $\mathbf{X}^*$. However, as was also discussed in Section III, for the types of systems that we are specially interested in analyzing, direct sampling strategies to estimate $\psi(\mathbf{X})$ offer a number of advantages. The mapping model that we employ, f, is similar to the one adopted for categorical y variables. A search is performed over all the available $(\mathbf{x}, y)$ data records, leading to the identification of a total of $n(\mathbf{X})$ pairs for which $\mathbf{x} \in \mathbf{X}$. The performance criterion estimate, $\psi^{\text{est}}(\mathbf{X})$, is the sample y average among these $n(\mathbf{X})$ pairs:

$$\psi^{\text{est}}(\mathbf{X}) = f(\mathbf{X}) = \frac{1}{n(\mathbf{X})} \sum_{j=1}^{n(\mathbf{X})} y_j \tag{14}$$

The corresponding confidence interval, CI, for $E(y|\mathbf{X})$, at a given significance level, α, is

$$\text{CI} = \left[f(\mathbf{X}) \pm t_{(\alpha/2, n(\mathbf{X})-1)} \cdot \frac{s_y(\mathbf{X})}{\sqrt{n(\mathbf{X})}} \right], \tag{15}$$

where $s_y(\mathbf{X})$ stands for the sample standard deviation:

$$s_y(\mathbf{X}) = \left\{ \frac{1}{n(\mathbf{X}) - 1} \sum_{j=1}^{n(\mathbf{X})} \left[y_j - f(\mathbf{X}) \right]^2 \right\}^{0.5}. \tag{16}$$

If one is interested only in finding the single feasible hyperrectangle (i.e., respecting minimum width constraints imposed due to control limitations) that minimizes $\psi^{\text{est}}(\mathbf{X})$, to find that hyperrectangle one may choose as search procedure, S, any optimization routine. However, our primary goal is to conduct an exploratory analysis of the decision space, leading to the identification of a set of particularly promising solutions, $\mathbf{X}^*$, that are presented to the decisionmaker, who is responsible for a final selection and a choice of the course of action to follow. The search procedure adopted in our learning methodology, S, reflects this goal, and will be described in a subsequent paragraph.

B. ALTERNATIVE PROBLEM STATEMENTS AND SOLUTIONS

Recognizing that, due to unavoidable variability in the decision variables, one has to operate within a zone of the decision space, and not at a single point, we might still believe that finding the optimal pointwise solution, $\mathbf{x}^*$, as usual, would be enough. The assumption behind such a belief is that centering the operation around $\mathbf{x}^*$ will correspond in practice to the adoption of a zone in the decision space to conduct the operation, $\mathbf{X}$ [with $\mathbf{m}(\mathbf{X}) = \mathbf{x}^*$], that is equivalent or close to the best possible zone, $\mathbf{X}^*$. However, because the evaluation of performance at a single point in the decision space, $\mathbf{x}$, completely ignores the system behavior around that point, the preceding assumption in general does not hold: *searching for an optimal hyperrectangle leads to a final solution, $\mathbf{X}^*$, that is likely to lie in a region of the decision space quite distant from $\mathbf{x}^*$, and $\mathbf{m}(\mathbf{X}^*) \neq \mathbf{x}^*$.* This observation emphasizes how critical it is to adopt the modified problem statements described at the beginning of Section V, where a direct and explicit search for the best zone to operate replaces the classical optimization paradigm, which ignores variability in the decision variables and seeks to identify as precisely as possible the optimal $\mathbf{x}^*$. No matter how accurately $\mathbf{x}^*$ is determined, targeting the operation around it can represent a quite suboptimal answer when the random nature of the decision variables is taken into account.

To illustrate how different $\mathbf{m}(\mathbf{X}^*)$ and $\mathbf{x}^*$ may happen to be, let's consider as a specific example (others can be found in Saraiva and Stephanopoulos, 1992c) a Kraft pulp digester. The performance metric y, that one wishes to minimize, is determined by the kappa index of the pulp produced and the cooking yield. Two decision variables are considered: *H-factor* (x_1), and *alkali charge* (x_2). Furthermore, we will assume as perfect an available deterministic empirical model (Saraiva and Stephanopoulos, 1992c), f, which expresses y as function of $\mathbf{x}$, i.e., that $y = f(x_1, x_2)$ is perfectly known.

If one follows the conventional optimization paradigm, adopting point-wise solution formats, the best feasible answer, $\mathbf{x}^*$, which minimizes $f(\mathbf{x})$, is $\mathbf{x}^* = (200; 17.9)$, as can be confirmed by examining the contour plots of f shown in Fig. 5a.

On the other hand, when the unavoidable variability in the decision space is considered explicitly, the goal of the search becomes the identification of the optimal hyperrectangle, $\mathbf{X}^*$, which solves the following problem:

$$\min_{\mathbf{X} \in \mathbf{I}^2} f(\mathbf{X}), \tag{17}$$

subject to

$$w(X_1) \geq \Delta x_1 = 300,$$

$$w(X_2) \geq \Delta x_2 = 1.0,$$

PEDRO M. SARAIVA

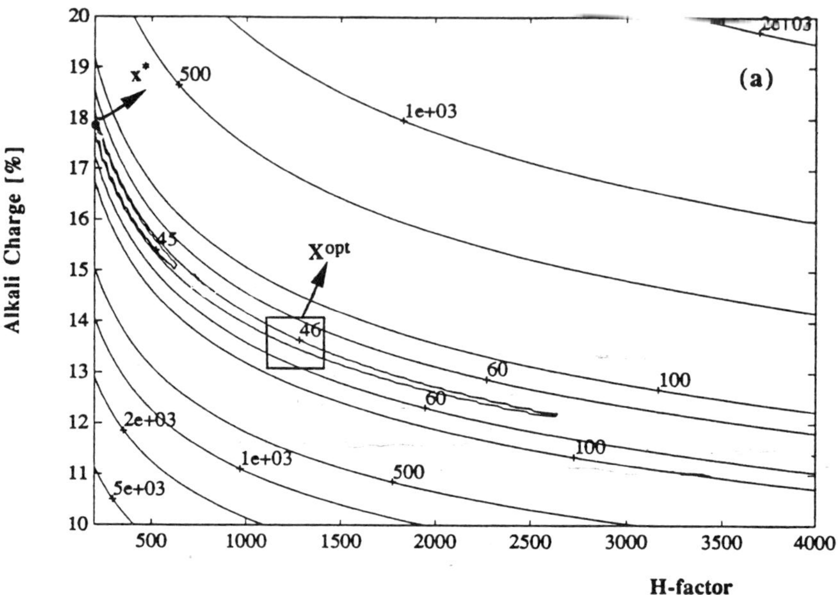

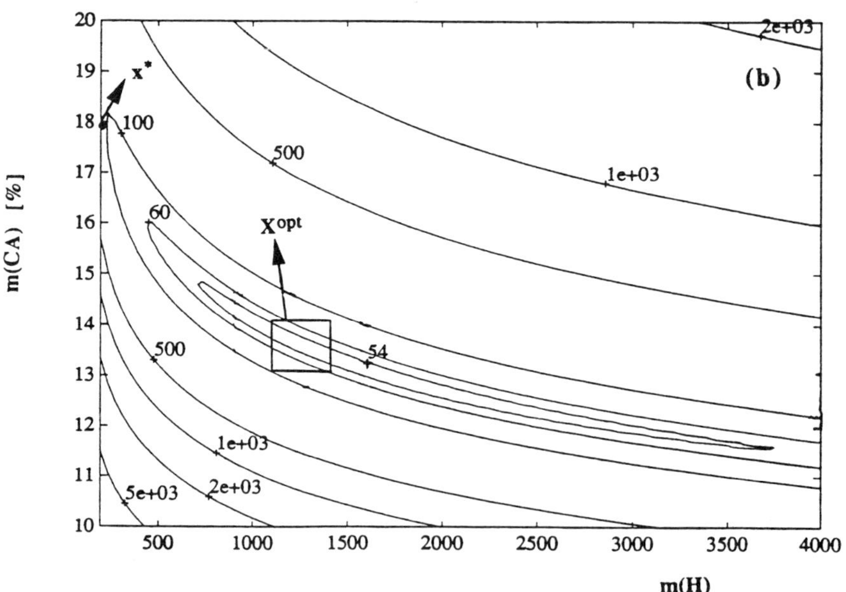

FIG. 5. Contour plots and optimal solutions for (a) $f(\mathbf{x})$ versus $\mathbf{x}$ and (b) $E(f(\mathbf{x})|\mathbf{X})$ versus $\mathbf{m}(\mathbf{X})$.

with

$$f(\mathbf{X}) = E(y|\mathbf{X}) = \frac{\int_{i(X_1)}^{s(X_1)} \int_{i(X_2)}^{s(X_2)} p(\mathbf{x}) f(\mathbf{x})\, dx_1\, dx_2}{\int_{i(X_1)}^{s(X_1)} \int_{i(X_2)}^{s(X_2)} p(\mathbf{x})\, dx_1\, dx_2},$$

and $p(\mathbf{x})$ representing the (x_1, x_2) joint probability density function (for this particular study we assumed that both decision variables are independent and uniformly distributed).

In Fig. 5b we present the contour plots for $E(y|\mathbf{X})$ as a function of $\mathbf{m}(\mathbf{X})$. The corresponding optimal rectangle, $\mathbf{X}^*$, is

$$\mathbf{X}^* = \{x_1 \in [1114; 1414] \wedge x_2 \in [13.1; 14.1]\}. \tag{18}$$

This solution lies in a zone of the decision space that is quite distant from $\mathbf{x}^* = (200; 17.9)$ and targeting the operation around $\mathbf{x}^*$ results in clearly suboptimal performance.

This discrepancy illustrates some of the dangers associated with looking exclusively for pointwise solutions and performances, while neglecting the decision variables' variability. Although the specific problem-solving strategies developed in this chapter are aimed at the analysis of systems for which no good quantitative $f(\mathbf{x})$ and $p(\mathbf{x})$ exist a priori, the example presented shows the more general nature of the benefits that may derive from adopting the suggested alternative problem statements even when such models are available and used to find the final optimal hyperrectangle, $\mathbf{X}^*$.

Finally, it should be added that the conventional problem statement and pointwise solution format can be interpreted as a particular degenerate case of our more general formulations. As the minimum acceptable size for zones in the decision space decreases, the different performance criteria converge to each other and $\mathbf{X}^*$ gets closer and closer to $\mathbf{x}^*$. Both approaches become exactly identical in the extreme limiting case where $\Delta x_m = 0$, $m = 1, \ldots, M$, which is the particular degenerate case adopted in traditional formulations.

C. Taguchi Loss Functions as Continuous Quality Cost Models

The development of most of the optimization and operations research techniques was motivated and focused on the minimization of operating costs, which are usually expressed on a quantitative basis. However, when

y represents a quality related measure, performance has been traditionally evaluated through a categorical variable, whose values depend on whether the product is inside or outside the range of desired specifications. But it is recognized today that just being within any type of specification limits is not good enough, and the idea that any product is equally good inside a given range of values and equally bad outside it must be revised (Deming, 1986; Roy, 1990). This points to the need for assuming continuous performance metrics even when y is quality related. One of the most powerful contributions of Taguchi (1986) to quality management was the proposition of loss functions as ways of quantifying and penalizing on a continuous basis any deviations from a desired nominal target (Phadke, 1989; Clausing, 1993). Given a quality functional characteristic z, with a nominal target z^*, any deviation from z^* has some quality cost associated with it, and this cost increases gradually as we move away from the target (Fig. 6). To operationalize and quantify this quality degradation process, Taguchi loss functions express quality costs on a monetary basis, commensurate with operating costs, and define the particular quality cost associated with a generic z value, $L(z)$, as

$$L(z) = k(z - z^*)^2, \qquad (19)$$

where k is a constant known as quality loss coefficient. The value of k is

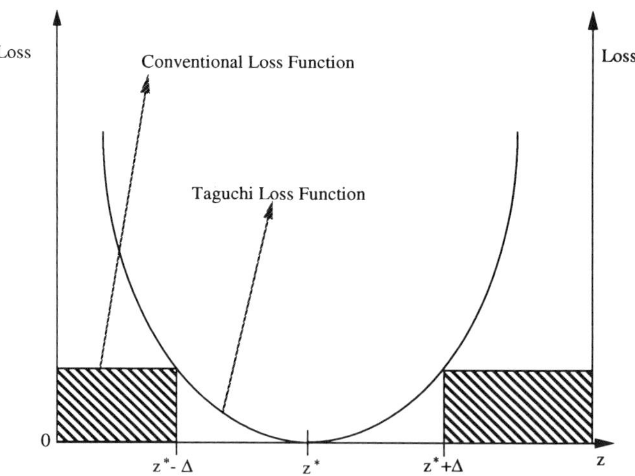

FIG. 6. Conventional and Taguchi quality cost models.

usually found by assigning a loss value to the specification limits, established at $z^* \pm \Delta$:

$$k = L(z^* \pm \Delta)/\Delta^2, \tag{20}$$

$$L(z) = k(z - z^*)^2 = L(z^* \pm \Delta) \cdot (z - z^*)^2/\Delta^2. \tag{21}$$

It is also important to realize that Taguchi loss functions not only bring into consideration both issues of location and dispersion of z but also provide a consistent format for combining them. By taking expectations on both sides of Eq. (19), and after a few algebraic rearrangements, we can show that the expected loss, $E[L(z)]$ is

$$E[L(z)] = k\left[\sigma_z^2 + (\mu_z - z^*)^2\right], \tag{22}$$

where σ_z^2 is the z variance and μ_z is the z expectation.

For a zone $\mathbf{X}$ in the decision space to lead to a small conditional expected loss, $E[L(z)|\mathbf{X}]$, it must achieve both precise (reduced σ_z) and accurate ($\mu_z \cong z^*$) performance with respect to z. Finding such robust zones and operating on them results in inoculating the process against the transmission of variation from disturbances and the decision space to the performance space (Taylor, 1991).

If, besides the quality-related measure, z, one also wishes to include operating costs, ζ, in the analysis, because quality loss functions express quality costs on a monetary basis, commensurate with operating costs, the final global performance metric, y, which reflects total manufacturing cost, is simply the sum of both quality and operating costs (Clausing, 1993),

$$y = L(z) + \zeta. \tag{23}$$

Consequently, the goal of our learning methodology is the identification of hyperrectangles in the decision space, $\mathbf{X}$, that minimize expected total manufacturing cost, $E(y|\mathbf{X})$, a performance measure that combines in a consistent form and a quantitative basis both operating and quality costs.

D. Learning Methodology and Search Procedure, S

In the previous paragraphs we defined the solution format ξ, performance criterion ψ, mapping procedure f, and performance metric y that characterize our learning methodology for systems with a quantitative metric y. Here we will assemble all these pieces together and briefly discuss the search procedure, S (further details can be found in Saraiva and Stephanopoulos, 1992c), that is employed to identify a set of final solutions, $\mathbf{X}^*$, which achieve low $\psi^{\text{est}}(\mathbf{X})$ scores. The two main stages of the learning procedure are examined in the following paragraphs.

1. Problem Formulation

The total loss function, y, given by Eq. (23), is not directly measured and has to be computed from information that is available and collected from the process, consisting of $(\mathbf{x}, z)$ pairs. After defining an adequate loss function, $L(z)$, and considering operating costs, ζ, one can identify the $(\mathbf{x}, y)$ pairs that correspond to each of the initial $(\mathbf{x}, z)$ data records.

Before starting the search for solutions, it is necessary to select among the M decision variables a subset of H variables, x_h, $h = 1, \ldots, H$, which influence significantly the system performance, and thus will be used by S and included in the definition of the final set of hyperrectangles, $\mathbf{X}^*$. For this preliminary choice of critical decision variables, other than his or her own specific process knowledge, the decisionmaker can count on a number of auxiliary techniques enumerated in Saraiva and Stephanopoulos (1992c).

This first stage of the learning procedure is concluded with the definition of a set of constraints related to

(a) Minimum acceptable width of operating windows $[w(X_h) \geq \Delta x_h, h = 1, \ldots, H]$
(b) Minimum number of data records, $N_{\min}$, that must be covered by any final solution, $\mathbf{X}^*[n(\mathbf{X}) \geq N_{\min}$ for $\mathbf{X}$ to be considered a feasible solution]. This constraint is identical to specifying a given maximum acceptable width for the $E(y|\mathbf{X})$ confidence interval that is obtained at a significance level, α, as defined by Eq. (15).

2. Search Procedure

Rather than finding the exact location of the single feasible hyperrectangle that optimizes $\psi^{\text{est}}(\mathbf{X})$, our primary goal is to conduct an exploratory analysis of the decision space, leading to the definition of a set of particularly promising solutions, $\mathbf{X}^*$, to be presented to the decisionmaker.

To identify this set of final feasible solutions, $\mathbf{X}^* \in \mathbf{I}^H$, with low $\psi^{\text{est}}(\mathbf{X})$ scores, we developed a greedy search procedure, S (Saraiva and Stephanopoulos, 1992c), that has resulted, within an acceptable computation time, in almost-optimal solutions for all the cases studied so far, while avoiding the combinatorial explosion with the number of $(\mathbf{x}, y)$ pairs of an exhaustive enumeration/evaluation of all feasible alternatives. The algorithm starts by partitioning the decision space into a number of isovolumetric and contiguous hyperrectangular seed cells, where for each cell the width associated with variable x_h is smaller than the corresponding Δx_h. Then, we gradually enlarge these seed cells, until they satisfy all imposed constraints, and further growth is found to degrade their estimated performance. Each initial cell is thus converted into a feasible solution candidate, $\mathbf{X}$, and the corresponding $\psi^{est}(\mathbf{X})$ score is evaluated. Those feasible solution candidates receiving the lowest $\psi^{est}(\mathbf{X})$ are included in the set of final solutions, $\mathbf{X}^*$, that is presented to the decisionmaker. It is the user's responsibility to analyze this set, make a selection among its elements, and thus choose the ultimate target zone to conduct the operation of the process.

E. CASE STUDY: PULP DIGESTER

In order to verify how close to a known true optimum the final solutions found by our learning methodology happen to be, we will describe here its application to a pulp digester, for which a perfect empirical model $f(\mathbf{x})$ is assumed to be available. Other applications are discussed in Saraiva and Stephanopoulos (1992c).

The original data format consists of (x_1, x_2, z, ω) records, where

x_1 stands for the *H-factor*.

x_2 is the *alkali charge*.

z is the pulp *kappa index*, with nominal target set at 30.0.

ω is the cooking yield, an indirect measure of operating cost that one wishes to maximize.

After defining the z loss function as

$$L(z) = 10(z - 30)^2 \quad \$/\text{ton of pulp} \tag{24}$$

(where $\$ = $ U.S. dollars), and combining it with a commensurate measure of operating cost, expressed as a very simple function of ω

$$\zeta = 100 - \omega \quad \$/\text{ton of pulp,} \tag{25}$$

one finally arrives at the identification of our total manufacturing cost performance metric,

$$y = 10(z - 30)^2 + 100 - \omega \quad \$/\text{ton of pulp,} \tag{26}$$

which leads to the conversion of the original data records into the usual $(\mathbf{x}, y)$ format.

Let's consider that under the current operating conditions the values of $\mathbf{x}$ fall within a rectangle $\mathbf{X}_{\text{current}} = \{x_1 \in [200; 4000] \wedge x_2 \in [10; 20]\}$. Furthermore, we will assume that the two decision variables (x_1 and x_2) are independent and have uniform probability distributions. Using the available model, $f(\mathbf{x})$, we computed the current average total manufacturing cost, $E(y|\mathbf{X}_{\text{current}}) = 743.5$, a reference value that can be used to estimate the savings achieved with the implementation of any uncovered final solutions, $\mathbf{X}^*$.

To support the application of the learning methodology, $f(\mathbf{x})$ was used to generate 500 $(\mathbf{x}, z, \omega)$ records of simulated operational data, transformed by Eq. (26) into an equivalent number of $(\mathbf{x}, y)$ pairs. Finally, the following constraints were imposed to the search procedure, S:

(a) $w(X_1) \geq \Delta x_1 = 300$
   $w(X_2) \geq \Delta x_2 = 1.0$
(b) $N_{\min} = 15$

Given the preceding problem definition, and after going through S, the final solution, $\mathbf{X}^*$, chosen for implementation is (Fig. 7):

$$\mathbf{X}^* = \{x_1 \in [910.2; 1566.3] \wedge x_2 \in [12.8; 14.6]\} \qquad E(y|\mathbf{X}^*) = 69.9. \tag{27}$$

Thus, $\mathbf{X}^*$ indeed leads to a quite significant average total cost reduction, because $E(y|\mathbf{X}^*) \ll E(y|\mathbf{X}_{\text{current}})$. Both $\mathbf{X}^*$ and $E(y|\mathbf{X}^*)$ are also close approximations (Fig. 7) to the true optimal solution given by Eq. (18), i.e., $\mathbf{X}_{\text{opt}}$ and $E(y|\mathbf{X}_{\text{opt}})$ are

$$\mathbf{X}_{\text{opt}} = \{x_1 \in [1114; 1414] \wedge x_2 \in [13.1; 14.1]\} \qquad E(y|\mathbf{X}_{\text{opt}}) = 52.2. \tag{28}$$

To benchmark our learning methodology with alternative conventional approaches, we used the same 500 $(\mathbf{x}, y)$ data records and followed the usual regression analysis steps (including stepwise variable selection, examination of residuals, and variable transformations) to find an approximate empirical model, $f^{\text{est}}(\mathbf{x})$, with a coefficient of determination $R^2 = 0.79$. This model is given by

$$y \approx f^{\text{est}}(\mathbf{x}) = a \ln(x_1) + b \ln(x_2) + cx_2 + dx_1^2 + ex_2^2 + gx_1 \cdot x_2, \tag{29}$$

whose parameters were fitted by ordinary least squares.

By employing $f^{\text{est}}(\mathbf{x})$ in Eq. (17), we used this approximate model to find a final solution, $\mathbf{X}_{\text{est}}$ (Fig. 7), that satisfies the (a) constraints and

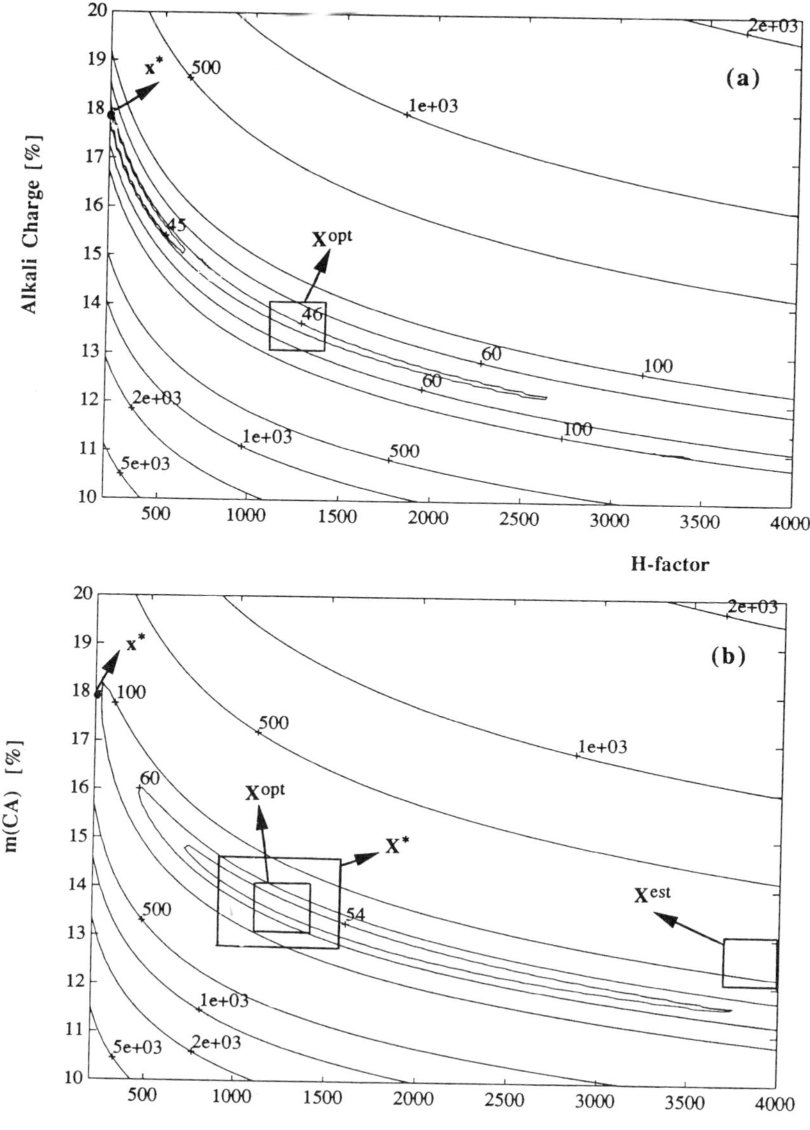

FIG. 7. Locations of $\mathbf{x}^*$, $\mathbf{X}^*$, $\mathbf{X}^{\mathrm{opt}}$, and $\mathbf{X}^{\mathrm{est}}$ in the decision space, and contour plots of (a) $y = f(\mathbf{x})$ versus $\mathbf{X}$, (b) $E(y|\mathbf{X})$ versus $\mathbf{m}(\mathbf{X})$.

minimizes $E[f^{est}(\mathbf{x})|\mathbf{X}]$:

$$\mathbf{X}_{est} = \{x_1 \in [3698; 3998] \wedge x_2 \in [12.1; 13.1]\} \qquad E(y|\mathbf{X}_{est}) = 145.6. \tag{30}$$

This solution leads to a considerably worse performance, $E(y|\mathbf{X})$, than $\mathbf{X}^*$, and it is also much more distant from the zone of the decision space where the true optimum, $\mathbf{X}_{opt}$, is located.

Results obtained with other $f^{est}(\mathbf{x})$ functional forms, for this and other similar problems, seem to indicate that even when only moderate amounts of data are available our direct sampling estimation procedure, f, and search algorithm, S, provide better final solutions than classical regression analysis followed by the use of Eq. (17) unless one is able to build from the data almost perfect $f^{est}(\mathbf{x})$ empirical models with the appropriate functional forms.

VI. Systems with Multiple Operational Objectives

Except for the combination of quality loss and operating costs given by Eq. (23), in the previous sections we assumed the system performance to be determined by a single objective. However, in the analysis of pieces of equipment or plant segments of nontrivial size/ complexity, a multitude of objectives has usually to be taken into account in order to evaluate the system's global performance, and find ways to improve it.

In this section we describe extensions of the basic learning methodologies introduced in Sections IV and V that, while preserving the same premises and paradigms, enlarge considerably their scope by adding the capability to consider simultaneously multiple objectives. As before, and without loss of generality, we will focus our attention on the coexistence of several quality-related objectives.

Both situations with categorical and continuous, real-valued performance metrics will be considered and analyzed. Since Taguchi loss functions provide quality cost models that allow the different objectives to be expressed on a commensurate basis, for continuous performance variables only minor modifications in the problem definition of the approach presented in Section V are needed. On the other hand, if categorical variables are chosen to characterize the system's multiple performance metrics, important modifications and additional components have to be incorporated into the basic learning methodology described in Section IV.

A case study on the operational improvement of a plasma etching unit in microelectronics fabrication ends the section. This case study illustrates that if similar preference structures are used in both types of formulation, identical final solutions are found when either categorical or continuous performance evaluation modes are employed.

A. Continuous Performance Variables

Instead of a single quality-related performance variable, z, as in Section V, let's suppose that one has to consider a total of P distinct objectives and the corresponding continuous performance variables, z_i, $i = 1, \ldots, P$, which are components of a performance vector $\mathbf{z} = [z_1, \ldots, z_P]^T$. In such case, one has to identify the corresponding Taguchi loss functions, $L(z_i)$, $i = 1, \ldots, P$, for each of the performance variables:

$$L(z_i) = k_i(z_i - z_i^*)^2. \tag{31}$$

Since these loss functions express quality costs on a common and commensurate basis, extending the learning methodology of Section V to a situation with P objectives is straightforward. All one has to do is replace the original definition of the y performance metric [Eq. (23)] by the following more general version:

$$y = \sum_{i=1}^{P} k_i(z_i - z_i^*)^2 + \zeta. \tag{32}$$

Except for this modification, all the procedures and steps discussed in Section V carry over to the solution of multiobjective problems.

B. Categorical Performance Variables

Rather than a single objective, y, as in Section IV, we now have a total of P distinct categorical performance variables, y_i, $i = 1, \ldots, P$, associated with an equivalent number of objectives. Consequently, each data record is now composed of a $(\mathbf{x}, \mathbf{y})$ pair, where $\mathbf{y}$ is a performance vector defined by

$$\mathbf{y} = [y_1, \ldots, y_i, \ldots, y_P]^T. \tag{33}$$

The most important changes and adaptations that were introduced in order to handle such multiobjective problems are summarized in Table III. The solution space remains the same as for the single objective case.

TABLE III

SINGLE AND MULTIPLE CATEGORICAL PERFORMANCE VARIABLES

	Single objective	Multiple objectives
ξ	$\mathbf{X} \in \mathbf{I}^M$	$\mathbf{X} \in \mathbf{I}^M$
ψ	$y(\mathbf{X})$ or $p(y = j \mid \mathbf{X})$	$y(\mathbf{X})$ or $p(y_i = j \mid \mathbf{X})$
f	$\dfrac{n_j(\mathbf{X})}{n(\mathbf{X})}$	$\dfrac{n_{i,j}(\mathbf{X})}{n(\mathbf{X})}$
S	Induction of decision trees	Multiple agents and lexicographic search

However, since now we have P different objectives, the performance criteria, ψ, must include all of them. They may assume one of the following formats:

1. $\mathbf{y}(\mathbf{X}) = [y_1(\mathbf{X}), \ldots, y_i(\mathbf{X}), \ldots, y_P(\mathbf{X})]^T$,
   where $y_i(\mathbf{X})$ stands for the most likely y_i value within the zone of the decision space defined by $\mathbf{X}$.
2. $p(y_i = j \mid \mathbf{X})$, $i = 1, \ldots, P$ and $j = 1, \ldots, K_i$,
   the set of conditional probabilities for getting any given value for each of the y_i variables, with K_i representing the total number of different possible values that y_i can assume.

The mapping procedure, f, is identical to the one adopted for a single objective:

$$p^{\text{est}}(y_i = j \mid \mathbf{X}) = \frac{n_{i,j}(\mathbf{X})}{n(\mathbf{X})}, \qquad j = 1, \ldots, K_i, \tag{34}$$

where $n_{i,j}(\mathbf{X})$ is the total number of $(\mathbf{x}, \mathbf{y})$ pairs for which $\mathbf{x} \in \mathbf{X}$, and $y_i = j$.

The search procedure, S, requires major modifications in order to account for multiple objectives. As we will see, S becomes now a highly interactive process, with progressive articulation/ elicitation of the user's preference structure, successive relaxation of aspiration levels and lexicographic construction of final solutions, $\mathbf{X}^*$, that lead to satisfactory joint performances according to all the P objectives. It includes P replicates of the basic learning methodology introduced in Section IV for a single objective, designated as *agents*, and a *coordination mechanism* that combines the results found by individual agents, in an attempt to identify conjunctions of zones uncovered by the agents that lead to the formation of the desired final solutions, $\mathbf{X}^*$. In the following paragraphs we will summarize the main steps of this search procedure (for details, see Saraiva and Stephanopoulos, 1992b).

1. Problem Definition

Besides the identification of the decision variables, x_m, $m = 1, \ldots, M$, and the performance variables, y_i, the user is asked to

(a) Rank the P objectives in order of decreasing relative importance.
(b) Provide constraints on minimum acceptable widths $[w(X_m) \geq \Delta x_m,$ $m = 1, \ldots, M]$, and coverage, $N_{\min}$.
(c) Identify minimum acceptability criteria or constraints in the performance space (e.g., no more than 10% of "very low" y_3 values will be tolerated) that must be satisfied irrespectively of how good or bad the corresponding performances for the other objectives may be.

2. Identification and Refinement of Initial Aspiration Levels

From past experience and an examination of the results provided by each of the P agents, a first tentative group of aspiration levels, $\mathbf{y}^*$, that one should aim to reach, is defined by the user ($y_1^* = $ "excellent$_1$," $y_2^* = $ "good-or-excellent$_2$," etc.):

$$\mathbf{y}^* = [y_1^*, \ldots, y_i^*, \ldots, y_P^*]^T. \tag{35}$$

These aspiration levels may be expressed either on an absolute ($y_1^* = $ "excellent$_1$") or on a probabilistic basis $[y_1^* \equiv p(y_1 = $ "excellent$_1$") $\geq 0.90]$.

Before beginning the search for feasible zones of the decision space where the preceding tentative aspiration levels can be achieved, a preliminary check for the possibility of existence of such a zone is conducted. If the perceived ideal, $\mathbf{y}^*$, does not pass this preliminary check, i.e., there is no commensurable solution to the multiobjective problem, the decisionmaker is asked to relax $\mathbf{y}^*$, in order to transform the problem into one with commensurable solutions. For instance, if the initial tentative perceived ideal were $\mathbf{y}^* = ($"excellent$_1$," "excellent$_2$," "excellent$_3$"), but there aren't any available $(\mathbf{x}, \mathbf{y})$ pairs with $\mathbf{y} = \mathbf{y}^*$, the user might adopt as a revised combination of aspiration levels $\mathbf{y}^* = ($"excellent$_1$," "excellent$_2$," "good-or-excellent$_3$"). This relaxation process, guided by the decisionmaker, continues until a final perceived ideal, $\mathbf{y}^*$, for which there are at least $N_{\min}(\mathbf{x}, \mathbf{y})$ pairs with $\mathbf{y} = \mathbf{y}^*$, can be identified. This is the set of aspiration levels that will be used to initiate the interactive search procedure for final solutions, $\mathbf{X}^*$.

3. Search for Solutions

The aspiration levels inherited from the previous step, $\mathbf{y}^*$, are used to guide the search process. Each agent i employs the corresponding aspiration level, y_i^*, and through the application of the learning methodology presented in Section IV tries to identify feasible hyperrectangles, $\mathbf{X}_i^*$, that lead to performance consistent with y_i^*.

Then, we try to combine the partial solutions uncovered by the several agents, $\mathbf{X}_i^*$, in order to find feasible final solutions, $\mathbf{X}^*$, that lead to joint satisfactory behavior in terms of all the objectives, and thus consistent with the current $\mathbf{y}^*$. This is achieved by building conjunctions of multiple $\mathbf{X}_i^*$, uncovered by different agents, according to a breadth-first type of search (Winston, 1984), that takes into account the relative importance assigned to the objectives. Following this lexicographic approach, final solutions $\mathbf{X}^*$ are gradually constructed, and partial paths expanded to accomplish less important goals only when the aspiration levels for the most important ones have already been satisfied. The construction process (whose details are given in Saraiva, 1993; Saraiva and Stephanopoulos, 1992b) relies heavily on the interaction with the decisionmaker to overcome dead-ends, examine arising conflicts, establish tradeoffs, and guide the procedure.

4. Validation of Results

After the search has been concluded, all the uncovered feasible final solutions, $\mathbf{X}^*$, leading to satisfactory joint performances, consistent with $\mathbf{y}^*$, are presented to the decisionmaker for close examination and for the selection of a particular hyperrectangle within this group for eventual implementation.

However, conflicts between the fulfillment of different objectives and aspiration levels may prevent any feasible zone of the decision space from leading to satisfactory joint performances. If the search procedure fails to uncover at least one feasible final solution, $\mathbf{X}^*$, consistent with $\mathbf{y}^*$, a number of options are available to the decisionmaker to try to overcome this impasse. Namely, the decisionmaker can revise the initial problem definition, by either

(a) Redefining any of the constraints originally imposed.
(b) Introducing further relaxations of aspiration levels, which result in new and less demanding perceived ideals, $\mathbf{y}^*$.
(c) Excluding from the search space of agents the particular decision variable, x_m, which creates the conflict among objectives.

Such revisions to the problem statement in order to overcome unsuccessful applications of the search procedure may have to be repeated a number of times before the problem possesses commensurable solutions, and one can find at least one final feasible solution, $\mathbf{X}^*$, that satisfies all the imposed constraints and achieves performances consistent with the current set of aspiration levels, $\mathbf{y}^*$.

C. CASE STUDY: OPERATIONAL ANALYSIS OF A PLASMA ETCHING UNIT

To conclude this section on systems with multiple objectives, we will consider a specific plasma etching unit case study. This unit will be analyzed considering both categorical and continuous performance measurement variables. Provided that similar preference structures are expressed in both instances, we will see that the two approaches lead to similar final answers. Additional applications of the learning methodologies to multiobjective systems can be found in Saraiva and Stephanopoulos (1992b, c).

1. System Characterization

This case study is based on real industrial data collected from a plasma etching plant, as presented and discussed in Reece *et al.* (1989). The task of the unit is to remove the top layer from wafers, while preserving the bottom one. Four different objectives and performance variables are considered:

(a) Maximize a measure of etching selectivity, z_1, expressed as the ratio of etching rates for the top and bottom layers.
(b) Minimize dispersion, z_2, of etching rate values across the wafer bottom-layer surface.
(c) Minimize dispersion, z_3, of etching rate values across the wafer, but now for the top layer.
(d) Maximize the average etching rate for the top layer, z_4.

A quantization of the z_i variables resulted in the definition of the following categorical performance variables, y_i:

Etching selectivity:

$$y_1 = \begin{cases} \text{``bad}_1\text{''} & \text{if} \quad z_1 \leq 3.4, \\ \text{``good}_1\text{''} & \text{if} \quad 3.4 < z_1 \leq 4.0, \\ \text{``excellent}_1\text{''} & \text{if} \quad 4.0 < z_1. \end{cases}$$

Bottom-layer etching dispersion:

$$y_2 = \begin{cases} \text{``bad}_2\text{''} & \text{if} \quad 7.7 < z_2, \\ \text{``good}_2\text{''} & \text{if} \quad 7.0 < z_2 \leq 7.7, \\ \text{``excellent}_2\text{''} & \text{if} \quad z_2 \leq 7.0. \end{cases}$$

Top-layer etching dispersion:

$$y_3 = \begin{cases} \text{``bad}_3\text{''} & \text{if} \quad 4.6 < z_3, \\ \text{``good}_3\text{''} & \text{if} \quad 3.0 < z_3 \leq 4.6, \\ \text{``excellent}_3\text{''} & \text{if} \quad z_3 \leq 3.0. \end{cases}$$

Average top-layer etching rate:

$$y_4 = \begin{cases} \text{``bad}_4\text{''} & \text{if} \quad z_4 \leq 1870, \\ \text{``good}_4\text{''} & \text{if} \quad 1870 < z_4 \leq 2000, \\ \text{``excellent}_4\text{''} & \text{if} \quad 2000 < z_4. \end{cases}$$

Similarly, for the case where a continuous performance metric, y, is employed, the following loss functions were defined (Saraiva and Stephanopoulos, 1992c):

(a) $L(z_1) = 1.924(4.321 - z_1)^2$;
(b) $L(z_2) = 0.033(z_2 - 4.50)^2$;
(c) $L(z_3) = 0.022(z_3 - 0.31)^2$;
(d) $L(z_4) = 5.1387 \cdot 10^{-7}(2595.0 - z_4)^2$;

and $y = \sum_{i=1}^{4} L(z_i)$.

The three decision variables, and the corresponding ranges of values in $\mathbf{X}_{\text{current}}$, are

x_1:   power at which the unit is operated, with values ranging between 75 and 150 W (watts)

x_2:   pressure in the apparatus, ranging from 200 to 255 mtorr (millitorr)

x_3:   flow of etchant gas, varying between 20 and 40 sccm (cubic centimeters per minute of gas flow at standard temperature and pressure conditions)

Since only 20 data records were collected from the system during the execution of the designed experiments conducted by Reece *et al.* (1989), we used their response surface models, deliberately contaminated with small Gaussian noise terms, to generate a total of 500 $(\mathbf{x}, \mathbf{z})$ pairs (assuming that the three variables, x_1, x_2, x_3, have independent and uniform probability distributions). Finally, the following constraints were considered:

(a) As minimum acceptable window sizes, values close to 10% of the $\mathbf{X}_{current}$ ranges are used, leading to $\Delta x_1 = 7.5$, $\Delta x_2 = 5.5$, and $\Delta x_3 = 2.05$.

(b) As minimum coverage, we set $N_{min} = 5$.

2. Categorical Performance Variables

The initial tentative perceived ideal, $\mathbf{y}^*$, was set at

$$\mathbf{y}^* = [\text{``excellent}_i\text{''}]^T, \qquad i = 1,\ldots,4.$$

After successive interactive relaxations, all of them leading to an insufficient number of $(\mathbf{x}, \mathbf{y})$ pairs that jointly satisfy the aspiration levels, we finally came down to the following revision of the perceived ideal:

$$\mathbf{y}^* = [\text{``good-or-excellent}_i\text{''}]^T, \qquad i = 1,\ldots,4.$$

After going through the complete search procedure, given the perceived ideal shown above, the following final solution, $\mathbf{X}_1^*$, was selected:

$$\mathbf{X}_1^* = \{x_1 \in [134.6, 149.1] \wedge x_2 \in [235.1, 243.2] \wedge x_3 \in [20.9, 25.4]\}. \tag{36}$$

A projection of this solution into the $(x_1 - x_2)$ plane is shown in Fig. 8, together with the available $(\mathbf{x}, \mathbf{y})$ pairs that verify the condition imposed over x_3 values. One can qualitatively confirm the validity of condition (36): $\mathbf{X}_1^*$ is indeed a good approximation of the zones in the $x_1 - x_2$ plane that by visual inspection one would associate with leading simultaneously to "good-or-excellent$_i$" performances for all the four objectives.

3. Continuous Performance Variables

The solution found when the plasma etching was analyzed in terms of continuous performance metrics is also presented in Fig. 8, and is given by

$$\mathbf{X}_2^* = \{x_1 \in [129.79; 140.39] \wedge x_2 \in [230.17; 243.22]$$

$$\wedge x_3 \in [20.22; 23.18]\}. \tag{37}$$

This solution is similar to the one found [see hyperrectangle defined by Eq. (36)] previously, when categorical performance evaluation variables were employed. Since the preference structures expressed under both

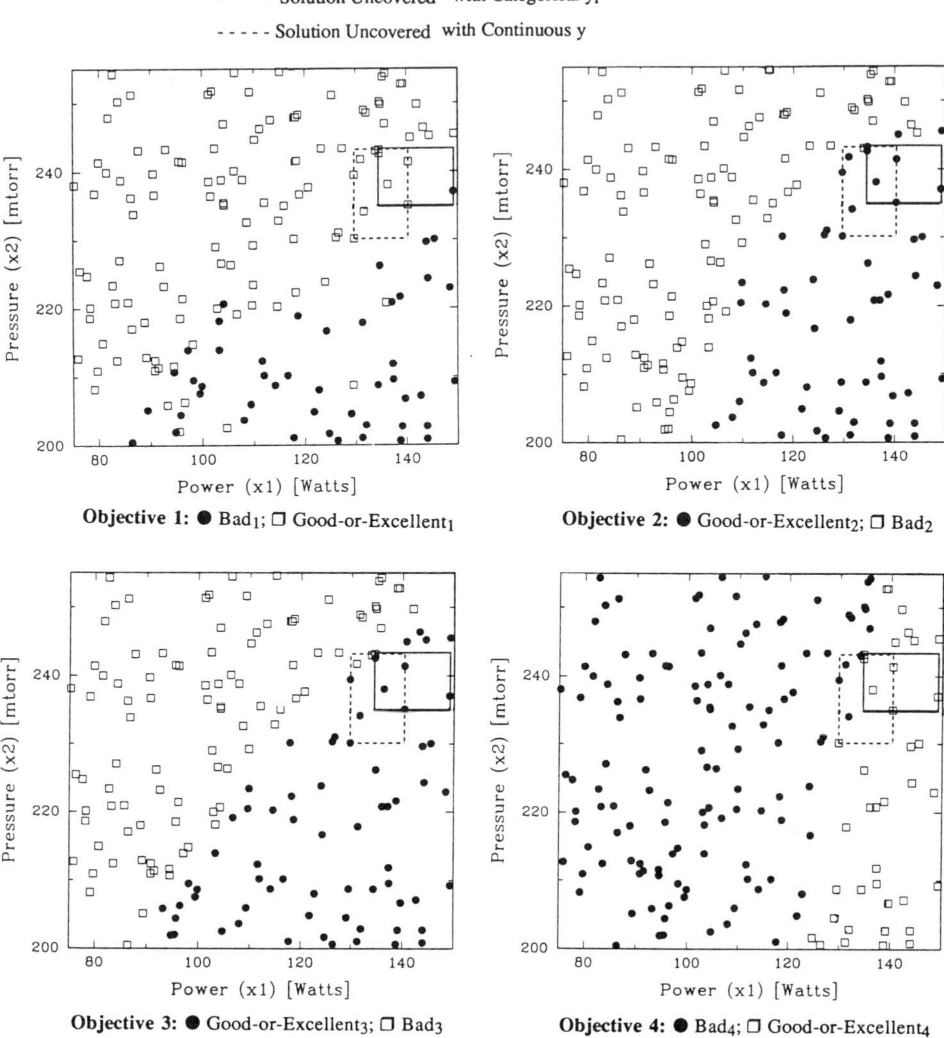

Fig. 8. Data records and final solutions.

formats were chosen to be consistent with each other, this is what one should expect and desire to happen.

Additional studies documented in Saraiva and Stephanopoulos (1992b, c) also illustrate how the introduction of changes in preference structures is translated into displacements of the final uncovered solutions in the decision space.

VII. Complex Systems with Internal Structure

In previous sections we covered a variety of possible applications, including single or multiple objectives, continuous or categorical performance variables. However, so far it has always been assumed that the systems studied are simple systems without any type of internal structure, isolated from the remaining world, and self-sufficient for decisionmaking purposes. In this section we discuss additional extensions of the basic learning methodologies, in order to address complex systems composed of a number of interconnected subsystems. Although situations with categorical performance variables can be treated in similar ways, requiring only minor changes and adaptations, we will consider here only continuous performance metrics.

First, we discuss the problem statements and key features of the learning architecture that are specific to complex systems. This is followed by a brief presentation of the search procedures that are used to build a final solution. The section ends with a summary of the application of the learning architecture to the analysis of a Kraft pulp mill.

A more detailed description of this section's contents can be found in Saraiva (1993) or Saraiva and Stephanopoulos (1992d).

A. PROBLEM STATEMENT AND KEY FEATURES

Complex manufacturing systems, such as an unbleached Kraft pulp plant (Fig. 9), are almost always characterized by some type of internal structure, composed of a number of interconnected subsystems with their own data collection and decisionmaking responsibilities. This raises a number of additional issues, not addressed in previous sections. For instance, if the learning methodology described in Section VI is applied to the digester module of a pulp plant (Fig. 9), it is possible for the final selected solution, $\mathbf{X}_{\text{digester}}$, to include ranges of desired values of sulfidity or other composition properties of the white liquor that enters the digester. This being the case, and since an adjustment of the liquor composition has to be achieved elsewhere in the plant, the request over white liquor sulfidity values has to be propagated backward, to the causticizing area, and eventually from this module to the one preceding it, before a final solution can be found.

In the development of a learning architecture able to extend the methodologies introduced in other sections to complex systems, we looked

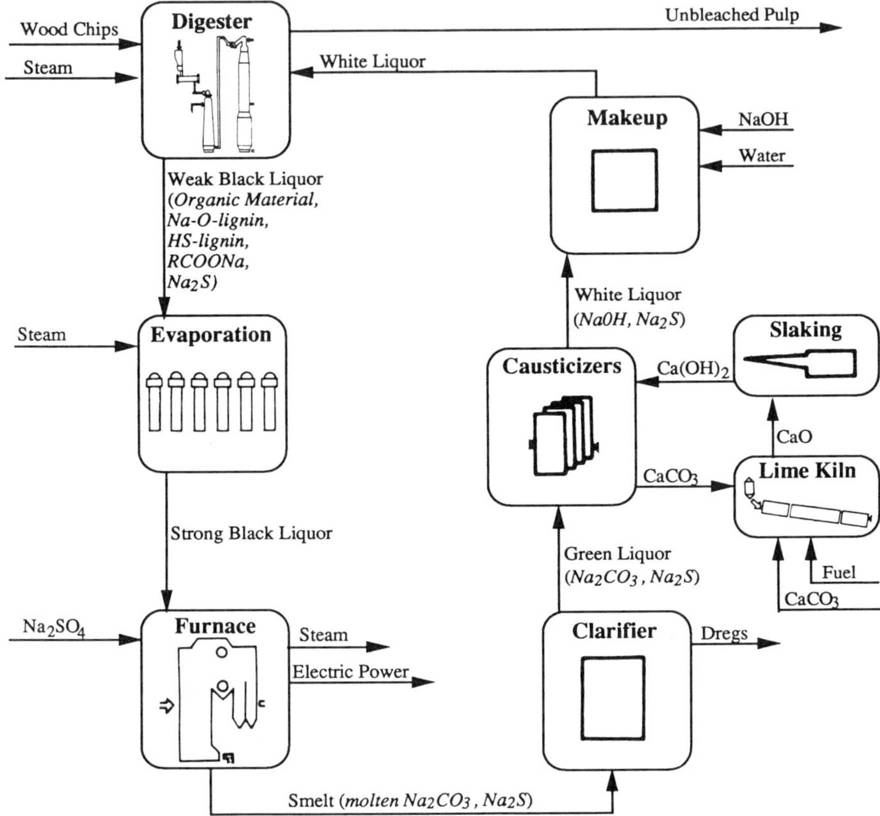

FIG. 9. General overview of unbleached Kraft pulp plant.

explicitly (for reasons presented and justified in Saraiva and Stephanopoulos, 1992c) for approaches that

(a) Are modular and decentralized, allowing each subsystem to take the initiative or make its own decisions, and assigning coordination roles to the upper hierarchical level.
(b) Support and reflect the existing organizational decisionmaking structures, responsibilities, data collection, and analysis activities.

Most of the existing tools to improve process operations fail to provide a systematic and formal process of handling complex systems, or do so in ways that do not fulfill the preceding set of requests. In this paragraph we provide a more formal characterization of a complex system and its several

FIG. 10. Schematic representation of a decision unit.

subsystems, which will then be used to introduce our problem statement, and compare it with analogous conventional formulations.

1. Complex Systems as Networks of Interconnected Subsystems

The basic building block in the definition of a complex system, as well as the key element in our learning architecture, is what we will designate as an *infimal decision unit* or subsystem (Mesarović *et al.*, 1970; Findeisen *et al.*, 1980), DU_k (Fig. 10). These decision units will in general correspond to a particular piece of equipment or section of the plant. The overall system is represented by a single *supremal decision unit* (Mesarović *et al.*, 1970; Findeisen *et al.*, 1980), DU_0, and contains a total of K interconnected infimal decision units (Fig. 11), DU_k, $k = 1, \ldots, K$.

For each of the infimal decision units, DU_k, one has to consider several groups of input and output variables (Fig. 10). Among the inputs are

(a) A vector of decision variables, $x_{d,k}$, which are variables that fall under the scope of authority and can be directly manipulated by DU_k.

(b) A vector of connection variables, $x_{i,k}$, containing those variables that link consecutive infimal decision units, because they are simultaneously inputs to DU_k (although not under its control) and outputs from the preceding decision unit, DU_{k+1}.

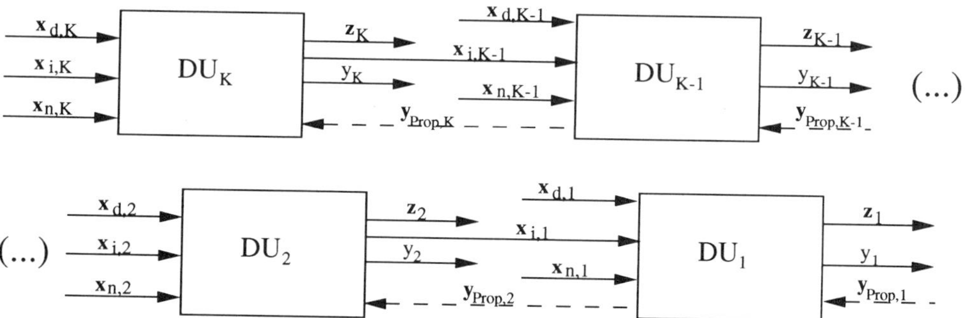

FIG. 11. Complex system as a sequence of infimal decision units.

(c) A vector of disturbance factors, $x_{n,k}$, including variables that reside within the boundaries of DU_k, but nonetheless over which neither DU_k nor any of the other decision units have any control.

As for possible outputs, one has

(a) A vector of performance variables, $z_k = [z_{1,k}, \ldots, z_{p(k),k}]$ [*note:* $p(k)$ stands for the total number of subsystem k objectives], that reflect operating costs, quality related measures and/or the violation of existing constraints on either the input or output spaces.
(b) A scalar global measure of performance for DU_k, y_k, which depends on z_k and results from the combination on a common basis of operating, quality, and constraint violation costs. This measure is global in the sense that it aggregates together all the operational objectives of the decision unit, but local in the sense that it is limited in scope to DU_k goals.
(c) A vector of temporary propagation loss functions, $y_{Prop,k}$, used to transmit the requests over the values of connection variables from one decision unit to the ones that precede it, during the backpropagation process that takes place in the construction of a final solution (for details about the procedure adopted to define propagation loss functions, see Saraiva, 1993; Saraiva and Stephanopoulos, 1992d). Once that solution has been found, these propagation loss functions cease to exist.

2. Conventional and Alternative Problem Definitions

In order to define and compare in a concise way the different problem formulations, let's designate as

n the total number of decision variables, distributed among the K subsystems, $x_{d,j}$, $j = 1, \ldots, n$.

x_{DP} a generic pointwise decision policy, i.e., a vector whose components are values of decision variables,

$$x_{DP} = [x_{d,1}, \ldots, x_{d,j}, \ldots, x_{d,n}].$$

$\dot{X}_{DP}$ a generic interval vector decision policy, whose components are ranges of decision variables

$$X_{DP} = [X_{d,1}, \ldots, X_{d,n}].$$

TABLE IV

Conventional Approaches and Suggested Alternative

	Conventional approaches	Suggested alternative
ξ	$\mathbf{x}_{DP} \in \mathscr{R}^M$	$\mathbf{X}_{DP} \in \mathbf{I}^M$
ψ	$y_0(\mathbf{x}_{DP})$ or $E(y_0|\mathbf{x}_{DP})$	$E(\mathbf{y}_k|\mathbf{X}_{DP})$ and $\sigma(\mathbf{y}_k|\mathbf{X}_{DP})$
f	Technique-dependent	Sample averages and standard deviations
S	Optimization	Top–down and bottom–up

$\mathbf{y}_k$ a performance vector whose components are all the unit k performance variables,

$$\mathbf{y}_k = [\mathbf{z}_k, y_k], \qquad k = 0, 1, \ldots, K.$$

The main differences between conventional approaches and our learning architecture are summarized in Table IV, and are discussed below:

1. As final solution formats, interval vector decision policies, $\mathbf{X}_{DP}$, replace their pointwise counterparts, $\mathbf{x}_{DP}$. Thus, a decision policy, $\mathbf{X}_{DP}$, in the context of this section is an interval vector whose components are intervals of decision variables associated with one or more of the infimal decision units. No connection variables or disturbance factors are involved in their definition;

2. As performance criteria, ψ, the traditional centralized optimization approach considers an overall objective function, y_0, as its single evaluation measure, and the role assigned to the search procedure is the identification of a pointwise decision policy that minimizes $y_0(\mathbf{x}_{DP})$ or $E(y_0|\mathbf{x}_{DP})$. However, any single aggregate measure alone, such as y_0, can not provide a complete and meaningful evaluation of performance for a complex plant that includes several subsystems, a large variety of objectives and local performance measurement variables. Although such aggregate measures may be used to facilitate the search for a promising solution, they should not be interpreted as leading to the very best possible answer, neither should the solutions found be implemented without a detailed analysis of how they will affect all subsystems and the achievement of their own local objectives. Thus, in our approach each subsystem is assumed to have its own goals, and these are taken into account explicitly in the definition of the performance criteria, ψ, which include the conditional expectations and standard deviations of the several goal achievement measures associated with the multiple subsystems, as

well as the system as a whole:

$$E(\mathbf{y}_k|\mathbf{X}_{\mathrm{DP}}) = \left\{E(z_{1,k}|\mathbf{X}_{\mathrm{DP}}),\ldots,E(z_{p(k),k}|\mathbf{X}_{\mathrm{DP}}),E(y_k|\mathbf{X}_{\mathrm{DP}})\right\}^T,$$
$$k = 0,1,\ldots,K, \quad (38)$$

$$\sigma(\mathbf{y}_k|\mathbf{X}_{\mathrm{DP}}) = \left\{\sigma(z_{1,k}|\mathbf{X}_{\mathrm{DP}}),\ldots,\sigma(z_{p(k),k}|\mathbf{X}_{\mathrm{DP}}),\sigma(y_k|\mathbf{X}_{\mathrm{DP}})\right\}^T,$$
$$k = 0,1,\ldots,K. \quad (39)$$

3. Traditional techniques use $f(\mathbf{x}_{\mathrm{DP}})$ quantitative models to perform the mapping from the solution to the performance space. These models essentially reduce the problem to a simple system, having the decision variables $x_{d,j}$ as inputs and y_0 as output, because they allow one to express the $y_0(\mathbf{x}_{\mathrm{DP}})$ or $E(y_0|\mathbf{x}_{\mathrm{DP}})$ estimates as a function of the decision variables, $\psi^{\mathrm{est}} = f(\mathbf{x}_{\mathrm{DP}})$. On the other hand, our mapping procedures, f, are based on the construction of direct sampling estimates. Given a decision policy, $\mathbf{X}_{\mathrm{DP}}$, both $E(\mathbf{y}_k|\mathbf{X}_{\mathrm{DP}})$ and $\sigma(\mathbf{y}_k|\mathbf{X}_{\mathrm{DP}})$ are estimated by identifying the available data records for which $\mathbf{x} \in \mathbf{X}_{\mathrm{DP}}$, and computing the corresponding sample averages and standard deviations for all the $z_{i,k}$ and y_k variables.

4. In conventional centralized approaches the goal of the search procedures is to find a final feasible decision policy, $\mathbf{x}^*_{\mathrm{DP}}$, that optimizes $f(\mathbf{x}_{\mathrm{DP}})$. This solution is then imposed to the several subsystems for implementation, although these were not directly involved in its construction. Other similar methodologies include problem decomposition strategies (Lasdon, 1970; Biegler, 1992) that explore particular properties of the complex system structure and dynamic programming (Bellman and Dreyfus, 1962; Roberts, 1964; Nemhauser, 1966). They share the same basic assumptions, problem statement and solution formats as centralized approaches, although, for the sake of computational efficiency gains, a multistage and sequential identification of the final result, $\mathbf{x}^*_{\mathrm{DP}}$, is adopted as the problem solving strategy.

Our search procedures represent a departure from the above type of paradigm. Rather than simply accepting and implementing a decision policy found by DU_0, that optimizes an overall measure of performance, the infimal subsystems and corresponding plant personnel play an active role in the construction and validation of solutions. One tries to build a consensus decision policy, $\mathbf{X}_{\mathrm{DP}}$, validated by all subsystems, DU_k, $k = 1,\ldots,K$, as well as by the whole plant, DU_0, and only when that consensus has been reached does one move toward implementation. Within this context, the upper-level decision unit, DU_0, assumes a coordination role, and does not have the power to impose solutions to the several subsystems. Two different search procedures, used to find such consensus decision policies, $\mathbf{X}_{DP}$, and designated respectively as *bottom–up* and *top–down*, are described in the following paragraphs.

3. Final Problem Statement

The ultimate goal of our learning architecture is to uncover at least one decision policy, $\mathbf{X}_{DP}$, that

(a) Is feasible, i.e., satisfies constraints imposed over the minimum acceptable size of operating windows $[w(X_{d,j}) \geq \Delta x_{d,j}, \; j = 1, \ldots, n]$ and coverage $[n(\mathbf{X}_{DP}) \geq N_{min}]$.

(b) Leads to a significant improvement over the current levels of performance for one or more of the decision units.

(c) Is accepted and validated by all decision units involved with or affected by the use of $\mathbf{X}_{DP}$ as the zone to conduct the operation.

A decision policy that satisfies all these requirements is designated as an active decision policy, $\mathbf{X}_{DP}^*$.

To declare a decision policy, $\mathbf{X}_{DP}$, as either unacceptable, acceptable or leading to a significant improvement, each decision unit compares its current levels of performance with the ones that are expected within $\mathbf{X}_{DP}$. The current levels of performance for unit k are provided by the $E(\mathbf{y}_k | \mathbf{X}_{current})$ and $\sigma(\mathbf{y}_k | \mathbf{X}_{current})$ estimates obtained from a sample of data records, just as in the case of $E(\mathbf{y}_k | \mathbf{X}_{DP})$ and $\sigma(\mathbf{y}_k | \mathbf{X}_{DP})$. A comparison of these reference performance values for $\mathbf{X}_{current}$ with the ones achieved by $\mathbf{X}_{DP}$ is made by each decision unit. As a result of this comparison, a final evaluation of $\mathbf{X}_{DP}$ by decision unit k leads to one of the three possible outcomes:

1. $f_k(\mathbf{X}_{DP}) \gg f_k(\mathbf{X}_{current})$, meaning that significant improvement is expected.

2. $f_k(\mathbf{X}_{DP}) \approx f_k(\mathbf{X}_{current})$, in case $\mathbf{X}_{DP}$ is accepted and validated by decision unit k, although no significant improvements are expected.

3. $f_k(\mathbf{X}_{DP}) \ll f_k(\mathbf{X}_{current})$, meaning that $\mathbf{X}_{DP}$ can not be accepted by unit k, because it would result in performance deterioration down to levels that fall below what the unit can tolerate for at least one of its performance variables.

Thus, an active decision policy, $\mathbf{X}_{DP}^*$, is a feasible decision policy such that

$$\exists_{k \in \{0, 1, \ldots, K\}} \mid f_k(\mathbf{X}_{DP}) \gg f_k(\mathbf{X}_{current});$$

$$\forall_{k \in \{0, 1, \ldots, K\}} \{ f_k(\mathbf{X}_{DP}) \approx f_k(\mathbf{X}_{current}) \vee f_k(\mathbf{X}_{DP}) \gg f_k(\mathbf{X}_{current}) \}.$$

B. Search Procedures

Two different search procedures (bottom–up and top–down) can be followed to build active decision policies, $\mathbf{X}_{\mathrm{DP}}^{*}$.

In the bottom–up approach the initiative to start the learning process is taken by one of the infimal decision units. Since solutions found at this unit may include connection variables, the request for given values of these variables is propagated backward, to unit $k + 1$, through temporary loss functions. After successive backpropagation steps, the participation of several other DU_k and the operators associated with them, a final decision policy, accepted and validated by all infimal decision units, is eventually found. Then, this policy is brought to the attention of the supremal decision unit, DU_0, who is responsible for detecting whether it leads to an improved performance of the system as a whole. If so, the uncovered policy is an active decision policy, and one can proceed with its implementation.

On the other hand, the top–down approach starts the learning process at the supremal decision unit, DU_0, and only on a second stage does it move down to the infimal decision units for approval and validation.

The next paragraphs provide a brief description of both the bottom–up and top–down search procedures (for further details, see Saraiva, 1993; Saraiva and Stephanopoulos, 1992d).

1. Bottom–Up Approach

The bottom–up approach contains two distinct stages. First, by successive backpropagation steps one builds a decision policy. Then, this uncovered policy is evaluated and refined, and its expected benefits confirmed before any implementation actually takes place. This two-stage process is conceptually similar to dynamic programming solution strategies, where first a decision policy is constructed by backward induction, and then one finds a realization of the process for the given policy, in order to check its expected performance (Bradley *et al.*, 1977).

a. Decision Policy Construction. Learning activities may be initiated by any of the infimal decision units, DU_k, to which one applies the **basic** learning methodology introduced in Sections V and VI, leading to the identification of a particular final solution, $\mathbf{X}_k$.

If $\mathbf{X}_k$ involves only ranges of decision variables attached to unit k, $x_{d,k}$, it defines a decision policy, and thus one can move directly to the validation and refinement phase.

However, $\mathbf{X}_k$, besides decision variables, may also include ranges of connection variables, $x_{i,k}$, that link units $k + 1$ and k. That being the case, the learning process has to be propagated backward, toward decision unit DU_{k+1}. To induce this propagation, one has first to identify temporary loss functions (Saraiva, 1993; Saraiva and Stephanopoulos, 1992d) for all connection variables present in the definition of $\mathbf{X}_k$. These temporary propagation loss functions are combined with other DU_{k+1} goals, and *basic* is now applied to decision unit DU_{k+1}. Further propagations, from unit $k+1$ to unit $k+2$, $k+2$ to $k+3$, etc., are identical to the one that occurred from DU_k to DU_{k+1}. This backpropagation through different decision units continues until one eventually reaches an infimal decision unit, DU_{k+j}, where a solution, $\mathbf{X}_{k+j}$, involving only ranges of decision variables, is found. When that happens, a final decision policy, $\mathbf{X}_{DP}$, can be immediately constructed by simply assembling all of its pieces together: it is the conjunction of the decision variable intervals, distributed among several decision units, that were uncovered during the upstream propagation process.

b. Validation and Refinement. Because the construction of $\mathbf{X}_{DP}$ resulted from the contributions of multiple infimal decision units, taking into consideration their specific goals and imposed constraints, it may be already an active decision policy. However, and even if that is the case, before proceeding to any implementation, it is necessary to evaluate the benefits that would derive from operating within $\mathbf{X}_{DP}$.

The process of $\mathbf{X}_{DP}$ validation and refinement starts with a final detailed analysis of its realization, through the computation of $E(\mathbf{y}_k|\mathbf{X}_{DP})$ and $\sigma(\mathbf{y}_k|\mathbf{X}_{DP})$, $k = 0, 1, \ldots, K$, estimates. If for some reason one or more infimal decision units judge the implementation of $\mathbf{X}_{DP}$ to be unacceptable, an additional attempt is made to refine the decision policy and find a revised version of it, $\mathbf{X}_{DP}^{final}$, that is accepted and validated by all infimal decision units, through additional applications of *basic* to the decision units in conflict with the initial $\mathbf{X}_{DP}$ version (see Saraiva, 1993; Saraiva and Stephanopoulos, 1992d, for details).

Once agreement and consensus have been reached by all infimal decision units, for $\mathbf{X}_{DP}^{final}$ to be declared an active decision policy, it is necessary to bring it to the attention of DU_0. The effects of a possible implementation of $\mathbf{X}_{DP}^{final}$ on the system as a whole are examined, to check that it also translates into improved performance from a global perspective. If this final validation test is passed, $\mathbf{X}_{DP}^{final}$ represents an active decision policy, and one can proceed to its implementation.

2. Top–Down Approach

In the top–down approach the supremal decision unit, DU_0, starts the learning process by itself, and identifies a decision policy, $\mathbf{X}_{DP}$. Then, in a second stage, one moves down to the infimal subsystems, seeking their support and validation for $\mathbf{X}_{DP}$.

Let's assume that the inputs to the supremal decision unit are a subset of all the decision variables attached to infimal decision units, consisting of those $x_{d,k}$ variables that are believed to be particularly influential with respect to the operation of the overall system. Then, an application of **basic** to DU_0 results directly in the identification of a decision policy, $\mathbf{X}_{DP}$. This decision policy is then passed down to the lower level in the hierarchy, where it is submitted to a process of validation and refinement by all infimal decision units that is identical to the one that takes place in the bottom–up approach.

C. Case Study: Operational Analysis of a Pulp Plant

The overall system that we will analyze comprises the unbleached Kraft pulp line, chemicals and energy recovery zones of a specific paper mill (Melville and Williams, 1977). We will employ a somewhat simplified but still realistic representation of the plant, originally developed in a series of research projects at Purdue University (Adler and Goodson, 1972; Foster *et al.*, 1973; Melville and Williams, 1977). The records of simulated operation data, used to support the application of our learning architecture, were generated by a reimplementation, with only minor changes, of steady-state models (for each individual module and the system as a

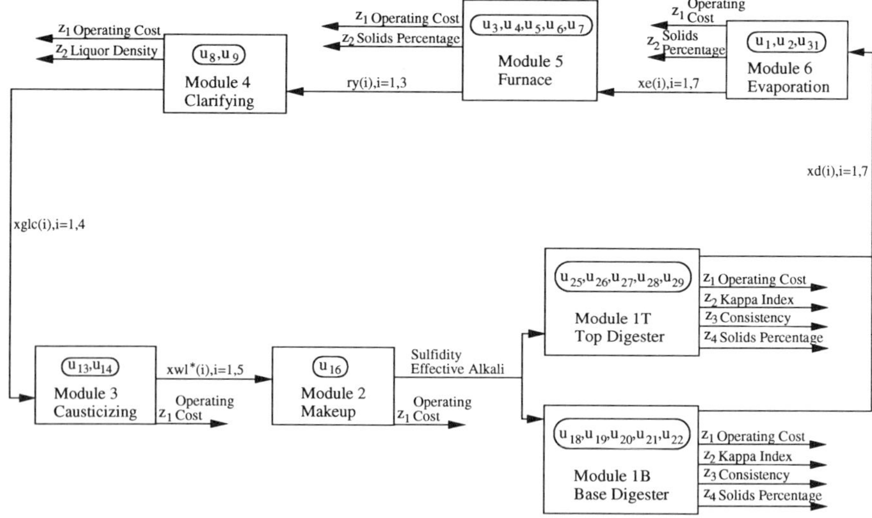

Fig. 12. Plant internal structure and modular representation.

whole) presented in the references cited above. A more detailed description of this case study can be found in Saraiva (1993), or Saraiva and Stephanopoulos (1992d).

The global structure of a Kraft pulp plant was illustrated in Fig. 9. The corresponding modular representation adopted in this study is shown schematically in Fig. 12. It includes 7 infimal decision units, with a total of 23 different decision variables distributed among them [to keep the notation consistent with Melville and Williams (1977), decision variables are designated as u_d]. A complete list of these variables, including their physical meaning, measurement units and ranges of values in $\mathbf{X}_{current}$, is provided in Table V. There are also 28 connection variables, linking successive infimal subsystems, and 16 local performance variables, $z_{i,k}$. All the performance measures (operating costs, quality losses, penalty functions) are expressed on a common basis of U.S. dollars per ton of air-dried pulp produced ($/ TADP).

TABLE V

LIST OF ALL 23 DECISION VARIABLES AND CORRESPONDING
WINDOWS UNDER CURRENT OPERATING CONDITIONS

u_d	u_d description	Units	Range
u_1	Oxidation tower efficiency	Adimensional	$[0.0; 0.9]$
u_2	Steam flow to evaporators	lb/h	$[0.9 \cdot 10^5; 1.25 \cdot 10^5]$
u_3	Sodium sulfate addition	lb/h	$[2000; 4000]$
u_4	Black liquor temperature at nozzles	°F	$[245; 250]$
u_5	Primary airflow to furnace	lb/h	$[3 \cdot 10^5; 4 \cdot 10^5]$
u_6	Secondary airflow to furnace	lb/h	$[1.75 \cdot 10^5; 2.75 \cdot 10^5]$
u_7	Furnace flue gas temperature	°F	$[550; 750]$
u_8	Steam/condensate added to smelt tank	lb/TADP	$[4000; 5000]$
u_9	Washing water flow to dregs filter	lb/TADP	$[0.0; 75.0]$
u_{13}	Water fraction in lime mud slurry	Adimensional	$[0.3; 0.4]$
u_{14}	Water spray flow to white liquor filter	lb/TADP	$[1000; 2000]$
u_{16}	White liquor effective alkali	lb/gal	$[0.8; 0.9]$
u_{18}	Lower heater temperature	°F	$[295; 310]$
u_{19}	Digester blow flow	gpm[a]	$[1100; 1250]$
u_{20}	Washing circulation temperature	°F	$[225; 280]$
u_{21}	Black liquor extraction flow	gpm	$[1150; 1250]$
u_{22}	White liquor added to digester	gpm	$[400; 500]$
u_{25}	Lower heater temperature	°F	$[295; 308]$
u_{26}	Digester blow flow	gpm	$[800; 950]$
u_{27}	Washing circulation temperature	°F	$[225; 280]$
u_{28}	Black liquor extraction flow	gpm	$[950; 1150]$
u_{29}	White liquor added to digester	gpm	$[400; 500]$
u_{31}	Steam flow to flash system	lb/h	$[0; 10000]$

[a] Gallons per minute.

1. Top–Down Approach

When examined at the supremal level, the system's primary goal is the production of pulp with the desired kappa indices. Consequently, as DU_0 performance variables we will consider the following:

(a) $z_{1,0}$, total operating cost, which is the sum of all the 7 subsystems operating costs, $z_{1,k}$, $k = 1, \ldots, 7$
(b) $z_{2,0}$, kappa index of the pulp produced at the base digester
(c) $z_{3,0}$, kappa index of the pulp produced at the top digester

Furthermore, we will consider the 23 different decision variables, u_d, as the DU_0 inputs.

Both operating costs and quality losses were combined together, leading to the following overall performance measure, y_0, for the supremal decision unit:

$$y_0 = z_{1,0} + (z_{2,0} - 100)^2 + 5(z_{3,0} - 60)^2. \tag{40}$$

A preliminary analysis of the available DU_0 data records showed that y_0 is currently dominated by the behavior of $z_{3,0}$, and its deviations from the target. After going through the search procedure described in Section VII, the following final active decision policy, $\mathbf{X}_{DP}^{final}$, was identified:

$$\mathbf{X}_{DP}^{final} = \{u_2 \in [112,000; 122,000] \wedge u_3 \in [2060; 2700]$$

$$\wedge u_4 \in [245.3; 247.3] \wedge u_{25} \in [304.2; 307.8]\}. \tag{41}$$

In Table VI we compare the performance achieved through the implementation of the above strategy with that defined by the current values. An implementation of $\mathbf{X}_{DP}^{final}$ results in a significant decrease of the y_0 average, primarily as a consequence of a reduction in the average cost associated with the operation of the top digester, y_{1T}. On its own hand, the y_{1T} average decrease derives from the fact that $\mathbf{X}_{DP}^{final}$ centers the average kappa index of the top-digester pulp much closer to its target of 60, while also reducing its standard deviation. All these results are consistent with the observation made earlier that pointed to $z_{3,0}$ as the key performance variable that conditions the current levels of overall system performance, $E(y_0|\mathbf{X}_{current})$.

2. Bottom–Up Approach

The learning process was initiated at the top-digester infimal decision unit, leading to a solution, $\mathbf{X}_{1T}$, that involves local decision variables and a range of white liquor sulfidity (fraction of active reactants in the white

TABLE VI

COMPARISON OF PERFORMANCE MEASURES BEFORE
AND AFTER IMPLEMENTATION OF $\mathbf{X}_{\text{DP}}^{\text{final}}$

	Average $(\mathbf{X}_{\text{current}})$	Average $(\mathbf{X}_{\text{DP}}^{\text{final}})$
y_0	1786.5	895.6
y_{1T}	1500.6	695.3
y_{1B}	281.5	187.8
y_2	8.8	9.2
y_3	7.4	7.8
y_4	185.6	168.8
y_5	120.0	50.1
y_6	5.8	4.8
Kappa index (top)	67.4	59.9
Kappa index (base)	102.9	102.0

liquor that are present as HS^- rather than as OH^-) values. Successive upstream propagations of this request had to be performed before a solution involving only decision variables, $\mathbf{X}_5$, was found at the furnace module, thus leading to the identification of a decision policy, $\mathbf{X}_{\text{DP}}$, which consists of the conjunction of all the decision variable ranges identified up to that point.

It is worth noticing that the path that was followed in the construction of $\mathbf{X}_{\text{DP}}$ is coherent and logical with respect to a physical understanding of the plant. It was found at the top digester that one of the most important variables, conditioning the location and dispersion of the pulp kappa index, is the amount of sulfur present in the white liquor added to the digester. But sodium sulfate enters the plant to compensate for sulfur losses at the recovery furnace, and thus it is basically within this infimal decision unit that the levels of sulfidity can be adjusted. Accordingly, the backpropagation process involved requests over the values of sulfur flows in several intermediate streams, and it stopped only at the furnace module, where the decision policy construction was concluded, leading to a $\mathbf{X}_{\text{DP}}$ that involves the amount of sodium sulfate added to the furnace.

After submitting $\mathbf{X}_{\text{DP}}$ through the validation and refinement stage, the final uncovered active decision policy, $\mathbf{X}_{\text{DP}}^{\text{final}}$, was given by

$$\mathbf{X}_{\text{DP}}^{\text{final}} = \{u_3 \in [3200; 3635] \wedge u_4 \in [247.5; 248.7] \wedge u_8 \in [4800; 5000]$$

$$\wedge u_{18} \in [306, 309] \wedge u_{25} \in [304.5; 308.0] \wedge u_{29} \in [460.0; 490.0]\} \quad (42)$$

Table VII summarizes the levels of performance that are achieved within $\mathbf{X}_{\text{DP}}^{\text{final}}$, and compares them with the performance corresponding to the current values.

TABLE VII

Comparison of Performance Measures Before
and After Implementation of $\mathbf{X}_{\mathrm{DP}}^{\mathrm{final}}$

	Average $(\mathbf{X}_{\mathrm{current}})$	Average $(\mathbf{X}_{\mathrm{DP}}^{\mathrm{final}})$
y_0	1786.5	489.1
y_{1T}	1500.6	387.4
y_{1B}	281.5	89.2
y_2	8.8	11.8
y_3	7.4	8.6
y_4	185.6	14.9
y_5	120.0	91.7
y_6	5.8	4.9
Kappa index (top)	67.4	60.7
Kappa index (base)	102.9	99.5

3. Brief Comparison of Final Decision Policies

In the previous paragraphs it was shown how by following the top–down and bottom–up approaches one arrived at the construction of two distinct and promising decision policies, Eqs. (41) and (42). Both of these decision policies include intervals of values for certain critical decision variables (e.g., u_3, u_4, u_{25}). Since there is some consistency between the infimal and supremal decision unit goals, this communality of variables should be expected. However, different infimal decision unit local goals were taken into account in the construction of these policies, and choices among several possible solutions were made by the user during that construction. Thus, it does not come as a surprise that the two final decision policies also involve ranges of different decision variables. Similarly, the performances achieved are not entirely identical.

A final comparison of the results obtained with each policy, as well as under the current operating conditions, is given in Fig. 13. Both the top–down and the bottom–up approaches uncovered promising decision policies, which lead to considerable improvements over the current plant performance levels. When one compares in a more detailed way the differences in infimal decision unit performances for the two policies, one can see that they reflect the participation of the corresponding subsystems in their construction. Finally, it should be added that, besides performance related issues, the upstream propagation associated with the bottom–up approach provided some important insight into understanding the main causes of dispersion and location for the top-digester pulp kappa index. Specifically, it was uncovered that the kappa index is highly dependent on

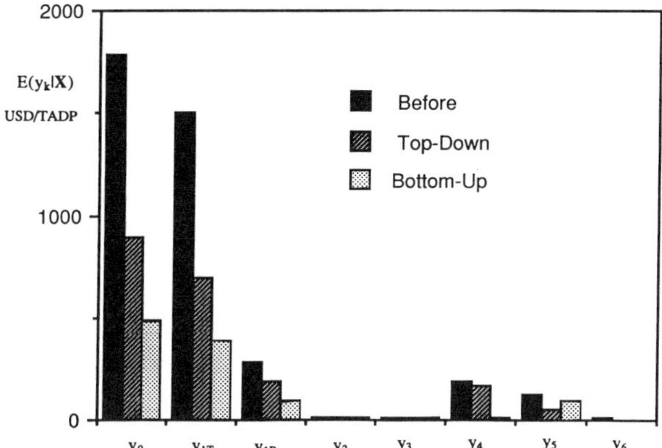

FIG. 13. Final comparison of performances for both decision policies.

the white liquor sulfidity, a relationship translated later on into the definition of intervals for decision variables associated with the recovery boiler, which is the module where one has the ability to manipulate the amounts of sulfur present in all the plant liquor streams. This type of knowledge would have been hard or impossible to acquire by merely examining the final decision policies or through an analysis of the top–down approach alone.

VIII. Summary and Conclusions

In this chapter we revisited an old problem, namely, exploring the information provided by a set of $(\mathbf{x}, y)$ operation data records and learn from it how to improve the behavior of the performance variable, y. Although some of the ideas and methodologies presented can be applied to other types of situations, we defined as our primary target an analysis at the supervisory control level of $(\mathbf{x}, y)$ data, generated by systems that cannot be described effectively through first-principles models, and whose performance depends to a large extent on quality-related issues and measurements.

We have introduced modified statements and solution formats for the preceding problem, with hyperrectangles in the decision space replacing the conventional pointwise results. The advantages and implications of adopting this alternative language to express final solutions were discussed, and it was also shown that traditional formulations can be interpreted as a particular degenerate case of the suggested more general problem definitions, where one searches for ranges of decision variables rather than single values.

To address the modified problem statements and uncover final solutions with the desired alternative formats, data-driven nonparametric learning methodologies, based on direct sampling approaches, were described. They require far fewer assumptions and a priori decisions on the part of the user than most conventional techniques. These practical frameworks for extracting knowledge from operating data present the final uncovered solutions to the decisionmaker in formats that are both easy to understand and implement.

We presented extensions and variations of the basic learning methodologies aimed at enlarging their flexibility and cover a number of different situations, including systems where performance is evaluated by categorical or continuous variables, with single or multiple objectives, simple or complex plants containing some type of internal structure and composed of a number of interconnected subsystems.

The potential practical capabilities of the described learning methodologies, and their attractive implementational features from an industrial point of view, were illustrated through the presentation of a series of case studies with both real-world industrial and simulated operating data.

Acknowledgments

The author would like to acknowledge financial an other types of support received from the Leaders for Manufacturing program at MIT, Fulbright Program, Rotary Foundation, Comissão Permanente da INVOTAN, Fundação Luso-Americana para o Desenvolvimento, and Comissão Cultural Luso-Americana. Special thanks also to Professor George Stephanopoulos, who always provided the right amount of support and guidance, while at the same time allowing me to have all the freedom that I needed to pursue my own research dreams and try to convert them into reality.

References

Adler, L., and Goodson, R., "An Economic Optimization of a Kraft Pulping Process," Laboratory for Applied Industrial Control, Report 48. Purdue University, West Lafayette, IN, 1972.

Alefeld, G., and Herzberger, J., "Introduction to Interval Computations." Academic Press, New York, 1983.

Bellman, R., and Dreyfus, S., "Applied Dynamic Programming." Princeton University Press, Princeton, NJ, 1962.

Biegler, L., Optimization strategies for complex process models. *Adv. Chem. Eng.* **18**, 197 (1992).

Bradley, S., *et al.*, "Applied Mathematical Programming." Addison-Wesley, Reading, MA, 1977.

Breiman, L., *et al.*, "Classification and Regression Trees." Wadsworth, Belmont, CA, 1984.

Clausing, D., "Total Quality Development." ASME Press, New York, 1993.

Daniel, C., and Wood, F., "Fitting Equations to Data." 2nd ed. Wiley, New York, 1980.

Deming, W., "Out of the Crisis." Massachusetts Institute of Technology, Center for Advanced Engineering Study, Cambridge, MA, 1986.

Duda, R., and Hart, P., "Pattern Classification and Scene Analysis." Wiley, New York, 1973.

Ellingsen, W., Implementation of advanced control systems. *AIChE Symp. Ser.* **159**, 150 (1976).

Findeisen, W., *et al.*, "Control and Coordination in Hierarchical Systems." Wiley, New York, 1980.

Foster, R., *et al.*, "Optimization of the Chemical Recovery Cycle of the Kraft Pulping Process," Laboratory for Applied Industrial Control, Report 54. Purdue University, West Lafayette, IN, 1973.

Fu, K., "Sequential Methods in Pattern Recognition and Machine Learning." Academic Press, New York, 1968.

Gaines, B., The trade-off between knowledge and data in knowledge acquisition. *In* "Knowledge Discovery in Databases" (G. Shapiro and W. Frawley, eds.), p. 491. MIT Press, Cambridge, MA, 1991.

Garcia, C., and Prett, D., Advances in industrial model-predictive control. *In* "Chemical Process Control, CPC-III." (Morari, M. and McAvoy, T. J., eds.). CACHE-Elsevier, New York, 1986.

Goodman, R., and Smyth, P., Decision tree design using information theory. *Knowl. Acquis.* **2**, 1 (1990).

Hayes, R., *et al.*, "Dynamic Manufacturing: Creating the Learning Organization." Free Press, New York, 1988.

Hunt, E., "Concept Learning: An Information Processing Problem." Wiley, New York, 1962.

James, M., "Classification Algorithms." Wiley, New York, 1985.

Juran, J., "Managerial Breakthrough." McGraw-Hill, New York, 1964.

Klein, J., "Revitalizing Manufacturing." R.D. Irwin, Homewood, IL, 1990.

Kodratoff, Y., and Michalski, R., eds., "Machine Learning: An Artificial Intelligence Approach." Vol. 3. Morgan Kaufmann, San Mateo, CA, 1990.

Lasdon, L., "Optimization Theory for Large Systems." Macmillan, New York, 1970.

Lasdon, L., and Baker, T., The integration of planning, scheduling and process control. *In* "Chemical Process Control, CPC-III." (Morari, M. and McAvoy, T. J., eds.). CACHE-Elsevier, New York, 1986.

Latour, P., Comments on assessment and needs. *AIChE Symp. Ser.* **159**, 161 (1976).

Latour, P., Use of steady-state optimization for computer control in the process industries. *In* "On-line Optimization Techniques in Industrial Control" (Kompass, E. J. and Williams, T. J., eds.). Technical Publishing Company, 1979.

Launks, U., *et al.*, On-line optimization of an ethylene plant. *Comput. Chem. Eng.* **16**, S213 (1992).

Melville, S., and Williams, T., "Application of Economic Optimization to the Chemical Recovery System of a Kraft Pulping Process," Laboratory for Applied Industrial Control, Report 107. Purdue University, West Lafayette, IN, 1977.

Mesarović, M., *et al.*, "Theory of Hierarchical, Multilevel Systems." Academic Press, New York, 1970.

Moore, R., "Methods and Applications of Interval Analysis." SIAM, Philadelphia, 1979.

Moore, R., What and who is in control. *In* "The Second Shell Process Control Workshop" (D.M. Prett, C.E. Garcia, and B.L. Ramaker, eds.). Butterworth, Stoneham, MA, 1990.

Moret, B., Decision trees and diagrams. *ACM Comput. Surv.* **14**(4), 593 (1982).

National Research Council, "The Competitive Edge. "National Academy Press, Washington, DC, 1991.

Nemhauser, G., "Introduction to Dynamic Programming." Wiley, New York, 1966.

Phadke, M., "Quality Engineering Using Robust Design." Prentice Hall, Englewood Cliffs, NJ, 1989.

Quinlan, J., Induction of decision trees. *Mach. Learn.* **1**, 81 (1986).

Quinlan, J., Simplifying decision trees. *Int. J. Man-Mach. Stud.* **27**, 221 (1987).

Quinlan, J., "C4.5: Programs for Machine Learning." Morgan Kaufmann, San Mateo, CA, 1993.

Reece, J., Daniel, D. and Bloom R., Identifying a plasma etch process window. *In* "Understanding Industrial Designed Experiments" (Schmidt S. and Launsby R., eds.), 2nd ed. AIR Academy Press, Colorado Springs, CO, 1989.

Roberts, S., "Dynamic Programming in Chemical Engineering." Academic Press, New York, 1964.

Roy, R., "A Primer on the Taguchi Method." Van Nostrand-Reinhold, Princeton, NJ, 1990.

Saraiva, P., Data-driven learning frameworks for continuous process analysis and improvement. Ph.D. Thesis, Massachusetts Institute of Technology, Dept. Chem. Eng., Cambridge, MA, 1993.

Saraiva, P., and Stephanopoulos, G., Continuous process improvement through inductive and analogical learning. *AIChE J.* **38**(2), 161 (1992a).

Saraiva, P., and Stephanopoulos, G., "Learning to Improve Processes with Multiple Pattern Recognition Objectives," Working paper. Massachusetts Institute of Technology, Dept. Chem. Eng., Cambridge, MA, 1992b.

Saraiva, P., and Stephanopoulos, G., "An Exploratory Data Analysis Robust Optimization Approach to Continuous Process Improvement," Working paper. Massachusetts Institute of Technology, Dept. Chem. Eng., Cambridge, MA, 1992c.

Saraiva, P., and Stephanopoulos, G., "Data-Driven Learning Architectures for Process Improvement in Complex Systems with Internal Structure," Working paper. Massachusetts Institute of Technology, Dept. Chem. Eng., Cambridge, MA, 1992d.

Sargent, R., The future of digital computer based industrial control systems. *In* "Industrial Computing Control After 25 Years" (Kompass, E. J. and Williams, T. J., eds.), p. 63. Technical Publishing Company, 1984.

Senge, P., "The Fifth Discipline." Currency, 1990.

Shannon, C., and Weaver, W., "The Mathematical Theory of Communication," 11th printing. University of Illinois Press, Urbana, 1964.

Shapiro, G., and Frawley, W., eds., "Knowledge Discovery in Databases." MIT Press, Cambridge, MA, 1991.

Shavlik, J., and Dietterich, T., eds., "Readings in Machine Learning." Morgan Kaufmann, San Mateo, CA, 1990.

Sheridan, T., "45 Years of Man-Machine Systems." Massachusetts Institute of Technology, Dept. Mech. Eng., Cambridge, MA, 1985.

Shiba, S., *et al.*, "The Four Revolutions of Management Thinking: Planning and Implementation of TQM for Executives." Productivity Press, 1993.

Sinnar, R., Impact of model uncertainties and nonlinearities on modern controller design. *In* "Chemical Process Control, CPC-III." (Morari, M. and McAvoy, T. J., eds.), p. 53. CACHE-Elsevier, 1986.

Sonquist, J., *et al.*, "Searching for Structure." University of Michigan, Ann Arbor, Michigan, 1971.

Taguchi, G., "Introduction to Quality Engineering." Asian Productivity Association, 1986.

Taylor, W., "What Every Engineer Should Know About Artificial Intelligence." MIT Press, Cambridge, MA, 1989.

Taylor, W., "Optimization and Variation Reduction in Quality." McGraw-Hill, New York, 1991.

Tukey, J., "Exploratory Data Analysis." Addison-Wesley, Reading, MA, 1977.

Turban, E., "Decision Support and Expert Systems." Macmillan, New York, 1988.

Utgoff, P., Perception trees: A case study in hybrid concept representations. *In* "Proceedings of AAAI88," Vol. 2, p. 601. Morgan Kaufmann, San Mateo, CA, 1988.

Winston, P., "Artificial Intelligence," 2nd ed. Addison-Wesley, Reading, MA, 1984.

EMPIRICAL LEARNING THROUGH NEURAL NETWORKS: THE WAVE-NET SOLUTION

Alexandros Koulouris, Bhavik R. Bakshi,[1]
and George Stephanopoulos

Laboratory for Intelligent Systems in Process Engineering
Department of Chemical Engineering
Massachusetts Institute of Technology
Cambridge, Massachusetts 02139

Empirical learning is an ever-lasting and ever-improving procedure. Although neural networks (NN) captured the imagination of many researchers as an outgrowth of activities in artificial intelligence (AI), most of the progress was accomplished when empirical learning through NNs was cast within the rigorous analytical framework of the *functional estimation problem*, or *regression*, or *model realization*. Independently of the

---

[1]Present address: Ohio State University, Department of Chemical Engineering, Columbus, OH 43210.

name, it has been long recognized that, due to the inductive nature of the learning problem, to achieve the desired accuracy and generalization (with respect to the available data) in a dynamic sense (as more data become available) one needs to seek the unknown approximating function(s) in functional spaces of varying structure. Consequently, a recursive construction of the approximating functions at multiple resolutions emerges as a central requirement and leads to the utilization of wavelets as the basis functions for the recursively expanding functional spaces. This chapter fuses the most attractive features of a NN: representational simplicity, ability for universal approximation, and ease in dynamic adaptation, with the theoretical soundness of a recursive functional estimation problem, using wavelets as basis functions. The result is the *Wave-Net* (wavelets, network of), a multiresolution hierarchical NN with localized learning. Within the framework of a *Wave-Net* where adaptation of the approximating function is allowed, we have explored the use of the L^{∞} error measure as the design criterion. Within the framework of a *Wave-Net* one may cast any form of data-driven empirical learning to address a variety of modeling situations encountered in engineering problems such as design of process controllers, diagnosis of process faults, and planning and scheduling of process operations. This chapter will discuss the properties of a *Wave-Net* and will illustrate its use on a series of examples.

I. Introduction

Estimating an unknown function from its examples is a central problem in statistics, where it is known as the problem of *regression*. Under a different name but following the same spirit, the problem of *learning* has attracted interest in the AI research since the discovery that many intelligent tasks can be represented through functional relationships and, therefore, learning those tasks involves the solution of a regression problem (Poggio and Girosi, 1989). The significance, however, of the problem is by no means limited to those two fields. *Modeling*, an essential part of science and engineering, is the process of deriving mathematical correlations from empirical observations and as such, obeys the same principles as the regression or learning problem. Many methods developed for the solution of these problems are extensively used as tools to advance understanding and allow prediction of physical behavior.

In recent years the merging between the statistical and AI points of view on the same problem has benefited both approaches. Statistical regression techniques have been enriched by the addition of new methods and learning techniques have found in statistics the framework under which their properties can be studied and proved. This is especially true for neural networks (NNs), which are the main product of AI research in the field and the most promising solution to the problem of functional estimation. Despite their origination as biologically motivated models of the brain, NNs have been established as nonlinear regression techniques and studied as such. Their main features are their capability to represent nonlinear mappings, their ability to learn from data and finally their inherent parallelism which allows fast implementation. Every application where an unknown nonlinear function has to be reconstructed from examples is amenable to a NN solution. This explains the wide spread of NNs outside the AI community and the explosion of their applications in numerous disciplines and for a variety of tasks. In chemical engineering and especially in process systems engineering, NNs have found ground for applications in process control as models of nonlinear systems behavior (Narendra and Parthasarathy, 1990; Ungar *et al.*, 1990; Ydstie, 1990; Hernandez and Arkun, 1992; Bhat and McAvoy, 1990; Psichogios and Ungar, 1991; Lee and Park, 1992), fault diagnosis (Hoskins and Himmelblau, 1988; Leonard and Kramer, 1991), operation trend analysis (Rengaswamy and Venkatasubramaniam, 1991), and many other areas. In all these applications, the exclusive task, NNs are required to perform, is *prediction*. From the interpretation point of view, NNs are blackbox models. They fail to provide explicitly any physical insight simply because their representation does not coincide and is not motivated by the underlying phenomenon they model. The only expectation from their use is accuracy of their predictions, and this is the criterion on which the merits of their use can be judged.

The mathematical point of view on the analysis of NNs did not only aim at depriving them from their mystery related to their biological origin, but was also supposed to provide guarantees, if any, on their performance and guide the user with their implementation. Many theorems have been recruited to support the approximation capabilities of NNs. The universal approximation property has been proved for many different activation functions used in NNs such as sigmoids (Hornik *et al.*, 1989), or radial basis functions (RBFs) (Hartman *et al.*, 1990). At the same time, it was shown that such a property is quite general and not difficult to prove for many other sets of basis functions not used in NNs. Rates of convergence to the real function have also been derived (Barron, 1994), but these are limited to special conditions on the space of functions where the unknown function belongs. The central theme of all these theorems is that given a large enough network and enough data points any unknown function can be approximated arbitrarily well. Despite these theorems, the potential user is still puzzled by issues such as what type of network to implement, how many nodes or hidden layers to use, and how to interconnect them. The issues to these questions are still derived on empirical grounds or through a trial-and-error procedure. The theorems have, however, clearly shown that it is exactly these choices related to the above questions that determine the approximating capabilities and accuracy of a given network.

Under the new light that the mathematical analysis and practical considerations have brought, new directions in NN research have been revealed. The list of basis functions used in NNs expands steadily with new additions, and methods for overcoming the empiricism in determining the network architecture are explored (Bhat and McAvoy, 1992; Mavrovouniotis and Chang, 1992). Following this spirit, a novel NN architecture, the wavelet network or *Wave-Net* has been recently proposed by Bakshi and Stephanopoulos (1993). *Wave-Nets* use wavelets as their activation functions and exploit their remarkable properties. They apply a hierarchical localized procedure to evolve in their structure, guided by the local approximation error. *Wave-Nets* are data-centered, and all important decisions on their architecture are decided constructively by the data and that is their main attractive property.

In this study the problem of estimating an unknown function from its examples is revisited. Its mathematical description is attempted to map as closely as possible the practical problem that the potential NN user has to face. The objective of the chapter is twofold: (1) to draw the framework in which NN solutions to the problem can be developed and studied, and (2) to show how careful considerations on the fundamental issues naturally lead to the *Wave-Net* solution. The analysis will not only attempt to justify the development of the *Wave-Net*, but will also refine its operational characteristics. The motivation for studying the functional estimation problem is the derivation of a modeling framework suitable for process control. The applicability of the derived solution, however, is not limited to control implementations.

The remainder of this chapter is structured as follows. In Section II the problem of deriving an estimate of an unknown function from empirical data is posed and studied in a theoretical level. Then, following Vapnik's original work (Vapnik, 1982), the problem is formulated in mathematical terms and the sources of the error related to any proposed solution to the estimation problem are identified. Considerations on how to reduce these errors show the inadequacy of the NN solutions and lead in Section III to the formulation of the basic algorithm whose new element is the pointwise presentation of the data and the dynamic evolution of the solution itself. The algorithm is subsequently refined by incorporating the novel idea of structural adaptation guided by the use of the L^∞ error measure. The need for a multiresolution framework in representing the unknown function is then recognized and the wavelet transform is proposed as the essential vehicle to satisfy this requirement. With this addition, the complete algorithm is presented and identified as the modified *Wave-Net* model. Modeling examples (Section IV) demonstrate the properties of the derived solution and the chapter concludes with some final thoughts (Section V).

II. Formulation of the Functional Estimation Problem

In its more abstract form, the problem is to estimate an unknown function $f(\mathbf{x})$ from a number of, possibly noisy, observations $(\mathbf{x}_i, y_i)$, where $\mathbf{x}$ and y correspond to the vector of *input* or *independent variables* and the *output variable* of that function, respectively. To avoid confusion later on, the terms *regression*, *functional estimation* or *modeling* will all equivalently refer to the process of obtaining a *solution* or *approximating function* or *model* from the set of available data. The function $f(\mathbf{x})$ will be referred as the *real* or *target* function. The objective is to use the derived estimate to make predictions on the behavior of the real function in new, unseen situations. In almost all cases, the function $f(\mathbf{x})$ provides a mathematical description of a physically sound, but otherwise unknown, relationship between $\mathbf{x}$ and y. However, the form of the approximating function sought is in no way motivated by the underlying physical phenomenon and, consequently, from the point of view of interpretability, the proposed solutions are "blackbox" models. The only basic requirements for $f(\mathbf{x})$ related to its physical origination are that (1) it exists and (2) it is continuous with respect to all its arguments and that (3) the dimensionality of the input space is known.

Before attempting to formalize and solve the problem in mathematical terms, it is instructive to recognize the characteristics that are inherent to the functional estimation task and not attributable to the particular tool used to solve it. This will also set the ground on which different solutions to the problem can be analyzed and compared.

First, it can be easily recognized that finite data are always available within a bounded region of the input space. *Extrapolation* of any approximating function beyond that region is meaningful only when the model is derived by physical considerations applicable in the entire input space. In any other case, extrapolation is equivalent to postulating hypotheses that cannot be supported by evidence and, therefore, are both meaningless and dangerous. For that reason, we will confine the search for the unknown function in the fraction of the input space where data are available.

A very important characteristic of the problem is its *inductiveness*. The task is to obtain globally valid models by generalizing partial, localized and, many times, incorrect information. Because of its inductive nature, the problem inherits some disturbing properties. No solution can ever be guaranteed and the validity of any model derived by any means is not amenable to mathematical proofs. The model validation can only be perceived as a *dynamic* procedure triggered by any available piece of data and leading to an endless cycle between model postulating and model testing. This cycle can potentially prove the inefficiencies of any proposed model, but can never establish unambiguously its correctness.

Additional complications result from the fact that the functional estimation problem is *ill-posed* (Poggio and Girosi, 1989). Our only way to check whether a given function is a potential solution, is by measuring the *accuracy* of the fit for the available data. It is, however, clear that, for every given set of data points, there exists an infinite number of, arbitrarily different from the real, functions that can approximate the data arbitrarily well. Indeed, the implementation of every available tool for data regression, including NNs, will result in an equal number of different approximating functions. All these functions are equally plausible solutions to the functional estimation problem. The question that naturally arises is whether there exists some criterion based on which potential solutions can be screened out. What is certain is that the requirement for accuracy (independently of how it is mathematically measured) is not adequate in defining a unique solution to the problem and definitely not an appropriate basis for comparing the performance of various regression methods.

Where mathematical logic fails, intuition takes over. Intuitively, not all approximating functions for a given data set are equally plausible solutions to the functional estimation problem. The attribute that distinguishes the intuitively plausible from the unfavorable solutions is the *smoothness* of the approximating curve (in this context, smoothness does not refer to the mathematical property of differentiability). It is natural to seek the "best" solution in the face of the simplest, smooth-looking function that approximates well the available data. This conforms with a celebrated philosophical principle known as *Occam's razor*, which has found wide applications in the AI area (Kearns and Vazirani, 1994) and which, in simple terms, favors the shortest hypotheses that can explain the observations. In our context, shortness is equivalent to smoothness where the latter has so far only an intuitive connotation. For example, the approximating curve in Fig. 1a is a perfectly reasonable model of the data points, despite the fact that it differs considerably from the real function that produced those points. It is also remarkable to notice that, for that example and the given data, the

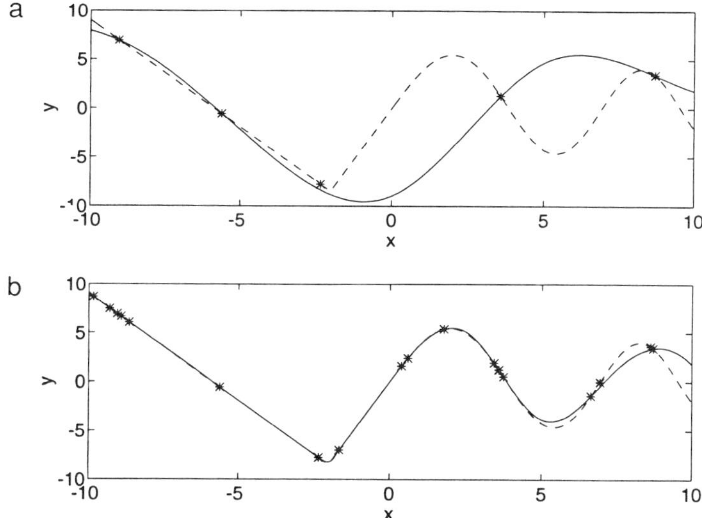

FIG. 1. Example of a functional estimation problem: evolution of the model (solid line) and comparison with the real function (dashed line) as more data (asterisks) become available.

real function is intuitively inferior (i.e., unnecessarily complicated) as an approximating function compared to its incorrect model. The real function becomes both the most accurate and smoothest (therefore, most favorable) solution when more reveal in data become available as it is the case in Fig. 1b. The fact that the model in Fig. 1a failed for additional data does not disqualify it as the "best" (in a weak, empirical sense) solution when only the initial points were available. The encouraging observation is that, by favoring intuitively simple approximating functions, the real function gradually emerges as it is evident by comparing the models in Figs. 1a and 1b.

The simple example in Fig. 1 indicates very clearly that the way the functional estimation problem should be attacked is by dynamically searching for the smoothest function that fits accurately the data. What we need is tools that will automatically produce approximating curves like the ones shown in Fig. 1. Of course, that will require us to first express the problem and its solution in mathematical terms and give precise mathematical meaning to the ambiguous attributes "accurate" and "smooth." The degree to which each proposed algorithm will prove able to mathematically represent those terms and appropriately satisfy the requirements for accuracy and smoothness will be the basis for judging its success.

A. Mathematical Description

Mathematically, the problem of deriving an estimate of a function from empirical data can be stated as follows (Vapnik, 1982):

1. Functional Estimation Problem

Given a set of data points $(\mathbf{x}_i, y_i) \in R^{k \times 1}$ $i = 1, 2, \ldots, l$ drawn with some unknown probability distribution, $P(\mathbf{x}, y)$, find a function $g^*(\mathbf{x})$: $X \to Y$ belonging to some class of functions G that minimizes the *expected risk functional*

$$I(g) = \int_{X \times Y} \mu[y, g(\mathbf{x})] P(\mathbf{x}, y) \, d\mathbf{x} \, dy, \tag{1}$$

where $\mu(\cdot, \cdot)$ is a metric in the space Y. We will restrict this definition to the case where y is given as a nonlinear function of $\mathbf{x}$ corrupted by some additive noise independent of $\mathbf{x}$:

$$y = f(\mathbf{x}) + d, \tag{2}$$

where d follows some unknown probability function $\mathbf{P}(d)$ and is also bounded: $|d| < \delta$. In that case, $P(\mathbf{x}, y) = P(\mathbf{x})P(y|\mathbf{x})$ and $P(y|\mathbf{x}) = \mathbf{P}[y - f(\mathbf{x})]$. The real function, $f(\mathbf{x})$, belongs to the space $C(X)$ of continuous functions defined in a closed hypercube $X = [a, b]^k$, where a and b are, respectively, the lower and upper bounds of the hypercube and k is the dimensionality of the input space. Therefore, $f(\mathbf{x})$ is bounded in both its L^2 and L^∞-norm:

$$\|f\|_2 = \int_X f^2(\mathbf{x}) \, d\mathbf{x} < \infty,$$

$$\|f\|_\infty = \sup_x |f(\mathbf{x})| < \infty.$$

No other a priori assumptions about the form or the structure of the function will be made. For a given choice of g, $I(g)$ in Eq. (1) provides a measure of the real approximation error with respect to the data in the entire input space X. Its minimization will produce the function $g^*(\mathbf{x})$ that is closest to G to the real function, $f(\mathbf{x})$ with respect to the, weighted by the probability $P(\mathbf{x}, y)$ metric μ. The usual choice for μ is the Euclidean distance. Then $I(g)$ becomes the L^2-metric:

$$I(g) = \int_{X \times Y} [y - g(\mathbf{x})]^2 P(\mathbf{x}, y) \, d\mathbf{x} \, dy \tag{3}$$

In an analogous way, $I(g)$ can be defined in terms of all L^n norms with

$1 \leq n < \infty$. With a stretch of the notation, we can make Eq. (1) correspond to the L^∞-metric:

$$I(g) = \sup_x |y - g(\mathbf{x})|, \tag{4}$$

which is a meaningful definition $[I(g) < \infty]$ only because we have assumed the noise to be bounded and, therefore, $y - g(\mathbf{x})$ is finite.

The minimization of the expected risk given by Eq. (1) cannot be explicitly performed, because $P(\mathbf{x}, y)$ is unknown and data are not available in the entire input space. In practice, an estimate of $I(g)$ based on the empirical observations is used instead with the hope that the function that minimizes the *empirical risk* $I_{emp}(g)$ (or *objective function*, as it is most commonly referred) will be close to the one that minimizes the real risk $I(g)$.

Let $(\mathbf{x}_i, y_i)i = 1, \ldots, l$ be the available observations. The empirical equivalents of Eqs. (3) and (4) are respectively

$$I_{emp}(g) = \sum_{i=1}^{l} [y_i - g(\mathbf{x}_i)]^2, \tag{5}$$

$$I_{emp}(g) = \max_{i=1,\ldots,l} |y_i - g(\mathbf{x}_i)|. \tag{6}$$

The magnitude of $I_{emp}(g)$ will be referred as the *empirical error*. All regression algorithms, by minimizing the empirical risk $I_{emp}(g)$, produce an estimate, $\hat{g}(\mathbf{x})$, which is the solution to the functional estimation problem.

a. Decisions Involved in the Functional Estimation Problem. As it was theoretically explained earlier, due to the inductive nature of the problem, the only requirements we can impose on the model, $\hat{g}(\mathbf{x})$ are accuracy and smoothness with respect to the data. Equations (5) and (6) provide two of the possible mathematical descriptions of the accuracy requirement. The smoothness requirement is more difficult to describe mathematically and, in fact, does not appear anywhere in the problem definition. We will postpone the mathematical description of smoothness for later. The important question now is what tools are available for controlling the accuracy and smoothness of the approximating function, $\hat{g}(\mathbf{x})$.

The mathematical description of the problem involves the following elements:

(a) The data, $(\mathbf{x}_i, y_i)i = 1, \ldots, l$
(b) The function space, G, where the solution is sought
(c) The empirical risk, $I_{emp}(g)$
(d) The algorithm for the minimization of $I_{emp}(g)$

The problem definition itself does not pose any restrictions on the decisions described above. For all practical purposes all these choices are arbitrary. That, in fact, demonstrates the ill-posedness of the problem, which results in a variety of solutions depending on the particular tool used and the specific choices that are forced.

The data are usually given a priori. Even when experimentation is tolerated, there exist very few cases where it is known how to construct good experiments to produce useful knowledge suitable for particular model forms. Such a case is the identification of linear systems and the related issue on data quality is known under the term *persistency of excitation* (Ljung, 1987).

The critical decisions in the modeling problem are related to the other three elements. The space G is most often defined as the linear span of a finite number, m, of basis functions, $\theta(\mathbf{x})$, each parametrized by a set of unknown *coefficients* $\mathbf{w}$ according to the formula

$$ G = \left\{ g(\mathbf{x}) | g(\mathbf{x}) = \sum_{k=1}^{m} c_k \theta_k(\mathbf{w}, \mathbf{x}), \mathbf{c} \in R^k, \mathbf{w} \in R^n \right\}, \tag{7} $$

where $\mathbf{c}$ is the vector of independent *weights*. All models we consider will be of the form given by Eq. (7). The number of basis functions in G will be referred as the *size of G*. The choice of the basis functions $\theta(\mathbf{x})$ is a problem of *representation*. Any a priori knowledge on the unknown function, $f(\mathbf{x})$, could effectively be exploited to reduce the space G where the model is sought. In most practical situations, however, it is unlikely that such a condition is available. Indeed, in the problem definition we have only assumed $f(\mathbf{x})$ to be continuous and bounded. That sets G equal to the L^∞ space, which, however, is not helpful at all, since the L^∞ space is already too large.

The important question is what restrictions the accuracy and smoothness requirements impose on the function space G, i.e., if smoothness and accuracy (a) can be achieved with any choice of G. (b) Can "optimally" be satisfied in *all* cases for a specific choice of G.
First, with respect to the type of basis functions used in G, smoothness is by no means restrictive. As it is intuitively clear and proved in practice, weird nonsmooth basis functions have to be excluded from consideration but beyond that, all "normal" bases are able to create smooth approximations of the available data. Accuracy is not a constraint either. Given enough basis functions, arbitrary accuracy for the prediction on the data is possible.

Both smoothness and accuracy are possible for given selection of the basis functions. It is, however, clear that a tradeoff between the two appears with respect to the size of G. Each set of basis functions resolves optimally this tradeoff for different sizes of G. This gives rise to an interesting interplay between the size of G and the number of data points. This is appropriately exemplified by the so-called *bias vs. variance* tradeoff in statistics (Barron, 1994), or *generalization vs. generality* tradeoff in the AI literature (Barto, 1991).

If G is "small," good generalization can be achieved (in the sense that with few degrees of freedom, the behavior of the approximating function in between the data points is restricted) but that introduces bias in the model selection. The solution might look smooth, but it is likely to exhibit poor accuracy of approximation even for the data used to derive it. This is, for example, usually the case when linear approximations are sought to nonlinear functions. Increased accuracy can be achieved by making G large enough. If that is the case, generality can also be achieved, since a large set of functions can potentially approximate well different sets of data. The problem is, however, that in a large G more than one and probably quite different functions in G are able to approximate well or even interpolate perfectly the data set. The variance of the potential solutions is large and the problem can be computationally ill-posed. With such a variety of choices, the risk of deriving a nonsmooth model is large. This is the problem most often encountered with the space of polynomials as basis functions and is also related to the problem of *overfitting*. Data are overfitted when smoothness is sacrificed for the sake of increased accuracy. Because overfitting usually occurs when a large number of basis functions are used for fitting the data, the two have become almost synonymous. It should be noted, however, that this might not always be the case. There can be sets of basis functions (like the piecewise linear Haar basis) that are so poor as approximation schemes, that many of them are required to fit decently (but not overfit) a given set of data. On the other hand, the newly developed fractal-like functions (see, for example, the wavelets created by Daubechies, 1992) provide examples of structures that even in small numbers give rise to approximating functions that make the data look definitely overfitted.

The conclusion is that for every particular set of basis functions and given data, there exists an "appropriate" size of G that can approximate both accurately and smoothly this data set. A decisive advantage would be if there existed a set of basis functions, which could probably represent *any* data set or function with *minimal complexity* (as measured by the number of basis functions for given accuracy). It is, however, straightforward to construct different examples that acquire minimal representations with respect to different types of basis functions. Each basis function for itself is the most obvious positive example. A Gaussian (or discrete points from a Gaussian) can minimally be represented by the Gaussian itself as the basis function and so on. It is, consequently, hopeless to search for the unique *universal approximator*. The question boils down to defining for each selected basis and data the appropriate size of G that yields both an accurate and smooth approximation of the data.

2. Sources of the Generalization Error

The model, $\hat{g}(\mathbf{x})$, that any regression algorithm produces will, in general, be different from the target function, $f(\mathbf{x})$. This difference is expressed by the *generalization error* $\|f(\mathbf{x}) - \hat{g}(\mathbf{x})\|$, where $\|\cdot\|$ is a selected functional norm. The generalization error is a measure of the prediction accuracy of the model, $\hat{g}(\mathbf{x})$, for unseen inputs. Although this error cannot in practice by measured (due to the inductive character of the problem), the identification of its sources is important in formulating the solution to the problem.

There are two sources to the generalization error (Girosi and Anzellotti, 1993):

1. The *approximation error* that stems from the "finiteness" of the function space, G
2. The *estimation error* due to the finiteness of the data.

The unknown function, $f(\mathbf{x})$ is an infinite-dimensional object requiring an infinite number of basis functions for an exact representation. In Fourier analysis, any square-integrable function is exactly represented by an infinite number of trigonometric functions. Similarly, a polynomial of infinite degree is needed for an exact reconstruction of any generic function. On the other hand, the approximating space G, for computations to be feasible, can be spanned only by a finite set of basis functions and characterized only by a finite number of adjustable parameters. Independently of the basis used, the size o G has to be finite. The selection of any function from G [including the "best" approximating function $g^*(\mathbf{x})$] will inevitably introduce some approximation error compared to the target function. Even worse, due to the finiteness of the available data that does not allow the minimization of $I(g)$, a suboptimal solution, $\hat{g}(\mathbf{x})$, rather than $g^*(\mathbf{x})$, will be derived as the model. The difference between $\hat{g}(\mathbf{x})$ and $g^*(\mathbf{x})$ corresponds to the estimation error which further increases the generalization error. To eliminate the approximation error, G has to be infinite-dimensional. Similarly, infinite number of data are required to avoid the estimation error. Both requirements are impractical and that forces us to an important conclusion that the expectation to perfectly reconstruct the real function has somehow to be relaxed.

B. Neural Network Solution to the Functional Estimation Problem

For an introduction to NNs and their functionality, the reader is referred to the rich literature on the subject (e.g., Rumelhart *et al.*, 1986; Barron and Barron, 1988). For our purposes it suffices to say that NNs represent nonlinear mappings formulated inductively from the data. In doing so, they offer potential solutions to the functional estimation problem and will be studied as such.

Mathematically, NNs are weighted combinations of activation functions (units), $\theta(\mathbf{x})$, parametrized by a set of unknown coefficients. In that respect, they fall exactly under the general model form given by Eq. (7). The use of all NNs, for deriving an approximating function $g(\mathbf{x})$ given a set of data, follows a common pattern summarized by the following general NN algorithm:

Step 1. Select a family of basis functions, $\theta(\mathbf{x})$.

Step 2. Use some procedure to define G (the number of basis functions, m, and possibly the coefficients, $\mathbf{w}$).

Step 3. Learning Step: Solve the objective function minimization problem with respect to the weights $\mathbf{c}$ (and the coefficients, $\mathbf{w}$, if not defined in the previous step).

Although the minimization of the objective function might run to convergence problems for different NN structures (such as backpropagation for multilayer perceptrons), here we will assume that step 3 of the NN algorithm unambiguously produces the best, unique model, $\hat{g}(\mathbf{x})$. The question we would like to address is what properties this model inherits from the NN algorithm and the specific choices that are forced.

1. Properties of the Approximations

In recent years some theoretical results have seemed to defeat the basic principle of induction that no mathematical proofs on the validity of the model can be derived. More specifically, the *universal approximation* property has been proved for different sets of basis functions (Hornik *et al.*, 1989, for sigmoids; Hartman *et al.*, 1990, for Gaussians) in order to justify the bias of NN developers to these types of basis functions. This property basically establishes that, for every function, there exists a NN model that exhibits arbitrarily small generalization error. This property, however, should not be erroneously interpreted as a guarantee for small generalization error. Even though there might exist a NN that could approximate well any given function, once the NN structure is fixed (step 2), the universality of the approximation is lost. Any given NN can approximate well only a limited set of functions, and it is only sheer luck that determines whether the unknown function belongs in that set. On the other hand, the universal approximation property is quite general and can be proved for different sets of basis functions, such as polynomials or sinusoids, even if they are not used in NNs. Basically, every nonlinearity can be used as a universal approximator (Kreinovich, 1991). This conforms with our earlier conclusion that all "normal" basis functions are eligible as approximation schemes, and no one can be singled out as an outstanding tool. As the study of the NN history reveals, the use of most bases in NNs has been motivated by other than mathematical reasons.

2. Error Bounds

Furthermore, there exist some theoretical results (Barron, 1994; Girosi, 1993) which provide a priori order of magnitude bounds on the generalization error as a function of the size of the network and the number of the data points. For all these results, however, an assumption on the function space, where the target function belongs, is invariably made. Since it is unlikely that the user will have a priori knowledge on issues like the magnitude of the first Fourier moment of the real function (Barron, 1994) or if the same function is of a Sobolev type (Girosi, 1993), the convergence proofs and error bounds derived under those assumptions have limited practical value. It can be safely concluded that NNs offer no more guarantees with respect to the magnitude of the generalization error than any other regression technique.

The next question is whether the construction of NNs at least guarantees accurate and smooth approximating functions with respect to the available data. The critical issue here is, given a specific selection of the basis functions, how their number is determined (step 2). The established practice reveals that the selection of the NN size is based on either empirical grounds or statistical techniques. These techniques use different aspects of the data, such as the size of the sample or its distribution in the input space, to guide the determination of the network structure. Some techniques, such as the k-means clustering algorithm (Moody and Darken, 1989), are essentially reflections of the empirical rule that the basis functions should be placed in proportion to the data density in the input space. That is an intuitive argument, but also easy to defy. Large density in the input space does not necessarily entail the existence of a complex feature of the underlying function requiring many basis functions to be represented.

It is interesting to notice that the decision on the network size is taken in the absence of any consideration on both the accuracy and the smoothness of the resulting model. On one hand, this might lead to unnecessary complexity and nonsmoothness of the model if the size is larger than needed. This is the case with the regularization approach to radial basis function networks (RBFNs) (Poggio and Girosi, 1989), which assigns a basis function to every data point available. On the other hand, because of the arbitrariness in structuring the network, the accuracy can be poor. The discovery of the network inefficiency, though some validation technique using additional empirical (testing) data, results to a repeated cycle of trial-and-error tests between steps 2 and 3 of the NN algorithm. This is a strong indication that the NN algorithm does not provide in the first place any guarantees of accuracy and smoothness for the approximating function.

Finally, it would be interesting to see if, with an increasing number of data and allowing the NN to find the optimal coefficient values, the real function can ever be closely approximated. Because of the arbitrariness of the NN structuring, the approximation error for the selected network can be arbitrarily large. As long as adaptation of the structure is not allowed, the gap between the model and the real function can never close. The adaptation of the coefficient values (reduction of the estimation error) is certainly not the remedy, if the initial selection of G is inappropriate. It should be noted that there exists some work (e.g., Bhat and McAvoy, 1992) on the problem of optimally structuring the network. However, the problem is dealt at the initial stage (step 1) and the proposed solutions do not include any considerations for structural adaptation at the learning stage (step 3). In most of the cases, the definition of the network architecture involves few initial data points and the solution of a considerable optimization problem that is difficult to duplicate in an on-line fashion when more data are available. As long as the network structuring becomes a static (with respect to the data) decision, the preceding considerations still hold independently of the method used for determining the NN structure.

III. Solution to the Functional Estimation Problem

A. FORMULATION OF THE LEARNING PROBLEM

The analysis of NNs has shown that, in order to assure both accuracy and smoothness for the approximating function, the solution algorithm will have to allow the model (essentially its size) to evolve *dynamically* with the data. This calls not only for a new algorithm but also for a new problem definition. The new element is the stepwise presentation of the data to the modeling algorithm and the dynamic evolution of the solution itself. All elements in the problem have to be indexed by a time-like integer l, which will correspond to the amount of data available and that, by abuse of notation, will be referred as the *instant*. The new problem is called the *learning problem* and is defined as follows:

Learning Problem.   At every instant l let $Z_l = \{(\mathbf{x}_i, y_i) \in R^{k \times 1} i = 1, 2, \ldots, l\}$ be a set of data drawn with some unknown probability distribution, $P(\mathbf{x}, y)$ and G_l, a space of functions. Find a function $\hat{g}_l(\mathbf{x}): X \to Y$ belonging to G_l that minimizes the empirical risk $I_{\text{emp}}(g)$.

There is a "time" continuity of all the variables in the problem, which is expressed by the following relations:

(a) $Z_l = Z_{l-1} \cup \{(\mathbf{x}_l, y_l)\}$
(b) $G_l \supseteq G_{l-1}$
(c) $G_l = A(G_{l-1})$, where A is the algorithm for the adaptation of the space, G_l
(d) $\hat{g}_l(\mathbf{x}) = M(Z_l, G_l)$, where M is the algorithm for the generation of the model, $\hat{g}_l(\mathbf{x})$

The first relation represents the data buildup, while the next two reassure that the space G_l evolves over its predecessor, G_{l-1} according to algorithm A. At every instant, the model results by applying algorithm M on the available data, Z_l, and the chosen input space, G_l.

With such an approach to the functional estimation problem, we escape from the cycle of deriving models and testing them. With every amount of data available, the simplest and smoothest model is sought. Every new data point builds constructively on the previous solution and does not defy it, as it is the case with statistical techniques such as cross-validation. The expectation is that eventually a function close to the real one will be acquired. The procedure is ever-continuing and is hoped to be ever-improving. Not unjustifiably, we call such a procedure *learning* and the computational scheme that realizes it, a *learning scheme*. The learning scheme consists of four elements:

1. A dynamic memory of previous data, Z_l
2. An approximating function or model, $\hat{g}_l(\mathbf{x})$
3. A built-in algorithm A for the adaptation of the space, G_l
4. A built-in algorithm M for the generation of the model, $\hat{g}(\mathbf{x})$

Independently of the amount of data and the way they are acquired, the definition of the learning problem simulates a continuous data flow. Such a framework is of particular importance for the cases where the data are indeed available sequentially in time. Process modeling for adaptive control is one important application, which fits perfectly into this framework. The very nature of feedback control relies on and presupposes the explicit or implicit measurement of the process inputs and outputs; therefore, a continuous *teacher* to the process model is available. Accurate process modeling is essential for control. However, process identification is a hard problem. The derivation of a suitable mechanism which can fully exploit the continuous flow of useful information has been our main motivation for developing the learning framework itself and its solution.

1. Derivation of Basic Algorithm

The solution of the learning problem requires the construction of an algorithm, A, for adaptation of the structure and that in turn presupposes the selection of a proper criterion, which will indicate the need for adaptation. Unlike NNs, such a criterion will have to be based on the empirical error to avoid the dangers of overdetermining the size of the network, which might cause overfitting, or underdetermining it, which will result in large approximation error. The framework calls for an appropriately defined *threshold* on the empirical error, as the criterion which will serve as a guide for the structural adaptation.

We are now ready to propose the basic algorithm for the solution of the learning problem.

Basic Algorithm

Step 1. Select a family of basis functions, $\theta(\mathbf{x})$.

Step 2. Select a form for the empirical risk, $I_{\mathrm{emp}}(g)$, and establish a threshold, ε, on the empirical error.

Step 3. Select a minimal space, G_0.

Step 4. Learning step: For every available data point, apply algorithm M to calculate the model and estimate the empirical error on all available data. If it exceeds the defined threshold, use algorithm A to update the structure of G until threshold is satisfied.

In the remaining sections, specific answers will be given for the selection of the

(a) Functional space, G.

(b) Basis functions $\theta(\mathbf{x})$.

(c) The empirical risk, $I_{\mathrm{emp}}(g)$, and the corresponding error threshold, ε.

With every specification of the above parameters, the basic algorithm will be refined and its properties will be studied, until the complete algorithm is revealed. However, each of the presented algorithms can be considered as a point of departure, where different solutions from the finally proposed can be obtained.

2. Selection of the Function Space

The solution to the learning problem should provide the flexibility to search for the model in increasingly larger spaces, as the inadequacy of the smaller spaces to approximate well the given data are proved. This immediately calls for a hierarchy in the space of functions. Vapnik (1982) has introduced the notion of *structure* as an infinite ladder of finite-dimensional nested subspaces:

$$S_0 \subset S_1 \subset S_2 \subset \cdots S_n \subset \cdots. \tag{8}$$

For reasons that will become clear later on, we will refer to the index j of each subspace as the *scale*. For continuity to be satisfied, the basis functions that span each subspace S_j have to be a superset of the set of basis functions spanning the space S_{j-1}. Vapnik solves the minimization problem in each subspace and uses statistical arguments to choose one as the final solution to the approximation problem for a given set of data. He calls his method *structural minimization*. Although we have adopted his idea of the structure given by Eq. (8), we have implemented it in a fairly different spirit. At every instant, the space G_l, where the model is sought, is set equal to one of the predefined spaces S_j. With more data available and depending on the prediction accuracy, the model is sought in increasingly larger spaces. In this way, we establish a strictly forward move into the structure and allow the incoming data dictate the pace of the move.

One example of a structure (8) is the space of polynomials, where the ladder of subspaces corresponds to polynomials of increasing degree. As the index j of S_j increases, the subspaces become increasingly more complex where complexity is referred to the number of basis functions spanning each subspace. Since we seek the solution at the lowest index space, we express our bias toward simpler solutions. This is not, however, enough in guaranteeing smoothness for the approximating function. Additional restrictions will have to be imposed on the structure to accommodate better the notion of smoothness and that, in turn, depends on our ability to relate this intuitive requirement to mathematical descriptions.

Although other descriptions are possible, the mathematical concept that matches more closely the intuitive notion of smoothness is the *frequency* content of the function. Smooth functions are sluggish and coarse and characterized by very gradual changes on the value of the output as we scan the input space. This, in a Fourier analysis of the function, corresponds to high content of low frequencies. Furthermore, we expect the frequency content of the approximating function to vary with the position in the input space. Many functions contain high-frequency features dispersed in the input space that are very important to capture. The tool used to describe the function will have to support *local* features of *multiple resolutions* (variable frequencies) within the input space.

Smoothness of the approximating function with respect to the data forces analogous considerations. First, we expect the approximating function to be smoother where data are sparse. Complex behavior is not justified in a region of the input space where data are absent. On the other hand, smoothness has to be satisfied in a dynamic way with every new data point. The model adaptation, triggered by any point, should not extend beyond the neighborhood of "dominance" of the new data point and, at the same time, should not be limited to a smaller area. It is not justifiable to change the entire model every time one point is not predicted accurately by the model, nor is it meaningful to induce so local changes that only a single point is satisfied. Again the need for *localization both in space and frequency* shows up. This is not surprising, since the very nature of the problem is a game between the localization of the data and the globality of the sought model. The problem itself forces us to think in multiple resolutions.

Here is how we can achieve a multiresolution decomposition of the input space. Let $I_0 = [a, b]^k$ be the entire bounded input space according to the problem definition. In each dimension, the interval $[a, b]$ is divided into two parts that are combined into 2^k subregions, $I_{1,\mathbf{p}}$, where $\mathbf{p}$ is k-dimensional vector that signifies the position of the subregion and whose elements take the values 1 or 2. Let I_1 signify the set of all these subregions: $I_1 = \{I_{1,\mathbf{p}} | \mathbf{p} = (q, r), q, r = 1 \text{ or } 2)$, which will be referred as the scale 1 *decomposition* of the input space. The union of all these subregions forms the entire input space: $\cup I_{1,\mathbf{p}} = I_0$. the subdivision of each subregion in I_1 follows the same pattern, resulting in a *hierarchy* of decompositions $I_0, I_1, I_2, \ldots$ each corresponding to a division of the input space in a different resolution. An example of the resulting ladder of subregions is given in Fig. 2 for the two-dimensional case.

There exists a one-to-one correspondence between the structure of subspaces and the hierarchy of input space decompositions. This already

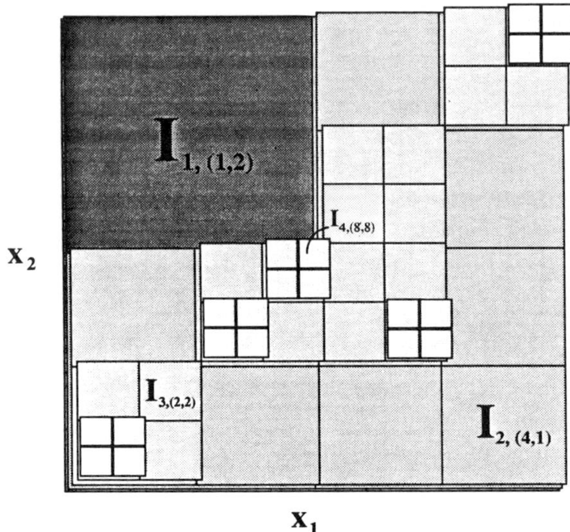

FIG. 2. Multiresolution decomposition of the input space.

poses a restriction on the basis functions that span the subspaces. They have to be local and have bounded support. Such a requirement excludes popular in regression functional bases such as polynomials and sigmoids that are global. Each subregion in I_j corresponds to the support of a basis function in S_j. In this way, the problem is decomposed into smaller subproblems where it is solved almost independently depending on the degree of overlap of the functions spanning those regions. The extreme case is when there is no overlap such as when piecewise approximation in each subregion is used. In this analysis, we have adopted for simplicity a predefined dyadic subdivision of the input space. However, this is not a unique choice. Different formulas can be invented to break the dyadic regularity and even formulate the grid of subregions not in a predefined way but in parallel with the data acquisition. The additional requirement for the basis functions used is the construction of the basis to support this localized multiresolution structure.

Given a multiresolution decomposition of the input space, the algorithm A for the adaptation of the model structure, identifies the region $I_{j+1,\mathbf{p}}$ in the next scale where the incoming data point belongs and adds the corresponding basis function that spans this region. With this addition, the basic algorithm can now be modified as follows.

Algorithm 1

Step 1. Select a family of basis functions, $\theta(\mathbf{x})$, supporting a multiresolution decomposition of the input space.

Step 2. Select a form for the empirical risk, $I_{emp}(g)$, and establish a
threshold, ε, on the empirical error.

Step 3. Select a minimal space, G_0, and input space decomposition, I_0.

Step 4. Learning step: For every available data point, apply algorithm
M to calculate the model and estimate the empirical error on
all available data. If it exceeds the defined threshold, use
algorithm A to update the structure of G, by inserting the
lowest scale basis function that includes the new point. Repeat
until threshold is satisfied.

3. Selection of the Error Threshold

Algorithm 1 requires the a priori selection of a threshold, ε, on the
empirical risk, $I_{emp}(g)$, which will indicate whether the model needs
adaptation to retain its accuracy, with respect to the data, at a minimum
acceptable level. At the same time, this threshold will serve as a termination criterion for the adaptation of the approximating function. When (and
if) a model is reached so that the generalization error is smaller than ε,
learning will have concluded. For that reason, and since, as shown earlier,
some error is unavoidable, the selection of the threshold should reflect our
preference on how "close" and in what "sense" we would like the model
to be with respect to the real function.

Given a space G, let $g^*(\mathbf{x})$ be the "closest" model in G to the real
function, $f(\mathbf{x})$. As it is shown in Appendix 1, if $f \in G$ and the L^∞ error
measure [Eq. (4)] is used, the real function is also the best function in G,
$g^* = f$, independently of the statistics of the noise and as long as the noise
is symmetrically bounded. In contrast, for the L^2 measure [Eq. (3)], the
real function is not the best model in G if the noise is not zero-mean. This
is a very important observation considering the fact that in many applications (e.g., process control), the data are corrupted by non-zero-mean
(load) disturbances, in which cases, the L^2 error measure will fail to
retrieve the real function even with infinite data. On the other hand, as it
is also explained in Appendix 1, if $f \notin G$ (which is the most probable
case), closeness of the real and "best" functions, $f(\mathbf{x})$ and $g^*(\mathbf{x})$, respectively, is guaranteed only in the metric that is used in the definition of
$I(g)$. That is, if $I(g)$ is given by Eq. (3), $g^*(\mathbf{x})$ can be close to $f(\mathbf{x})$ only in
the L^2-sense and similarly for the L^∞ definition of $I(g)$. As is clear, L^2

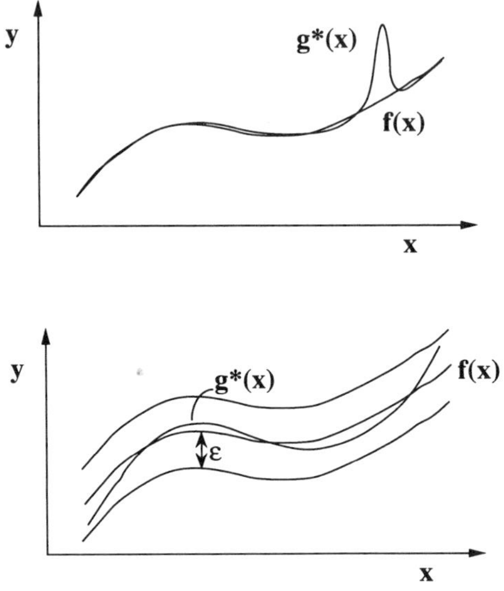

FIG. 3. L^2 vs. L^∞ closeness of functions.

closeness does not entail L^∞ closeness and vice versa. The question now is what type of closeness is preferable.

Figure 3 gives two examples of L^2 and L^∞ closeness of two functions. The L^2 closeness leaves open the possibility that in a small region of the input space (with, therefore, small contribution to the overall error) the two functions can be considerably different. This is not the case for L^∞ closeness, which guarantees some minimal proximity of the two functions. Such a proximity is important when, as in this case, one of the functions is used to predict the behavior of the other, and the accuracy of the prediction has to be established on a pointwise basis. In these cases, the L^∞ error criterion (4) and its equivalent [Eq. (6)] are superior. In fact, L^∞ closeness is a much stricter requirement than L^2 closeness. It should be noted that whereas the minimization of Eq. (3) is a quadratic problem and is guaranteed to have a unique solution, by minimizing the L^∞ expected risk [Eq. (4)], one may yield many solutions with the same minimum error. With respect to their predictive accuracy, however, all these solutions are equivalent and, in addition, we have already retreated from the requirement to find the one and only real function. Therefore, the multiplicity of the "best" solutions is not a problem.

There is one additional reason why the L^∞ empirical risk is a better objective function to use. With the empirical risk given by Eq. (6), which is by definition a pointwise measure, it is clear how to define in practice the numerical value of the threshold that will be applied to every data point to assess the model accuracy. That would not be the case if the L^2 error were used whose absolute magnitude is difficult to correlate with the goodness of the approximation.

In light of the previous discussion and contrary to the established practice, we propose the use of the maximum absolute error (6) as the empirical risk to be minimized because it offers the following advantages:

1. As long as the noise is symmetrically bounded, the L^∞ error is minimized by the true function, $f(\mathbf{x})$, independently of the statistics of the noise.
2. Minimization of the L^∞ error ensures L^∞ closeness of the approximating function to the real function.
3. It is straightforward to define a numerical value for the threshold on the L^∞ error.

The first two points raise the question of convergence of the model to the real function, or its best model, $g^*(\mathbf{x})$. In the following analysis we will address the convergence question, first assuming that the model with the desired accuracy lies in a known space G and then when this space is attempted to be reached with the hierarchical adaptation procedure of Algorithm 1.

Given a choice for G, the question is, if and under which conditions the model, $\hat{g}(\mathbf{x})$, converges to $g^*(\mathbf{x})$ as the number of points increases to infinity, or, in other words, if it is possible to completely eliminate the estimation error. The answer will emerge after addressing the following questions.

4. As the Sample Size Increases, Does $I_{emp}(g)$ Converge to $I(g)$?

Under very general conditions, it is true that the empirical mean given by Eq. (5), converges to the mathematical expectation (3), as $l \to \infty$ (Vapnik, 1982). On the other hand, it is clear that

$$\max_{i=1,\ldots,l} |y_i - g(\mathbf{x}_i)| \to \sup_x |y - g(\mathbf{x})| \quad \text{as} \quad l \to \infty$$

and convergence of $I_{emp}(g)$, given by Eq. (4) to $I(g)$, given by Eq. (6), is also guaranteed.

5. Does Convergence of $I_{emp}(g)$ to $I(g)$ Entail Convergence of $\hat{g}$ to g^*?

If the answer were positive, that would ensure that, as we minimize the $I_{emp}(g)$ and more data become available, we approach the best approximating function, $g^*(\mathbf{x})$. Unfortunately, in general, this is not true and that means that as more data become available the estimation error does not necessarily decrease. Convergence can, however, be proved in a weaker sense for the L^∞ case, if we retreat from the requirement to acquire $g^*(\mathbf{x})$ at infinity and be satisfied with *any* solution that satisfies an error bound in its L^∞-norm distance from the real function $f(\mathbf{x})$.

Theorem 1.   *Let $\varepsilon > 0$ and $G_\varepsilon = \{g \in G \mid I(g) < \varepsilon\}$ be a nonempty subset of G and ε, a small positive number. The minimization of the empirical risk [Eq. (6)] will converge in G_ε as $l \to \infty$.*

The proof is given in Appendix 2. The theorem basically states that if there exists a function in G that satisfies $I(g) < \varepsilon$, then by minimizing the L^∞ error measure, this function will be found. The generalization error for this function is guaranteed to be less than $\varepsilon - \delta$. The theorem, however, presupposes the existence of the function with the desired accuracy. This, in general, cannot be guaranteed a priori. In the following, we will prove that the adaptation methodology developed for Algorithm 1 will converge to a subspace where the existence of such a function is guaranteed. For the proof, we have to make the following assumptions for the ladder of subspaces defined by Eq. (8):

Assumption 1.   At every subspace S_{j2} there exists a better approximating function than any function in S_{j1} with $j_2 > j_1$:

$$\forall j_2 > j_1, \exists g' \in S_{j2} \quad \text{such that} \quad I(g') < I(g) \ \forall g \in S_{j1}.$$

Assumption 2.   There exists a subspace S_{j*} with $j^* < \infty$ where a function with the desired approximation accuracy exists:

$$\exists j^* \quad \text{and} \quad g^* \in S_{j*} \quad \text{such that} \quad I(g^*) < \varepsilon,$$

or, in other words, G_ε in S_{j*} is nonempty.

On the basis of these assumptions we can prove the following theorem:

Theorem 2.   *Algorithm 1 will converge to S_{j*}, and no further structural adaptation will be needed for any additional data points.*

The proof is given in Appendix 3. Theorem 2 indicates that by forcing $I_{\text{emp}}(g)$ to be less than ε at all times and by following Algorithm 1, we are guaranteed to reach a subspace S_{j*} where a solution with $I(g) < \varepsilon$ exists. According to Theorem 1, the algorithm will converge into the set G_ε and all subsequent solutions to the minimization of $I_{\text{emp}}(g)$ will obey $I(g) < \varepsilon$ which is the criterion we want to be satisfied. The generalization error will be $\varepsilon - \delta$. Therefore, by defining an error threshold $\varepsilon + \delta$, we will eventually reach a solution with a generalization error less than ε. The signficance of Algorithm 1 according to Theorem 2, is that it provides a formal way to screen out functional spaces where solutions with the desired accuracy are not present, and that it eventually concludes on the smaller subspace where such a solution exists. This result is possible by the use of the L^∞ error measure and the structural adaptation procedure. It should be reminded that if the space G were to be determined a priori and statically (as it is the case for NNs), such a convergence would not be possible unless, luckily, that space already contained a function with the desired accuracy.

It should be noted, however, that the convergence result does not entail a monotonic decrease of the generalization error. The algorithm moves to spaces where potentially better solutions can be obtained, but the solution that it chooses is not necessarily better than the previous one. This is too much to ask and indeed no approximation scheme, whether it applies structural adaptation or not, can guarantee strict improvement as the number of data points increases to infinity. On the other hand, no reassurance on the number of data needed for convergence in S_{j*} and G_ε are provided. This issue is related to the data quality, i.e., the degree to which the incoming data reveal the inadequacies of the existing model. If the data are not "good enough," convergence can be delayed beyond the point where data are available. In this case, however, the solution is still satisfactory as long as, by construction, it serves properly the accuracy—smoothness tradeoff, which, as stated earlier, is the only criterion in assessing the goodness of the approximation.

We can now revise Algorithm 1 with the definition of the error threshold.

Algorithm 2

    Step 1. Select a family of basis functions, $\theta(\mathbf{x})$, supporting a mutliresolution decomposition of the input space.

    Step 2. Select a threshold ε, with $\varepsilon > 0$ and $\varepsilon > \delta$, on the empirical error defined by Eq. (6).

    Step 3. Select a minimal space, G_0, and input space decomposition, I_0.

    Step 4. Learning step: For every available data point, apply algorithm M to calculate the model by solving the minimization problem, $\hat{g} = \min_g I_{emp}(g) = \min_g \max_{i=1,\ldots,l} |y_i - g(x_i)|,\ g_j \in S_j$, and estimate the empirical error on all available data, $I_{emp}(g)$. If $I_{emp}(g) > \varepsilon$, use algorithm A to update the structure of G, by inserting the lowest scale basis function that includes that point. Repeat until threshold is satisfied.

6. Representation of the Functional Spaces

The only element of the learning algorithm that remains undetermined is the basis functions that span the subspaces according to Eq. (8). The basic requirement for the basis functions are to be local, smooth and to have adjustable support to fit the variable size subregions of the input space. The Gaussian function, for example, with variable mean and variance has all the required properties. The convergence theorems stated in the previous section, however, show that the basis used is also required to exhibit well-defined properties as an approximation scheme. This stems from the requirements (implied by Assumptions 1 and 2) that, as we move on to higher index subspaces, we increase our ability to fit better not only the data (which is the easy part) but the unknown function as well. This is a major difference of this approach with other algorithms, such as the resource-allocating network (RAN) algorithm (Platt, 1991) where on-line adaptation of the structure is performed but not in a well-defined framework that will guarantee improved approximation capabilities of the augmented network. The additional approximation properties are sought in the framework of *multiresolution analysis* and the *wavelet representation*. For details on both these tools, the reader is referred to the work of Mallat (1989), Daubechies (1992), Strang (1989), and also Bakshi and Stephanopoulos (1993) for a NN-motivated treatment of wavelets. Here, we will state briefly the basic facts that prove that the wavelet transform indeed fits the developed framework and will emphasize more its properties as an approximation scheme.

a. Multiresolution Analysis. Let $f(x) \in L^2(R)$ be a one-dimensional function of finite energy. Following Mallat (1989), a multiresolution analysis of $f(x)$ is defined as a sequence of successive approximations of $f(x)$ resulting from the projection of $f(x)$ on finer and finer subspaces of functions, S_j. Le $f(x)$ be the approximation of $f(x)$ at resolution 2^j or scale j. As the scale decreases, $f^j(x)$ looks coarser and smoother and devoid of any detail of high frequency. Inversely, with the scale increasing, the approximation looks finer and finer until the real function is recovered at infinite scale:

$$\lim_{j \to \infty} f^j(x) = f(x). \tag{9}$$

Multiresolution analysis conforms with the definition of structure (8). More importantly, it guarantees that by moving to higher subspaces (scales), better approximations of the unknown functions can potentially be obtained, which is the additional property sought.

Regarding the representation of the subspaces, S_j, there exists a unique function $\phi(x)$, called the *scaling function*, whose translations and dilations span those subspaces:

$$S_j = \left\{ \sum_n c_{jn}\phi_{jn}(x) \,|\, j, n \in Z, c_{jn} \in R \right\}, \quad \text{where}$$

$$\phi_{jn}(x) = \sqrt{2^j}\,\phi\!\left(2^j x - n\right) \tag{10}$$

The scaling function, $\phi(x)$, has either local support or decays very fast to zero. For all practical purposes, it is a local function. By translating and dilating that function we are able to cover the entire input space in multiple resolutions, as it is required.

The framework, however, as introduced so far is of little help for our purpose since the shift from any subspace to its immediate in hierarchy would require to change entirely the set of basis functions. Although $\phi_{jn}(x)$ are all created by the same function, they are different functions and, consequently, the approximation problem has to be solved from scratch with any change of subspace. The theory of wavelets and its relation to multiresolution analysis provides the ladder that allows the transition from one space to the other.

It can be proved that the approximation $f^{j+1}(x)$ of $f(x)$ in scale $j+1$ can be written as a combination of its approximation at the lower scale j with additional detail represented by the wavelet transform at the same scale:

$$S_{j+1} = \left\{ \sum_n c_{jn}\phi_{jn}(x) + \sum_n d_{jn}\psi_{jn}(x) \,|\, j, n \in Z, c_{jn}, d_{jn} \in R \right\}, \quad \text{where}$$

$$\psi_{jn}(x) = \sqrt{2^j}\,\psi\!\left(2^j x - n\right) \tag{11}$$

and $\psi(x)$ is the *wavelet function* determined in unique correspondence to the scaling function. With this addition, the shift to higher scales for improved approximation involves the addition of a new set of basis functions that are the wavelets at the same scale and whose purpose is to capture the high-frequency detail ignored in the previous scale. This analysis can be easily extended to dimensions higher than one.

There exist different pairs of wavelets and scaling functions. One such pair is shown in Fig. 4. This is the "Mexican hat" pair (Daubechies, 1992), which draws its name by the fact that the scaling function looks like the

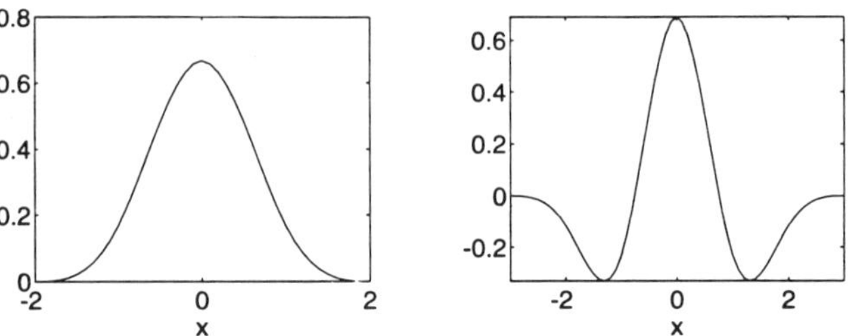

FIG. 4. "Mexican hat" scaling function and wavelet.

Gaussian and the wavelet like the Gaussian second derivative ("Mexican hat" function).

b. Wavelet Properties. All wavelets and scaling functions share the property of *localization in space and in frequency*, as required. Although such a property (in the weak sense that we are using it as a means to represent the smoothness requirement) can be satisfied with other functions as well, it is only with wavelets that it finds an exact mathematical description. The Fourier transform of the wavelets is a bandpass filter. The higher the scale, the more the passband extends over higher frequencies. That explains how the wavelets of high scales are able to capture the fine, high-frequency detail of any function they represent. The combination of space and frequency localization is a unique property for bases of $L^2(R)$ and it is what made wavelets a distinguished mathematical tool for applications where a multiresolution decomposition of a function or a signal is required.

As approximation schemes, wavelets trivially satisfy the Assumptions 1 and 2 of our framework. Both the L^2 and the L^∞ error of approximation is decreased as we move to higher index spaces. More specifically, recent work (Kon and Raphael, 1993) has proved that the wavelet transform converges uniformly according to the formula

$$\sup_x |f^j(x) - f(x)| \le C2^{-jp}, \tag{12}$$

where j is the scale, C a constant, and p the order of the approximation. The latter corresponds to the number of vanishing moments of the wavelet (Strang, 1989):

$$\int x^m \psi(x)\, dx = 0 \qquad \text{for } m = 0, \ldots, p - 1, \tag{13}$$

and is an important indicator of the accuracy of each wavelet as an approximator. As can be easily verified by Eq. (12), the larger the number of vanishing moments, the faster the wavelet transform converges to the real function. The approximation error decreases with the scale and for any arbitrary small number ε there exists a scale $j^* = -(1/p)\log_2(\varepsilon/C)$, where the approximation error is less than ε. To use the NN terminology, wavelets are universal approximators since any function in $L^2(R^k)$ can be approximated arbitrarily well by its wavelet representation. Wavelets, however, offer more than any other universal approximator in that, they provide, by construction, a systematic mechanism (their hierarchical structure) to achieve the desired accuracy.

The different pairs of wavelet and scaling functions are not equivalent as approximators. Besides the number of vanishing moments, another distinguishable property is that of orthogonality. Orthogonality results in considerable simplifications in the structural adaptation procedure, since every new orthogonal wavelet inserted in the approximating function introduces independent information. From the approximation point of view, orthogonality is equivalent to more compact and, therefore, more economical representations of the unknown functions. There is, however, a price to pay. With the exception of the discontinuous Haar wavelet, there do not exist wavelets that are compactly supported, symmetrical and orthogonal. In addition, all orthogonal and compactly supported wavelets are highly nonsmooth and, therefore, are not suitable for approximation problems. For this reason, *biorthogonal* and *semiorthogonal* wavelets have been constructed (Cohen, 1992; Feauveau, 1992) that retain the advantages of orthogonality and, at the same time, satisfy the smoothness requirement.

The space–frequency localization of wavelets has lead other researchers as well (Pati, 1992; Zhang and Benveniste, 1992) in considering their use in a NN scheme. In their schemes, however, the determination of the network involves the solution of complicated optimization problem where not only the coefficients but also the wavelet scales and positions in the input space are unknown. Such an approach evidently defies the on-line character of the learning problem and renders any structural adaptation procedure impractical. In that case, those networks suffer from all the deficiencies of NNs for which the network structure is a static decision.

B. Learning Algorithm

With the selection of wavelets as the basis functions the learning algorithm can now be finalized.

Algorithm 3

Step 1. Select a family of scaling functions and wavelets.
Step 2. Select a threshold, ε, with $\varepsilon > 0$ and $\varepsilon > \delta$, on the empirical error defined by Eq. (6).
Step 3. Select an initial scale, j_0.
Step 4. Learning step: For every available data point, apply algorithm M to calculate the model by solving the minimization problem, $\hat{g} = \min_g I_{\mathrm{emp}}(g) = \min_g \max_{i=1,\ldots,l} |y_i - g(\mathbf{x}_i)|, \; g_j \in S_j$, and estimate the empirical error on all available data, $I_{\mathrm{emp}}(g)$. If $I_{\mathrm{emp}}(g) > \varepsilon$, use algorithm A to update the structure of G, by inserting the lowest scale wavelet that includes that point. Repeat until threshold is satisfied.

The algorithm is schematically presented in Fig. 5. The implementation of the algorithm presupposes the a priori specification of

(a) The scaling function and wavelet pair.
(b) The initial scale j_0.
(c) The value of the error threshold, ε.

For the moment, there are no guidelines for the selection of the particular basis functions for any given application. The important issue here is that the properties of the wavelets will be inherited by the approximating

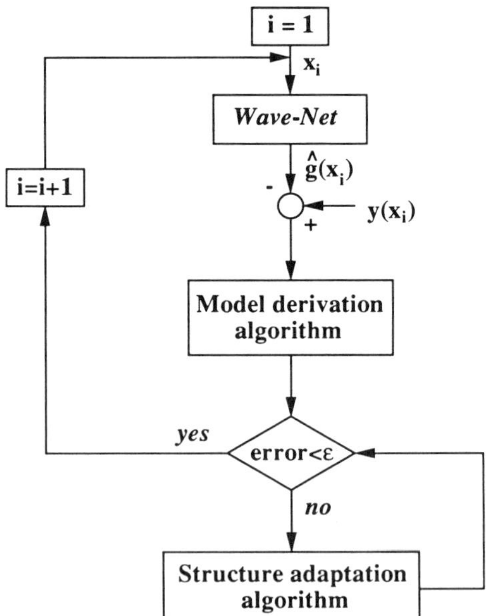

FIG. 5. Learning algorithm.

function and, therefore, if there exist prerequisites on the form of the approximating function (such as the number of existing derivatives), they should be taken into account before the selection of the wavelet. The initial scale, where the first approximation will be constructed, can be chosen as the scale in which few scaling functions (even one) fit within the input range, I_0, of the expected values of $\mathbf{x}$. Finally, the selection of the error bound depends on the accuracy of the approximation sought for the particular application. In any case, it should be equal or greater than the bound on the expected noise, δ.

1. The Model and Its Derivation

The approximating function constructed by the previous algorithm is of the form

$$g(x) = \sum_{k=1}^{m} c_k \theta_k(\mathbf{x}), \tag{14}$$

where $\theta(\mathbf{x})$ stands for both the scaling functions at scale j_0 and wavelets at the same or higher scales. Equation (14) can be easily identified as the mathematical representation of the *Wave-Net* (Bakshi and Stephanopoulos, 1993). With the use of the L^∞ error measure, the coefficients are calculated by solving the minimization problem:

$$\mathbf{c} = \arg\min_{\mathbf{c}} \max_{i=1,\dots,l} \left| y_i - \sum_{k=1}^{m} c_k \theta_k(\mathbf{x}_i) \right|. \tag{15}$$

The problem [Eq. (15)] is a minimax optimization problem. For the case (as it is here) where the approximating function depends linearly on the coefficients, the optimization problem [Eq. (15)] has the form of the *Chebyshev approximation* problem and has a known solution (Murty, 1983). Indeed, it can be easily shown that with the introduction of the dummy variables z, z_i, z_i^* the minimax problem can be transformed to the following linear program (LP):

$$\min_{\mathbf{c}} z,$$

$$z_i = -y_i + \sum_{k=1}^{m} c_k \theta_k(\mathbf{x}_i) + z \geq 0, \tag{16}$$

$$z_i^* = y_i - \sum_{k=1}^{m} c_k \theta_k(\mathbf{x}_i) + z \geq 0.$$

The transformation to this LP program is graphically depicted in Fig. 6 for the case when $\mathbf{c}$ is a scalar. For each data pair $(\mathbf{x}_i, y_i)$ the term $|y_i - \Sigma c_k \theta_k(\mathbf{x}_i)|$ represents two $(m + 1)$-dimensional hyperplanes, $z = y_i - \Sigma c_k \theta_k(\mathbf{x}_i)$ and $z = -y_i + \Sigma c_k \theta_k(\mathbf{x}_i)$. For the scalar case, these correspond

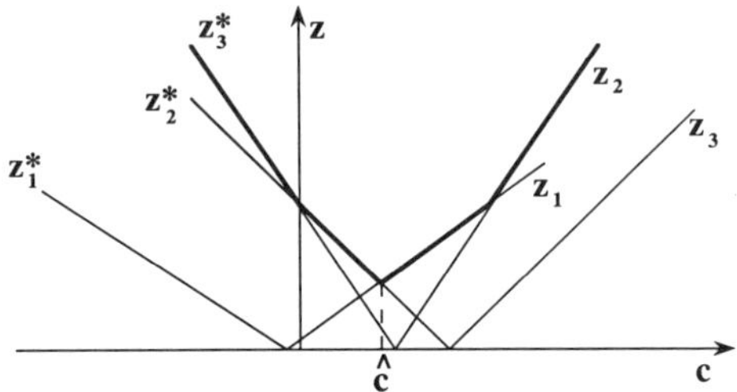

FIG. 6. Geometric interpretation of the Chebyshev approximation problem.

to two lines of opposite slope that meet on the c axis as shown in Fig. 6. For every value of c, according to problem (15), we need to identify the data pair whose corresponding hyperplane has the maximum value, z. The locus of these points is the polyhedron that inscribes the common space above each hyperplane. If an LP formulation, the space above every hyperplane corresponds to an inequality constraint according to Eq. (16), and the intersection of all these constraints is the feasible region. The bottom of the convex polyhedron that inscribes the feasible region is the value of c that minimizes both Eqs. (15) and (16). Its distance from c axis is the minimum value of z, or the minimum possible error. The optimization problem is convex and is guaranteed to have a solution. It can be easily solved using standard LP techniques based on the dual simplex algorithm.

The selection to minimize absolute error [Eq. (6)] calls for optimization algorithms different from those of the standard least-squares problem. Both problems have simple and extensively documented solutions. A slight advantage of the LP solution is that it does not need to be solved for the points for which the approximation error is less than the selected error threshold. In contrast, the least squares problem has to be solved with every newly acquired piece of data. The LP problem can effectively be solved with the dual simplex algorithm, which allows the solution to proceed recursively with the gradual introduction of constraints corresponding to the new data points.

2. Variations on the Structural Adaptation Algorithm

The multiresolution framework allows us to reconsider more constructively some of the features of the structure adaptation algorithm. First, a strictly forward move in the ladder of subspaces [Eq. (8)] is not necessary. Due to localization, the structural correction can be sought in higher spaces before all functions in the previous space are exhausted. This results in an uneven distribution of variable scale features in the input space, which, in turn, means considerable savings in the number of basis functions used.

In a k-dimensional space, one way to construct the scaling function and the wavelets is by taking the k-term tensor products of the one-dimensional parts. For example, in two-dimensional (2D) space the scaling function is $\phi(x_1)\phi(x_2)$ and there are three wavelets: $\phi(x_1)\psi(x_2)$, $\psi(x_1)\phi(x_2)$, and $\psi(x_1)\psi(x_2)$. This result in $2^k - 1$ wavelets all sharing the same support, $I_j, \mathbf{p}$, in the input space. Every time adaptation of the structure is needed at a given position and scale, there are $2^k - 1$ choices and a selection problem arises. By solving locally a small optimization problem, we can identify the wavelet whose insertion in the model minimizes the empirical error and use this one to update the structure. The fact, however, that we are not seeking a complete wavelet representation allows us to consider alternatively empirical selection methods that are not computationally expensive. For example, we might choose to use only the one symmetrical wavelet $\psi(x_1)\psi(x_2)\ldots\psi(x_k)$ at every scale and ignore all the others. On the other hand, there exist methods for constructing high-dimensional wavelets that do not use the tensor products (*nonseparable* wavelets; Kovačević and Vetterli, 1992) and, therefore, automatically overcome this selection problem.

Another variation on the algorithm can be applied by allowing the freedom to select the initial space S_0 outside the multiresolution framework. In many cases, a model of the unknown function is already available by first-principles modeling, or there might exist some a priori knowledge (or bias) on the structure of the unknown function (linear, polynomial, etc.). In all these cases, we would like to combine the available knowledge with the NN model to reduce the blackbox character of the latter and increase its predictive power. There has been some recent work in this direction (Psichogios and Ungar, 1992; Kramer *et al.*, 1992). The multiresolution framework allows us to naturally unify the a priori model with the NN by allowing us to select the initial space S_0 in accordance to our preference. Then, the model in S_0 provides an initial global representation of the unknown function, and the learning algorithm is used to expand and correct locally this model. This procedure is in agreement with the notion of learning as a way to extend what is already known, and is also of considerable practical value.

3. Derivation of Error Bounds

As shown earlier, by imposing a threshold on the L^∞ empirical error and applying the learning algorithm, an approximating function with generalization error less that the threshold will eventually be reached. Therefore, the selected threshold can be regarded as a *global* measure of the approximating capability of the derived model. The multiresolution decomposition of the input space, however, allows us to impose *local* error thresholds and, therefore, seek for variable approximation accuracy in different areas of the input space according to the needs of the problem.

The error threshold can be allowed to vary not only on a spatial but also on a temporal basis as well. Sometimes, the bound on the magnitude of the noise, δ, is not easy to be known, in which cases, the value of the threshold (which must be greater than δ) cannot be determined. In some other situations, it might be risky to impose a tight threshold from the beginning when relatively few data points are available because this might temporarily lead to overfitting. An empirical but efficient way to alleviate those problems is to enforce the error threshold gradually. In other words, we can devise some sequence of error thresholds that satisfy the relationship

$$\varepsilon_1 > \varepsilon_2 > \cdots > \varepsilon_n = \varepsilon$$

and apply them sequentially in time. With this approach, conservative models with respect to the prediction accuracy are initially derived. But as the threshold tightens and data become more plentiful, the model is allowed to follow the data more closely. Since the final value for the threshold is still ε, the theoretical results derived in previous section still hold.

The value of the threshold provides a global upper bound on the expected error of approximation. As stated earlier, guaranteeing any error bound is impossible unless some a priori assumption is imposed on the real function. Such an assumption could effectively be an upper and lower bound on the magnitude of the first derivative of the function. These bounds naturally arise when the unknown function describes a physical phenomenon, in which case, no steep gradients should be expected. If such bounds are available, the value of function between two neighboring data points is restricted between the intersection of two conic regions whose boundaries correspond to the upper and lower values of the first derivative. This conic region can effectively be considered as a guaranteed bound on the error, which, however, might or might not satisfy the desired accuracy as expressed by the value of the selected error threshold.

A tighter and local *estimate* on the generalization error bound can be derived by observing locally the maximum *encountered* empirical error. Consider a given dyadic multiresolution decomposition of the input space and, for simplicity, let us assume piecewise constant functions as approximators. In a given subregion of the input space, $I_{j,\mathbf{p}}$, let $Z_{l,\mathbf{p}}$ be the set of data at instant l that are present into this region. The maximum encountered empirical error in this area is equal to

$$\eta_{l,\mathbf{p}} = \frac{\max(y_{l,\mathbf{p}}) - \min(y_{l,\mathbf{p}})}{2},$$

and also satisfies the relationship: $\eta_{l,\mathbf{p}} < \varepsilon$. This value constitutes a local error bound on the expected accuracy of the approximation unless otherwise proved by the incoming data. By observing the evolution of the local error bound, we can draw useful conclusions on the accuracy of the approximation in the different areas of the input space.

IV. Applications of the Learning Algorithm

In this section, the functionality and performance of the proposed algorithm are demonstrated through a set of examples. In all examples, the "Mexican hat" wavelet (Fig. 4) was used. An illustration of the effectiveness of the algorithm was given with the example of Fig. 1. The approximating curves were indeed derived by applying the algorithm using an error threshold equal to .5. The example was introduced in Zhang and Benveniste (1992) and its solution using *Wave-Nets* is more thoroughly explained in Bakshi *et al.* (1994).

A. EXAMPLE 1

The first example is the estimation of the 2D function (from Narendra and Parthasarathy, 1990)

$$f(x_1, x_2) = \frac{x_1}{(1 + x_1^2)} + x_2^3. \tag{17}$$

A graph of the real function is presented in Fig. 7. A set of examples was created by random selections of the pairs (x_1, x_2) in the range $I = [0, 1]^2$ and the error threshold, ε, was set equal to .1.

Figure 8 presents the evolution of the model during training. More specifically, the size of the network (number of basis functions) and the generalization error are plotted as a function of data presented to the learning algorithm. It can be observed that although the number of data is small, the generalization error is high and the network adaptation proceeds in high rates to compensate for the larger error. The generalization

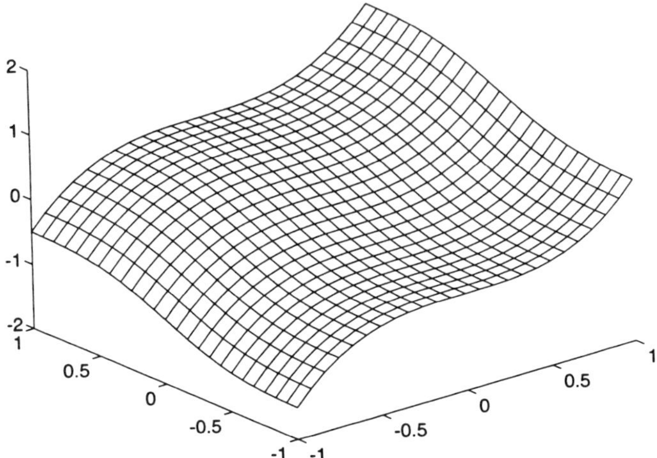

Fig. 7. Real function for Example 1.

error does not decrease monotonically, but it exhibits a clear descending trend. Large changes in the error coincide with data points that trigger structural adaptation (for example, after about 100 and 160 data points). That means that at those points the algorithm shifts to subspaces where better solutions are available and, most importantly, picks one of them resulting in drastic reductions of the error. After almost 160 points, no

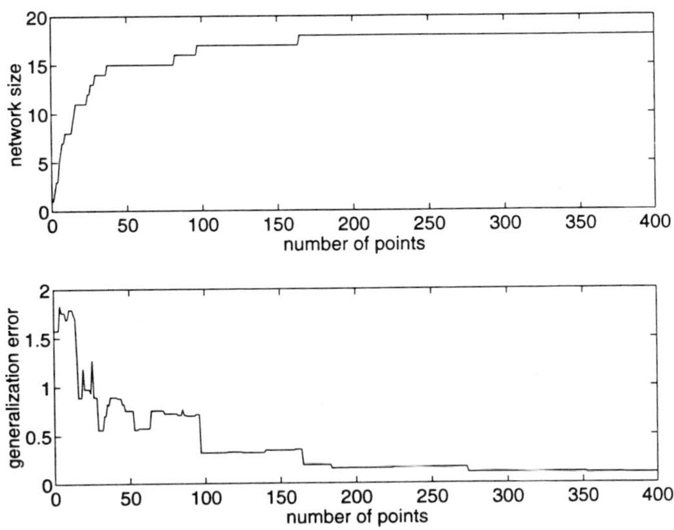

Fig. 8. Evolution of learning for Example 1.

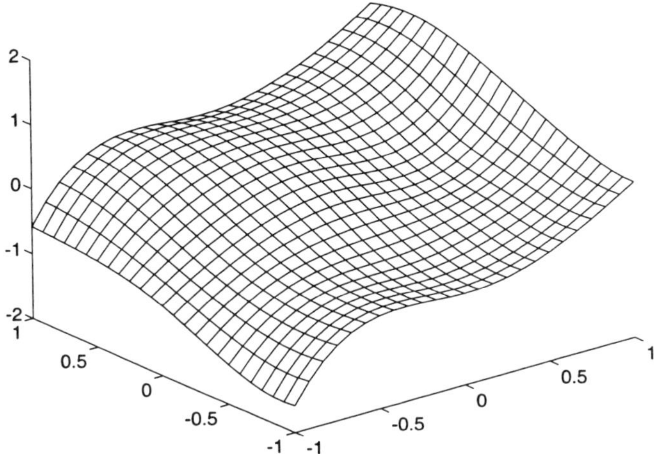

FIG. 9. Model surface for Example 1 after training with 400 points.

further structural adaptation is required and the network size reaches a steady state. The flat sections in the generalization error correspond to periods where no model adaptation is needed. After the 160th data point, coefficient adaptation is required only for three points until the 350th data when the generalization error drops for the first time below .1. At this point, learning has concluded, and a model with the specified accuracy has

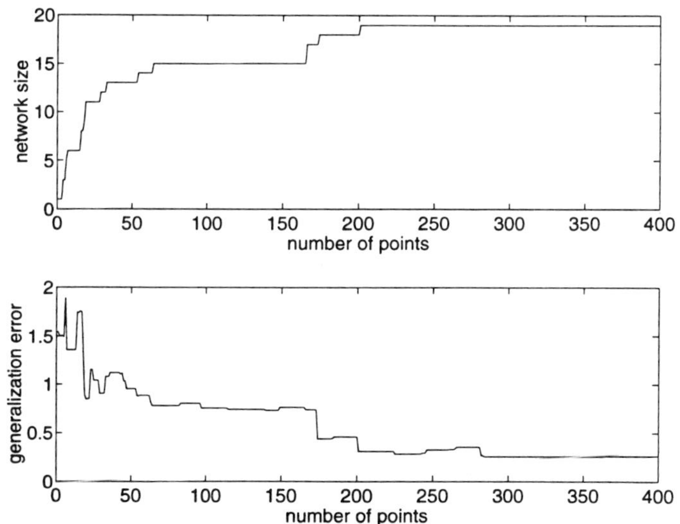

FIG. 10. Evolution of learning for Example 1: data corrupted with noise.

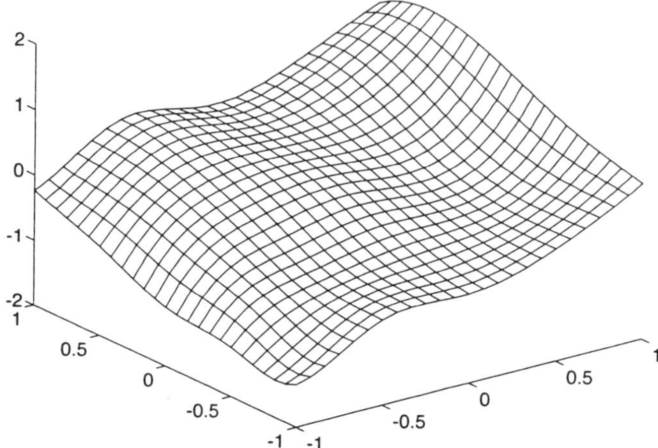

FIG. 11. Model surface for Example 1 after training with 400 noisy points.

been found. This model is shown in Fig. 9. It should be noticed, as is clear by comparing Fig. 7 and 9, that the error of .1 is the worst accuracy achieved. The approximation is much closer in the largest portion of the input space.

The same example was solved for the case when noise is present at the output, y. The maximum amplitude of the noise, δ, was .1 and the error threshold, ε, was set to .2 to reflect our desire to achieve a generalization error of .1($= \varepsilon - \delta$). The results (shown in Fig. 10) are similar with the no-noise case, except that the decrease of the generalization error and the derivation of the final model are slower. This is expected, since the presence of noise degrades the quality of information carried by the data as compared to equal amount of noise-free data. After 400 points the generalization error is .26 and, therefore, a model with the desired accuracy is yet to be achieved. The model (shown in Fig. 11), although not as accurate as before, provides a decent approximation of the real function. More importantly, the use of a larger threshold allows greater tolerance for the observed empirical error and, in this way, allows the model to avoid overfitting the noisy data.

B. EXAMPLE 2

Consider a continuous-stirred-tank reactor (CSTR) with cooling jacket where a first order exothermic reaction takes place. It is required to derive a model relating the extent of the reaction with the flowrate of the heat transfer fluid. This is a nonlinear identification example very popular in chemical engineering literature. In this study we adopt the approach followed by Hernandez and Arkun (1992). The dimensionless differential equations that describe the material and energy balances in the reactor are

$$\frac{dx_1}{dt} = -x_1 + Da(1 - x_1)\exp\left(\frac{x_2}{1 + x_2/\gamma}\right),$$

$$\frac{dx_2}{dt} = -x_2 + BDa(1 - x_1)\exp\left(\frac{x_2}{1 + x_2/\gamma}\right) + b(u - x_1), \qquad (18)$$

$$y = x_1$$

where x_1 (or y) is the measured extent of the reaction, and x_2 the dimensionless temperature of the reactor. The input, u, is the dimensionless flowrate of the heat transfer fluid through the cooling jacket. The input (shown in Fig. 12) was constructed as a concatenation of step changes and random signals created by adding a pseudo random binary signal (PRBS) signal between -1 and 1 and a random variable with uniform distribution between $-.5$ and $.5$. The set of differential equations [Eq. (18)] was solved numerically with the initial conditions corresponding to

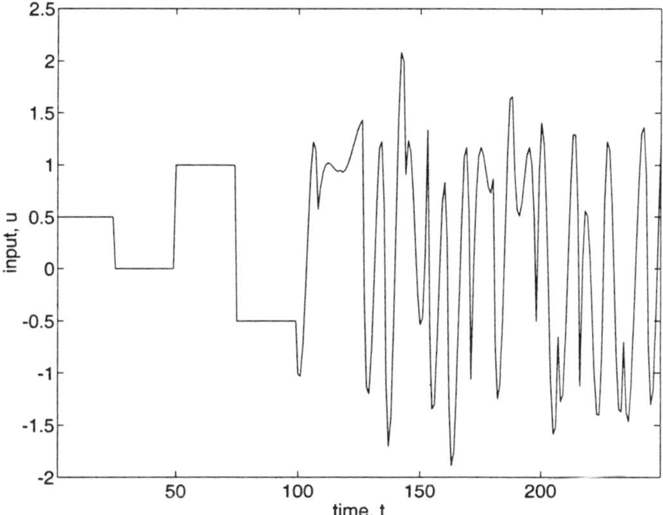

FIG. 12. Input data for identification example.

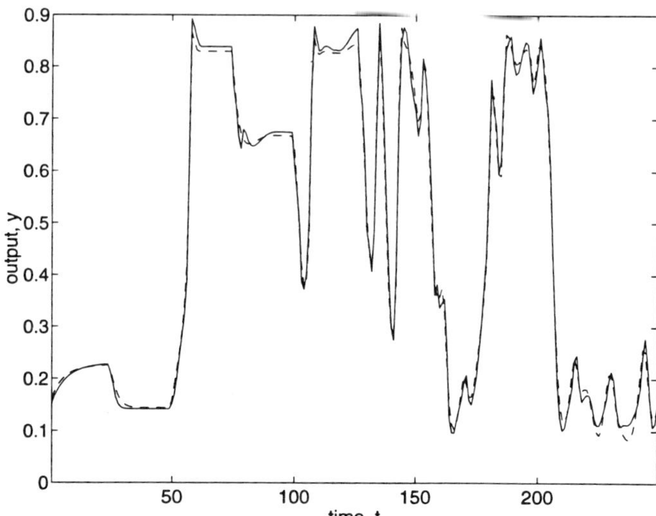

Fig. 13. Comparison of model predictions (solid line) with training data (dashed line) for identification example.

the equilibrium point $x_1 = .144$, $x_2 = .855$, and $u = .0$. The values of the output, created in this way, are represented by the dashed line in Fig. 13. The derived data were used to construct an input/output process model of the form

$$y(t+1) = f[(y(t), y(t-1), u(t))] \qquad (19)$$

by using the *Wave-Net* algorithm to learn the unknown function $f(\cdot, \cdot, \cdot)$. The objective was to use the derived model to predict the behavior of the system in the future for different input values.

In Fig. 13, the model predictions are compared with the real output after training the network with 250 data points. The error threshold was set to .03 and the resulting model contained 37 basis functions with equal number of unknown coefficients. This model was used without further training to predict the system output for the next 750 time instants. The results are shown in Fig. 14, where a fairly good accuracy can be observed. The maximum error associated with the predictions is .11. It is greater than the defined threshold but it is much better than the maximum error $(= .28)$ that would result by predicting the value of $y(t+1)$ equal to $y(t)$. This shows that the model has indeed captured the underlying dynamics of the physical phenomenon and is not merely predicting the future as a zero-order extrapolation of the present.

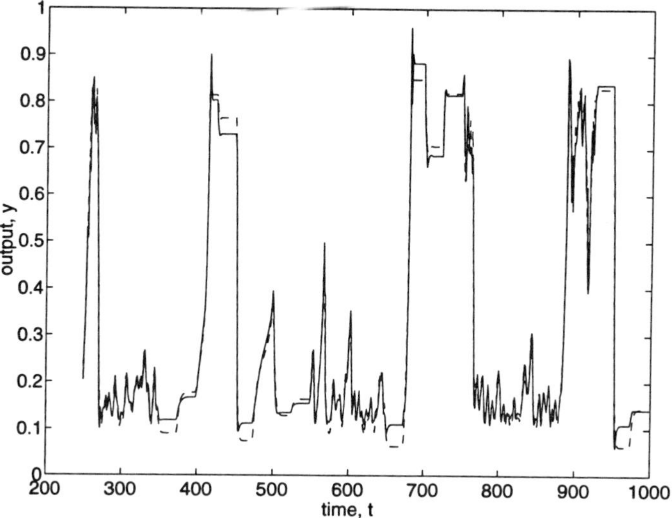

FIG. 14. Comparison of model predictions (solid line) with testing data (dashed line) for identification example.

C. EXAMPLE 3

The final example introduced in Zhang and Benveniste (1992) is a two-dimensional function with distinguished localized features. It is analytically represented by the formula

$$f(x_1, x_2) = \left(x_1^2 - x_2^2\right)\sin(5x_1) \tag{20}$$

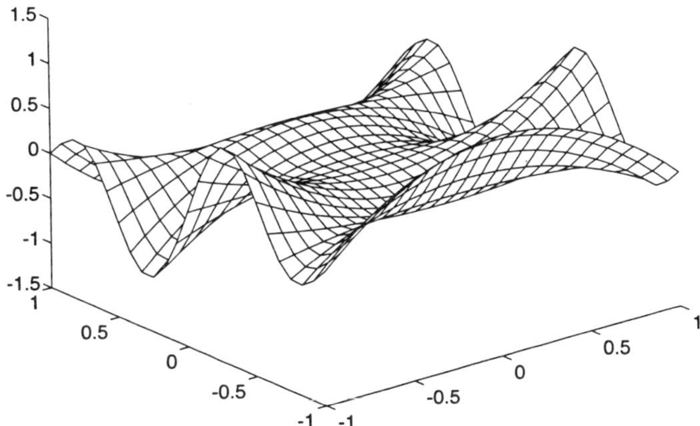

FIG. 15. Real function for example 3.

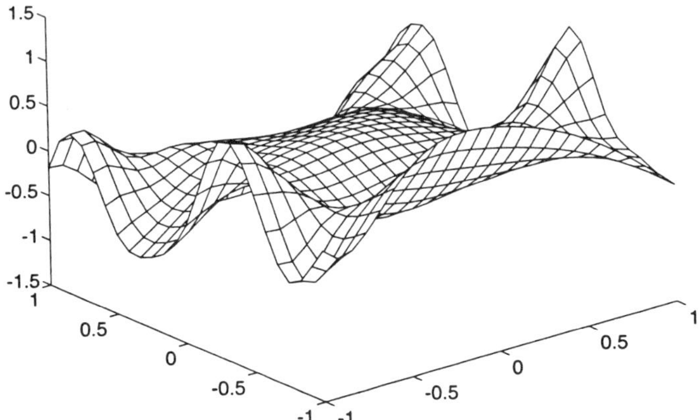

FIG. 16. Model surface for example 3 after training with 500 points.

and graphically represented in Fig. 15. After setting $\varepsilon = .2$ and using 500 data points for training, the resulting model exhibited a generalization error of .22. The model, shown in Fig. 16, has, within the predefined accuracy, captured the important features of the unknown function. This was possible by including during training more high-scale wavelets along the x_1 axis where the significant features lie. The coverage of the input space by the basis functions is shown in Fig. 17, where the circles

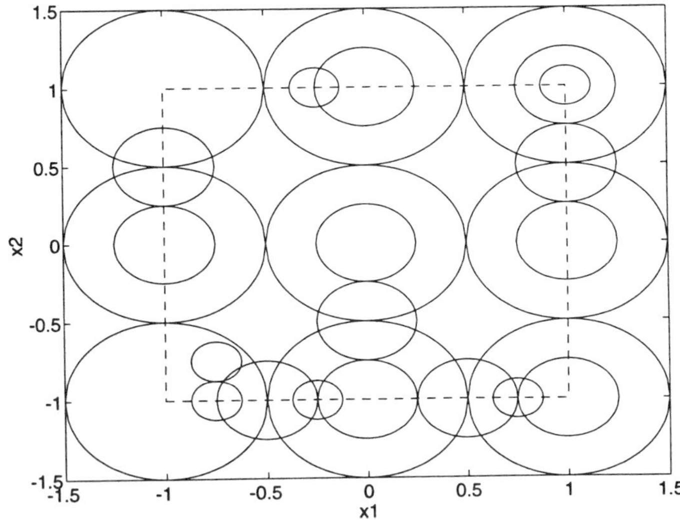

FIG. 17. Coverage of the input space (square) by the support (circles) of the basis functions for the model of Example 3.

correspond to the "dominant" support of each basis function and the square bounds the area where data were available. It is not surprising that most of the basis functions lie in the boundaries of the input space because this is where the biggest errors are most often encountered.

V. Conclusions

In this chapter the problem of estimating an unknown function from examples was studied. We emphasized the issues that stem from the practical point of view in posing the problem. In practice, one starts with a set of data and wants to construct an approximating function, for which, almost always, nothing is known a priori. The main point of the chapter is that the unknown function cannot be found but can be learned. Learning is an ever-lasting and ever-improving procedure. It requires the existence of a teacher that is the data and an evaluative measure. To ensure continuous improvement, the model has to be susceptible to new data and structural adaptation is an essential requirement to achieve this. NNs are neither data-centered nor flexible enough to enable one to learn from them since the important decisions for a given application are made in the absence of data. The derived approximating function has to be coarse and conservative where data are sparse and to follow closely the fluctuations of the data that might indicate the presence of a distinguished feature; localization and control of smoothness are important elements of the approximating scheme.

All the issues raised by these practical considerations have found rigorous answers within the *Wave-Net* framework. The chapter advocates the following points:

(a) The need for on-line model adaptation in parallel with the data acquisition
(b) The use of the local approximation error measured by the L^∞ error to guide the model adaptation
(c) The use of the multiresolution framework to formalize the intuitive preference to simple and smooth approximating functions
(d) The use of bases, like wavelets, with well-defined properties as approximation schemes

The derivation of process models for adaptive control falls exactly within the framework of the estimation problem studied in this chapter. Control-related implementation are natural extensions to the current work and are studied in another publication (Koulouris and Stephanopoulos, 1995). Some computational details and variations of the adaptation algorithm have also to be evaluated to improve the efficiency of the computations.

VI. Appendices

A. APPENDIX 1

1. Case 1: $f \in G$.

When $I(g)$ is given by Eq. (3), the solution to the minimization problem is the *regression function* (Vapnik, 1982):

$$g^* = \int yP(y|\mathbf{x})\, dy. \tag{21}$$

Since $y = f(\mathbf{x}) + d$, $P(y|z\mathbf{x}) = P[y - f(\mathbf{x})] = \mathbf{P}(d)$, and Eq. (21) yields

$$g^* = \int [f(\mathbf{x}) + d] P[y - f(\mathbf{x})]\, dy = \int f(\mathbf{x}) \mathbf{P}[y - f(\mathbf{x})]\, dy + \int d\mathbf{P}(d)\, dd$$

$$= f(x) + \bar{d}.$$

That is, the real function $f(\mathbf{x})$ is the solution to the minimization of Eq. (3) only in the absence of noise ($d = 0$) or when the noise has zero mean ($\bar{d} = 0$). This is, in fact, true for all L^n norms with $2 < n < \infty$ under the condition that n moments of the noise are zero. When the L^∞ metric given by Eq. (4) is used, we will prove that $f(\mathbf{x})$ yields the minimal value of $I(g)$, independently of the statistics of the noise and as long as the noise is symmetrically bounded.

$$\text{If } g = f, I(f) = \sup_{x} |y - f(\mathbf{x})| = \sup|d| = \delta,$$

$$\text{If } g \neq f, I(g) = \sup_{x} |y - g(\mathbf{x})| = \sup_{x} |f(\mathbf{x}) - g(\mathbf{x}) + d|$$

$$= \sup_{x} |f(\mathbf{x}) - g(\mathbf{x})| + \delta > \delta. \tag{22}$$

The reason for the last inequality is that for every $\mathbf{x}$ the maximum value of $|f(\mathbf{x}) - g(\mathbf{x}) + d|$ is $|f(\mathbf{x}) - g(\mathbf{x})| + \delta$ and the supremum value for all $\mathbf{x}$ is $\sup_{x} |f(\mathbf{x}) - g(\mathbf{x})| + \delta$.

2. Case 2: $f \notin G$.

It can be easily proved that when $f(\mathbf{x})$ is the regression function (Vapnik, 1982)

$$\int [y - g(x)]^2 P(\mathbf{x}, y)\, d\mathbf{x}\, dy = \int [y - f(\mathbf{x})]^2 P(\mathbf{x}, y)\, d\mathbf{x}, dy$$

$$+ \int [f(\mathbf{x}) - g(\mathbf{x})]^2 P(\mathbf{x})\, d\mathbf{x},$$

$$I(g) = I(f) + \mu_2(g, f),$$

$$\mu_2(g, f) = I(g) - I(f),$$

which implies L^2 closeness [$\mu_2(g, f)$ is small] if $I(g) - I(f)$ is small [$I(g) \geq I(f)$]. Similarly for the L^∞ case,

$$(22) \quad \Rightarrow \quad \sup_x |f(\mathbf{x}) - g(\mathbf{x})| = I(g) - I(f) \qquad [\text{since } I(f) = \delta].$$

Again, the smaller the value of $I(g)[I(g) \geq I(f)]$, the closer $g(\mathbf{x})$ is to $f(\mathbf{x})$ in the L^∞ sense.

B. Appendix 2

The proof follows three intermediate steps.

(a) Let $I_{\text{emp}}^{(l)}(g) = \max_{i=1,\dots,l} |y_i - g(\mathbf{x}_i)|$. Then

$$I_{\text{emp}}^{(l)}(g) \leq I_{\text{emp}}^{(m)}(g) \text{ for } m > l \text{ and } \forall g \in G, \qquad (23)$$

which essentially states that every new point reveals for every function in G an empirical error that is equal or worse than the one already encountered. Equation (23) also conforms with the fact that $I_{\text{emp}}(g) < I(g)$ and is the basis of convergence of $I_{\text{emp}}(g)$ to $I(g)$.

(b) $(23) \Rightarrow \quad I(g) - I_{\text{emp}}^{(l)}(g) \geq I(g) - I_{\text{emp}}^{(m)}(g), \qquad m > l,$

$$\sup_g \left[I(g) - I_{\text{emp}}^{(l)}(g) \right] \geq \sup_g \left[I(g) - I_{\text{emp}}^{(m)}(g) \right],$$

$$\lim_{l \to \infty} \sup_g \left[I(g) - I_{\text{emp}}^{(l)}(g) \right] = 0.$$

This basically proves uniform convergence of $I_{\mathrm{emp}}^{(l)}(g)$ to $I(g)$ and that, in turn, means

$$\forall k > 0, \exists N > 0 \quad \text{such that} \quad \sup_g \left[I(g) - I_{\mathrm{emp}}^{(l)}(g) \right] < \kappa \quad \forall l > N.$$

(24)

(c) Let $\varepsilon > 0$ be a given positive number and $G_\varepsilon = \{g \in G \,|\, I(g) < \varepsilon\}$, which we assume to be nonempty. Let also $g^* = \mathrm{argmin}\, I(g)$. Obviously, $g^* \epsilon G$ and $I(g^*) < \varepsilon$. If, in Eq. (24), we set $\kappa = \varepsilon\text{-}I(g^*)$, then

$$\exists N > 0 \quad \text{such that} \quad I(g) - I_{\mathrm{emp}}^{(m)}(g) < \varepsilon - I(g^*) \quad \forall m > N$$

$$I_{\mathrm{emp}}^{(m)}(g) > I(g) - \varepsilon + I(g^*) \qquad (25)$$

We want to prove that, if this is the case, then only solutions with $I(g) < \varepsilon$ will be produced by the minimization of the empirical risk, and convergence in this weak sense will be guaranteed. Let g' be a function such that $I(g') > \varepsilon$. Then from Eq. (25)

$$I_{\mathrm{emp}}^{(m)}(g) > I(g') - \varepsilon + I(g^*) > I(g^*),$$

but

$$I(g^*) > I_{\mathrm{emp}}^{(m)}(g^*), \quad \text{so} \quad I_{\mathrm{emp}}^{(m)}(g') > I_{\mathrm{emp}}^{(m)}(g^*),$$

and, therefore, g' can never be the solution to the minimization of $I_{\mathrm{emp}}^{(m)}(g)$ since there will always be at least be at least the best approximation g^*, with guaranteed lower value of the empirical risk.

C. APPENDIX 3

We first claim that if in some subspace $j < j^*$, $I_{\mathrm{emp}}(g) > \varepsilon \,\forall g \in S_j$, there does not exist $g \in S_j$ so that $I(g) < \varepsilon$. That is straightforward since $I_{\mathrm{emp}}(g) < I(g)$. By forcing a threshold on I_{emp}, we are guaranteed to find a data point for which no solution in S_j can satisfy the bound. The algorithm will then look for the solution at the immediate subspace S_{j+1} where, by Assumption 1, a better solution exists. Following this procedure, eventually subspace S_j^* is reached. Its existence is guaranteed by Assumption 2. The algorithm will never move to higher subspaces $j > j^*$ since $I_{\mathrm{emp}}(g) < I(g)$ $< \varepsilon \,\forall g \in G_\varepsilon$, and the requirement $I_{\mathrm{emp}}(g) < \varepsilon$ will always be satisfied.

References

Bakshi, B., and Stephanopoulos, G., Wave-Net: A multiresolution, hierarchical neural network with localized learning. *AIChE J.* **39**, 57 (1993).

Bakshi, B., Koulouris, A., and Stepanopoulos, G., Learning at multiple resolutions: Wavelets as basis functions in artificial neural networks and inductive decision trees. *In* "Wavelet Applications in Chemical Engineering" (R. L. Motard and B. Joseph, eds.) Kluwer Academic Publishers, Dordrecht/Norwell, MA, p. 139 (1994).

Barron, A. R. Approximation and estimation bounds for artificial neural networks. *Mach. Learn.* **14**, 115 (1994).

Barron, A. R., and Barron, R. L., Statistical learning networks: A unifying view. *In* "Symposium on the Interface: Statistics and Computing Science." p. 192. Reston, VA, 1988.

Barto, A. G., Connectionist learning for control. *In* "Neural Networks for Control." (W. T. Miller, R. S. Sutton and P. J. Werbos, eds.) p. 5. MIT Press, Cambridge, MA, 1991.

Bhat, N. V., and McAvoy, T. J., Use of neural nets for dynamic modeling and control of chemical process systems. *Comput. Chem. Eng.* **14**, 573 (1990).

Bhat, N. V., and McAvoy, T. J., Determining the structure for neural models by network stripping. *Comput. Chem. Eng.* **16**, 271 (1992).

Cohen, A., Biorthogonal wavelets. *In* "Wavelets—A Tutorial in Theory and Applications," (C. K. Chui, ed.), Academic Press, San Diego, CA, p. 123. 1992.

Daubechies, I., "Ten Lectures on Wavelets." SIAM Philadelphia, 1992.

Feauveau, J. C., Nonorthogonal multiresolution analysis using wavelets. *In* "Wavelets—A Tutorial in Theory and Applications" (C. K. Chui, ed.), Academic Press, San Diego, CA, p. 153. 1992.

Girosi, F., "Rates of Convergence of Approximation by Translates and Dilates," AI Lab Memo, Massachusetts Institute of Technology, Cambridge, MA, 1993.

Girosi, F., and Anzellotti, G., Rates of convergence for radial basis functions and neural networks. "Artificial Neural Networks with Applications in Speech and Vision," (R. J. Mammone, ed.), p. 97. Chapman & Hall, London, 1993.

Hartman, E., Keeler, K., and Kowalski, J. K., Layered Neural Networks with Gaussian hidden units as universal approximators. *Neural Comput.* **2**, 210 (1990).

Hernandez, E., and Arkun, Y., A study of the control relevant properties of backpropagation neural net models of nonlinear dynamical systems. *Comput. Chem. Eng.* **16**, 227 (1992).

Hornik, K., Stinchcombe, M., and White, H., Multi-layer feedforward networks are universal approximators. *Neural Networks* **2**, 359 (1989).

Hoskins, J. C., and Himmelblau, D. M., Artificial neural network models of knowledge representation in chemical engineering. *Comput. Chem. Eng.* **12**, 881 (1988).

Kearns, M., and Vazirani, U., "An Introduction to Computational Learning Theory." MIT Press, Cambridge, MA. 1994.

Kon, M., and Raphael, L., "Convergence Rates of Wavelet Expansions," preprint 1993.

Koulouris, A., and Stepanopoulos, G., On-line empirical learning of process dynamics with Wave-Nets, submitted to *Comput. Chem. Eng.* (1995).

Kovačević, J., and Vetterli, M., Nonseparable multiresolutional perfect reconstruction banks and wavelet bases for R^n. *IEEE Trans. Inf. Theory*, **38**, 533 (1992).

Kramer, M. A., Thompson, M. L. and Bhagat, P. M., Embedding theoretical models in neural networks. *Proc. Am. Control Conf.* 475 (1992).

Kreinovich, V. Y., Arbitrary nonlinearity is sufficient to represent all functions by neural networks: A theorem. *Neural Networks* **4**, 381 (1991).

Lee, M., and Park, S., A new scheme combining neural feedforward control with model predictive control. *AIChE J.*, **38**, 193 (1992).

Leonard, J. A., and Kramer, M. A., Radial basis function networks for classifying process faults. *IEEE Control Syst.* **11**, pp. 31–38, (1991).

Ljung, L., "System Identification: Theory for the User." Prentice-Hall Englewood Cliffs, NJ. (1987).

Mallat, S. G., A theory for multiresolution signal decomposition: The wavelet representation. *IEEE Trans. Pattern Anal. Mach. Intell.* **PAMI-11**, 674 (1989).

Mavrovouniotis, M. L., and Chang, S., Hierarchical Neural Networks. *Comput. Chem. Eng.* **16**, 347 (1992).

Moody, J., and Darken, C. J., Fast learning in networks of locally-tuned processing units. *Neural Comput.* **1**, 281 (1989).

Murty, K. G., "Linear Programming." Wiley, New York, 1983.

Narendra, K. S., and Parthasarathy, K., Identification and control of dynamical systems using neural networks. *IEEE Trans. Neural Networks* **1**, (1990).

Pati, Y. C., Wavelets and time-frequency methods in linear systems and neural networks. Ph.D. Thesis, University of Maryland, College Park (1992).

Platt, J., A resource-allocating network for function interpolation. *Neural Comput.* **3**, 213 (1991)

Poggio, T., and Girosi, F., "A Theory of Networks for Approximation and Learning," AI Lab. Memo. No. 1140. Massachusetts Institute of Technology, Cambridge, MA, 1989.

Psichogios, D. C., and Ungar, L. H., Direct and indirect model based control using artificial neural networks, *Ind. Eng. Chem. Res.* **30**, 2564 (1991).

Psichogios, D. C., and Ungar, L. H., A hybrid neural network-first principles approach to process modeling. *AIChE J.* **38**, 1499 (1992).

Rengaswamy, R., and Venkatasubramaniam, V., Extraction of qualitative trends from noisy process data using neural networks. *AIChE Annu. Meet. Los Angeles* (1991).

Rumelhart, D. E., McClelland, J. L., and the PDP Research Group, "Parallel Distributing Processing." MIT Press, Cambridge, MA, 1986.

Strang, G., Wavelets and dilation equations: A brief introduction. *SIAM Rev.* **31**, 614 (1989).

Ungar, L. H., Powell, B. A., and Kamens, S. N., Adaptive Networks for fault diagnosis and process control. *Comput. Chem. Eng.* **14**, 561 (1990).

Vapnik, V., "Estimation of Dependences Based on Empirical Data." Springer-Verlag, Berlin, 1982.

Ydstie, B. E., Forecasting and control using adaptive connectionist networks. *Comput. Chem. Eng.* **14**, 583 (1990).

Zhang, Q., and Benveniste, A., Wavelet networks. *IEEE Trans. Neural Networks* **3**, 889 (1992).

REASONING IN TIME: MODELING, ANALYSIS, AND PATTERN RECOGNITION OF TEMPORAL PROCESS TRENDS

Bhavik R. Bakshi

Department of Chemical Engineering
Ohio State University
Columbus, Ohio 43210

George Stephanopoulos

Laboratory for Intelligent Systems in Process Engineering
Department of Chemical Engineering
Massachusetts Institute of Technology
Cambridge, Massachusetts 02139

The plain record of a variable's numerical values over time does not invoke appreciable levels of *cognitive* activity to a human. Although it can cause a fervor of numerical computations by a computer, the levels of cognitive appreciation of the variable's temporal behavior remain low. On the other hand, if one presents the human with a graphical depiction of the variable's temporal behavior, the level of cognition increases and a wave of reasoning activities is unleashed. Nevertheless, when the human is presented with scores of graphs, depicting the temporal behavior of interacting variables, that person's reasoning abilities are severely tested. In such case, the computer will happily continue crunching numbers without ever rising above the fray and thus developing a "mental" model, interpreting correctly the temporal interactions among the many variables.

Reasoning in time is very demanding, because time introduces a new dimension with significant levels of additional freedom and complexity. While the real-valued representation of variables in time is completely satisfactory for many engineering tasks (e.g., control, dynamic simulation, planning and scheduling of operations), it is very unsatisfactory for all those tasks, which require decision making via logical reasoning (e.g., diagnosis of process faults, recovery of operations from large unsolicited deviations, "supervised" execution of startup for shutdown operating procedures).

To improve the computer's ability to reason efficiently in time, we must first establish new forms for the representation of temporal behaviors. It is the purpose of this chapter to examine the engineering needs for temporal decisionmaking and to propose specific models that encapsulate the requisite temporal characteristics of individual variables and composite processes. Through a combination of analytical techniques, such as *scale-space filtering* and *wavelet-based, multiresolution decomposition of functions*, and modeling paradigms from *artificial intelligence* (AI), we have developed a concise framework that can be used to model, analyze, and synthesize the temporal trends of process operations. Within this framework, the modeling needs for logical reasoning in time can be fully satisfied, while maintaining consistency with the numerical tasks carried out at the same time. Thus, through the modeling paradigms of this chapter one may put together intelligent systems that use consistent representations for their logical reasoning and numerical tasks.

I. Introduction

Present-day computer-aided monitoring and control of chemical plants has caused an explosion in the amount of process information that can be conveyed to process operators and engineers. The real-time history of thousands of variables can be displayed and monitored. However, whereas a simple visual inspection of scores of displayed trends is sufficient to allow the operator confirm the process' status during normal, steady-state operations, when the process is in significant transience or crises have occurred the displayed trends and alarms can confound even the best of the operators. When the process variables change with different rates, or are affected by varying transportation lags, or inverse the response dynamics, or information is "suspicious" or lost because of sensors' malfunctioning, it is very difficult for a human operator to carry out routine tasks, such as the following:

- Distinguish normal from abnormal operating conditions (Bastl and Fenkel, 1980; Long and Kanazava, 1980).
- Identify the causes of process trends, e.g., external load disturbances, equipment faults, operational degradation, operator-induced mishandling.
- Evaluate current process trends and anticipate future operational states.
- Plan and schedule sequences of operating steps to bring the plant at the desired operating level, e.g., recover from safety fallback position, return to feasible operation after a fault.

The key recognitive skill required to carry out the above tasks is the formation of a "mental" model of the process operations that fits the current facts about the process and enables the operators to correctly assess process behavior and predict the effects of possible control actions. Correct "mental" models of process operations have allowed operators to overcome the weakness of "lost" sensors and conflicting trends, even under the pressure of an emergency (Dvorak, 1987), whereas most of the operational mishandlings are due to an erroneous perception as to what is going on in the process (O'Shima, 1983).

In order to develop intelligent, computer-aided systems with systematic and sound methodologies for the automatic creation of "mental" models of process operations, we need to resolve the following two and interrelated issues:

- What is the appropriate representational model for describing the "true" process trends, and how is it generated from the process data?
- How does one generate relationships among process trends in order to provide the desired "mental" model of process operations?

In the subsequent paragraphs and sections of this chapter we will see that these two issues impose requirements that transgress the abilities of simple "smoothing" filters and conventional regression techniques.

A. THE CONTENT OF PROCESS TRENDS: LOCAL IN TIME AND MULTISCALE

The time-dependent behavior of measured variables in a chemical process reflects the composite effect of many distinct contributions, coming from the underlying physicochemical phenomena and the status of processing equipment, sensors, and control valves. Thus, basic process dynamics, sensor noise, actuator dynamics, parameter drifts, equipment faults, external load disturbances, and operator-induced actions combine their contributions in some unknown way to form the temporal behavior of the measured operational data. As an example, consider the process signal shown in Fig. 1 (Cheung, 1992). It reflects the composite effect of contributions from five distinct sources; as a slow drift caused by a fouling process (source 1), equipment faults (sources 2 and 3), a periodic disturbance (source 4), and changing sensor noise (source 5). Ideally, we would like to have analytical techniques that can take the observed signal apart and render the exact temporal reproduction of each contributing signal. We know that this is theoretically impossible, but practically acceptable representations of the individual components is feasible and very valuable for the correct interpretation of process trends.

A systematic analysis of a process signal over (1) different segments of its time record and (2) various ranges of frequency (or *scale*) can provide a *local* (in time) and *multiscale* hierarchical description of the signal. Such description is needed if an intelligent computer-aided tool is to be constructed in order to (1) localize in time the "step" and "spike" from the equipment faults (Fig. 1), or the onset of change in sensor noise characteristics, and (2) extract the slow drift and the periodic load disturbance.

The engineering context of the need for multiscale representation of process trends can be best seen within the framework of the hierarchical

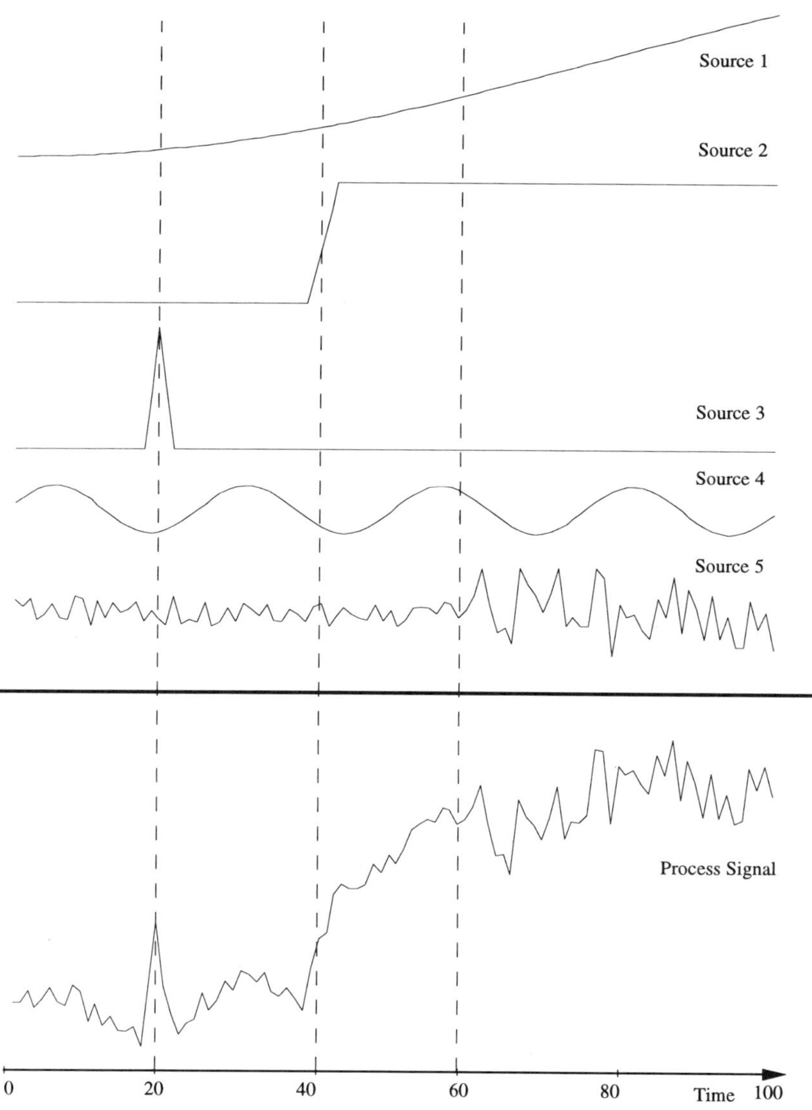

FIG. 1. A process signal and its component.

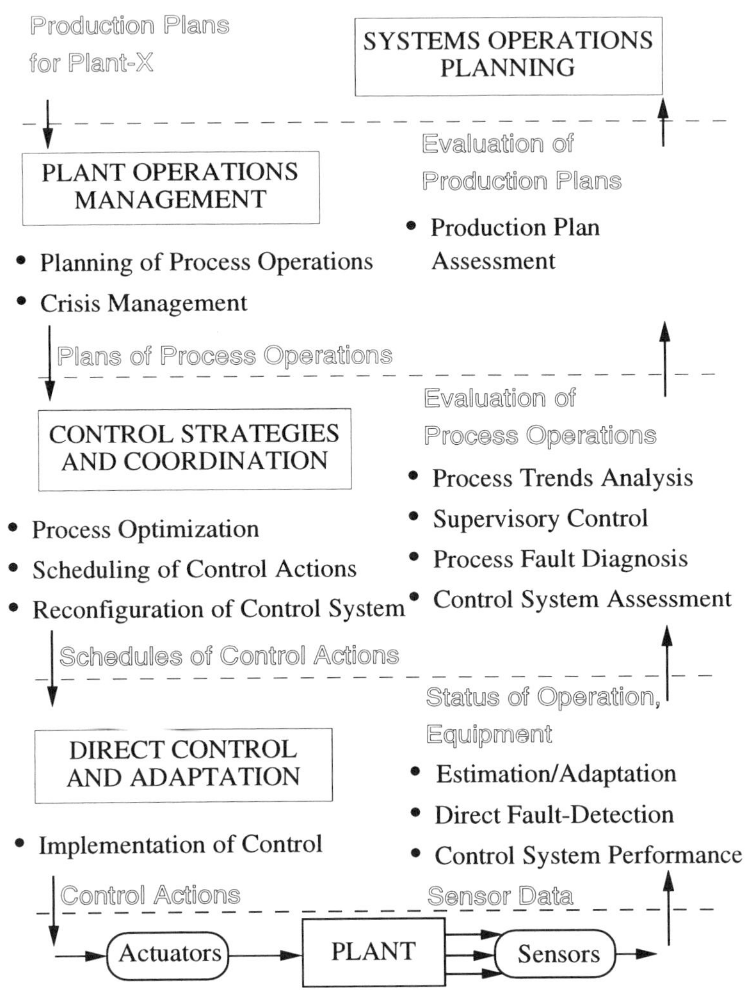

FIG. 2. Hierarchy of process operational tasks.

stratification of operational tasks, shown in Fig. 2 (Stephanopoulos, 1990). At the lowest level of abstraction, process data at a scale of seconds or minute are used to carry out a variety of numerical and logical tasks.

B. THE AD HOC TREATMENT OF PROCESS TRENDS

The term, *process trend*, undoubtedly carries an intuitive meaning about how process behavior changes over time. However, the exact meaning has never been formalized to allow one to articulate it in a general and concrete manner. For a long time, a process trend has been considered just as a smooth representation of a noisy signal, and consequently belonged to the province of digital signal processing community. However, the paradigms advanced are mostly ad hoc in character, and rarely can they sustain adequate fundamental physical justification. For example, during the design of digital filters, *the frequency response is selected* and the essential design task is to determine the coefficients of the filter that match the response. How do you come up, though, with the correct frequency response for the discovery of the "true" process trends in a signal? The answers have been essentially empirical and ad hoc, and have led to an explosive proliferation of design techniques. For example, the Fourier approach and over 100 windowing techniques (e.g., the uniform, von Hann, Hamming, and Kaiser windows) can be used for the design of nonrecursive filters, whereas the *impulse-invariant* method, the *bilinear transform*, and various computer-aided, trial-and-error techniques can be used for the design of recursive filters.

Furthermore, the current understanding of temporal process trends, produced by digital filters, is heavily founded on statistical considerations. Strong assumptions are often made about the statistical nature of the trend, in order to obtain "rigorously" optimal detection schemes. For example, the optimal Bayesian detection, the moving-average filters with adaptive controlled width, and the Wiener filters are only *theoretically optimal* (i.e., their practical implementation could yield grossly inadequate results) for (1) known signals with additive Gaussian noise (2) unknown signals with narrow band relative to the noise, and (3) random signals with known spectra, respectively (Van Trees, 1968; Papoulis, 1977). Although it is commonly agreed that a trend will appear if the random elements are removed from the signal, no filtering technique has incorporated a formal notion of the trend itself by eliciting its fundamental properties and its relation to the physicochemical process it describes.

Once a smooth signal has been constructed, how is the trend represented? Most of the available techniques do not provide a framework for the representation (and thus, interpretation) of trends, because their representations (in the frequency or time domains) do not include primitives that capture the salient features of a trend, such as continuity, discontinuity, linearity, extremity, singularity, and locality. In other words, most of the approaches used to "represent" process signals are in fact *data compaction techniques*, rather than trend representation approaches. Furthermore, whether an approach cmploys a frequency or a time-domain representation, it must make several major decisions before the data are compacted. For frequency-domain representations, assumptions about the origin of the trend must be made, e.g., whether it is stationary (i.e., frequency content independent of time), quasistationary (i.e., frequency content varies slowly with time), or nonstationary. For the curve-fitting representations in the time domain, the selection of the particular functional model and the thresholds for fitting errors are ad hoc decisions.

It is clear from the preceding discussion that *the deficiencies of the existing frequency- and time-domain representations of process trends stem from their procedural character* (Cheung and Stephanopoulos, 1990), i.e. they represent trends as the outputs of a computational process, which quite often bears no relationships to the process physics and chemistry. What is needed is a *declarative* representation, which can capture explicitly all the desirable characteristics of process trends.

C. Recognition of Temporal Patterns in Process Trends

The correct interpretation of measured process data is essential for the satisfactory execution of many computer-aided, intelligent decision support systems that modern processing plants require. In supervisory control, detection and diagnosis of faults, adaptive control, product quality control, and recovery from large operational deviations, determining the "mapping" from process trends to operational conditions is the pivotal task. Plant operators "skilled" in the extraction of real-time patterns of process data and the identification of *distinguishing features* in process trends, can form a "mental" model on the operational status and its anticipated evolution in time.

A formal induction of mappings from measured operating data to process conditions is composed of the following three tasks (Fig. 3):

Task 1. Extraction of pivotal, temporal features from process data.

Task 2. Inductive learning of the relationship between the features of process trends and process conditions.

Task 3. Adaptation of the relationship utilizing future operating data.

Linear, polynomial, or statistical discriminant functions (Fukunaga, 1990; Kramer, 1991; MacGregor *et al.*, 1991), or adaptive connectionist networks (Rumelhart *et al.*, 1986; Funahashi, 1989; Vaidyanathan and Venkatasubramanian, 1990; Bakshi and Stephanopoulos, 1993; third chapter of this volume, Koulouris *et al.*), combine tasks 1 and 2 into one and solve the corresponding problems simultaneously. These methodologies utilize a priori defined general functional relationships between the operating data and process conditions, and as such they *are not inductive*. Nearest-neigh-

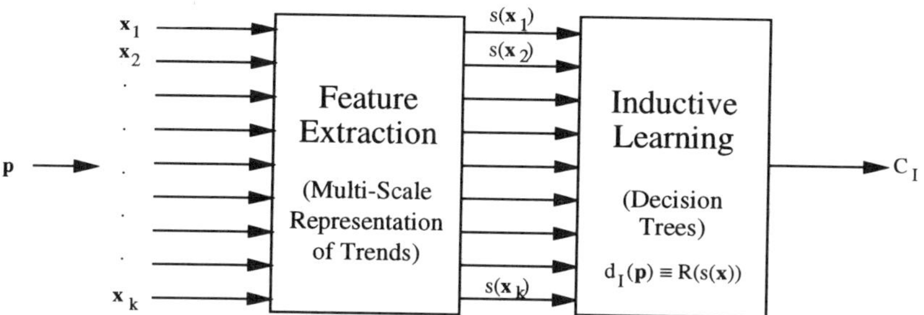

Fig. 3. Inductive generalization of pattern-based relations between process variables and operating conditions.

bor classifiers (Silverman, 1986; Kanal and Dattatreya, 1985), case-based analogical reasoners (Jantke, 1989), or inductive decision trees (Saraiva and Stephanopoulos, 1992), do not assume a functional model and produce truly inductive strategies.

The extraction, though, of the so-called pivotal features from operating data, encounters the same impediments that we discussed earlier on the subject of process trends representation: (1) localization in time of operating features and (2) the multiscale content of operating trends. It is clear, therefore, that any systematic and sound methodology for the identification of patterns between process data and operating conditions can be built only on formal and sound descriptions of process trends.

D. COMPRESSION OF PROCESS DATA

The expression "a mountain of data and an ant hill of knowledge" can be used to signify, besides its obvious message, the significance of the task of maintaining and sorting the vast amounts of data accumulated by present-day, digital process data acquisition systems. Efficient compression, storage, and recovery of historical process data is essential for many engineering tasks. Data compression methodologies must store with minimum distortion of qualitative and quantitative features, independently of whether these features represent behavior *localized in time* or a particular range of frequencies (i.e., *scale*). In other words, the successful compression of process data must necessarily go through an efficient representation of process trends (see Section I, A). Data compression has received surprisingly little attention by the chemical engineering practice, despite the clear articulation of the many benefits that can be drawn from it (Hale and Sellars, 1981; Bader and Tucker, 1987a, b). The *boxcar* and *backward slope* methods are fast, and produce piecewise linear interpolations of the data within predefined, acceptable error bounds, but are unable to capture small process transients and upsets. More techniques (Feehs and Arce, 1988; Bakshi and Stephanopoulos, 1995) have been inspired by technologies used for the compression of speech, images, and communications data, e.g., *vector quantization* (Gray, 1984), and *wavelet decomposition* theory (Mallat, 1989).

The primary objective of any data compression technique is to transform the data to a form that requires the smallest possible amount of storage space, while retaining all the relevant information. The desired qualities of a technique for efficient storage and retrieval of chemical process data are as follows:

1. Compacted data should require minimum storage space.
2. Compaction and retrieval should be fast, often in real time.
3. A clear and explicit measure of the quality of the signal, obtained after retrieval, should be available, to be used as a criterion for guiding the compression.
4. The retrieved signal should have minimum distortion and should contain all the desired features.
5. The compaction should be based on physically intuitive criteria and should require minimum a priori assumptions.

None of the practiced compression techniques satisfies all of these requirements. In addition, it should be remembered that compression of process data is not a task in isolation, but it is intimately related to the other two subjects of this chapter: (1) description of process trends and (2) recognition of temporal patterns in process trends. Consequently, we need to develop a common theoretical framework, which will provide a uniformly consistent basis for all three needs. This is the aim of the present chapter.

E. OVERVIEW OF THE CHAPTER'S STRUCTURE

Section II introduces the formal framework for the definition and description of process trends at all levels of detail: qualitative, order-of-magnitude, and analytic. A detour through the basic concepts of *scale-space filtering* is necessary in order to see the connection between the concept of process trends and the classical material on signal analysis. Within the framework of scale-space filtering we can then elucidate the notions of "episode," "scale," "local filtering," "structure of scale," "distinguished features," and others.

In Section III we introduce the theory of the multiresolution analysis of signals using *wavelet decomposition*, which is used to provide the scale-space image of a function with correct local characteristics. This localized description of a signal's features allows the correct extraction of distinguished attributes from a signal, a task that forms the basis for the inductive generation of pattern-based logical relationships among input and output variables, or the efficient compaction of process data. Of particular value is the construction of *translationally invariant* wavelet decompositions, which allow the correct formation of temporal patterns in process variables. Using translationally invariant decomposition of signals, Section III contains specific methodologies for the construction of the *wavelet interval–tree of scale*, the extraction of distinguished features and the generalization of process trends.

The ideas presented in Section III are used to develop a concise and efficient methodology for the compression of process data, which is presented in Section IV. Of particular importance here is the conceptual foundation of the data compression algorithm; instead of seeking noninterpretable, numerical compaction of data, it strives for an explicit retention of distinguished features in a signal. It is shown that this approach is both numerically efficient and amenable to explicit interpretations of historical process trends.

In Section V we discuss how the temporal distinguished features of many input and output signals can be correlated in propositional forms to provide logical rules for the diagnosis and control of processes, which are difficult to model. *Inductive decision trees* are used to capture the knowledge contained in previous operating data. The explicit description of process trends and the explicit statement of the knowledge "mined" from the data, overcome many of the real-world obstacles in applying these ideas at the manufacturing floor.

Throughout the remaining four sections of this chapter, we will provide illustrations on the use of the various techniques, using real-world case studies.

II. Formal Representation of Process Trends

Although the term "process trend" emulates a certain intuitive understanding in the minds of the speaker and the listener, this understanding may not be the same. Certainly, we do not have a clear, sound, and unambiguous definition of the term "trend," and this must be the first issue to be addressed.

A. The Definition of a Trend

In order to represent and reason with temporal information, we need to represent time explicitly and concisely. The discrete-time character of computer-aided data acquisition and control dictates that time should be represented as a sequence of strictly increasing time points:

$$t = \left\{ t_{-\infty} \cdots t_i \cdots t_j \cdots t_{\infty} \right\},$$

where

$$t_i < t_j \Leftrightarrow i < j \qquad \text{for all integer } i, j.$$

Furthermore, a *time interval*, I_{ij}, is defined to be an open interval of time as follows:

$$I_{ij} = (t_i, t_j), \qquad \text{where} \quad t_i < t_j.$$

Thus, we may represent time as a sequence of open intervals separated by time points, or a sequence of time intervals. We will adopt the second interpretation, considering a time point as an interval of zero duration.

1. From the Quantitative to the Qualitative Representation of a Function

Consider a real-valued function, $x(t)$, with the following properties over a time interval $[a, b]$:

1. $x(t)$ is continuous over $[a, b]$ but is allowed to have a finite number of discontinuities in its value or/ and first derivative.
2. $x'(t)$ and $x''(t)$ are continuous in (a, b), and their one-sided limits exist at a and b.
3. $x(t)$ has a finite number of extrema and inflexion points in $[a, b]$.

Variables that satisfy the above requirements will be called *reasonable variables* (Cheung and Stephanopoulos, 1990). In defining the "reasonableness" of a function, we are concerned only with the properties of the function's value, first and second derivatives. Such definition is less restrictive (it does not require existence of all derivatives), but it is completely general and allows the characterization of a function at different levels (Cheung and Stephanopoulos, 1990). All the physical variables encountered in the operation of a plant are reasonable.

Since we elected to represent time as a sequence of time intervals, we consider that the *state* of a reasonable variable is completely known, if we know the value and the derivative of the variable over a time interval. As the duration of the defining time interval approaches zero, we take the following definition of the state (continuous) of the variable at a given time point (Cheung and Stephanopoulos, 1990):

a. State (Continuous). The state (continuous), $CS(x, t)$ of a reasonable variable, $x : [a, b] \in R \; \forall t \in [a, b]$, is the point value $PtVl(x, t)$ defined by a triplet as follows:

1. If x is continuous at t, then

$$CS(x, t) \equiv PtVl(x, t) = \langle x(t), x'(t), x''(t) \rangle.$$

2. If x is discontinuous at t, then

$$CS(x, t) \equiv PtVl(x, t)$$

$$= \langle \langle x_L(t), x_R(t) \rangle, \langle x_L(t), x_R(t) \rangle, \langle x_L''(t), x_R''(t) \rangle \rangle,$$

where the subscripts L and R denote the left- and right-side limits at the discontinuity.

Consequently, the *trend* (continuous) of a variable can be defined as follows

b. Trend (Continuous). The continuous trend of a reasonable variable, $x : [a, b] \to R$, is given by the continuous sequence of states over $[a, b]$.

As we go from continuous to discrete-time representations, the time is represented by a strictly increasing sequence of points. In such case, the state (discrete) of a reasonable function and the associated trend are given by the following definitions:

c. State (Discrete). The state (discrete), $DS(x, t)$ of a reasonable function, $x : [a, b] \to R$, over a set of strictly increasing time points $T = \{a = t_0, \ldots, t_j, \ldots, t_n = b\} \subseteq [a, b]$, is defined as follows:

1. If $t \in T$, then $DS(x, t) = PtVl(x, t)$.
2. If $t_i < t < t_{i+1}$ for $i = 0, 1, \ldots, n$, then $DS(x, t) = \langle PtVl(x, t_i), PtVl(x, t_{i+1}) \rangle$.

d. Trend (Discrete). The discrete trend of a reasonable variable, $x : [a, b] \to R$, is given by the set of discrete states corresponding to the strictly increasing time points of the time interval $[a, b]$.

Clearly, the quantitative description of the discrete state and of the discrete trend must be declaratively explicit, since we cannot perform differentiation at the single points defining the intervals of $[a, b]$. This is a strict requirement and can be theoretically met only if we know the underlying continuous function that provides the values of the derivatives at the time points of a discrete representation. The availability, though, of such a continuous function is based on a series of ad hoc decisions on the character and properties of the functions, and if one prefers to avoid them, then one must accept a series of approximations for the evaluation of first and second derivatives. These approximations provide a sequence of representations with increasing abstraction, leading, ultimately, to qualitative descriptions of the state and trend as follows (Cheung and Stephanopoulos, 1990):

e. State (Qualitative). Let $x:[a, b] \to R$ be a reasonable function. $QS(x, t)$, the qualitative state of x at $t \in [a, b]$, is defined as the triplet of qualitative values as follows:

$$QS(x, t) = \begin{cases} \text{undefined} & \text{if } x \text{ is discontinuous at } t \\ \langle [x(t)], [\partial x(t)], [\partial \partial x(t)] \rangle & \text{otherwise,} \end{cases}$$

where

$$[x(t)] = \begin{cases} + & \text{if } x(t) > 0, \\ 0 & \text{if } x(t) = 0, \\ - & \text{if } x(t) < 0; \end{cases}$$

$$[\partial x(t)] = \begin{cases} + & \text{if } x'(t) > 0, \\ 0 & \text{if } x'(t) = 0, \\ - & \text{if } x'(t) > 0; \end{cases}$$

$$[\partial \partial x(t)] = \begin{cases} + & \text{if } x''(t) > 0, \\ 0 & \text{if } x''(t) = 0, \\ - & \text{if } x''(t) > 0. \end{cases}$$

f. Trend (Qualitative). The qualitative trend of a reasonable variable, $x:[a, b] \to R$, is the continuous sequence of qualitative states over $[a, b]$.

2. Episodes and Trends

The practical value of the qualitative state and trend lies in the fact that both are very close to the intuitive notions employed by humans in interpreting the temporal behavior of signals. But humans capture the trend as a finite sequence of ordered segments with constant qualitative state over each segment. To emulate a similar notion we introduce the concept of *episode* through the following definition (Cheung and Stephanopoulos, 1990):

a. Episode. Let $x:[a,b] \to r$ be a reasonable function. For any time interval $I = (t_i, t_j) \subseteq [a,b]$ such that the qualitative state, $QS(x,t)$ is constant for $\forall t \in (t_i, t_j)$, an episode, E, of x over (t_i, t_j) is the pair, $E = \langle I, QS(x,I) \rangle$, with (1) I signifying the temporal extent of an episode and (2) $QS(,I) = QS(x,t) \ \forall t \in (t_i, t_j)$ characterizing the constant qualitative state over I. Whenever two episodes, defined over adjacent time intervals, have the same qualitative state values, they can be combined to form an episode with broader temporal extent. On the other hand, if a qualitative state is constant over a time interval, then it is also constant over any of its time subintervals. These observations lead to the following definition of a *maximal episode.*

b. Maximal Episode. An episode, E_1, is maximal if there is no episode, E_2, such that (1) E_1 and E_2 have the same qualitative state and (2) $I_1 \subseteq I_2$, i.e., the temporal extent of E_1 is contained in the temporal extent of E_2.

From the definition of the maximal episode we conclude that the maximal episodes occur between adjacent time points, at which $x(t)$, $x'(t)$ or/ and $x''(t)$ change qualitative value. We will call these points *distinguished time points*, and from now on we will employ the following definition of a *trend*:

c. Trend. The trend of a reasonable function, $x:[a,b] \to R$ is a sequence of maximal episodes, defined over time intervals whose distinguished points are strictly ordered in time.

3. Triangular Episode: A Geometric Language to Describe Trends

If a trend is to be described by an ordered sequence of maximal episodes, then we only need to generate a language that contains the declarative value of an episode. Such a language was proposed by Cheung and Stephanopoulos (1990). It is composed of seven primitives (Fig. 4a) and possesses the following properties:

Completeness. Every trend can be represented by a legal sequence of triangular episodes (Fig. 4b).

Correctness. The recursive refinement of triangular episodes allows the description of a trend at any level of detail, converging to the real-valued description of a signal.

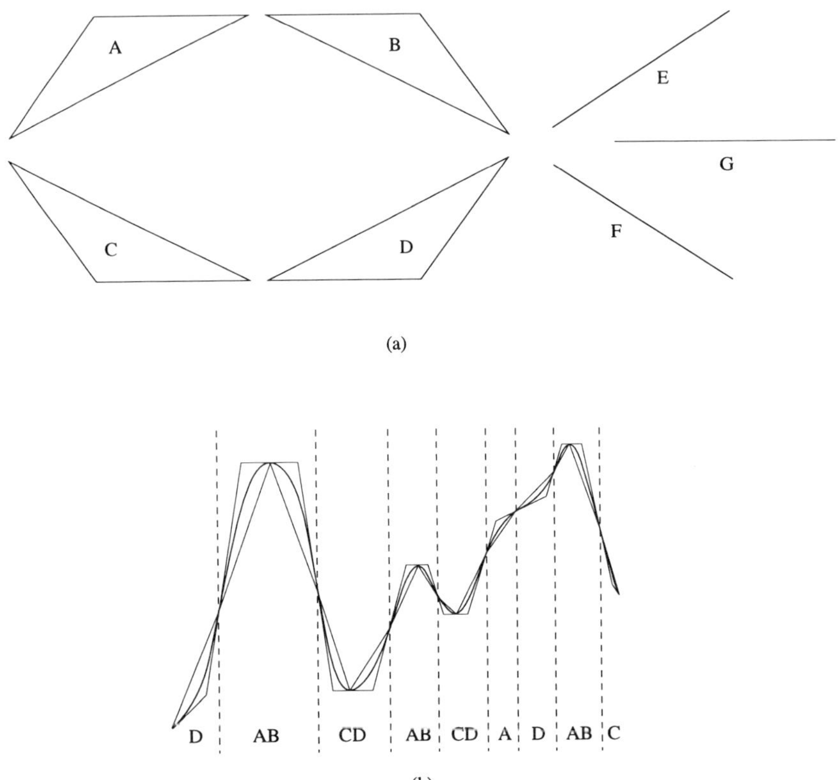

FIG. 4. (a) The seven primitives of the triangular description of trends; (b) sequence of triangular episodes describing a specific "trend" of a signal.

Robustness. The relative ordering of the triangular episodes in a trend is invariant to scaling of both the time axis and the function value. It is also invariant to any linear transformation (e.g., rotation, translation). Finally it is quite robust to uncertainties in the real value of the signal (e.g., noise), provided that the extent of a maximal episode is much larger than the period of noise.

B. Trends and Scale-Space Filtering

Consider the continuous function shown in Fig. 5a. Points 1 through 13 are the inflextion points and constitute a subset of all the distinguished points over the indicated period of time, and define the following

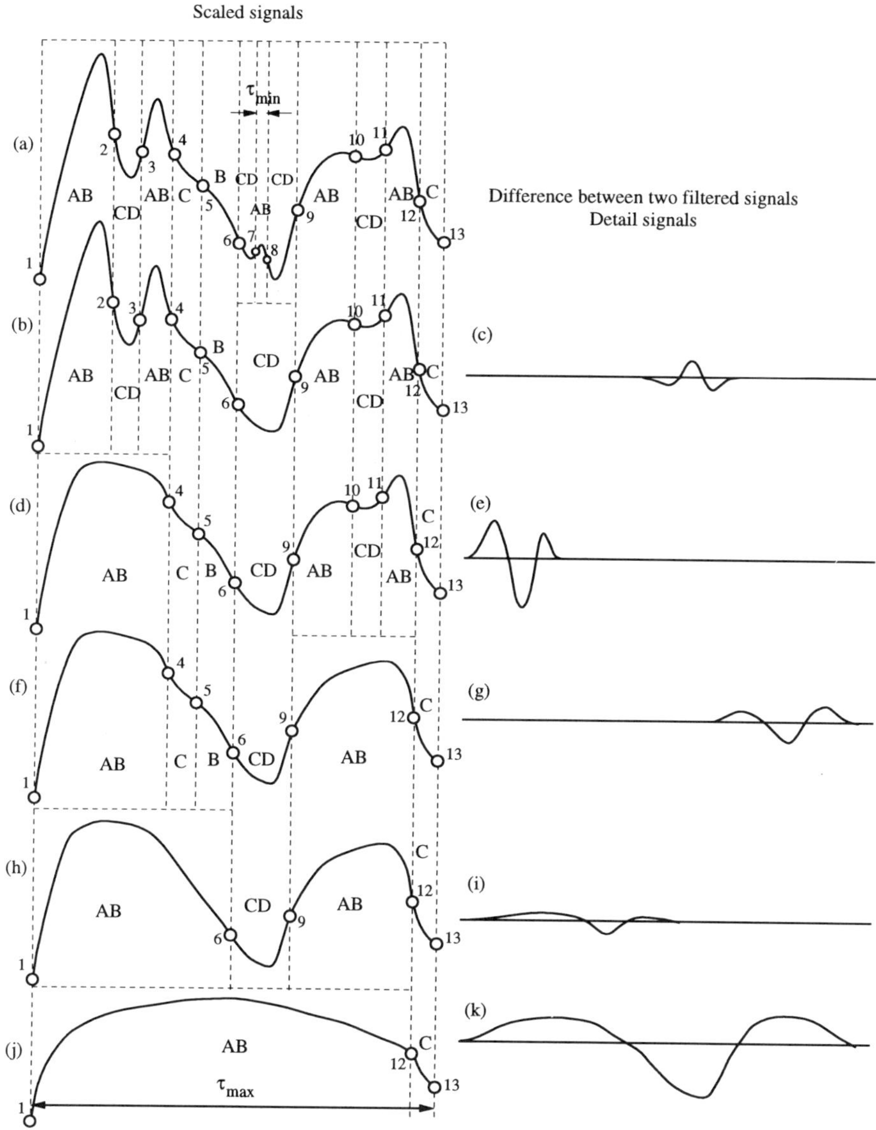

FIG. 5. Multiscale representation of process trends.

sequence of groupings of the triangular episodes:

$$AB(1,2) - CD(2,3) - AB(3,4) - C(4,5) - B(5,6)$$

$$- CD(6,7) - AB(7,8) - CD(8,9) - AB(9,10)$$

$$- CD(10,11) - AB(11,12) - C(12,13).$$

Let us filter the signal of Fig. 5a with a compact filter of variable width of compactness. The first distinguished feature to disappear from the trend of Fig. 5a is the sequence, $CD(6,7) - AB(7,8) - CD(8,9)$ (defined by the inflexion points, 6–9), which is replaced by the $CD(6,9)$ grouping of triangular episodes (see Fig. 5b). The trend of the filtered function (Fig. 5b) differs from the trend of the original one (Fig. 5a) over the time segment defined by the inflexion points 6–9. In an analogous manner, as the width of compactness of the local filter increases, additional features of the trend are replaced with more abstract descriptions. Examples of this process from Fig. 5 are the following:

(a) $\{AB(1,2) - CD(2,3) - AB(3,4)\}$ replaced by $\{AB(1,4)\}$
(Fig. 5d)

(b) $\{AB(9,10) - CD(10,11) - AB(11,12)$ replaced by $\{AB(9,12)]$
(Fig. 5f)

(c) $\{AB(1,4) - C(4,5) - B(5,6)\}$ replaced by $\{AB(1,6)\}$
(Fig. 5h)

(d) $\{AB(1,6) - CD(6,9) - AB(9,12)\}$ replaced by $\{AB(1,12)\}$
(Fig. 5j)

From Fig. 5 we conclude that the original function can be represented by six (6) distinct trends (Fig. 5a, b, d, f, h, j), *each with its own sequence of triangular episodes*. Each successive trend of Fig. 5 contains information at a coarser resolution (scale). The differences among two successive trends are shown in Figs. 5c, 5e, 5g, 5i, and 5k, assuming for presentation purposes perfectly local filters.

The procedure described above is a pictorial approximation of a process called *scale-space filtering* of a function, proposed by Witkin (1983). The surface (e.g., Fig. 6) swept out by a filtered signal as the Gaussian filter's standard deviation is varied, is called *scale-space image* of the signal and is given by

$$F(t,\sigma) = f(t)*g(t,\sigma) = \int_{-\infty}^{+\infty} f(u) \left\{ \frac{1}{\sigma\sqrt{2\pi}} \exp\left[\frac{-(t-u)^2}{2\sigma^2} \right] \right\} du, \quad (1)$$

and gives the time-dependent behavior of the signal, $f(t)$, at different scales. As σ increases certain distinguishing features disappear, replaced

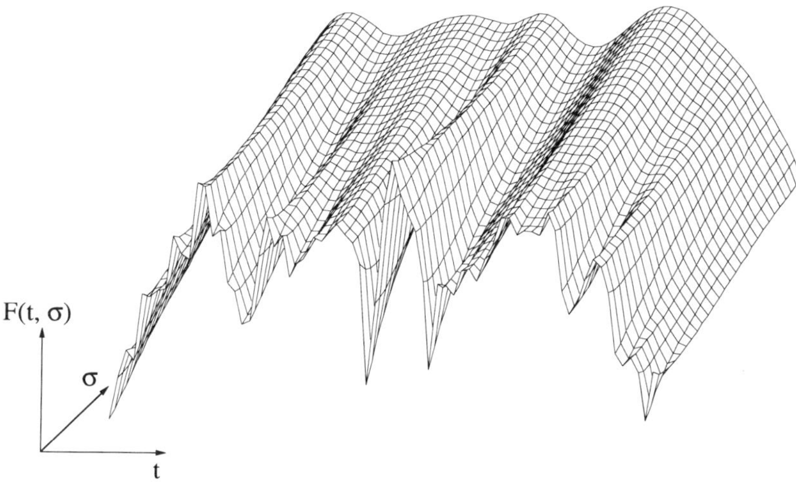

FIG. 6. Scale-space image of a function $F(t)$.

by coarser characteristics, very much in the spirit described by the pictorial examples of Fig. 5.

1. Structure of Scale

A close examination of the scale-space image of Fig. 6 reveals some interesting features:

1. As σ increases, pairs of inflexion points disappear and give rise to a smoother representation of the signal.
2. As σ decreases, a smooth segment of the signal goes through a singular point and gives rise to a "ripple" with the generation of a pair of inflexion points.
3. Some of the inflextion points persist over a large range of σ values, while others disappear. Clearly, the former correspond to *dominant features of the signal* while the latter indicate weak features, e.g., noise, which disappear readily with filtering.
4. The position of the inflexion points at higher values of σ, i.e., higher scales, shifts as a result of the increasingly global effect of the Gaussian filter.
5. If we connect the positions of the same inflexion points over various values of σ by straight lines, we create an *interval tree of scales*, as shown in Fig. 7 for the signals of Fig. 5. The interval tree allows us to generate two very important pieces of information about the trends of a measured variable:

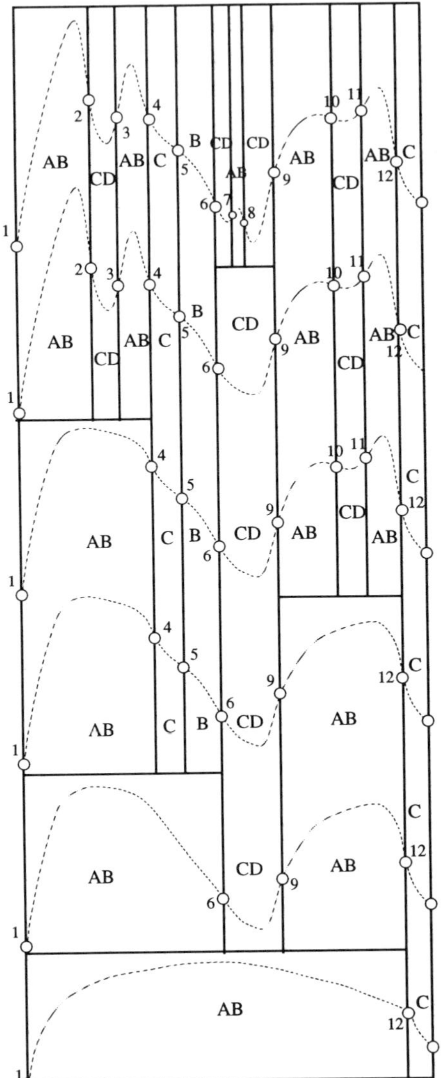

FIG. 7. The interval tree of scales for the signal of Fig. 5.

a. Number of Distinct Trends at Multiple Scales. The disappearance of inflexion points through filtering, or their generation through refinement, changes the sequence of triangular episodes needed for the description of a measured variable. This, in turn, implies that a new trend has been identified. Having the interval tree of scales, we can easily generate the complete set of trends that can be used to describe a measured signal at various scales of abstraction (or, detail). For example, the signal of Fig. 5 can be described by any of six (6) distinct trends.

b. Dominant Features in a Signal. The inflexion points which persist over extended ranges of σ (or scale), bound dominant features of a signal. For example, consider the signal of Fig. 5. The pairs of inflexion points $(1, 6)$ and $(9, 12)$ bound the features, $AB(1, 6)$ and $AB(9, 12)$ (see Fig. 5h), respectively, and have "survived" four levels of filtering. They correspond to dominant features of the original signal and as such characterize profoundly the signal itself.

2. Scaling Episodes and the Second-Order Zero Crossings

Scale-space filtering provides a multiscale description of a signal's trends in terms of its inflexion points (second-order zero crossings). The only legal sequences of triangles between two adjacent inflexion points are (in terms of triangular episodes):

AB

CD

A (if it is preceded and succeeded by D episodes)

B (if it is preceded and succeeded by C episodes)

C (if it is preceded and succeeded by B episodes)

D (if it is preceded and succeeded by A episodes)

Any of these legal sequences will be called a *scaling episode.*

If a signal is represented by a sequence of triangular episodes, scale-space filtering manipulates the sequences of triangular episodes with very concrete mechanisms. Here is the complete list of syntactic manipulations carried out by scale-space filtering:

1. $AB - CD - AB \quad \Rightarrow \quad AB$
2. $CD - AB - CD \quad \Rightarrow \quad CD$
3. $AB - C - B \quad \Rightarrow \quad AB$
4. $CD - A - D \quad \Rightarrow \quad CD$
5. $D - A - D \quad \Rightarrow \quad D$
6. $C - B - C \quad \Rightarrow \quad C$
7. $B - C - B \quad \Rightarrow \quad B$
8. $A - D - A \quad \Rightarrow \quad A$

The reader should note the following properties:

(a) Filtering occurs over a segment of *three* scaling episodes, producing a segment with *one* scaling episode.
(b) The syntax of the resulting scaling episode is determined by the type of the first and last episodes in the sequence of the three scaling episodes, e.g., $\underline{A}B - CD - A\underline{B} \Rightarrow AB$, $\underline{B} - C - \underline{B} \Rightarrow BB \equiv B$.

So, within the context of scale-space filtering, it is more convenient to express *a trend as a sequence of scaling episodes*, rather than as a sequence of episodes.

The utility of representing trends in terms of scaling episodes, or equivalently in terms of *second-order zero crossings* (i.e., inflexion points) increases even more, if we can reconstruct a signal from these zero crossings at multiple scales. Marr and Hildreth (1980), using the Laplacian of a Gaussian had surmised that the reconstruction was stable and complete, but gave no proof. Yuille and Poggio (1984) proved that the scale map of almost all signals filtered by a Gaussian of varying σ determines the signal uniquely, up to a constant scaling. Hummel and Moniot (1989) on the other hand have shown that reconstruction using second-order zero crossings alone is unstable, but together with gradient values at the zero crossings is possible and stable, although it uses a lot of redundant information. Mallat and Zhong (1992) have developed an algo-algorithm that converges quickly to stable reconstructions using second-order zero crossings of a signal, derived from the wavelet decomposition of the signal. Nevertheless, the completeness and stability of the reconstruction from second-order zero crossings remains an open theoretical question.

3. Properties of Scale-Space Filtering

The interval tree of scale allows us to extract the temporal features contained in a signal and systematically establish the trends at different scales. Witkin (1983) defined *the stability of a temporal episode as the number of scales over which it persists*. The most conspicuous features of a signal are also the most stable. He also developed a heuristic procedure through which he would construct representations of a signal by maintaining the most stable episodes. The stability criterion, though is entirely heuristic, and other criteria may be desired. In Section III we will see that wavelet decomposition of signals offers a sound and much more powerful approach in constructing stable representations of dominant trends. For the time being, it is very important to identify several important properties and shortcomings of the traditional scale-space filtering.

a. Distortion of Signal's Features. As a signal is filtered through a Gaussian, its features become progressively quantitatively more and more distorted. This is a natural effect of the averaging of the smoothing process over increasingly longer segments of a signal. Such distortion inhibits the extraction of accurate distinguished features. Feature extraction by selecting stable episodes from the interval tree involves fitting piecewise quadratic segments to the raw data covered by the stable episodes. This procedure is ad hoc, and creates discontinuities at the inflexion points.

b. Creation of Fictitious Features. Gaussian filtering has two very important properties:

1. Inflexion points never disappear as we move from more abstract to more detailed descriptions.
2. Fictitious inflexion points are not generated as the scale of filtering increases.

Taken together, these properties guarantee that Gaussian filtering never generates fictitious features. For many other filters, this is not the case. It is important that any filter we use maintains the integrity of the original signal.

III. Wavelet Decomposition: Extraction of Trends at Multiple Scales

The wavelet decomposition of continuous and discrete functions has emerged as a powerful theoretical framework over the last 10 years, and has led to significant technical developments in data compression, speech and video recognition and data analysis, fusion of multirate data, filtering techniques, etc. Starting with Morlet's work (Goupillaud *et al.* 1984) on the analysis of seismic data, wavelet decomposition of signals has become extremely attractive for two main reasons: (1) it offers excellent localization of a signal's features in both time and frequency and (2) it is numerically far more efficient than other techniques such as fast Fourier transform of signals (Bakshi and Stephanopoulos, 1994a). It is for the first reason that we discuss in this section the use of wavelet decomposition as the most appropriate framework for the extraction and representation of trends of process operating data.

A. The Theory of Wavelet Decomposition

A family of wavelets is a family of functions with all its members derived from the translations (e.g., in time) and dilations of a single, mother function. If $\psi(t)$ is the mother *wavelet*, then all the members of the family are given by

$$\psi_{su}(t) \equiv \frac{1}{\sqrt{s}} \psi\left(\frac{t-u}{s}\right) \qquad \text{for} \quad (s,u) \in R^2. \tag{2}$$

the parameters s and u that label the members of a wavelet family are known as the *dilation* and *translation* parameters, respectively. A wavelet function, $\psi(t)$, belongs to the set of square-integrable functions, i.e., $\psi(t) \in L^2(t)$, and satisfies the following admissibility condition

$$C_\psi = \int_0^\infty \frac{|\hat{\psi}(\omega)|}{\psi} \, d\omega < +\infty, \tag{3}$$

with $\hat{\psi}(\omega)$ signifying the Fourier transform of $\psi(t)$. Condition (3) implies that as $\omega \to 0$, $\hat{\psi}(\omega)$ approaches zero faster and thus represents a *band-pass filter*. $\psi(t)$ is either compactly supported in time, or has some fast decay at infinity implying that condition (3) is equivalent to the vanishing of the first p moments (see Section III of third Chapter in this volume):

$$\int_{-\infty}^\infty t^m \psi(t) \, dt = 0, \qquad m = 0, 1, 2, \ldots, p - 1 \tag{3a}$$

where p is the order of approximation considered for $\psi(t)$. Figure 8 shows examples of various wavelet functions.

1. Resolution in Time and Frequency

In the time domain, a wavelet placed at a translation point, u, has a standard derivation, σ_u, around this point. Similarly, in the frequency domain a wavelet is centered at a given frequency, ω (determined by the value of the dilation parameter) and has a standard deviation, σ_ω, around the specific frequency. The values of σ_u and σ_ω are given by

$$\sigma_u^2 = -\int_{-\infty}^{+\infty} t^2 |\psi(t)|^2 \, dt \qquad \sigma_\omega^2 = \int_{-\infty}^{+\infty} \omega^2 |\hat{\psi}(\omega)|^2 \, d\omega \tag{4}$$

As the dilation parameter, s, increases, the wavelets have significant values over a broader time segment and the resulting resolutions in the time and frequency domains change as $s\sigma_u$ and σ_ω/s, respectively. Thus, at large

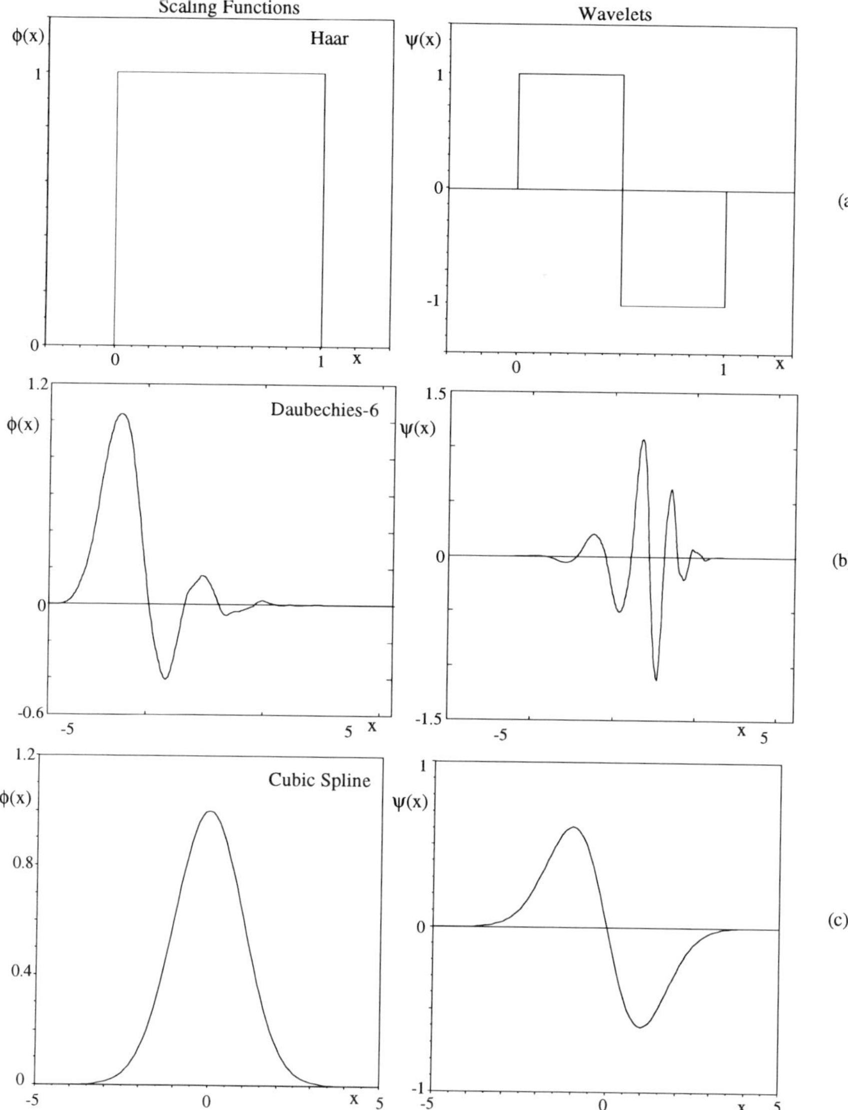

FIG. 8. Typical wavelets and scaling functions: (a) Haar, (b) Daubechies-6, (c) cubic spline.

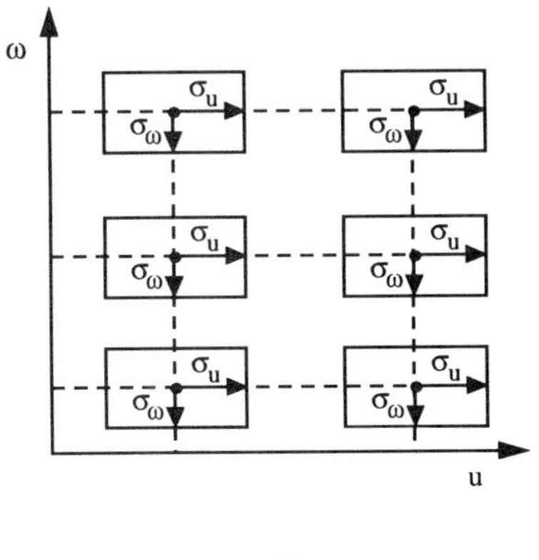

(a)

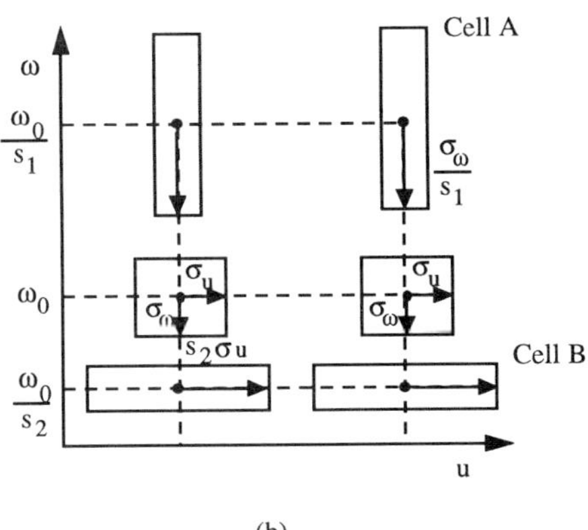

(b)

FIG. 9. Resolution in scale space of (a) window Fourier transform and (b) wavelet transform.

dilations (i.e., *scales*) the resolution in the time domain becomes coarser, while in the frequency domain, it becomes finer. The dimensions of the resolution cell (Fig. 9b) in the *scale-space* is equal to

$$[u_0 - s\sigma_u, u_0 + s\sigma_u] \times \left[\frac{\omega_0}{s} - \frac{\sigma_\omega}{s}, \frac{\omega_0}{s} + \frac{\sigma_\omega}{s}\right],$$

where u_0 is the translation point of a wavelet and consequently the center of its energy's concentration in the time domain, and ω_0/s is the frequency where the wavelet's energy is concentrated in the frequency domain. The variable dimensions of the resolution cell (Fig. 9b) are characteristic of the wavelet's ability to provide a variable-window framework for the localization in both time and frequency (i.e., scale), of the features contained in a signal. By comparison, the resolution cell of the windowed Fourier transform (Fig. 9a) maintains constant dimensions and thus produces inefficient representation of the signal's features.

Decomposition and Reconstruction of Functions

Let $F(t)$ be a square-integrable function, i.e., $F(t) \in L^2(t)$. The projection of $F(t)$ on a wavelet, $\psi_{su}(t)$, at dilation s and translation point, u, is given by

$$\alpha_{su} \equiv [W_{su} F(t)] = \int_{-\infty}^{+\infty} F(t)\psi_{su}(t)\, dt. \tag{5}$$

Using these projections over all wavelets with $(s, u) \in R^2$, we can reconstruct the function by the following equation:

$$F(t) = \int_{-\infty}^{+\infty} \int_{-\infty}^{+\infty} \alpha_{su}\psi_{su}(t)\, ds\, du. \tag{6}$$

If we consider a wavelet with compact support, then the projection given by Eq. (5) reflects the information content of $F(t)$ only in the resolution cell of the corresponding wavelet. For example, projections of $F(t)$ on wavelets with large dilation, s, represent the large-scale (i.e., low-frequency) components of $F(t)$. Consequently, the reconstruction of $F(t)$ through Eq. (6) indicates that we can put the functions together from "pieces," each of which represents the content of a specific segment of $F(t)$ over time [determined by $(u_0 - s\sigma_u, u_0 + s\sigma_u)$], and in terms of frequencies in a specific range [determined by $(\omega_0/s) - (\sigma_\omega/s), (\omega_0/s) + (\sigma_\omega/s)$. For example, if a function contains a sharp feature, e.g., spike, then the bulk of the informational content carried by the spike will be reflected by the projection of the function on a "narrow" wavelet, whose resolution cell is narrow in u and long in ω (e.g., cell A in Fig. 9b). On the other hand, the content of a function reflected by a slowly rising trend will be reflected by the projection of the function on a "wide" wavelet with a resolution cell such as that of cell B in Fig. 9b.

3. Discretization of Scale

When we analyze the scale-space image of a function (see Section II, B) in order to extract the trends of process variables, we are interested only in a finite number of distinct trends, as they are defined by the interval tree of scale.

Usually, the dilation parameter s is discretized along the dyadic sequence,

$$s = 2^m \qquad m \in Z,$$

where Z is the space of integer numbers. The resulting family of wavelets with a discretized dilation parameter is represented as follows:

$$\psi_{2^m u}(t) = \frac{1}{\sqrt{2^m}} \psi\left(\frac{t-u}{2^m}\right), \qquad m \in Z. \tag{2a}$$

The projection of $F(t)$ to all wavelets of the above form with $m \in Z$ and $u \in R$, yields the so-called *dyadic wavelet transform* of (t), with the following components:

$$\alpha_{2^m u} \equiv \left[W_{2^m u} F(t)\right] = \int_{-\infty}^{+\infty} F(t)\left\{\frac{1}{\sqrt{2^m}} \psi\left(\frac{t-u}{2^m}\right)\right\} dt,$$

$$m \in Z, \quad u \in R. \tag{5a}$$

The discretization of the dilation parameter need not be dyadic. Discretization schemes with integer or noninteger factors are possible and have been suggested. The reconstruction of $F(t)$ is given by

$$F(t) = \sum_{m=-\infty}^{+\infty} \int_{-\infty}^{+\infty} \alpha_{2^m u} \psi_{2^m u}(t)\, du. \tag{6a}$$

To ensure that no information is lost on $\hat{F}(\omega)$ as the dilation is discretized, the scale factors 2^m for $m \in Z$ must cover the whole frequency axis. This can be accomplished by requiring the wavelets to satisfy the following condition:

$$\sum_{m=-\infty}^{+\infty} |\hat{\psi}(2^m \omega)|^2 = 1. \tag{7}$$

Equation (6a) implies that the scale (dilation) parameter, m, is required to vary from $-\infty$ to $+\infty$. In practice, though, a process variable is measured at a finite resolution (sampling time), and only a finite number of distinct scales are of interest for the solution of engineering problems. Let $m = 0$ signify the finest temporal scale (i.e., the sampling interval at which a variable is measured) and $m = L$ be coarsest desired scale. To capture the information contained at scales $m > L$, we define a *scaling function*, $\phi(t)$, whose Fourier transform is related to that of the wavelet, $\psi(t)$, by

$$|\hat{\phi}(\omega)|^2 = \sum_{m=1}^{+\infty} |\hat{\psi}(2^m \omega)|^2. \tag{8}$$

Since $\hat{\psi}(\omega)$ must satisfy condition (3), then from Eqs. (7) and (8) we conclude that

$$\lim_{\omega \to 0} |\hat{\phi}(\omega)| = 1.$$

In other words, the energy of $\phi(t)$ can be seen as a lowpass filter.

Let us now create a family of scaling functions through the dilation and translation of $\phi(t)$

$$\phi_{2^m u}(t) = \frac{1}{\sqrt{2^m}} \phi\left(\frac{t-u}{2^m}\right), \quad m \in Z, \quad u \in R \qquad (9)$$

Consequently, the projection of $F(t)$ on $\phi_{2^m u}(t)$ produces a function whose content is completely stripped of high frequencies and includes information only at scales higher than m. As m increases, more and more details are removed from $F(t)$.

Through the employment of the scaling function we can reconstruct $F(t)$ using a finite number of scales, as follows:

$$F(t) = \sum_{m=0}^{L} \int_{-\infty}^{+\infty} \alpha_{2^m u} \psi_{2^m u}(t) \, du + \int_{-\infty}^{+\infty} \beta_{2^L u} \phi_{2^L u}(t) \, du, \qquad (10)$$

where, $\beta_{2^L u}$ is the projection of $F(t)$ on the scaling function, $\phi_{2^L u}$, and is given by

$$\beta_{2^L u} \equiv \left[S_{2^L u} F(t) \right] = \int_{-\infty}^{+\infty} F(t) \phi_{2^L u}(t) \, dt. \qquad (11)$$

4. Dyadic Discretization of Time

Having completed the decomposition and reconstruction of a function at a finite number of discrete values of scale, let us turn our attention to the discretization of the translation parameter, u, dictated by the discrete-time character of all measured process variables. The classical approach, suggested by Meyer (1985–1986), is to discretize time over dyadic intervals, using the sampling interval, τ, as the base. Thus, the translation parameter, u, can be expressed as

$$u = k(2^m \tau), \quad \text{with} \quad (m,k) \in Z^2, \qquad (12)$$

where k is the translation parameter. At the initial dilation level $m = 0$, and $u = k\tau$, $k = 0, 1, 2, \ldots$, i.e., the translation points coincide with the sampling instants, i.e., $0, \tau, 2\tau, \ldots$. At the next dilation level, $u = (2\tau)k$, i.e., the distance between two adjacent wavelets double, and the translation points are $0, 2\tau, 4\tau, \ldots$.

As a result of the dyadic discretization in dilation and translation, the members of the wavelet family are given by

$$\psi_{mk}(t) \equiv \frac{1}{\sqrt{2^m}} \psi\left(\frac{t}{2^m} - k\tau\right), \qquad (m,k) \in Z^2.$$

A discrete-time signal is decomposed into the following set of projections:

$$\alpha_{mk} \equiv \left[W_{2^m, k(2^m\tau)}F(t)\right] = \sum_{k=-\infty}^{+\infty} F(t)\psi_{mk}, \qquad 1 \le m \le L, \qquad (5b)$$

$$\beta_{Lk} \equiv \left[S_{2^L, k(2^L\tau)}F(t)\right] = \sum_{k=-\infty}^{+\infty} F(t)\phi_{Lk}, \qquad (11a)$$

from which it can be completely reconstructed:

$$F(t) = \sum_{m=1}^{L} \sum_{k=-\infty}^{+\infty} \alpha_{mk}\psi_{mk} + \sum_{k=-\infty}^{+\infty} \beta_{Lk}\phi_{Lk}. \qquad (6b)$$

For practical purposes, the wavelet decomposition can only be applied to a finite record of discrete-time signals. If N is the number of samples in the record, and $\tau = 1$, then the maximum value of the translation parameter can be found from Eq. (12), by setting $u = N$, and is equal to $k_{\max} = N/2^m$. Consequently, the decomposition and reconstruction relations [Eqs. (5b), (11a), (6b)] take the following form:

$$\alpha_{mk} \equiv \left[W_{2^m, k2^m}F(t)\right] = \sum_{k=1}^{k_{\max}} F(t)\psi_{mk}, \qquad 1 \le m \le L, \qquad (13)$$

$$\beta_{Lk} \equiv S_{2^m, k2^m}F(t) = \sum_{k=1}^{k_{\max}} F(t)\phi_{Lk}, \qquad (14)$$

$$F(t) = \sum_{m=1}^{L} \sum_{k=1}^{k_{\max}} \alpha_{mk}\psi_{mk} + \sum_{k=1}^{k_{\max}} \beta_{Lk}\phi_{Lk}. \qquad (15)$$

Equation (15) provides a discrete, complete and nonredundant representation of the function $F(t)$, since it requires the computation of N coefficients, if N is the number of discrete-time samples describing $F(t)$.

5. Uniform Discretization of Time

An alternative approach to the discretion of the translation parameter u involves uniform sampling of the measured signal at all scales, i.e., $u = k\tau$, with $k \in Z$. The resulting decomposition algorithm is of complexity $O(N \log N)$, and the associated reconstruction requires the computation of $N \log N$ coefficients, i.e., it contains redundant information.

Nevertheless, uniform discretization of time at all scales leads to representations that are highly suitable for feature extraction and pattern recognition. More on this subject in a subsequent paragraph.

6. Practical Considerations

Let us now see how the theory of the wavelet-based decomposition and reconstruction of discrete-time functions can be converted into an efficient numerical algorithm for the multiscale analysis of signals. From Eq. (6b) it is easy to see that, given a discrete-time signal, $F_0(t)$ we have

$$F_0(t) = D_1(t) + F_1(t), \tag{16}$$

where

$$F_1(t) = \sum_{k=-\infty}^{+\infty} \beta_{1k}\phi_{1k}(t) \quad \text{and} \quad D_1(t) = \sum_{k=-\infty}^{+\infty} \alpha_{1k}\psi_{1k}(t). \tag{17}$$

$F_1(t)$ is called the *scaled signal* and is derived from the filtering of $F_0(t)$ with the lowpass scaling function. It represents a smoother version of $F_0(t)$. $D_1(t)$ is called the *detail signal* and is derived from the filtering of $F_0(t)$ with the bandpass wavelet functions. It represents the information that was filtered out of $F_0(t)$ in producing $F_1(t)$.

Generalizing Eqs. (16) and (17), we take

$$F_m(t) = D_{m+1}(t) + F_{m+1}(t), \qquad m \in Z, \tag{18}$$

with

$$F_{m+1}(t) = F_m(t){}^*H_m \quad \text{and} \quad D_{m+1}(t) = F_m(t){}^*G_m, \tag{19}$$

where the operator, $*$, signifies the convolution operation. Filter H_m is a lowpass filter, emulating the effects of the scaling function, and G_m is a bandpass filter affecting the influence of the wavelet function. Equations (18) and (19) imply that the wavelet-based decomposition and reconstruction of a discrete-time signal can be carried out through a cascade of convolutions with filters H and G (see Fig. 10). Figure 11 shows the scaled and detail signals generated from the wavelet decomposition of a given signal using dyadic and uniform sampling. The detailed methodological aspects of the cascaded convolution of signals, described above, can be found in Mallat (1989) for the case of dyadic sampling of time and in Mallat and Zhong (1992) for uniform sampling.

For real-world signals with finite records, the convolution processes leading to decomposition and reconstruction of a signal, require data in regions beyond the signal's endpoints. Assuming a mirror image of the

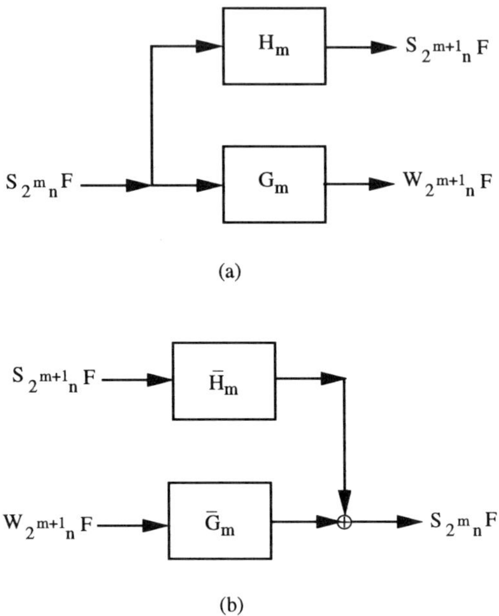

FIG. 10. Methodology for multiscale (a) decomposition and (b) reconstruction, using wavelets, with uniform sampling $(m, n) \in Z^2$.

original signal beyond both ends (Mallat, 1989), causes numerical inaccuracies in the computation of the wavelet coefficients. Bakshi and Stephanopoulos (1995) have studied various approaches for handling endpoint effects and they have suggested certain practical solutions in this problem.

B. Extraction of Multiscale Temporal Trends

In Section II we defined the trend of a measured variable as a strictly ordered sequence of scaling episodes. Since each scaling episode is defined by its bounding inflexion points, it is clear that the extraction of trends necessitates the localization of inflexion points of the measured variable at various scales of the scale-space image. Finally, the interval tree of scale (see Section II) indicates that there is a finite number of distinct sequences of inflexion points, implying a finite number of distinct trends. The question that we will try to answer in this section is, "How can you use the wavelet-based decomposition of signals in order to identify the distinct sequences of inflexion points and thus of the signal's trends?"

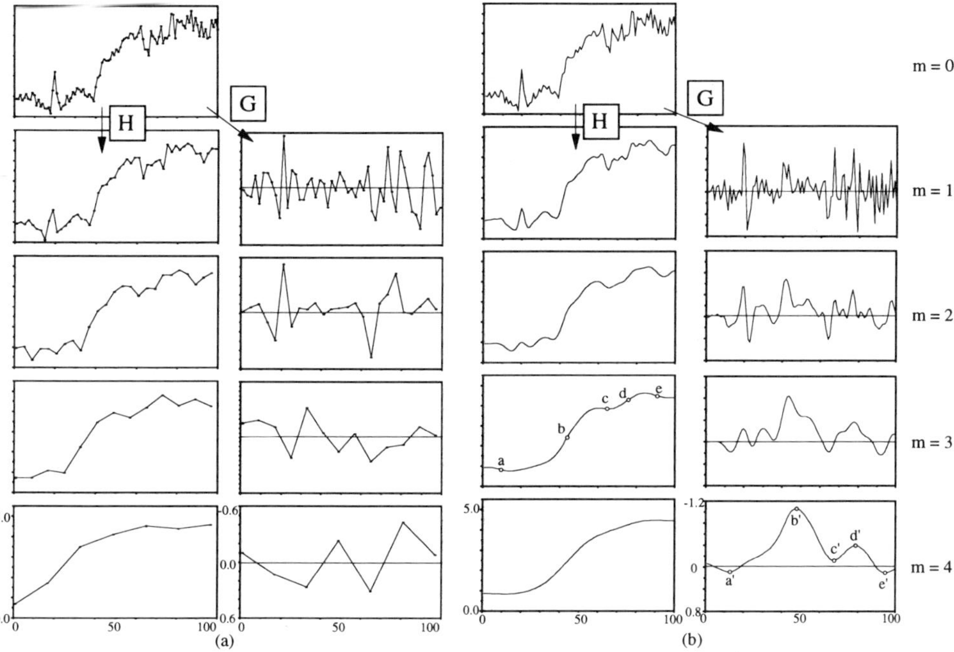

FIG. 11. Wavelet decomposition (a) dyadic sampling using Daubechies-6 wavelet; (b) uniform sampling using cubic spline wavelet.

1. Translationally Invariant Representation of Measured Variables

Measured variables may have identical features in different segments of the time record. Any analysis method should be able to identify and extract such common features, independently of the time segment in which they appear. Consequently, the wavelet decomposition of a time-translated signal should produce wavelet coefficients that are translated in time but equal in magnitude to the wavelet coefficients of the original (untranslated) signal. For example, if $F(t) \in L^2$ is the original signal and $F_\theta(t)$ is the same function translated in time by θ units, i.e., $F_\theta(t) = F(t - \theta)$, then

$$\left[W_{2^m u} F_\theta(t) \right] = \left[W_{2^m u} F(t - \theta) \right], \qquad m \in Z, \quad u \in R. \qquad (20)$$

Since the translation variable, u, is also discretized (dyadically, or uniformly at each scale), relationship (20) will hold only if the signal is translated in time by a period that is an integer multiple of the discretiza-

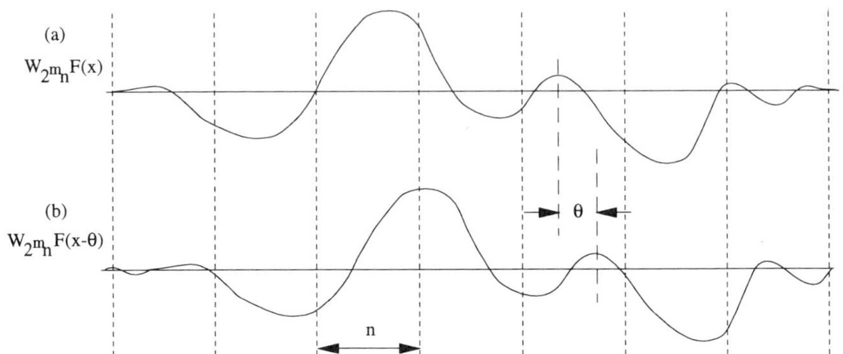

FIG. 12. Translation of the wavelet transform of $F(x)$ by θ, with sampling rate, n (Mallat, 1991).

tion interval at all scales. For example, in the case of dyadic sampling of time, translational invariance of the wavelet transform requires that $\theta = r(2^m \tau)$, $r \in Z$, whereas uniform sampling necessitates that $\theta = r\tau$, $r \in Z$, where τ is the sampling period of the original signal.

Unfortunately, the requirements for translational invariance of the wavelet decomposition are difficult to satisfy. Consequently, for either discretization scheme, comparison of the wavelet coefficients for two signals may mislead us into thinking that the two trends are different, when in fact one is simply a translation of the other.

On the other hand, the value of (zero-order) zero crossings and local extrema (first-order zero crossings) of a dyadic wavelet transform, $W_{2^m_u}F(t)$, do translate when the original signal is translated (Fig. 12). Therefore, we can generate translationally invariant representations of a signal if we employ zero-order and first-order zero crossings of the signal's wavelet transform.

2. Detection of Inflexion Points

Within the framework of scale-space filtering, inflexion points of $F(t)$ appear as extrema in $\partial F(t)/\partial t$ and zero crossings in $\partial^2 F(t)/\partial t^2$. Thus, filtering a signal by the Laplacian (second derivative) of a Gaussian will generate the inflexion points at various scales (Marr and Hildreth, 1980). In the same spirit, if the wavelet is chosen to be the first derivative of a scaling function, i.e., $\psi(t) = d\phi(t)/dt$, then from Eqs. (5a) and (11) we can easily show (Mallat and Zhong, 1992) that

$$[W_{2^m u} F(t)] = 2^m \frac{d}{dt}[S_{2^m u} F(t)].\tag{21}$$

A wavelet defined as above is called a *first-order wavelet*. From Eq. (21) we conclude that the extrema points of the first-order wavelet transform provide the position of the inflexion points of the scaled signal at any level of scale. Similarly, if $\psi(t) = d^2\phi(t)/dt^2$, then the zero crossings of the wavelet transform correspond to the inflexion points of the original signal smoothed (i.e., scaled) by the scaling function, $\psi(t)$ (Mallat, 1991).

An obvious choice for the scaling function is the Gaussian, but it is not a compactly supported function, and the corresponding filter requires a large number of coefficients. The scaling function and associated wavelet chosen in this chapter to support the extraction of process trends were suggested by Mallat and Zhong (1992). The scaling function is a cubic spline with compact support of size 4 (i.e., the corresponding filter requires four coefficients), whereas the associated wavelet is not compactly supported but decreases exponentially at infinity (Fig. 8c). Using these functions, we can generate efficient representations of signals. The scaled signals at various scales are equivalent to the cubic spline fits of the data at the various scales. Unfortunately, since the scaling function is not a Gaussian, we cannot guarantee that spurious oscillations with fictitious inflexion points will not be generated at coarser scales. Nevertheless, such fictitious inflexion points, if generated, are short-lived, i.e., disappear quickly with increasing scale, and thus they will never be stable and distinguished features of any process trend. It is also worth noting that the variation in the magnitude of local extrema of the wavelet decomposition across several scales provides useful information about the nature of the inflexion points, and thus offers a measure of the signal's regularity.

3. The Wavelet Interval–Tree of Scale

Consider the given signal and its wavelet decomposition, as shown in Fig. 11. Since the wavelet is a first-order wavelet, the inflexion points of the original function appear as extrema of the detail signal at the corresponding scale (Fig. 11). If we connect the extrema of the detailed signal across various scales, we generate a structure similar to the interval tree of scale, discussed in Section II, B. We will call this tree, the *wavelet interval–tree of scale* (Fig. 13). It gives the evolution of the features of a signal in the scale space, and can be used to generate the trends of measured variables with various combinations of features. The wavelet

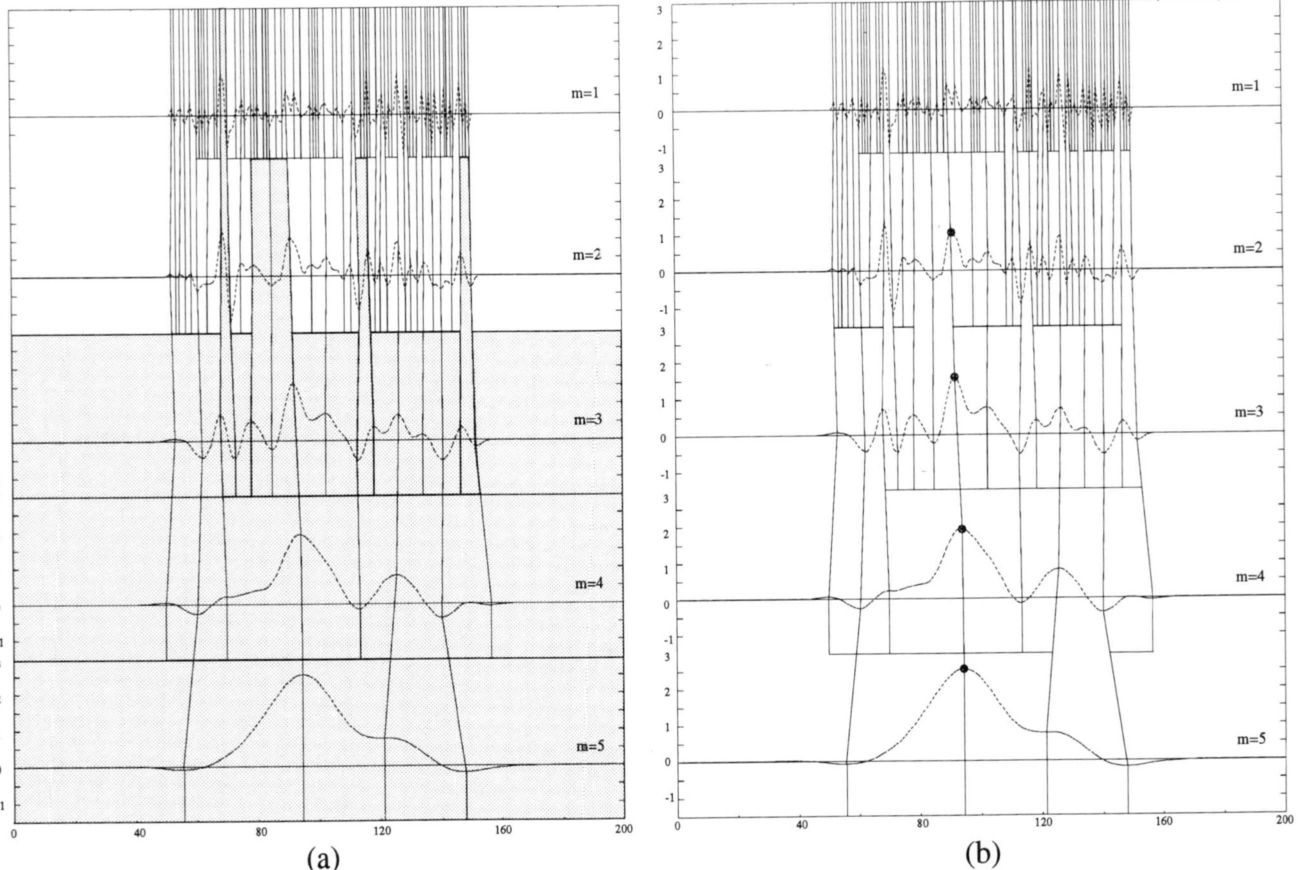

FIG. 13. Wavelet interval–tree of scales for the signal of Fig. 1; (a) extrema in shaded region reconstruct stable trend at $m = 3$ (see Fig. 15b); (b) encircled extrema reconstruct trend with accurate step change (see Fig. 15d).

interval–tree of scale presents certain advantages over the classical scale-space filtering, which are very important in the reconstruction of process trends. These advantages are (Bakshi and Stephanopoulos, 1994a):

1. The wavelet interval–tree of scale is constructed from $\log_2 N$ distinct representations, where N is the number of points in the record of measured data. This is a far more efficient representation than that of scale-space filtering with continuous variation of Gaussian σ.
2. In scale-space filtering, trends consisting of combinations of features at various scales are constructed by approximating the stable features through piecewise continuous segments, like parabolas. Such an approach leads to discontinuities in the trends. On the contrary, the trends generated from the wavelet interval-tree of scale with a mixture of features from different scales, are continuous up to the degree of continuity of the scaling function, and are based on formal and sound analysis.
3. The detail signals at the various scales of the tree provide the features of the trend that are distinct to the corresponding range of scales.
4. The evolution of the inflexion points, as given by the local extrema of the detail signals, also characterize the regularity of the original signal's inflexion points.

4. The Algorithm for the Extraction of Trends

Let us recall that *a trend is a strictly ordered sequence of scaling episodes*, whose bounding inflexion points can be identified by the local extrema of the detail signals, as discussed in the previous paragraph. Clearly, the more stable a scaling episode (over a range of scales), the more distinguished is the corresponding feature in describing the trend. Consider, for example, the noise-corrupted pulse shown in Fig. 14 along with the associated detail signals, resulting from its decomposition with a first-order wavelet. The extrema, A and B, of the detail signal at the smallest scale persist at all scales, indicating the presence of a very stable scaling episode. This scaling episode is bounded by the inflexion points on the left and right edges of the pulse and reveals the most dominant feature of the original signal, i.e., the pulse itself. On the contrary, the inflexion points corresponding to the fluctuations of the noise in the original signal disappear very quickly and do not represent distinguished scaling episodes, i.e., features. The range of each extremum point is equal to the range of the G filter and is equal to $2^{m+1}q + 1$, where m is the scale of the corresponding detail signal and $2q + 1$ is the length of the filter. By

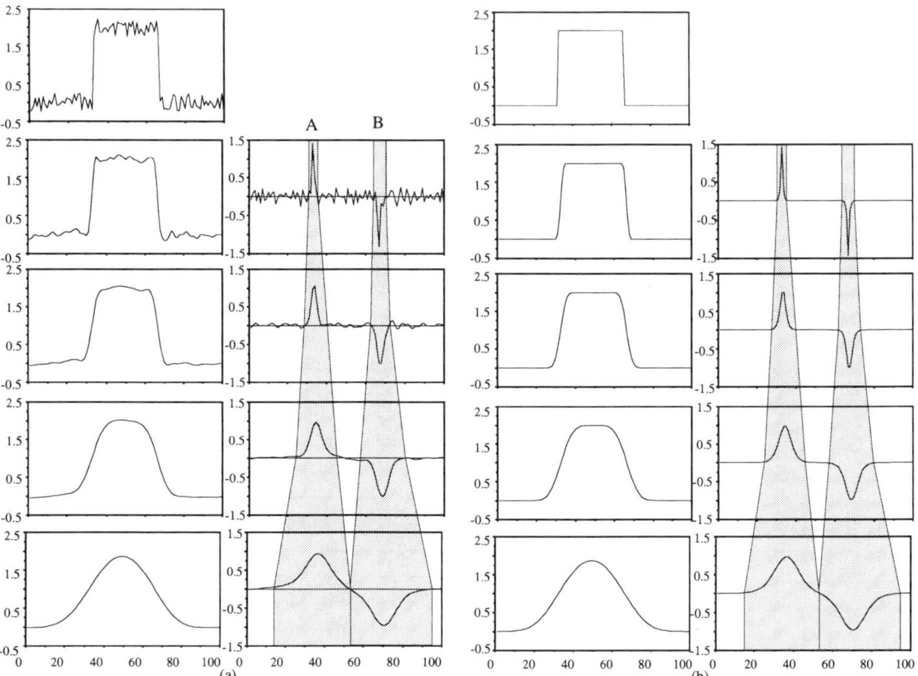

FIG. 14. Extracting distinguishing features from noise pulse signal. Wavelet coefficients in shaded regions represent stable extrema. (a) Wavelet decomposition of noisy pulse signal; (b) wavelet decomposition of pulse signal. (Reprinted from Bakshi and Stephanopoulos, "Representation of process trends, Part III. *Computers and Chemical Engineering*, **18**(4), p. 267, Copyright (1994), with kind permission from Elsevier Science Ltd., The Boulevard, Langford Lane, Kidlington OX5 1GB, UK.)

utilizing the wavelet coefficients of the stable extrema we can reconstruct a signal that contains the distinguished features at several scales. In Fig. 14, by utilizing the wavelet coefficients of the shaded region, we can completely reconstruct the pulse with minimum quantitative distortion.

The algorithm for extracting the most stable description of trends at each level of scale, proceeds as follows:

Step 1. Generate the finite, discrete dyadic wavelet transform of data using Mallat and Zhong's (1992) cubic spline wavelet (Fig. 8c).

Step 2. Generate the wavelet interval–tree of scale.

Step 3. For each scaling level, m, and for each scaling episode, (a) determine the range of stability and (b) collect the wavelet coefficients at all scales in the range of the episode's stability.

Step 4. Reconstruct the signal using the last scaled signal and the wavelet coefficients collected in step 3.

Once the stable reconstruction of a signal has been accomplished, its subsequent representation can be made at any level of detail, i.e. qualitative, semi-quantitative, or fully real-valued quantitative. The triangular episodes (described in Section I, A) can be constructed to offer an explicit, declarative description of process trends.

5. Illustrative Examples

a. Example 1: Generating Multiscale Descriptions. Consider the process signal and its wavelet decomposition shown in Fig. 11. Quantitatively accurate representations at each of the dyadic scales may be generated by selecting the wavelet transform extrema for the most stable episodes at each scale. For example, the extrema selected for reconstructing the process trend at scale, $m = 3$, are shown shaded in Fig. 13a. Note that wavelet coefficients at scales < 3 are needed to provide a stable representation. The reconstructed signal is shown in Fig. 15b, and is qualitatively equivalent to the scaled signal at $m = 2$, but is quantitatively much more accurate, since features such as the spike at $t = 20$ are undistorted. Similarly, the stable process trend at scale $m = 4$ is shown in Fig. 15c. The step change at $t = 40$ is not brought out accurately in the trends in Figs. 15b and 15c, due to the presence of other extrema in the range of the extremum corresponding to the step change. Most of the extrema in its range are much smaller than the extremum at $t = 40$. An empirical criterion, based on the relative magnitudes of wavelet transform extrema may be used for selecting large extrema, resulting in the reconstructed signal shown in Fig. 15d. The extremum at $t \approx 40$, at scales $m = 2$–5 used for this reconstruction, as shown in Fig. 13b. The step change is brought out quite accurately in this trend at $m = 5$.

Another powerful heuristic for extracting process trends is Witkin's stability criterion (Witkin, 1983). Starting with the episodes at the desired scale, lower-scale descriptions are considered only if the lower-scale episodes have a higher mean stability than their parent episode. The episodes selected using this criterion at scale, $m = 5$, and the reconstructed signal are shown in Fig. 16. The reconstructed signal contains all the conspicuous features in the raw data, without quantitative distortion. Witkin's stability criterion is very effective for extracting the most relevant features from a process signal.

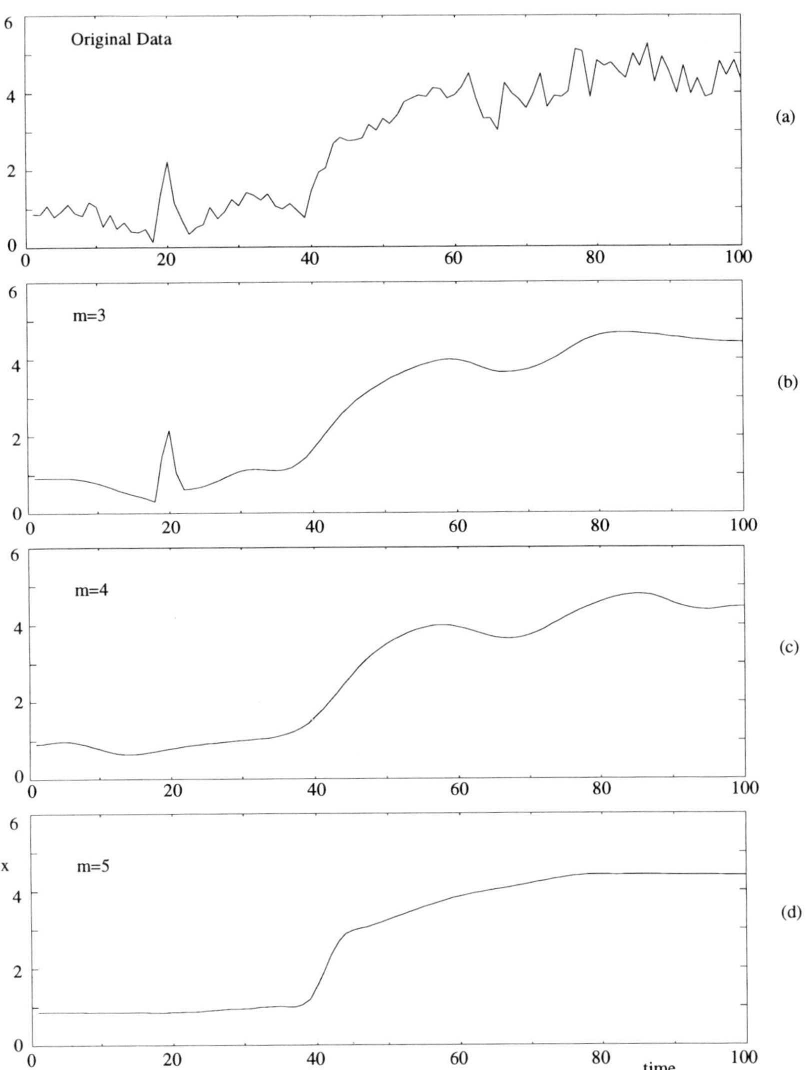

FIG. 15. Process trends of the signal in Fig. 1, extracted at various scales: (a) original data; (b) stable trend at $m = 3$ (see Fig. 13); (c) stable trend at $m = 4$; (d) stable trend at $m = 5$, neglecting small extrema.

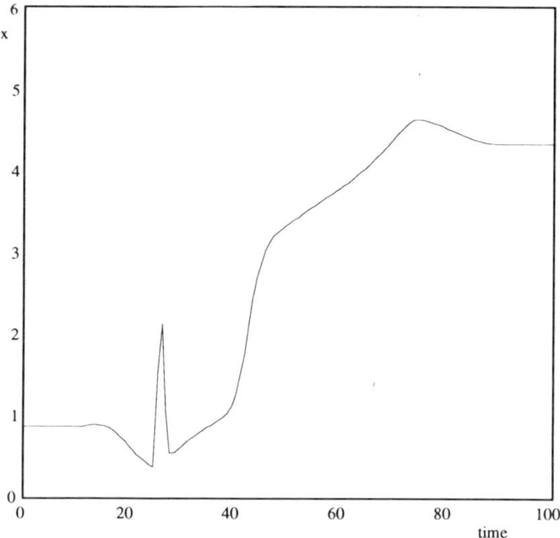

FIG. 16. Stable trend of the signal in Fig. 1, generated through the use of Witkin's stability criterion, and the wavelet coefficients shown in Fig. 13b.

b. Example 2. Generalization of process trends. Comparison of process signals obtained from several examples representing different process conditions is essential for evaluating the relevance of the extracted features. The representation of process trends at multiple scales provides a convenient, hierarchical technique for evaluating the features. Consider the signals shown in Fig. 17a obtained from three different batches of the fed-batch fermentation process described in Section V. At the level of the raw data, it is difficult to compare the three signals. On representing the signal at multiple scales, the differences in the signals disappear, and at a coarse-enough scale, the signals become qualitatively identical, as shown in Figs. 17–17g. The discovery of a generalized description provides a means of comparing the qualitative and quantitative features contained in the signals, and allows matching of features in the trends, facilitating easy extraction of qualitative differences. If the information in the signals at coarse scales is inadequate for distinguishing between the signals, then information in trends at finer scales is considered. As lower scales are considered, only the matched features need to be compared with each other. This provides a natural decomposition to the learning problem, and simplifies it significantly. The utility of generalized descriptions of process signals will be exploited further in the learning of input/output mappings, as described in Section V. The process trends obtained by applying Witkin's stability criterion at the coarsest scale are shown in Fig. 17h. These process trends are also quantitatively very similar, and the extracted features are physically meaningful and thus interpretable, as we will see in the illustrations of Section V (Figs. 21–23).

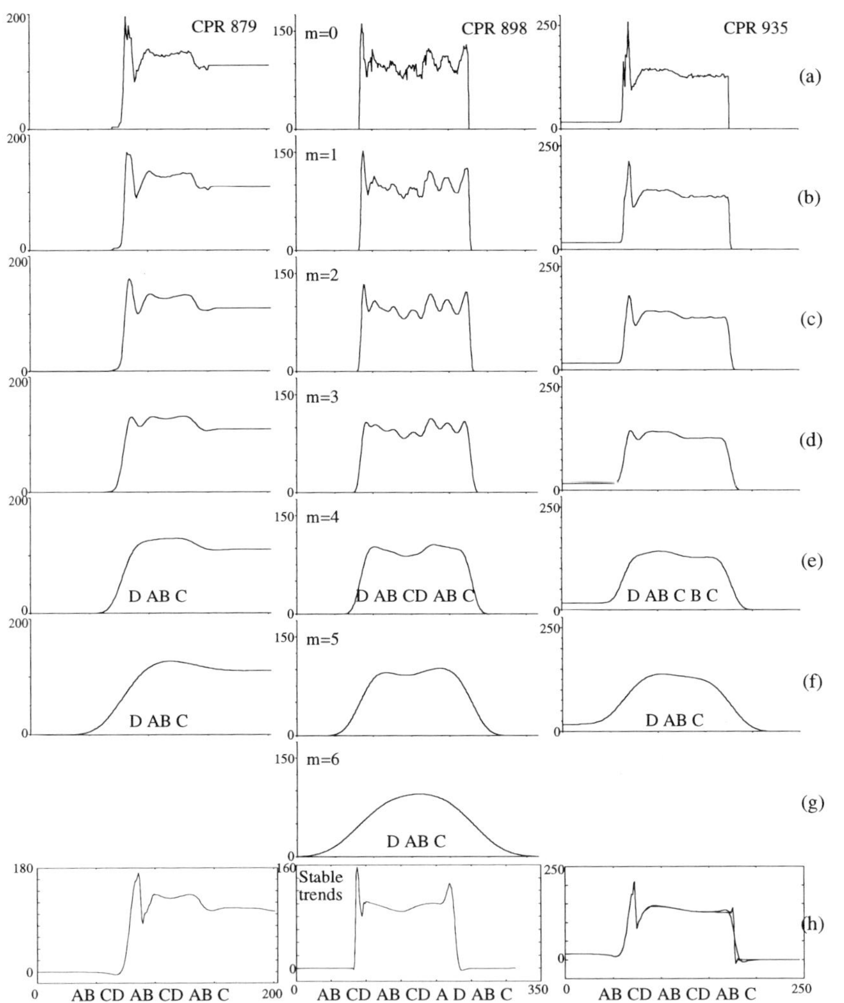

Fig. 17. Generalization of process trends for three distinct records; (a) raw data; (b)–(g) scaled signals; (h) stable trends.

IV. Compression of Process Data through Feature Extraction and Functional Approximation

With increasing computerization and improvements in sensor technology, it is relatively easy and inexpensive to collect large quantities of process data. Since measured data are useful for performing a variety of analytical and decision support engineering tasks, it is essential to store the data in historical records for future use. Efficient storage techniques are needed for two primary reasons: (1) to reduce the space required for the historical records and (2) to retrieve the data in a manner than renders the data easily interpretable for the execution of engineering tasks. In this section we will examine how both of these needs can be satisfied within the same theoretical framework of wavelet decomposition for the representation of measured signals.

A. DATA COMPRESSION THROUGH ORTHONORMAL WAVELETS

Data compaction involves representation of the measured signal as an approximation that requires less storage space, at the cost of losing some information from the original signal. The data compaction problem may be stated as an approximation problem (Bakshi and Stephanopoulos, 1995) as follows:

1. Definition: Data Compaction Problem

Determine the approximate representation $\tilde{F}$ of a discrete-time function, $F(t)$ so as to either (1) minimize the error of the approximation given by

$$e_{L_p} = \frac{\|F - \tilde{F}\|_{L_p(I)}}{N} = \frac{\left(\int_I |F(t) - \tilde{F}(t)|^P \, dt\right)^{1/p}}{N}, \qquad 0 < p < \infty,$$

where N is the number of data points contained in the original signal, for a given compression ratio, $C_R = (N/\tilde{N})$, where $\tilde{N}$ is the number of points stored in the compacted representation; or (2) maximize the compression rate, C_R, for a given error of approximation, $e_{L_p}^2$.

A popular technique for approximation consists of representing the data as a weighted sum of a set of basis functions. As described in Section III, A, wavelets form a convenient set of basis functions to represent signals consisting of a variety of features. A signal decomposed on an orthonormal basis, using dyadic discretization of the translation and dilation parameters may be represented in terms of its wavelet coefficients and the coefficients of the last scaled signal, as given by Eq. (15). Since the wavelets at all translations and dilations, and scaling functions at a given dilation are orthonormal, the total energy in the signal is equal to the sum of the square of the coefficients:

$$\|F(t)\|_{L_2}^2 = \sum_{m=1}^{L}\sum_{k=1}^{k_{max}} \alpha_{mk}^2 + \sum_{k=1}^{k_{max}} \beta_{Lk}^2.$$

Compression may be achieved if some regions of the time–frequency space in which the data are decomposed do not contain much information. The square of each wavelet coefficient is proportional to the least-squares error of approximation incurred by neglecting that coefficient in the reconstruction;

$$e_N^2 = \frac{\|F(t) - \tilde{F}(t)\|^2}{N^2} = \frac{\sum_{m=1}^{L}\sum_{k=1}^{k_{max}} \alpha_{mk,\,\text{neglected}}^2 + \sum_{k=1}^{k_{max}} \beta_{Lk,\,\text{neglected}}^2}{N^2},$$

$$(22)$$

where $\alpha_{mk,\,\text{neglected}}$ and $\beta_{mk,\,\text{neglected}}$ are the coefficients that are not stored in the compacted representation. Similarly, the local error in a region consisting of $(2r + 1)$ points, bounded by the interval $[l - r, l + r]$ is given by

$$e_{2r+1}^2 = \frac{\|F(t) - \tilde{F}(t)\|^2}{(2r+1)^2} = \frac{\sum_{m=1}^{L}\sum_{k=l-r}^{l+r} \alpha_{mk,\,\text{neglected}}^2 + \sum_{k=l-r}^{l+r} \beta_{Lk,\,\text{neglected}}^2}{(2r+1)^2}.$$

$$(23)$$

As $r \to 0$, the error of approximation tends to the L^∞ norm. Equations (22) and (23) are useful for data compaction with orthonormal wavelets.

Several complete orthonormal bases may be used for data compression. The selection of the best basis may be performed by utilizing several different criteria suggested by Coifman, Wickerhauser, and coworkers (Coifman and Wickerhauser, 1992; Wickerhauser, 1991). Some of the most interesting basis selection criteria include those discussed in the following paragraphs.

a. Number above a Threshold. The set of wavelets with the minimum number of coefficients above a threshold, ε, is selected as the best basis. This gives the best basis to represent a signal to precision ε. This measure is similar in principle to the error measure for the box car and backward slope methods.

b. Entropy. The statistical thermodynamical entropy is given by

$$H(\alpha) = -\sum_j p_j \log p_j,$$

where

$$p_j = \frac{|\alpha_j|^2}{\|\alpha_j\|^2} \quad \text{and} \quad p \log p = 0 \quad \text{for} \quad p = 0.$$

Coefficients of the wavelet expansion are indicated by α_j. Entropy is not an additive cost function, but it can be written as

$$H(\alpha) = \|\alpha\|^{-2}\lambda(\alpha) + \log\|\alpha\|^2,$$

where

$$\lambda(\alpha) = -\sum_j |\alpha_j|^2 \log|\alpha_j|^2$$

is additive, and can be minimized. The entropy measure provides a physically meaningful criterion for selecting the appropriate coefficients for data compression because $\exp[H(\alpha)]$ is proportional to the number of coefficients needed to represent the signal to a fixed mean square error.

c. Number Capturing Given Percentage of Signal Energy. This measure explicitly selects the smallest number of coefficients necessary to represent the signal with a given least-squares error. The cost function is the number of coefficients, N_e, with the largest absolute value that capture a percentage, e, of signal energy. These N_e coefficients represent the signal with the least-squares error, e, in the local region covered by the wavelet in the most compact form:

$$\sum_{j=1}^{N_e} \alpha_j^2 \geq \frac{e}{100} \sum_{j=1}^{N} \alpha_j^2,$$

where N is the total number of coefficients in the given wavelet decomposition. This measure evaluates the number of coefficients necessary to approximate the signal with a given least-squares error of approximation.

B. Compression through Feature Extraction

Compression of process data through feature extraction requires

1. Techniques for representing features in the process data in an explicit manner to allow selection of the relevant features.
2. Criteria for determining what features in a process signal are relevant and worth storing, and what features may be lost due to compression.

Multiresolution representation of a process signal allows satisfaction of both these requirements. Decomposition of a signal using derivative wavelets and uniform sampling of the translation parameter provides a technique for representing the dominant features in a signal at various scales, as discussed in Section III, B. The relevance of features in the process signal may be determined based on their persistence over multiple scales. These properties of derivative wavelets are based on the work of Mallat and Zhong (1992), and have been exploited for extracting features from process data by Bakshi and Stephanopoulos (1994a) and were discussed in Section III, B.

C. Practical Issues in Data Compression

The practical, implementational issues that arise during the utilization of the data compression techniques, presented in the previous two subsections, are discussed in the following paragraphs.

1. Compression in Real Time

The speed with which the data need to be compressed depends on the stage of data acquisition at which compression is desired. In intelligent sensors it may be necessary to do some preliminary data compression as the data are collected. Often data are collected for several days or weeks without any compression, and then stored into the company data archives. These data may be retrieved at a later stage for studying various aspects of the process operation.

In order to compress the measured data through a wavelet-based technique, it is necessary to perform a series of convolutions on the data. Because of the finite size of the convolution filters, the data may be decomposed only after enough data has been collected so as to allow convolution and decomposition on a wavelet basis. Therefore, point-by-point data compression as done by the boxcar or backward slope methods is not possible using wavelets. Usually, a window of data of length 2^m, $m \in Z$, is collected before decomposition and selection of the appropriate features for storage are carried out. The optimal window size depends on the nature of the signal, the compression parameter used, and the decomposing wavelet. The window size should be greater than the duration of the longest discardable or irrelevant even in the signal. For a measured variable, the best basis may be different for compressing data in different windows. Usually, the time–frequency characteristics of the signal from a given measured variable do not vary significantly with time. Therefore, a best basis may be selected from a few sets of data and then used for new data. The feasibility of the selected best basis may be evaluated at regular intervals, if necessary.

2. Inaccuracies Due to End Effects

A practical problem in decomposing and reconstructing signals using wavelets is the errors introduced due to boundary effects. Both wavelet decomposition and reconstruction with limited amounts of data require assumptions about the signal's behavior beyond its endpoints. Usually, the signal is assumed to have its mirror image beyond its boundaries. This assumption often introduces an unacceptable error in the reconstructed signal, particularly near its endpoints. For processes, where the signal is compressed by analyzing windows of finite sizes, this error may be eliminated by augmenting the signal beyond its endpoints by segments of constant value. Thus, for a signal of length 128 points, the decomposition is performed on a signal that is augmented by constant segments of length equal to 64 points on either end. The resulting decomposition has $L + 1$, i.e., one additional scale that is created because of the augmentation. For data compression, this additional scale is disregarded, and the scaling function and wavelet coefficients are selected from L scales only. This simple procedure results in highly accurate reconstruction by elimination of the boundary effects. The quality of the compression is also unaffected since the wavelet coefficients for the augmented portions are constant or zero. The computational complexity of decomposition and reconstruction of the augmented signal is $O(2N)$, and still linear in the length of the original signal.

3. Selecting the Mother Wavelet and Compression Criteria

Several types of wavelets have been developed, but no formal criteria exist for selecting the mother wavelet for compressing a given signal. Some qualitative criteria and experience-based heuristics are normally used. In approximating a signal by wavelet or derivative wavelet transform extrema, the smoothness of the reconstructed signal depends directly on the nature of the basis functions. Several orthonormal wavelets with different degrees of smoothness have been designed. The orthonormal wavelets with compact support designed by Daubechies (1988) are reasonably smooth for orders greater than 6. The Haar wavelets provide piecewise constant approximation which may be adequate for some process signals, such as those of manipulated variables. Among derivative wavelets, quadratic and cubic wavelets are described by Mallat and Zhong (1992). The first derivative of a Gaussian is an infinitely differentiable wavelet. A priori knowledge of the nature of the signal may be used to select the mother wavelet based on its smoothness.

The accuracy of the error equations (Eqs. (22) and (23)] also depends on the selected wavelet. A short and compactly supported wavelet such as the Haar wavelet provides the most accurate satisfaction of the error estimate. For longer wavelets, numerical inaccuracies are introduced in the error equations due to end effects. For wavelets that are not compactly supported, such as the Battle–Lemarie family of wavelets, the truncation of the filters contributes to the error of approximation in the reconstructed signal, resulting in a lower compression ratio for the same approximation error.

D. AN ILLUSTRATIVE EXAMPLE

A typical example of half a day's process data from a distillation tower, and its wavelet decomposition using an orthonormal wavelet (Daubechies, 1988), with dyadic sampling are shown in Fig. 18. The raw process data represent pressure variation measured every minute (Takei, 1991). The raw signal is augmented by a constant segment of half the length of the original signal to avoid boundary effects. The augmentation results in an additional level of the decomposition, which is disregarded in the selection of coefficients, and the signal reconstruction. The wavelet coefficients are normalized to be proportional to their contribution to the overall signal. From Fig. 18 it is clear that many of the wavelet coefficients have very small values, which may be neglected without significant loss of information. The reconstructed signal with compression ratios of 2, 4, and 24 are shown in Fig. 19. The performance of orthonormal wavelet-based data compaction is compared with that of the conventional techniques such as backward slope and boxcar methods. As shown in Fig. 20, the wavelet-based method outperforms both the boxcar, and the backward slope methods. The quality of the reconstructed signal is significantly better for the wavelet-based method for similar compression ratios. Process data compaction using biorthogonal wavelets, wavelet packets and wavelet trans-

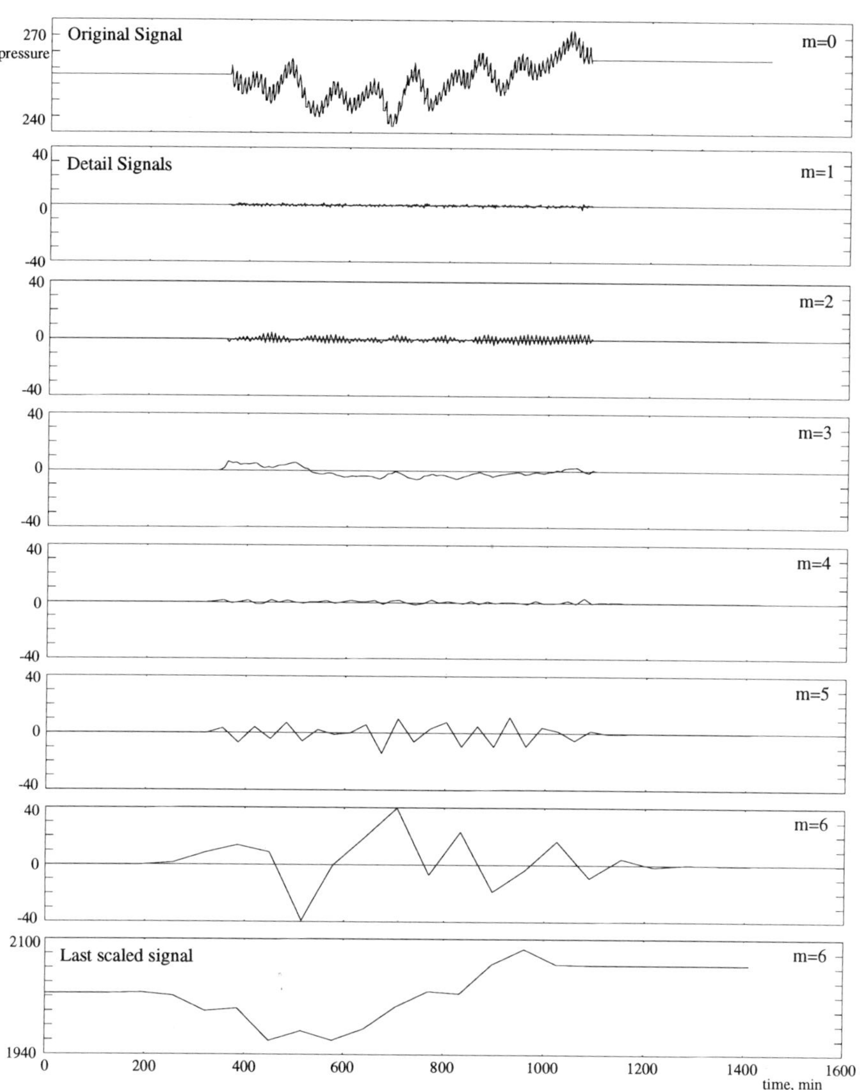

FIG. 18. Wavelet decomposition of a pressure signal, using Daubechies-6 wavelet.

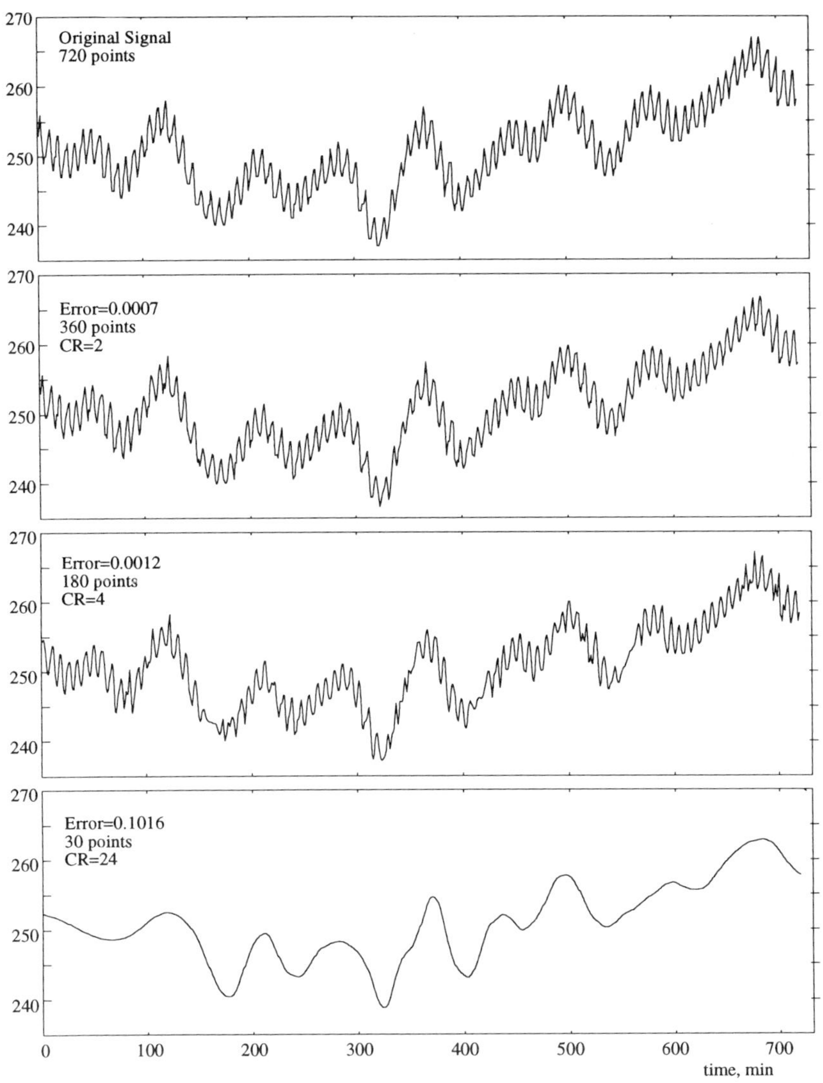

FIG. 19. Reconstruction of compressed signal from the wavelet decomposition of Fig. 18.

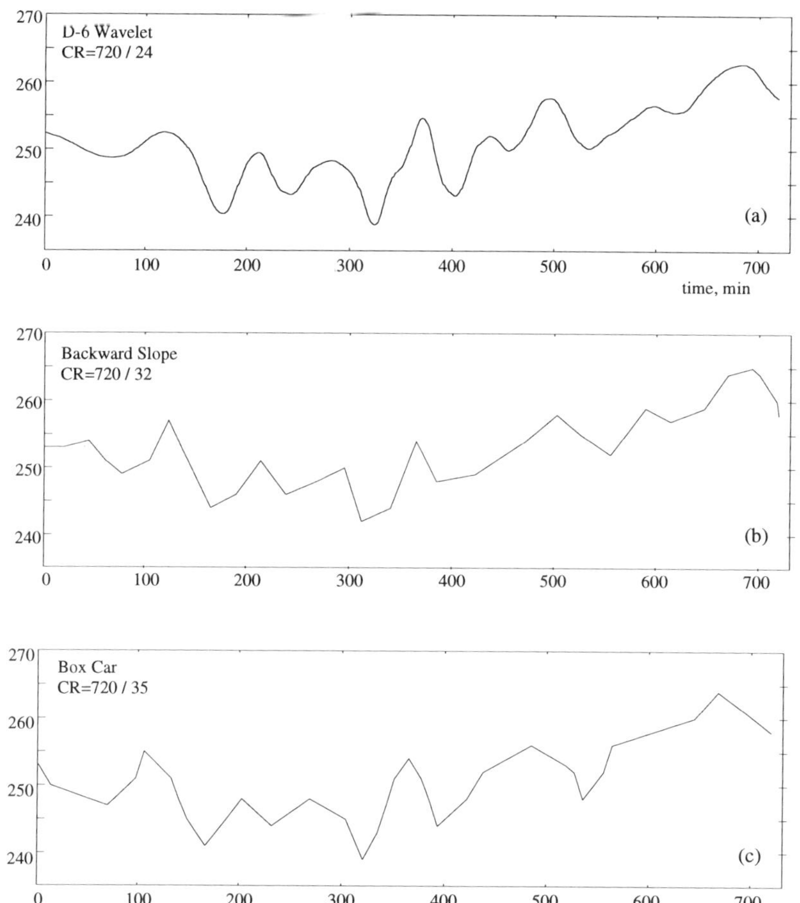

Fig. 20. Performance of data compression techniques: (a) orthonormal wavelet; (b) backward slope; (c) boxcar.

form extrema, and several more examples are presented in Bakshi and Stephanopoulos (1995).

V. Recognition of Temporal Patterns for Diagnosis and Control

The pivotal component of many engineering tasks—e.g., process fault diagnosis and product quality control—is the recognition of certain distinguishing temporal patterns (i.e., features) during the operation of the plants, and the development of associations between process operating patterns and process conditions. This two-step process is depicted in Fig. 3. Pattern recognition is the process through which a given pattern, $\mathbf{p}_i$, is assigned to the correct output class, C_I. A *pattern*, $\mathbf{p}$, is designated as the N-dimensional vector of inputs $(x_1, x_2, \ldots, x_N)$ in the input space, which may be partitioned into regions indicated by $(C_1, C_2, \ldots, C_K)$. The inputs may represent the value of manipulated and measured process variables, measured external disturbances, violation of output constraints, and/or set points at a given time point, or over a time interval. The feature extraction phase transforms the pattern $\mathbf{p}$, to a feature space, S_x, where only the most relevant parts of the inputs are retained. The extracted features may include the values of computed variables such as derivatives, integrals, or averages over time of the measured variables. Any pattern $\mathbf{p}_i = (x_1, x_2, \ldots, x_N)_i^{\mathrm{T}}$ corresponds to a particular class of operating situations, such as sensor or actuator failure, process equipment failure, process parameter changes, activation of unmodeled dynamics, and effect of unmodeled disturbances. During inductive learning, the feature space S_x is partitioned into K mutually exclusive regions, $S_x^{(I)}$, with $I = 1, 2, \ldots, K$. Thus

$$S_x^{(I)} \cap S_x^{(J)} = 0, \qquad I, J = 1, 2, \ldots, K, \quad \text{but} \quad I \neq J,$$

and

$$\bigcup_{I=1}^{K} S_x^{(I)} = S_x.$$

This mapping from S_x to the classes C_I is determined by the discriminant functions that define the boundaries of regions $S_x^{(I)}$, $I = 1, 2, \ldots, K$ in S_x. Let $d_I(\mathbf{p})$ be the discriminant function associated with the Ith class of operating situations, where $I = 1, 2, \ldots, K$. Then, a pattern of measurements $\mathbf{p}$, implies the operating situation C_I iff

$$d_I(\mathbf{p}) > d_J(\mathbf{p}) \qquad \text{for all } J \neq I.$$

Also, the boundary of the region $S_x^{(I)}$ with another region $S_x^{(J)}$ is given by

$$d_I(\mathbf{p}) - d_J(\mathbf{p}) = 0.$$

The inductive learning process determines the discriminant functions, using prior examples of $(\mathbf{p}, C_I)$ associations.

For solving the pattern recognition problem encountered in the operation of chemical processes, the analysis of measured process data and extraction of process trends at multiple scales constitutes the feature extraction, whereas induction via decision trees is used for inductive learning. A formal framework for the multiscale analysis of process data and the extraction of qualitative and quantitative features at various scales, based on the mathematical theory of wavelets, has already been developed (see Section III, B). The formal and efficient framework of wavelet decomposition provides a translationally invariant representation, and enables the extraction of qualitative and quantitative features at various scales and temporal locations with minimum distortion.

Having represented measured process data at multiple scales, we can develop consistent models for the various operational tasks. Learning input/output mappings between features in measured data and process conditions involves determining features that are most relevant to the process conditions. Inductive learning via decision trees provides explicit input/output mappings by identifying the most relevant qualitative and quantitative features from the measured variables. The mapping is easily expressed as *if-then rules* and may even be physically interpretable.

Several techniques from statistics, such as partial least-squares regression, and from artificial intelligence, such as artificial neural networks have been used to learn empirical input/output relationships. Two of the most significant disadvantages of these approaches are the following:

1. If the input to the learning procedure consists of raw process data from, e.g., several production batches, and the corresponding product yields, the learning technique has to overcome the "curse of dimensionality." The raw data consists of information, much of which may be irrelevant to the learning task, since the sensor data are measured at a scale much smaller than that of the events that may be relevant to the process yield. Such extraneous information increases the complexity of the learning process and may necessitate a large number of training examples to achieve the desired error rate.
2. The model learned is usually a "blackbox" and does not provide any insight into the physical phenomena and events influencing the process outputs.

These disadvantages are overcome by the methodology we will describe in the subsequent paragraph developed by Bakshi and Stephanopoulos. Effects of the curse of dimensionality may be decreased by using the hierarchical representation of process data, described in Section III. Such a multiscale representation of process data permits hierarchical development of the empirical model, by increasing the amount of input information in a stepwise and controlled manner. An explicit model between the features in the process trends, and the process conditions may be learned by *inductive learning* using *decision trees* (Quinlan, 1986; Breiman *et al.*, 1984), as described later on in this section.

A. Generating Generalized Descriptions of Process Trends

Consider a measured operating variable, $x(t)$, and its M distinct measurement records, $[x(t)]_i$, $i = 1, 2, \ldots, M$ over the same range of time. Using the multiscale decomposition of measured variables, discussed in Section III, we can represent each measurement record, $[x(t)]_i$, $i = 1, 2, \ldots, M$ by a finite state of trends, where each trend is a pattern of triangular episodes;

$$[p]_i^k = \{T_1, T_2, \ldots, T_{m_{i,k}}\}_i^k, \qquad k = 1, 2, \ldots, l_i,$$

where superscript k indicates the kth representation of the record at some temporal scale, and l_i is the number of distinct ranges of scale generated by the wavelet interval-tree of scale. $T_1, T_2, \ldots, T_m$ denote the primitive triangles, describing the monotonic temporal behavior of $[x(t)]_i$ over a certain period of time, and could be any of the seven types shown in Fig. 4a;

$$T \in \{A, B, C, D, E, F, G\}.$$

1. Qualitatively Equivalent Patterns

Two patterns are qualitatively equivalent if and only if their corresponding sequences of triangular episodes are qualitatively equivalent episode by episode; i.e., the condition

$$\{T_1, T_2, \ldots, T_{m_{i,k}}\}_i^k \langle QE \rangle \{T_1, T_2, \ldots, T_{m_{i,j}}\}_i^j$$

implies that

$$\{T_r\}_i^k \langle QE \rangle \{T_r\}_i^j, \qquad r = 1, 2, \ldots, m_{i,k}; \qquad m_{i,k} = m_{i,j},$$

where $\langle QE \rangle$ denotes the qualitative equivalence.

2. Generalized Description of Trends

Consider M distinct batch records of the same measured variable. Let $[p]_i^k$; $i = 1, 2, \ldots, M$ be the appropriate representation of the ith record at some scale of abstraction. If $[p]_1^{k_1} \langle QE \rangle [p]_2^{k_2} \langle QE \rangle \ldots [p]_i^{k_i} \langle QE \rangle \ldots [p]_M^{k_M}$, then the common qualitative trend of all records is called the *generalized description* of trends contained in M records of a measured variable.

Figure 17 shows the raw data and the scaled signals for three records (CPR 879, 898, and 935) of carbon dioxide (CO_2) production rate (CPR) from a fed-batch fermentor. The description of the three records at the coarsest scale is $\{DABC\}$ and is identical for the three batches, thus leading to a general description of the variation of CPR. Representation at scales coarser than that of the generalized description will also be qualitatively equivalent. Figure 21 shows the generalized description of six (6) operating variables. These generalized descriptions were generated

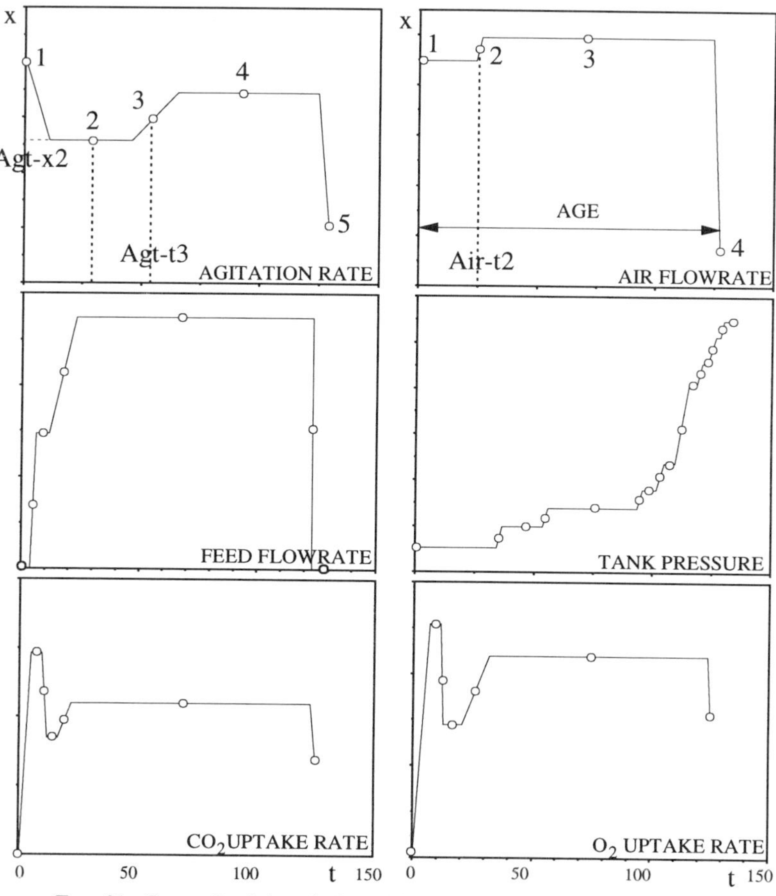

FIG. 21. Generalized description of fed-batch fermentation process data.

from 32 records of each variable, using Witkin's stability criterion to extract the most stable trends.

3. Pattern Matching of Multiscale Descriptions

Extraction of the most relevant parts of a process trend requires pattern matching of features that arise from the same physical phenomena, and are qualitatively identical. Matching of qualitatively identical features allows comparison of the qualitative and quantitative features at each scale in an organized, hierarchical manner, and helps fight the curse of dimensionality.

Consider the process data shown in Fig. 17a. These data represent the CPR for the fed-batch fermentation process for three batches giving different yields, as shown in the figure. The raw data are quite different, both in their qualitative and quantitative features. At coarser representations, the three signals start looking similar, and finally, at the coarsest scale, they are qualitatively identical, as shown in Figs. 17f and 17g, resulting in a unique generalized description: $\{D - AB - C\}$, and matching qualitatively identical features is straightforward.

The descriptions obtained using Witkin's stability criterion are shown in Fig. 17h. Pattern matching of the features in these descriptions results in one of the following matches:

Match 1: $AB - CD - AB - CD - A \qquad\qquad B - C$ CPR 879, CPR 935
$$AB - CD - AB - CD - A - (D - A)$$
$$B - C \quad \text{CPR 898}$$

Match 2: $AB - CD - AB - C \qquad\qquad D - AB - C$ CPR 879 CPR 935
$$AB - CD - AB - C(D - A) - D - AB - C \text{ CPR 898}$$

The qualitative feature, $\{D - A\}$, distinguishes the variable CPR 898 from the other two, and should be evaluated for its relevance to solving the classification problem. The pattern matching task may not be straightforward, especially at lower scales, due to greater detail in the trends. Heuristics may then be used to organize the matches in terms of their likelihood of being physically meaningful. Often, domain-specific information about the process is available for matching features in trends from different examples. For example, information about the sequence and type of features in a trend under normal operation, may be available. Such information simplifies the pattern matching problem, and eliminates several infeasible matches. In the absence of domain-specific information, all matches arc equally likely, and need to be evaluated for their relevance to solving the learning problem. For more details on the technical aspects of pattern matching see Bakshi and Stephanopoulos (1994b).

TABLE I

DATA FOR ILLUSTRATING INDUCTION USING DECISION TREES[a]

Example #	Input Features		Output Features	
	Pressure	Temperature	Color	Production Quality
1	N	99	N	Good
2	N	105	A	Bad
3	H	108	N	Good
4	L	92	N	Bad
5	L	106	N	Bad
6	N	106	N	Good
7	L	104	A	Bad
8	N	95	A	Bad

[a]Key: H—high; L—low; N—normal; A—abnormal. *Source*: reproduced from Bakshi and Stephanopoulos (1994b), by permission.

B. INDUCTIVE LEARNING THROUGH DECISION TREES

Once the several records of a process variable have been generalized into a pattern, as indicated in the previous paragraph, we need a mechanism to induce relationships among features of the generalized descriptions. In this section we will discuss the virtues of inductive learning through decision trees.

Inductive learning by decision trees is a popular machine learning technique, particularly for solving classification problems, and was developed by Quinlan (1986). A decision tree depicting the input/output mapping learned from the data in Table I is shown in Fig. 22. The input information consists of pressure, temperature, and color measurements of

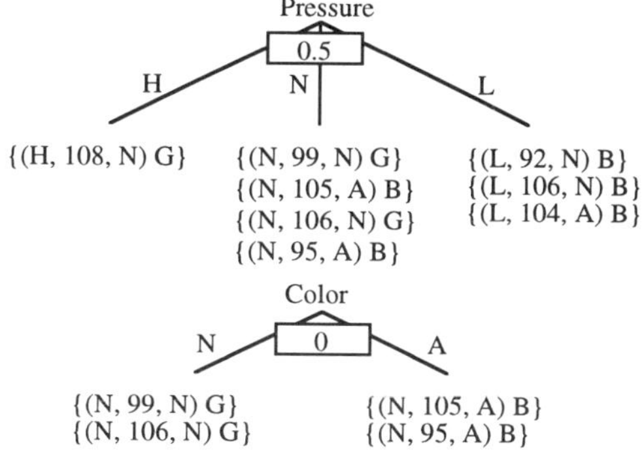

FIG. 22. Decision tree for data in Table I.

a chemical process. The output variable is the product quality corresponding to each set of measured variables. The decision tree in Fig. 22 provides the following information for obtaining product of good quality:

1. If pressure is normal and color is normal, then quality is good.
2. If pressure is low, then quality is bad.
3. If pressure is normal and color is abnormal, then quality is bad.

The model induced via the decision tree is not a blackbox, and provides explicit and interpretable rules for solving the pattern classification problem. The most relevant variables are also clearly identified. For example, for the data in Table I, the value of the temperature are not necessary for obtaining good or bad quality, as is clearly indicated by the decision tree in Fig. 22.

The procedure for generating a decision tree consists of selecting the variable that gives the best classification, as the root node. Each variable is evaluated for its ability to classify the training data using an information theoretic measure of entropy. Consider a data set with K classes, C_I, $I = 1, 2, \ldots, K$. Let M be the total number of training examples, and let M_{C_I} be the number of training examples in class C_I. The information content of the training data is calculated by Shannon's entropy:

$$I(M_{C_1}, M_{C_2}, \ldots, M_{C_K}) = -\sum_{I=1}^{n} \frac{M_{C_I}}{M} \log_2\left(\frac{M_{C_I}}{M}\right). \tag{24}$$

Equation (24) provides a measure of the variety of classes contained in the data set. If all examples belong to the same class, then the entropy is zero. Smaller entropy implies less variety of classes (more order) in the data set. If the data set is split into groups, G_1 and G_2, with M_{C_I,G_J} being the number of examples belonging to class, C_I, that are present in group, G_J, then the total information content is

$$E(G_1, G_2) = \frac{M_{G_1}}{M} I(M_{C_1,G_1}, \ldots, M_{C_K,G_1}) + \frac{M_{G_2}}{M} I(M_{C_1,G_2}, \ldots, M_{C_K,G_2}). \tag{25}$$

Equations (24) and (25) are adequate for designing decision trees. The feature that minimizes the information content is selected as a node. This procedure is repeated for every leaf node until adequate classification is obtained. Techniques for preventing overfitting of training data, such as cross validation are then applied.

Induction via decision trees is a greedy procedure and does not guarantee optimality of the mapping, but works well in practice, as illustrated by successful applications in several areas. The attractive features of learning by decision trees are listed below.

1. The model is easy to understand, and interpret physically. Concise rules may be developed.
2. Only the most relevant features are selected as nodes. Redundant, or unnecessary features are clearly identified through the maximization of entropy change.
3. Both continuous and discrete-valued features can be handled in a uniform framework. This permits easy combination of quantitative and structural methods.
4. No a priori assumptions about the distribution of data or class probability are required.
5. The technique is robust to noisy examples.

The discriminant hypersurface is approximated in a piecewise constant manner. This may result in decision trees that are very large and complicated, and deriving meaningful rules may not be very easy. This problem is alleviated by the hierarchical learning procedure, since the number of features evaluated is increased gradually, only if necessary. Techniques for overcoming some of the disadvantages of decision trees are described by Saraiva and Stephanopoulos (1992).

C. Pattern Recognition with Single Input Variable

Consider the three distinct records (examples) of measurements for the same operating variable, shown in Fig. 23. Let the records $[x(t)]_1$ and $[x(t)]_2$ correspond to operating conditions of class C_A, while the third, $[x(t)]_3$ corresponds to operating conditions of class C_B. At the lowest scale, the high-frequency components of the three records make all of them look very different. The first scale at which the first two records have the same qualitative description, yields the following common trend: $AB - C - B - C$. On the other hand, the most stable representation of the third record is $AB - C - B - CD - AB - C$. In this particular example, syntactic representation of trends has been sufficient to provide complete classification of the three records. Using inductive learning through decision trees leads to the following two cases for classifying trends of class C_A

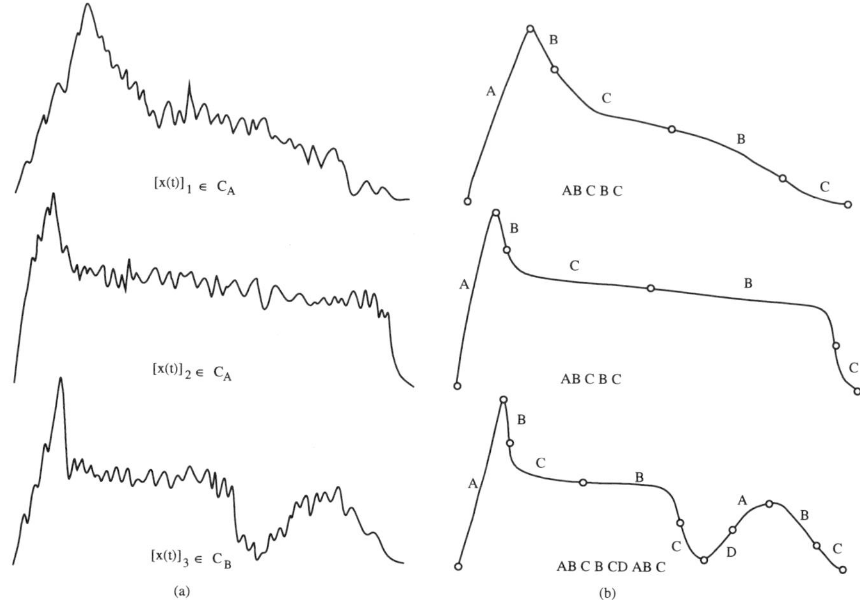

FIG. 23. (a) The raw data of three distinct records and (b) their corresponding syntactic generalizations. (Reprinted from Bakshi and Stephanopoulos, Representation of Process Trends, Part IV. *Computers and Chemical Engineering*, **18**(4), p. 303, Copyright (1994), with kind permission from Elsevier Science Ltd., The Boulevard, Langford Lane, Kidlington OX5 1GB, UK).

from those of class C_B:

Case 1. The syntactic difference is the string $(D - AB - C)$ at the end of the trend, since

$$AB - C - B - C \qquad \text{(description of first two records)}$$

$$AB - C - B - C(D - AB - C) \qquad \text{(description of third record)}$$

Case 2. The syntactic difference is the string $(B - CD - A)$ in the middle of the trend, since

$$AB - C - \qquad\qquad B - C \qquad \text{(for first two records)}$$

$$AB - C - (B - CD - A)B - C \qquad \text{(for third record)}$$

D. PATTERN RECOGNITION WITH MULTIPLE INPUT VARIABLES

The inductive classification of multiple-dimensional trends involves the mapping between the distinguishing features of several input-variables and the corresponding classes of a single output. It is carried out in two phases, as follows:

1. Phase 1: Generate Generalized Descriptions and Extract Distinguishing Features for Each Variable

The procedure for generating generalized trends was described earlier. The features of the generalized trends are given by the triangular episodes at any level of required detail, i.e., qualitative, semiquantitative, real-valued analytic.

2. Phase 2: Learn Mapping between Extracted Features and Output Classes

Once the distinguishing features for each input variable have been extracted through the generalization of descriptions of the available records, these features become the inputs of the inductive learning procedure through decision trees.

Bakshi and Stephanopoulos (1994b) have applied the above procedure to a fed-batch fermentation process. The problem involved 41 sets of batch records on 24 measured variables. Of these variables only very few were found by the decision tree to be relevant, and yield rules such as the following for guiding the diagnosis or control of a fermentor.

Diagnostic Rule-1 If the duration of the high bottom dissolved oxygen (BDO) phase is > 3.6 h and the level of CO_2 generation during the production phase is > 4.3 units, then the quality of the fermentation is excellent.

Control rule-1. To achieve fermention of excellent quality, keep (a) the agitation rate in the first scaling episode (growth phase) > 37 units and (b) the duration of the first episode in the air flowrate ≤ 25 h.

For the detailed discussion on the inductive learning of diagnostic and control rules around the fed-batch fermentor system, the reader should refer to the work of Bakshi and Stephanopoulos (1994b).

VI. Summary and Conclusions

The wavelet decomposition of measured data provides a natural framework for the extraction of temporal features, which characterize operating process variables and their trends. Such characterization, local in frequency and time, provides valuable insight as to what is going on in a process, and thus it can explicitly support a number of engineering methodologies, such as data compression, diagnosis of process upsets, and pattern recognition for quality control. Nevertheless, a number of interesting issues arise that can shape future developments in the area process operations and control.

1. *Multiscale modeling of process operations.* The description of process variables at different scales of abstraction implies that one could create models at several scales of time in such a way that these models communicate with each other and thus are inherently consistent with each other. The development of multiscale models is extremely important and constitutes the pivotal issue that must be resolved before the long-sought integration of operational tasks (e.g., planning, scheduling, control) can be placed on a firm foundation.

2. *Multiscale process identification and control.* Most of the insightful analytical results in systems identification and control have been derived in the frequency domain. The design and implementation, though, of identification and control algorithms occurs in the time domain, where little of the analytical results in truly operational. The time-frequency decomposition of process models would seem to offer a natural bridge, which would allow the use of analytical results in the time-domain deployment of multiscale, model-based estimation and control.

3. *Integration of process operational tasks.* The industrial deployment of computer-aided systems that can integrate planning-scheduling-diagnosis-control rests on two pillars; multiscale process models (see paragraph 1, above) and multiscale depiction of process data. Although the wavelet decomposition offers the theoretical answer to the second, the industrial implementation requires the solution of the following problems: (a) integration of quantitative, qualitative and semiquantitative descriptions of process operations through the establishment of an "appropriate" language; and (b) integration of planning, scheduling, diagnosis, and control methodologies around the common language. These issues require creative modeling of process data around the wavelet decomposition and need to be addressed soon by researchers.

References

Bader, F. P., and Tucker, T. W., Data compression applied to a chemical plant using a distributed historian station. *ISA Trans.* **26**(4), 9–14 (1987a).

Bader, F. P., and Tucker, T. W., Real-time data compression improves plant performance assessment. *InTech.* **34,** 53–56 (1987b).

Bakshi, B. R., and Stephanopoulos, G., Wave-net: A multi resolution, hierarchical neural network with localized learning. *AIChE J.* **39**(1), 57–81 (1993).

Bakshi, B. R., and Stephanopoulos, G., Representation of process trends. Part III. Multi-scale extraction of trends from process data. *Comput. Chem. Eng.* **18**, 267 (1994a).

Bakshi, B. R., and Stephanopoulos, G., Representation of process trends. Part IV. Induction of real-time patterns from operating data for diagnosis and supervisory control. *Comput. Chem. Eng.* **18** 303 (1994b).

Bakshi, B. R., and Stephanopoulos, G., Compression of chemical process data through functional approximation and feature extraction. *AIChE J.*, accepted for publication (1995).

Bastl, W., and Fenkel, L., Disturbance analysis systems. *In* "Human Diagnosis of System Failures," (Rasmussen and Rouse, eds.) Nato Symp. Denmark, Plenum, New York, 1980.

Breiman, L., Friedman, J. H., Olshen, R. A., and Stone, C. J., "Classification and Regression Trees." Wadsworth, Belmont, CA (1984).

Cheung, J. T.-Y., Representation and extraction of trends from process data. Sc.D. Thesis, Massachusetts Institute of Technology, Dept. Chem. Eng., Cambridge, MA (1992).

Cheung, J. T.-Y., and Stephanopoulos, G., Representation of process trends. Part I. A formal representation framework. *Comput. Chem. Eng.* **14**, 495–510 (1990).

Coifman, R. R., and Wickerhauser, M. V., Entropy-based algorithms for best basis selection. *IEEE Trans. Inf. Theory* **38**(2), 713–718 (1992).

Daubechies, I., Orthonormal Bases of Compactly Supported Wavelets, *Comm. Pure Appl. Math.*, **XLI**, 909–996 (1988).

Dvorak, D. L., "Expert Operations Systems," Tech. Rep. Dept. of Computer Science, University of Texas at Austin, 1987.

Feehs, R. J., and Arce, G. R., "Vector Quantization for Data Compression of Trend Recordings," Tech. Rep. 88-11-1, University of Delaware, Dept. Elect. Eng., Newark, 1988.

Fukunaga, K., "Introduction to Statistical Pattern Recognition." Academic Press, Boston, 1990.

Funahashi, K. I. On the approximate realization of continuous mappings by neural networks. *Neural Networks* **2** 183–192 (1989).

Goupillaud, P., Grossmann, A., and Morlet, J., Cycle-octave and related transforms in seismic signal analysis. *Geoexploration* **23**, 85–102 (1984).

Gray, R. M., Vector quantization. *IEEE ASSP Mag.*, April, pp. 4–29 (1984).

Hale, J. C., and Sellars, H. L., Historical data recording for process computers. *Chem. Eng. Prog.*, November, pp. 38–43 (1981).

Hummel, R., and Moniot, R., Reconstructions from zero crossings in scale space. *IEEE Trans. Acoust., Speech, Signal Processing* **ASSP 37**(12), 2111–2130 (1989).

Jantke, K., "Analogical and Inductive Inference." Springer-Verlag, Berlin, 1989.

Kanal, L. N., and Dattatreya, G. R., Problem-solving methods for pattern recognition. *In* Handbook of Pattern Recognition and Image Processing," (T. Y. Young, and K.-S. Fu, eds.) Academic Press, New York, 1985.

Kramer, M. A., Nonlinear principal component analysis using autoassociative neural networks. *AIChE J.* **37**, 233–243 (1991).

Long, A. B., and Kanazava, R. M., "Summary and Evaluation of Scoping and Feasibility Studies for Disturbance Analysis and Surveillance Systems," EPRI Report NP-1684. Electr. Power Res. Inst., Palo Alto, CA, 1980.

MacGregor, J. F., Marlin, T. E., Kresta, J. V., and Skagerberg, B., Multivariate statistics methods in process analysis and control. *In* "Chemical Process Control, CPCIV," (Y. Arkun and W.H. Ray, eds.). CACHE, AIChE Publishers, New York, 1991.

Mallat, S. G., A theory for multiresolution signal decomposition: The wavelet representation. *IEEE Trans. Pattern Anal. Mach. Intell.* **PAMI-11**(7), 674–693 (1989).

Mallat, S. G., Zero crossing of a wavelet transform. *IEEE Trans. Inf. Theory* **IT-37**(4), 1019–1033 (1991).

Mallat, S., and Zhong, S., Characterization of signals from multiscale edges, *IEEE Trans. Pattern Anal. Mach. Intell.* **PAMI-14**(7), 710–732 (1992).

Marr, D., and Hildreth, E., Theory of edge detection. *Proc. R. Soc. London B Ser.* **207**, 187–217 (1980).

Meyer, Y., Principle d'incertitude, bases hilbertiennes et algebres d'operateurs. *Bourbaki Sem.* No. 662 (1985–1986).

O'Shima, E., Computer-aided plant operation. *Comput. Chem. Eng.* **7**, 311 (1983).

Papoulis, A., "Signal analysis." McGraw-Hill, New York, 1977.

Quinlan, J. R., Induction of decision trees. *Mach. Learn.* **1**(1), 81–106 (1986).

Rumelhart, D. E., McClelland, J. L., "Parallel Distributed Processing," Vol. 1. MIT Press, Cambridge, MA, 1986.

Saraiva, P., and Stephanopoulos, G., Continuous process improvement through inductive and analogical learning. *AIChE J* **38**(2), 161–183 (1992).

Silverman, B. W., "Density Estimation for Statistics and Data Analysis." Chapman & Hall, New York, 1986.

Stephanopoulos, G., Artificial intelligence...'What will its contributions be to process control?' *In* "The Second Shell Process Control Workshop," (D.M. Prett, C. E. Garcia, and B. L. Ramaker, eds.) Butterworth, Stoneham, MA, 1990.

Takei, S., Multiresolution analysis of data in process operations and control. M.S. Thesis, Massachusetts Institute of Technology, Dept. Chem. Eng., Cambridge, MA, 1991.

Vaidyanathan, R., and Venkatasubramanian, V., Process fault detection and diagnosis using neural networks: II. Dynamic processes. *AIChE Ann. Meet.*, Chicago, IL (1990).

Van Trees, H., "Detection, Estimation and Modulation Theory." Wiley, New York, 1968.

Wickerhauser, M. V., "INRIA Lectures on Wavelet Packet Algorithms." Yale University, New Haven, CT, 1991.

Witkin, A. P., Scale space filtering: A new approach to multi-scale description. *In* "Image Understanding" (S. Ullman and W. Richard, eds.), pp. 79–95. Ablex, Norwood, NJ, 1983.

Yuille, A.L., and Poggio, T., Fingerprints theorems. *Proc. Natl. Conf. on Artif. Intell.*, pp. 362–365 (1984).

INTELLIGENCE IN NUMERICAL COMPUTING: IMPROVING BATCH SCHEDULING ALGORITHMS THROUGH EXPLANATION-BASED LEARNING

Matthew J. Realff[1]

School of Chemical Engineering
Georgia Institute of Technology
Atlanta, Georgia 30332

[1]The work reported in this chapter was carried out while the author was a Ph.D. student in the Laboratory for Intelligent Systems in Process Engineering, Department of Chemical Engineering, Massachusetts Institute of Technology, Cambridge, MA 02139.

Learning comes from reflection on accumulated experience and the identification of patterns found among the elements of previous experience. All numerical algorithms used in scientific and engineering computing are based on the same paradigm: *Execute a predetermined sequence of calculational tasks and produce a numerical answer*. The implementation of the specific numerical algorithm is oblivious to the experience gained during the solution of a specific problem, and the next time a different, or even the same, problem is solved through the execution of exactly the same sequence of calculational steps. The numerical algorithm makes no attempt to reflect on the structure and patterns of the results it produced, or reason about the structure of the calculations it has performed. This chapter shows that this need not be the case. By allowing an algorithm to reflect on and reason with aspects of the problems it solves and its *own structure of computational tasks, the algorithm can learn* how to carry out its tasks more efficiently. This form of *intelligent numerical computing* represents a new paradigm, which will dominate the future of scientific and engineering computing. But, in order to unlock the computer's potential for the implementation of truly intelligent numerical algorithms, the *procedural* depiction of a numerical algorithm must be replaced by a *declarative* representation of the algorithmic logic. Such a requirement upsets an established tradition and imposes new educational challenges which most educators and educational curricula have not, as yet, even recognized. This chapter shows how one can take a branch-and-bound algorithm, used to identify optimal schedules of batch operations, and endow it with the ability to learn to improve its own effectiveness in locating the optimal scheduling policies for flowshop problems. Given that most batch scheduling problems are NP-hard, it becomes clear how important it is to improve the effectiveness of algorithms for their solution. Using the framework of Ibaraki (1978) a branch-and-bound algorithm is declaratively modeled as a *discrete decision process*. Then, *explanation-based machine learning* strategies can be employed to uncover patterns of generic value in the experience gained by the branch-and-bound algorithm from solving specific instances of scheduling problems. The logic of the uncovered patterns (i.e., new knowledge) can be incorporated into the control strategy of the branch and bound algorithm when the next problem is to be solved.

I. Introduction

The main goal of this chapter is to introduce the concept of *intelligent numerical computing* within the context of solving optimization problems of relevance to chemical engineering. Before focusing on this area, it is worth noting that intelligent scientific computing has been the subject of research within the context of other problems such as bifurcation analysis and nonlinear dynamics, and chemical reaction kinetic schemes (Abelson et al., 1989; Yip, 1992), both of which are clearly subjects of interest to chemical engineers.

Our first task is to define intelligent numerical computing and to indicate what features it possesses to make it a distinct subclass of general numerical computing. For us, the essence of the distinction is the difference between *reasoning* and *calculation* (Simon, 1983). Implementations of numerical algorithms, such as the simplex method for linear programming (Dantzig, 1963), or gradient descent methods for more general nonlinear optimization problems (Avriel, 1976), take a problem formulated as a set of algebraic relationships and calculate an answer. Following performance of the calculation task, the implementation is completely oblivious to the *experience*, and the next time a different, or even the same, problem is solved exactly, the same set of steps will be performed. The implementation makes no attempt to reflect on its experience, or reason about the calculation it performed; it is completely oblivious to its own structure, which has not been declaratively represented, but that is procedurally embodied in the set of calculations performed. *Intelligent numerical computing* attempts to incorporate reasoning into the calculation procedure by allowing the computer to reason and reflect on various aspects of the problem, and its own structure. To do this we need to adopt new representations of information pertinent to problem solving, in particular symbolic representations, and to introduce new algorithms, which compute with symbolic as opposed to numerical information, such as natural deduction and resolution (Robinson, 1965) for logic-based reasoning.

To illustrate these concepts we will focus on a particular numerical algorithm, branch and bound (Nemhauser and Wolsey, 1988), which has been a workhorse for solving optimization problems with discrete structure, such as chemical batch scheduling (Kondili et al., 1993; Shah et al., 1993). This is a generic problem-solving strategy, and its successful application relies on the identification of effective *control information*, often in (1) the form of a lower-bound function, which characterizes how good an incomplete solution can be, and (2) as dominance and equivalence conditions, which use information about one partial solution to terminate another. In this chapter, we will present a methodology for *automating* the acquisition of this control knowledge for branch-and-bound algorithms, using flowshop scheduling as an illustration. Here, automation means that the *computer itself* acquires new control information, unaided by the human user, save for the specification of the problem. The acquisition is carried out by analyzing the experience gained by the computer during problem solving, illustrating the incorporation of one facet of reasoning, *learning from experience.*

The rest of the chapter will focus on solving four goals, engendered by the problem of automatically improving problem-solving experience.

1. Before we can learn new control information, we must formally specify what it is we are trying to learn and how what we learn is to be applied within the original problem-solving framework (Section II).
2. Having established the formal specification of the problem-solving framework, the next goal is to establish how the solution of example problems can be turned into relevant problem-solving experience (Section III).
3. To convert the problem-solving experience into a useful form, we need to be able to represent it to the computer, along with the other information necessary to reason about it efficiently (Section IV).
4. Finally, having set up the learning problem, we need to employ a learning method that will guarantee preservation of the correctness of the branch-and-bound algorithm and make useful additions to the control information we have about the problem (Section V).

These four goals are addressed sequentially in the next four sections. The flowshop problem will be used as an illustration throughout, because of its practical relevance, difficulty of solution, and yet relative simplicity of its mathematical formulation.

A. Flowshop Problem

Many chemical batch production facilities are dedicated to producing a set of products that require for their manufacturing a common set of unit operations. The unit operations are performed in the same sequence for each product. This type of production problem is often solved by configuring the available equipment so that each unit operation is carried out by a fixed set of equipment items that are disjoint from those used in any other operation. If one unit is assigned to each step, this is called a *flowshop* (Baker, 1974).

The flowshop problem has been widely studied in the fields of both operations research (Lagweg et al., 1978; Baker, 1975) and chemical engineering (Rajagopalan and Karimi, 1989; Wiede and Reklaitis, 1987). Since the purpose of this chapter is to illustrate a novel technique to synthesize new control knowledge for branch-and-bound algorithms, we will adopt the simplest, and most widely studied, form of the flowshop problem. The assumptions with respect to the problem structure are given below.

1. *Storage policy.* Much attention has been paid to exploring the implications of different storage policies, such as having no intermediate storage, finite intermediate storage, running the plant with a zero-wait policy, or combinations of the policies discussed above. We will assume unlimited intermediate storage is available to simplify the constraints between batches.
2. *Clean out, set up, and transfer policies.* In general, these operations could be dependent on the order in which the products are routed through the equipment. We will assume that these operations are *sequence independent* and can be factored into the processing times for each step.

With these simplifications, we can now formulate the mathematical model, which describes the flowshop. Let

$p_{ik} \equiv$ the processing time of job i on machine k,

$c(\sigma, k) \equiv$ the end-time of the last job of sequence σ on machine k.

Then, to calculate the end-time of an operation of job a_i on machine k, when a_i has been scheduled after a sequence σ, we use the following equations:

Machine 1:

$$c(\sigma a_i, 1) = c(\sigma, 1) + p_{a_i 1}. \tag{1}$$

Machine k:

$$c(\sigma a_i, k) = \max\{c(\sigma, k), c(\sigma a_i, k - 1)\} + p_{a_i k} \quad k = 2 \ldots m. \tag{2}$$

To generate a specific instance of a flowshop problem we will assume that the plant produces a fixed set of products. In addition to allowing the type of product to vary, we will also allow the size of the batch to be one of a fixed set of sizes. Further details of the formulation are given in Section III, A.

B. Characteristics of Solution Methodology

To solve the flowshop scheduling problem, or indeed most problems with significant discrete structure, we are forced to adopt some form of enumeration to find, and verify, optimal solutions (Garey and Johnson, 1979). To perform this enumeration, we must solve two distinct problems.

1. *Representation*. The solution space is composed of discrete combinatorial alternatives of batch production schedules. For example, in the permutation flowshop problem, where the batches are assumed to be executed in the same order on each unit, there are $N!$ number of solutions, where N is the number of batches. We must find a way to compactly represent this solution space, in such a way that significant portions of the space can be characterized with respect to our objective as either "poor" or "good" without explicitly enumerating them.

2. *Control strategy*. Having found an appropriate representation for the space of solutions, we must then take advantage of this representation to explore only as small a fraction of the space as possible. This involves the following:
   (a) Finding a method to systematically enumerate subsets of the solution space without explicitly enumerating their members.
   (b) Finding ways to prove that certain subsets of the space cannot contain optimal solutions, or if they do, that we have other subsets of the space that will produce equivalent solutions.

To solve the problems of representation and control, we will employ the framework of the branch-and-bound algorithm, which has been used to solve many types of combinatorial optimization problems, in chemical engineering, other domains of engineering, and a broad range of management problems. Specifically, we will use the framework proposed by Ibaraki (1978), which is characterized by the following features:

1. The combinatorial problem is represented by a *discrete decision process* (DDP) (Ibaraki, 1978) where the underlying information in the problem is captured by an explicit *state-space model* (Nilsson, 1980).

2. The control is divided into three parts.
   (a) Elimination of alternative solutions through the use of a *lower bound* on the value of the objective function.
   (b) Elimination of alternative solutions through *dominance* conditions.
   (c) Elimination of alternative solutions through *equivalence* conditions.

The next section will highlight these features of the branch-and-bound framework, within the context of the flowshop scheduling problem. Then we will give an abstract description of the algorithm, followed by the paraphrasing of an important theorem from Ibaraki (1978), which characterizes the relative efficiency of different branch-and-bound algorithms in terms of their control strategies.

II. Formal Description of Branch-and-Bound Framework

If an algorithm is to reflect on its own computational structure and thus *learn* how to improve its performance, then this algorithm needs to have a *declarative* representation of its components. Thus, a branch-and-bound algorithm used to generate optimal schedules of batch operations should possess declarative representations of (1) the schedules of batch operations (2) the predicates that determine the feasibility of a scheduling policy, (3) the objective function, and (4) the conditions that determine the control strategy of the branch-and-bound algorithm. The material in this section will provide such declarative representations for the various components of branch-and-bound algorithms, and will introduce a metric that can be used to measure improvements in the efficiency of such algorithms.

A. SOLUTION SPACE REPRESENTATION—DISCRETE DECISION PROCESS

The first step in solving a combinatorial optimization problem is to model the solution space itself. Such a model should be declarative in character, if it is to be independent of the characteristics of the specific algorithm that will be used to find the solution within the solution space. The model we have adopted for the scheduling of flowshop operations is the *discrete decision process* (DDP) introduced originally by Karp and Held (1967). As defined by Ibaraki (1978) a DDP, Y, is a triple (Σ, S, f) with its elements defined as follows:

Σ is a finite nonempty alphabet whose symbols are used to build the description of solutions. Let Σ^* denote the set of finite strings generated from the concatenation of symbols present in Σ.

S is a subset of Σ^* whose members satisfy a set of feasibility predicates over Σ^*. The members of Σ^* will represent the "feasible" solutions.

f is a cost function over S, mapping S into the set of real numbers.

For the flowshop problem, one choice of Σ is an alphabet with as many symbols as the number of distinct batches to be scheduled, i.e., one symbol for each batch, then, each discrete schedule of batches is represented by a finite string, σ, of symbols drawn from the alphabet, Σ. Only strings which contain each alphabetic element once and only once represent acceptable schedules of batch operations. We will use the notation σ to denote the set of symbols from Σ, which are contained in the solution string, Σ^*. Also, with $O(Y)$ we will denote the set of optimal solutions of the DDP, Y.

The DDP is a formalism more general than the customary formalisms of the combinatorial optimization problems and offers an excellent framework for unifying the formalization of very broad classes of such problems (Karp and Held, 1967; Ibaraki, 1978). It also exhibits many common features with the *state-space representation* of problems commonly employed by researchers in artificial intelligence (AI) (Kumar and Kanal, 1983; Nilsson, 1980). This is a feature that we will find very convenient in subsequent sections as we try to integrate machine learning algorithms, which use the state space representation, with branch-and-bound algorithms solving combinatorial optimization problems.

Thus, it can be effectively argued that the DDP formalism is superior to other formalisms since it can uniformly accommodate

1. A variety of problems seeking the set of feasible solutions, the set of optimal solutions, or one member of either set.
2. Various types of alphabets (used to describe solutions), predicates (used to describe feasibility of solutions), and cost functions.

1. Illustration: Flowshop Example

Let

$\Sigma \equiv \{\text{all batches}\}$.

$S \equiv$ Set of all σ_x, if $\{\sigma_x\}$ contains all symbols of Σ once and only once.

$f(x) \equiv C(\sigma_x, m)$, completion time of the last operation of the final batch.

In this alphabet, each batch is assigned its own symbol. The problem formulation allows for the same combination of product and size to be selected multiple times. Hence the schedules that have the same batch type in two or more different positions will be enumerated multiple times, even though they represent schedules which are indistinguishable from one another.

A more compact alphabet for this situation is one for which we create a unique symbol for each batch type, and allow the multiple occurrence of this symbol to stand for scheduling the batch type more than once. The feasibility predicate would be suitably modified to check to see when enough of the type had been added to a given branch. This gives the following DDP:

$\Sigma \equiv \{N$ distinct symbols: $N =$ all (product size) combinations$\}$.

$S \equiv$ Set of all σ_x, such that $\{\sigma_x\}$ contains all the symbols of Σ a number of times equal to the number of batches for each product-size combination that was selected for production.

$f \equiv C(\sigma_i, m)$ completion time of the last operation on the final machine.

The elimination of spurious equivalent solutions is important computationally, because either we will have to enumerate the spurious equivalent solutions, doubling the effort for each equivalent pair of solutions, or we will have to introduce rules that can detect the equivalence explicitly. For the purpose of illustrating the ideas of this chapter we will continue to use the naive alphabet, although the method is not restricted to such a choice.

A scheduling policy, x, is optimal if and only if the following conditions are satisfied:

Condition 1. Scheduling policy is feasible, i.e., $\sigma_x \in S$.

Condition 2. Scheduling policy has cost not higher than that of an other policy, y, i.e., $f(x) \le f(y)$ for any y such that $\sigma_y \in S$.

B. The Branch-and-Bound Strategy

1. Branching in General

The large size of the solution space for combinatorial optimization problems forces us to represent it *implicitly*. The branch-and-bound algorithm encodes the entire solution space in a *root node*, which is successively expanded into branching nodes. Each of these nodes represents a subset of the original solution space specialized to contain some particular element of the problem structure.

In the flowshop example, the subsets of the solution space consists of subsets of feasible schedules. We can organize these subsets in a variety of ways; for example, fixing any one position of the N available positions in the schedule to be a particular batch creates N subsets of size $(N - 1)!$. Subsequently, as we fix more and more of the positions, the sets will include fewer and fewer possibilities, until all the positions are fixed, and we have a single element in the set corresponding to a single, feasible schedule.

2. *Formal Statement of Branching*

The branching, or specialization of the solution subsets, should obey certain constraints to avoid potential problems of inefficiency, nontermination, or incorrectly omitting solutions.

1. *Mutual exclusivity.* In branching from one node to its descendants, we should ensure that none of the subsets overlap with one another; otherwise we could potentially explore the same solution subsets in multiple branches of the tree. Formally, if X is the original solution space and x_i is the ith subset, then

$$x_i \cap x_j = \varnothing \; \forall i, j > i$$

2. *Inclusiveness.* In branching to the descendants, we should ensure that no solution of the parent set is omitted from the child subsets. If we fail to have inclusiveness, the procedure may not correctly find the optimum solution. Formally, this requirement implied that

$$\bigcup_i x_i = X.$$

We must now link these two properties, and the notion of branching to our problem representation, i.e., the DDP formalism. To do this, we introduce a variant on our earlier notation; let

$$Y(x) \equiv \left[\Sigma, S(x), f \right],$$

where $S(x)$ is the feasibility predicate over all strings that begin with the partial sequence of symbols, x.

If ε denotes the empty strings, then

$$Y(\varepsilon) \equiv (\Sigma, S, f)$$

and $Y(\varepsilon)$ is equivalent to the original problem, since $S(\varepsilon)$ includes all the original set of solutions. Thus $Y(\varepsilon)$ is the problem associated with the root node and branching from the root node creates problems $Y(a_1)$, $Y(a_2) \cdots Y(a_n)$, one for each of the n elements of the alphabet. Each problem, $Y(a_i)$ captures a subset of the original problem, specialized by the inclusion of the alphabet symbol. If $\Sigma = \{a_1, a_2, \ldots, a_n\}$, then a branch-and-bound algorithm decomposes $Y(\varepsilon)$ into $|\Sigma|$ problems, $\{Y(a_1), Y(a_2), \ldots, Y(a_n)\}$, where the solution strings are required to start with the respective alphabetic element. We will also extend the definition of the objective function values as follows:

$$f(x) = \begin{cases} \inf\{f(xy): \sigma = xy \in S\} \\ \infty \text{ otherwise, i.e., } Y(x) \text{ infeasible} \end{cases}$$

According to this extension of the cost function definition, the following two conditions are always satisfied:

(a) $f(xy) \geq f(x)$ for $x, y \in \Sigma^*$.
(b) $f(x) = \min\{f(xa): a \in \Sigma\}$ for $x \notin S$.

3. Illustration: Flowshop Problem

With our alphabet equal to the set of batches to be scheduled, we will interpret ε, the empty string, to be the empty schedule. $Y(a_j)$ is then the problem $Y(\varepsilon)$ specialized to have schedules that "begin with" alphabet element a_j in the first position of the schedule. Thus, the discrete decision process $Y(a_j)$ is equivalent to solving the original scheduling problem with the first batch fixed to be a_j. The set $S(a_j)$ includes all feasible solutions that have a_j as their first batch, and, in our naive formulation, $S(a_j)$ differs from $S(\varepsilon)$ by prohibiting a_j to appear again in the string.

We can see that, in general, the branching process satisfies the property of inclusiveness, since we generate a subset for each possible specialization of the original set. We cannot, however, claim exclusiveness unless we are guaranteed that each string maps to a unique solution subset. For example, if an alphabet element corresponds to assigning a batch to a specific position, then the order in which batches are assigned their positions is irrelevant, hence, with that alphabet, $Y(a_i a_j) \equiv Y(a_j a_i)$.

Having defined the process of branching, we must now formally define the mechanisms for controlling the expansion of the subsets. The basic intuition behind each of these mechanisms is that we can "measure" the quality not just of a single feasible solution, but of the entire subset represented by the partial solution string.

4. Lower-Bound Function

Having formally defined the branching structure, we must now make explicit the mechanisms by which we can eliminate subsets of the solution space from further consideration. Ibaraki (1978) has stated three major mechanisms for controlling the evolution of the branch-and-bound search algorithms, by eliminating potential solution through

(a) Comparison with an estimate of the objective function's lower bound.
(b) Dominance conditions.
(c) Equivalence conditions.

In this section we will discuss the characteristics of the first mechanism, leaving the other two for the subsequent two sections.

In our scheduling example, a partial solution represents all the possible completions of a partial schedule. To estimate the quality of the partial solution, we could assume the best scenario, and assign it a lower-bound value based on the lowest possible value of its makespan. There are many ways to estimate the makespan; for example, we could simply ignore the remainder of the batches, and report the makespan of the current partial schedule, clearly a lower bound on the final makespan. Stated formally, the lower-bound function $g(x)$ satisfies the following requirements (Ibaraki, 1978):

(a) $g(x) \leq f(x)$
(b) $g(x) = f(x)$ $x \in S$
(c) $g(x) \leq g(xy)$ for $x, y \in \Sigma^*$ $x \in \Sigma^*$

If $g(x)$ satisfies these conditions, we can use the following *lower-bound elimination* criterion to terminate the solution of a discrete decision process, $Y(y)$.

Condition: If $x \in S$ and $g(y) \geq f(x)$, then all solutions of $Y(y)$ can be discarded.

Clearly, since $g(y) \leq f(y)$, it follows that $f(y) \geq f(x)$, and hence the solution of $Y(y)$ cannot lead to a better objective function value.

The disadvantage of the lower-bound elimination criterion is that it cannot eliminate solutions with lower bounds better than the optimal solution objective function value.

Since

$$f(x^*) \leq f(x) \; \forall x \; x^* \in \text{optimal solution set}, \; O[Y(\varepsilon)]$$

and

$$\text{If } g(y) < f(x^*)$$

$$\text{then } g(y) < f(x) \; \forall x$$

and the condition for lower-bound elimination cannot be satisfied.

The efficiency of branch-and-bound algorithms is profoundly influenced by the *strength* of a lower-bounding scheme or function. In general, the closer a lower-bound function value of a node is to the true objective function value, the fewer nodes will be expanded, and the more efficient the algorithm will be. For example, in solving mixed-integer linear programs (MILP) via branch and bound, if the lower-bound scheme employed is a linear programming (LP) relaxation of the integer variables, then the efficiency of the solution technique is a strong function of how closely the

LP objective function value approximates the best integer solution value that could be generated by solutions emanating from that node.

If using a stronger lower-bound function leads to the expansion of fewer nodes, why not always use the strongest lower bound? There are two potential problems with this. First, in many cases, it is impossible to prove that one lower-bounding function, f_1, yields a higher value for minimization than another, f_2, for every node in the tree. Second, stronger lower bounds tend to require more computational effort to evaluate; thus, even though fewer nodes are evaluated, they take longer to evaluate. This tradeoff can lead to weaker lower bounds, yielding better overall computational efficiency.

In addition to the elimination of partial solutions on the basis of their lower-bound values, we can provide two mechanisms that operate directly on pairs of partial solutions. These two mechanisms are based on *dominance* and *equivalence conditions*. The utility of these conditions comes from the fact that we need not have found a feasible solution to use them, and that the lower-bound values of the eliminated solutions do not have to be higher than the objective function value of the optimal solution. This is particularly important in scheduling problems where one may have a large number of equivalent schedules due to the use of equipment with identical processing characteristics, and many batches with equivalent demands on the available resources.

5. Dominance Test

The intuitive notion behind a *dominance condition*, D, is that by comparing certain properties of partial solutions x and y, we will be able to determine that for every solution to the problem $Y(y)$ we will be able to find a solution to $Y(x)$ which has a better objective function value (Ibaraki, 1977). In the flowshop scheduling problem several dominance conditions, sometimes called elimination criteria, have been developed (Baker, 1975; Szwarc, 1971). We will state only the simplest:

Condition: If $\{x\} = \{y\} \wedge \forall i \ C(x,i) < C(y,i)$ then $x.D.y$.

For example, if the completion times of the partial schedule x on each machine i are less than those of the partial schedule y on each machine i, and x and y have scheduled the same batches, then, if we complete the partial schedules x and y in the same way, the completion time of any schedule emanating from x will be less than the completion time of the same schedule emanating from y; that is,

$$\forall u \ C(xu,i) < C(yu,i),$$

and thus, specifically

$$C(xu, m) < C(yu, m),$$

and the makespan of schedule, xu, will be less than that of yu.

Formally, we define the dominance condition by specifying a set of necessary and sufficient conditions, which it has to satisfy (Ibaraki, 1978):

Definition. A dominance relation, D, is a partial ordering of the partial solutions of the discrete decision processes in Σ^*, which satisfies the following three properties for any partial solutions, x and y:

1. Reflexivity $x.D.x$
2. Transitivity $x.D.y \wedge y.D.z \rightarrow x.D.z$
3. Antisymmetry $x.D.y \rightarrow \neg(y.D.x)$

Furthermore, a dominance relationship is constructed in such a way that it satisfies the following properties:

4. $x.D.y$ and $x \neq y \Rightarrow f(x) < f(y)$
5. $x.D.y$ and $x \neq y \Rightarrow g(x) < g(y)$
6. $x.D.y$ and $x \neq y \Rightarrow \forall u \in \Sigma^* \exists v \in \Sigma^*$ such that $xu.D.yv$ and $xv \neq yu$
7. For each $x \in \Sigma^* \; \exists y \in \Sigma_D^*$ such that $y.D.x$

Note that the set Σ_D^* is defined such that it contains all solutions that have not been dominated by any other solution. Property 4 guarantees that we will not miss an element of the optimal set, since the objective function value of x is lower than that of y.

6. Equivalence Test

The final relationship between solution subsets, which we can use to curtail the enumeration of one subset, is an *equivalence condition*, EQ. Intuitively, equivalence between subsets means that for every solution in one subset, y, we can find a solution in the subset x, which will have the same objective function value, and that plays a similar role in the execution of the algorithm.

In the case of the flowshop example, we have the following equivalence condition

Condition: If $\{x\} = \{y\} \wedge \forall i \; C(x, i) = C(y, i)$ then $x.EQ.y$.

This condition implies that any complete schedule, xu, emanating from the partial schedule, x, will have the same end-times on all machines as the completions, yu, of y, not just for the completed schedule, but also for all intermediate partial schedules.

Formally, we define the equivalence condition by specifying a set of necessary and sufficient conditions, which it has to satisfy (Ibaraki, 1978).

Definition. An equivalence condition, EQ, between two partial solutions, $x \in \Sigma^*$ and $y \in \Sigma^*$ is a binary relationship, $x.EQ.y$, which has the following properties:

1. Reflexivity $x.EQ.x$
2. Transitivity $x.EQ.y \wedge y.EQ.z \rightarrow x.EQ.z$
3. Symmetry $x.EQ.y \rightarrow y.EQ.x$

and satisfies the following conditions.

Condition a:

$$x.EQ.y \;\Rightarrow\; \forall u \in \Sigma^* \begin{cases} xu \in S \Leftrightarrow yu \in S, \\ f(xu) = f(yu). \end{cases}$$

Condition b:

$$x.EQ.y \;\Rightarrow\; \forall u,w \in \Sigma^* \begin{cases} g(xu) = g(yu), \\ xu.EQ.w \Leftrightarrow yu.EQ.w, \\ xu.D.w \Leftrightarrow yu.D.w, \\ w.D.xu \Leftrightarrow w.D.yu. \end{cases}$$

These two definitions of dominance and equivalence have been stated in the context of seeking the optimal solution set $O(Y(\epsilon))$. If we are trying to seek only a single optimal solution, we can merge the definitions of dominance and equivalence into a single relationship, since we no longer need to retain solutions that might potentially have equal objective function values, but that are not strictly equivalent. In our flowshop example, this is equivalent to allowing the end-times of x be less than *or equal* to those of y, instead of strictly less than.

C. Specification of Branch-and-Bound Algorithm

The preceding definitions allow us to explicitly characterize a branch-and-bound algorithm by

1. Its representation as a discrete decision process, Y denoted by the triple of (Σ, S, f).
2. Its control information, which is represented by the lower-bound function, g, the dominance conditions, D, and the equivalence conditions, EQ.

To complete the specification of the algorithm, we require one additional decision parameter: *how to select the next problem $Y(x)$, which we will solve*, or equivalently, *which node in the branching structure to expand*. We will define a *search function*, s, which allows us to select a node from the currently unexpanded nodes for expansion. In this chapter, as in Ibaraki (1978), we consider only *best bound search*, where we select the node with the minimum $g(x)$ value for expansion. Thus our branch-and-bound algorithm, A, is explicitly specified by

$$A = (Y, g, D, EQ, s),$$

and procedurally is given by the following steps:

Step-1. **Initialize**
     Let $A \leftarrow \{\epsilon\}$, $N \leftarrow \{ \ \} \ z \leftarrow \infty$ where $\{\epsilon\}$ denotes the set of null strings (solutions of length zero).

Step 2. **Search**
     If $A = \varnothing$, exit; else $x \leftarrow s(A)$ goto Step-3.

Step-3 **Test for feasibility**
     If $x \in S$, let $z \leftarrow min(z, f(x))$. Goto Step-4

Step-4 **Test through the lower bound**
     If $g(x) > z$ goto Step-8; else goto Step-5.

Step-5 **Test for relative dominance**
     If $y.D.x$ for some $y(\neq x) \in N$; goto Step-8; else goto Step-6.

Step-6 **Test for equivalence**
     If $y.EQ.x$ for some $y \in N$ which has already been tested; goto Step-8; else goto Step-7

Step-7 **Decompose active node**
     $A \leftarrow A \cup \{xa | a \in \Sigma\} - \{x\}$
     $N \leftarrow N \cup \{xa | a \in \Sigma\}$; return to Step-2.

Step-8 **Terminate**
     $A \leftarrow A - \{x\}$ and return to Step-2.

D. Relative Efficiency of Branch-and-Bound Algorithms

One measure of efficiency is the number of nodes of the branching tree that are expanded. Ibaraki (1978) has proved that the number of nodes expanded can be linked to the strength of the control knowledge, and proves that if one branch-and-bound algorithm has a better lower-bounding function, dominance and equivalence rules, then it will expand fewer nodes.

Let us briefly discuss the theoretical results providing the basis for the improved efficiency of branch-and-bound algorithms. Let $F = \{x \in \Sigma^*: g(x) \leq f(x^*)\}$ be the set of solutions that cannot be terminated by the lower-bound test. Then, the set L_A, defined by $L_A = F \cap \Sigma_D^*$, contains all the partial solutions, which can be terminated only by an equivalence relation. Recall that, by definition, no node in Σ_D^* can be terminated by a dominance rule.

Theorem 1

Let $A = (Y, g, D, EQ, s)$ be a branch-and-bound algorithm. The algorithm terminates after decomposing exactly $|L_A / EQ|$ nodes, provided that $|L_A / EQ| < \infty$, where L_A / EQ denotes the set equivalent classes of solutions induced by the equivalence conditions EQ.

For the proof of this theorem see Ibaraki (1978) as a result of Theorem 1, a natural measure of the efficiency of a branch-and-bound procedure is the number of the resulting equivalence classes under EQ. Furthermore, the following theorem (Ibaraki, 1978) allows a direct comparison of the efficiencies of two distinct branch-and-bound algorithms:

Theorem 2

Assume that the following two branch-and-bound algorithms

$$A = (Y, g, D, EQ, s) \quad \text{and} \quad A' = (Y, g', D', EQ', s')$$

terminate and satisfy the following conditions:

$$
\begin{aligned}
g(x) \leq g'(x) \qquad &\text{for} \quad x \in \Sigma^* \\
x.D.y \Rightarrow x.D'.y \qquad &\text{for} \quad x, y \in \Sigma^* \\
x.EQ.y \Rightarrow x.EQ'.y \qquad &\text{for} \quad x, y \in \Sigma^*
\end{aligned}
$$

Then

$$T_A' = |L_A/EQ'| < |L_A/EQ| = T_A,$$

i.e., algorithm A' is more efficient than algorithm A.

The importance of Theorem 2 is that it provides a *theoretical basis* for improving the control knowledge we have available to solve a given problem class. If we can derive new dominance and/or equivalence rules —or in the single optimal solution case, the enhanced dominance rules—we will be able to reduce the number of nodes enumerated. The focus of this chapter is the description of a methodology for achieving this goal, so that the *computer itself*, can carry out the process of acquiring new control information. The acquisition is based on the analysis of problem-solving activity, in the light of the formal specifications laid out for dominance and equivalence rules, and the generalization of this "experience" to apply it to new problems within the same problem class.

E. Branching as State Updating

Branching from one partial solution to another involves more than just concatenating an alphabet element to a string; it also changes the underlying *state* of the problem. In our flowshop example the state of the partial schedule is conveniently and parsimoniously represented by the start- and end-times of the last operation on each unit. If the problem involved sequence-dependent equipment cleaning, we would have to include information about the identity of the batches as well. The choice of state variables for a problem is critical in determining the ease with which additional state properties and new state properties are calculated. For example, we could record just the list of batches scheduled in a given state and each time we wanted to calculate a new start- or end-time recalculate it from the list of batches. This would be computationally expensive, but our state information would be less complex. Thus, what we choose to record in a state addresses the issue of time/space tradeoff explicitly. In general, the more state variables we have the easier it will be to update each one during branching, but the more we will have to compute and store.

In addition to having to assign state variables to the strings of the DDP, we also have to assign properties to the alphabet symbols. In our flowshop example, the alphabet symbols can be interpreted as batches to be executed with a series of processing times. Thus, if we use the notation, $\Theta(x)$, to denote the state of partial solution, x, then

$$Y(x) \rightarrow Y(xa_i) \qquad \Theta(xa_i) = H\big[\Theta(x), I(a_i)\big],$$

where I returns the interpretation of the alphabet symbol and H is a vector of functions h_j; $j = 1, 2, \ldots, k$, which compute the properties, p_j; $j = 1, 2, \ldots, k$, characterizing $\Theta(xa_i)$:

$$p_j(xa_i) = h_j\big[\Theta(x), I(a_i)\big].$$

The state updating functions combine information about the constraints on the state variables with the objective function minimization. The feasibility predicate forces the state variables to obey certain constraints, such as the nonoverlap of batches, forcing the start-times of successive operations to be greater than the end-times of the previous operation. The constraints do not force the start-times to be equal to the previous end-times; this comes as the effect of the objective function minimization. Consequently, in addition to the update functions, we also need to represent the underlying set of constraints:

Update functions:

$$s_{i,\sigma_{j+1}} = \max(e_{i,\sigma_j}, e_{i-1,\sigma_{j+1}}),$$

$$e_{i,\sigma_{j+1}} = s_{i,\sigma_{j+1}} + p_{ij}.$$

Constraints:

$$s_{i,\sigma_{j+1}} \geq e_{i,\sigma_j}, \tag{3}$$

$$s_{i,\sigma_{j+1}} \geq e_{i-1,\sigma_{j+1}}, \tag{4}$$

$$e_{i,\sigma_{j+1}} = s_{i,\sigma_{j+1}} + p_{ij}. \tag{5}$$

Figure 1 gives the structure of the constraint set. We can now classify these constraints into two classes. The first class of constraints includes the *intersituational constraints*, which have variables of the next state as their potential output variables, and the variables of the current state as their input variables. We can see that these correspond to constraints (3). We then have the *intrasituational constraints*, which relate variables of the next state together and correspond to constraints (4) and (5). We will similarly define those variables that appear in intersituational constraints to be *intersituational variables*, and those that appear in intrasituational constraints to be *intrasituational*. Note that these classes are not mutually exclusive. Thus, we will refine our classes to reflect whether the variable in a specific constraint is a potential output of that constraint or is an input to the constraint. In essence this input/output distinction separates the variables of the current state from those of the next state. In the flowshop example, the end-times of the current state are the input intersituational variables and the start-times of the next state are the output intersituational variables. The start-times also belong to the class of input intrasituational variables, and the end-times of the next state are the output intrasituational variables.

Our ability to make these distinctions rests on the fact that we know the direction that the branching generation imposes on the updating of the variables. If we were not solving the problem in such a way that all the variables are explicitly determined by the branching, then these distinctions would not be so clear. For example, if some variable values were the result of solving an auxiliary linear program that involved these constraints, we could not classify the variables this way.

	$s_{i+1,1}$	$e_{i+1,1}$	$s_{i+1,2}$	$e_{i+1,2}$	$s_{i+1,3}$	$e_{i+1,3}$	$s_{i+1,4}$	$e_{i+1,4}$	$s_{i+1,5}$	$e_{i+1,5}$	$e_{i,1}$	$e_{i,2}$	$e_{i,3}$	$e_{i,4}$	$e_{i,5}$
Inter	✕										✕				
Intra	✕	✕													
Intra		✕	✕												
Inter			✕									✕			
Intra			✕	✕											
Intra				✕	✕										
Inter					✕								✕		
Intra					✕	✕									
Intra						✕	✕								
Inter							✕							✕	
Intra							✕	✕							
Intra								✕	✕						
Inter									✕						✕
Intra									✕	✕					

Inter	Intersituational Constraints
Intra	Intrasituational Constraints
s	Output Intersituational Variables, Input Intrasituational Variables
$e_{i+1,j}$	Output Intrasituational Variables
$e_{i,j}$	Input Intersituational Variables

Fig. 1.

The purpose of defining these classes of constraints and variables is that this will enable us to conveniently express general notions of equivalence and dominance, without explicit reference to the flowshop domain, but at a more abstract level.

F. Flowshop Lower-Bounding Scheme

To define the lower-bound function for flowshop scheduling, we will use the lower-bounding schemes proposed in Lagweg et al. (1978). The lower-bound schemes are organized into a hierarchy that reflects the *strength* of the lower-bounding scheme.

The *strength* of a lower-bounding scheme is a binary relation, over two lower-bounding schemes, which indicates that if $\Omega' \to_D \Omega$, then the lower-bounding scheme Ω' will always yield an equal or lower value for any σ, and hence will always lead to the expansion of fewer nodes, by the theorem of Section II, D.

We will use the simplest of lower-bounding schemes proposed in Lagweg et al. (1978), where all but one of the flowshop units are relaxed to be nonbottleneck machines, A nonbottleneck machine is a machine in one which the capacity constraint has been relaxed; i.e., it can process all the jobs simultaneously. Thus, to compute the lower bound, assuming that the bottleneck machine is u, and that we have completed the partial schedule σ, we need to schedule a single machine knowing that each unscheduled job, $i \in \phi$ will be released after a time $q_{i.u}$, representing the time it spends on the nonbottleneck machines up to u processed on u for a time p_{iu}, and then completed after a further period $q_{iu.}$, representing the time it spends on the nonbottleneck machines after u. We can calculate $q_{i.u}$ by adding the sum of the processing times of i on machines up to u to the partially completed schedule end-time where $q_{i.u} \equiv$ release time before reaching machine u and $q_{iu.} \equiv$ time spent on machines after u,

$$q_{i.u} = \max_{l} \left\{ C(\sigma,l) + \sum_{k=l}^{u-1} p_{ik} | l = 1 \ldots u \right\};$$

$q_{iu.}$ is simply the sum of the remaining processing time of i on the machines following u.

$$q_{iu} = \sum_{k=u+1}^{m} p_{ik}.$$

At this point the lower-bounding scheme consists of solving a single machine scheduling problem where for each job i, the release time is $q_{i.u}$, the due date is $-q_{iu.}$, and the processing time is p_{iu}. This nonbottleneck scheme can be further simplified in two steps. First, we can assume that $q_{i\alpha\beta} = \min_{i \in \phi} \{q_{\alpha\beta}\}$ for $\alpha\beta = .uu.$, avoiding the need to consider release times of $q_{i.u}$, and due dates of $-q_{iu.}$, which turns an NP-complete problem, into one solvable in polynomial time. If only one of these were to be relaxed, the schedule can still be found in polynomial time by Jackson's rule (Jackson, 1955). Second, we can avoid the computation of $q_{i.u}$ completely, by assuming that the maximum is obtained at $l = u$ for all values of i.

$$q_{i.u} = C(\sigma,u) \; \forall i.$$

Thus, our final lower-bound scheme

$$g(\sigma) = \max_{u} \left\{ C(\sigma,u) + \sum_{j \in \phi} p_{ju} + q_{u.} \right\},$$

where $q_{u.} = \min_{j} \{q_{ju.}\}$.

The reason we call this a scheme is because it is still parameterized by, u, the machine that we choose to retain as our bottleneck. The simplest choice of u is $u = m$, for which we get

$$g(\sigma) = C(\sigma, m) + \sum_{j \in \phi} p_{jm}.$$

The computational complexity of the lower bound is presented in Lagweg et al. (1978).

1. At the root node, we can calculate $q_{iu}.$ for all (i, u) in $O(mn)$ steps, for $u = m$ we avoid this calculation.
2. At each node during the branching, we calculate the lower bound, by taking the minimum of the remaining unscheduled batches, $q_{iu}..$ This takes $o(|\phi|)$ for each machine or $O(m|\phi|)$ in all. The final calculation is then of order 1, provided we update and store $\sum_{j \in \phi} p_{ju}$ at each node. For $u = m$, the calculation is $O(1)$.

III. The Use of Problem-Solving Experience in Synthesizing New Control Knowledge

The purpose of this section is to illustrate the role that the problem-solving experience plays in improving problem-solving performance, via learning, and examine the question; *"exactly what is it that we are trying to learn?"*

A. An Instance of a Flowshop Scheduling Problem

To make the ideas of this section more concrete we will use a specific example of a flowshop problem. The problem is a small one, only five batches will be considered, since this enables us to examine the enumeration tree by hand.

Consider the flowshop in Fig. 2. The unit operations being performed are mixing, reaction, distillation, cooling, and filtration. We have made the following assumptions about the way the flowshop operates:

1. Processing times will be similar for each product type on a given processing step, reflecting the fact that technologically identical operations have the same, or similar, equations governing their behavior, with only parametric variation between products.

2. Processing times have some proportionality to batch size. We have chosen to allow three batch sizes (500, 1000, 1500 mass units). The reactor

Flowshop Example Plant:

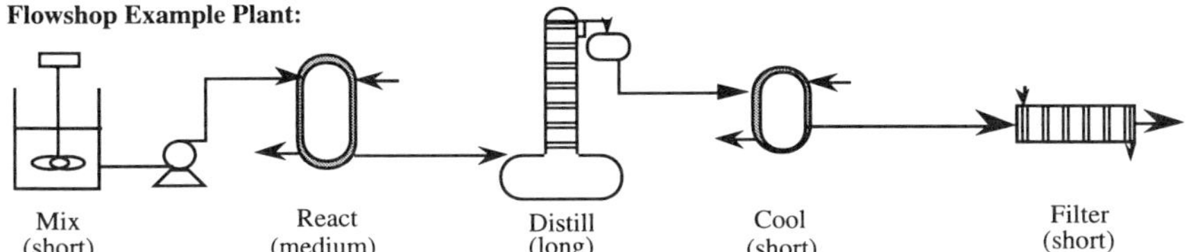

FIG. 2. Example configuration of unit operations in flowshop production.

processing time is the least sensitive to batch size, because the reaction time is assumed proportional to concentration and the increased time is due only to a constant pump rate assumption. The distillation processing time is the most sensitive, due to the assumption of constant boilup rate. The filtration period is roughly proportional to the size, because we have assumed a constant filter cake size and divided the larger batches to achieve this size.

3. The processing parameters have been chosen so that the distillation step is, in general, the longest step. In fact, processing periods for each operation have been roughly ordered as follows: D_1 (distillation) $> R_1$ (reaction) $> F_1$ (filtration) $> C_1$ (cooling) $> M_1$ (mixing). Obviously, the size considerations may alter this ranking for batches of the same type but of different sizes.

The example problem was generated by picking the type and size at random, while ensuring that no size and type combination was repeated. This last requirement was imposed to avoid making the equivalence condition appear to perform better due to spurious equivalences.

In particular, consider the batches shown in Table I. The first two columns of the table indicate the product type and batch size. The next column gives the processing times for each operation. The fourth column gives a unique identifier to each batch composed of a prefix "B-P," the

TABLE I

PRODUCTS IN THE SCHEDULING FLOWSHOP

Product	Size	Processing times					Code	Alphabet symbol
		M_1	R_1	D_1	C_1	F_1		
2	500	2	17	24	4	10	"B-P-2-0"	a_1
1	1000	4	20	26	8	12	"B-P-1-1"	a_2
3	1500	5	26	36	15	19	"B-P-3-2"	a_3
3	500	2	24	12	5	6	"B-P-3-3"	a_4
1	1500	6	21	40	11	19	"B-P-1-1"	a_5

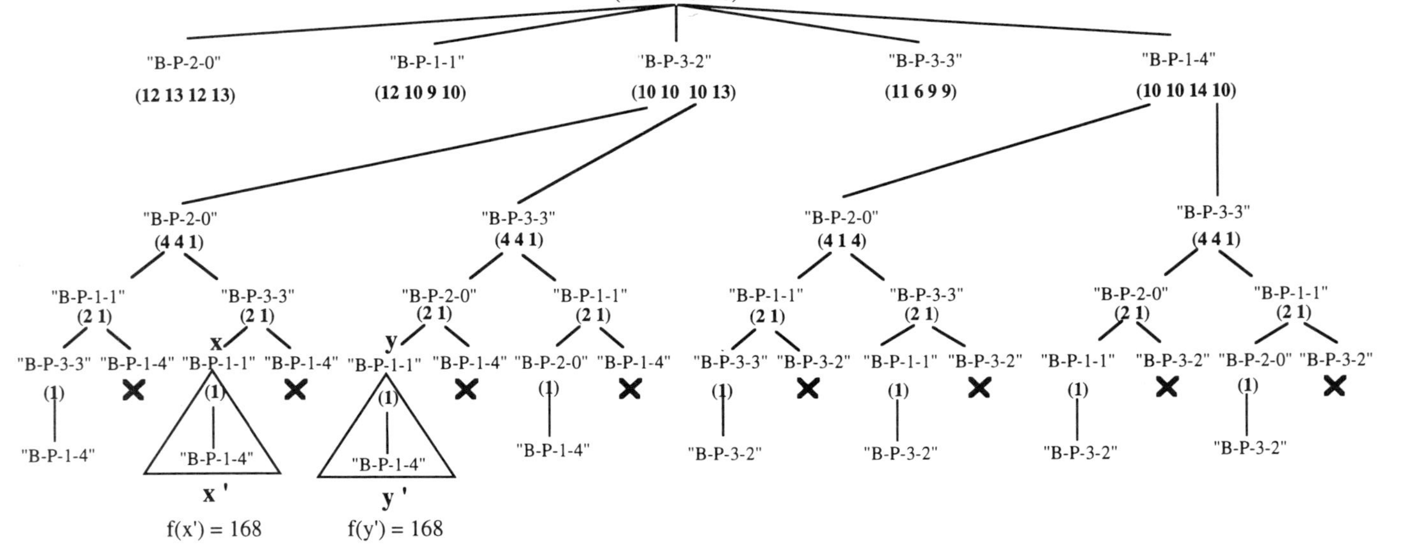

x', y' indentified on the basis of equal objective function values

x,y identified by identical sets of alphabetic elements

FIG. 3. Enumeration without any equivalence conditions; 218 nodes enumerated out of 326 possible.

product type and a counter. The fifth column lists the alphabet symbol associated with each batch. The total number of feasible schedules is 5! or 120. The total number of nodes in the tree is 326 (including the root node).

Utilizing only the simple lower bounding scheme with the bottleneck machine fixed to the last machine, we generate the branching tree given in Fig. 3, where much of the detail has been suppressed. The boldface numbers indicate the size of the subtrees (i.e., the number of nodes in the subtree) beneath the node.

We can now pose the following question:

> Do any of the nodes, which we have enumerated, satisfy the necessary and sufficient conditions for dominance or equivalence?

If we could answer the question affirmatively, we might be able to extract out the features of the nodes that cause them to participate in the relationship, and use these features to identify similar nodes in future problem solving activities. *This is the essence of the goal of learning.*

B. Definition and Analysis of Problem-Solving Experience

For branch-and-bound algorithms the notion of "problem-solving experience" is defined in term of the branching structure generated during the solution, and the various relationships between the nodes of the branching structure that are engendered by existing control knowledge.

1. Identification of Nodes for Dominance and Equivalence Relationships

To identify pairs of nodes that might be capable of being proved to participate in dominance or equivalence relationships, we will employ a strategy of *successive filtering* of the branching structure. We will concentrate on identifying equivalent nodes (for purposes of illustration), although a similar methodology applies to the case of nodes that participate in a dominance relationship.

The necessary and sufficient conditions for equivalence are repeated here for convenience.

Condition-a:

$$x.EQ.y \;\Rightarrow\; \forall u \in \Sigma^* \begin{cases} xu \in S \leftrightarrow yu \in S, \\ f(xu) = f(yu) \end{cases}$$

Condition-b:

$$x.EQ.y \Rightarrow \forall u,w \in \Sigma^* \begin{cases} g(xu) = g(yu), \\ xu.EQ.w \leftrightarrow yu.EQ.w, \\ xu.D.w \leftrightarrow yu.D.w, \\ w.D.xu \leftrightarrow w.D.yu \end{cases}.$$

First, let us concentrate on the meaning of *Condition-a*. Essentially, we are required to ensure that all the solutions, which lie below nodes x and y, have the same objective function value, and that x and y engender the same set of solutions. A problem involving a finite set of objects, each of which must be included in the final solution, is termed a *finite-set problem*. For finite-set problems, we can be confident that $x\omega \in S \leftrightarrow y\omega \in S$ if x and y contain the same set of alphabet symbols. In fact, if x and y do not contain the same set of alphabet symbols, they cannot be equivalent, the proof of which is trivial. Thus our first step in identifying candidates is to find two solutions $x', y' \in S$ that have equal objective function values. We then try to identify their ancestors x, y, such that the set of alphabetic symbols contained in x and y are equal. If in the process of finding this "common stem" we find that the stem is the same for both x', y', then we have the situation depicted in Fig. 4, and $x = y$. This is of no value; a solution is trivially equivalent to itself. We term this *ancestral-equality*. So far, all that has been established for x and y is that there exist two

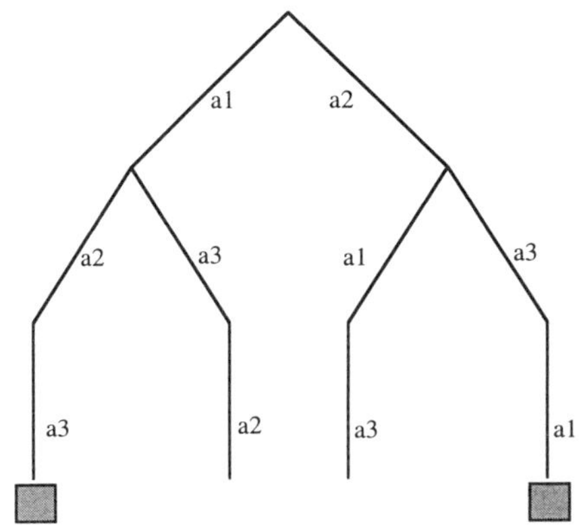

Two solutions can have different tails which lead to solutions with equal objective function values but which cannot produce equivalence relations because they have the same ancestor.

FIG. 4. Schematic depicting ancestral equality.

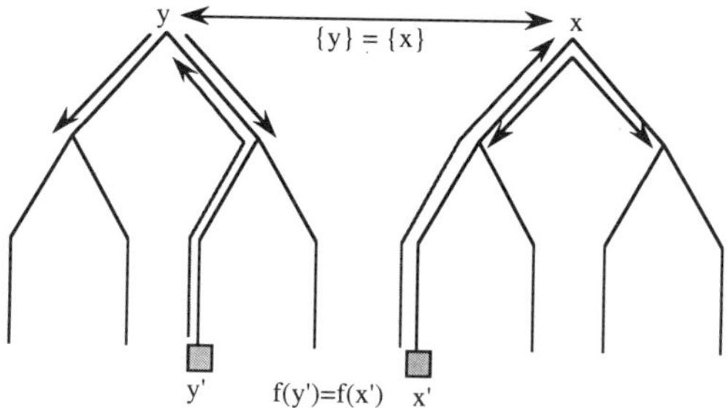

FIG. 5. Pattern of reasoning required to identify examples of equivalence or dominance by the syntactic criterion.

feasible solutions, x' and y', for which $f(x') = f(y')$ and x', y' are the descendants of x, y. We must now reverse the direction of reasoning, and verify that for all the feasible children of x and y that $f(x\omega) = f(y\omega)$, as in Fig. 5.

If we are successful, then we have verified that for the conditions prevailing in the example, the partial solutions, x and y would indeed be equivalent, as far as *Condition-a* is concerned. We now move to *Condition-b*, which ensures that the equivalent node will play an equivalent role in the enumeration as the one that was eliminated. Thus, as we examine the children of x, y, regardless of whether they are members of the feasible set, we would verify that their lower-bound values were equal, and that if we had any existing dominance, or equivalence conditions that the equivalent descendant of x, i.e., xu, participates in the same relationships as does yu.

To make the above discussion more concrete, consider the example branching structure of Fig. 3. In this structure, we have identified an x' and y', which have the same objective function values, and their ancestors, (x, y) which are characterized by the same set of symbols, (i.e., batches). Furthermore, we can see that the children of x, y do indeed satisfy the requirements of *Condition-a* and *Condition-b* and hence, (x, y) would be considered as candidates to develop a new equivalence relationship. If we examine the partial schedules (x, y) as depicted in Fig. 6, our knowledge

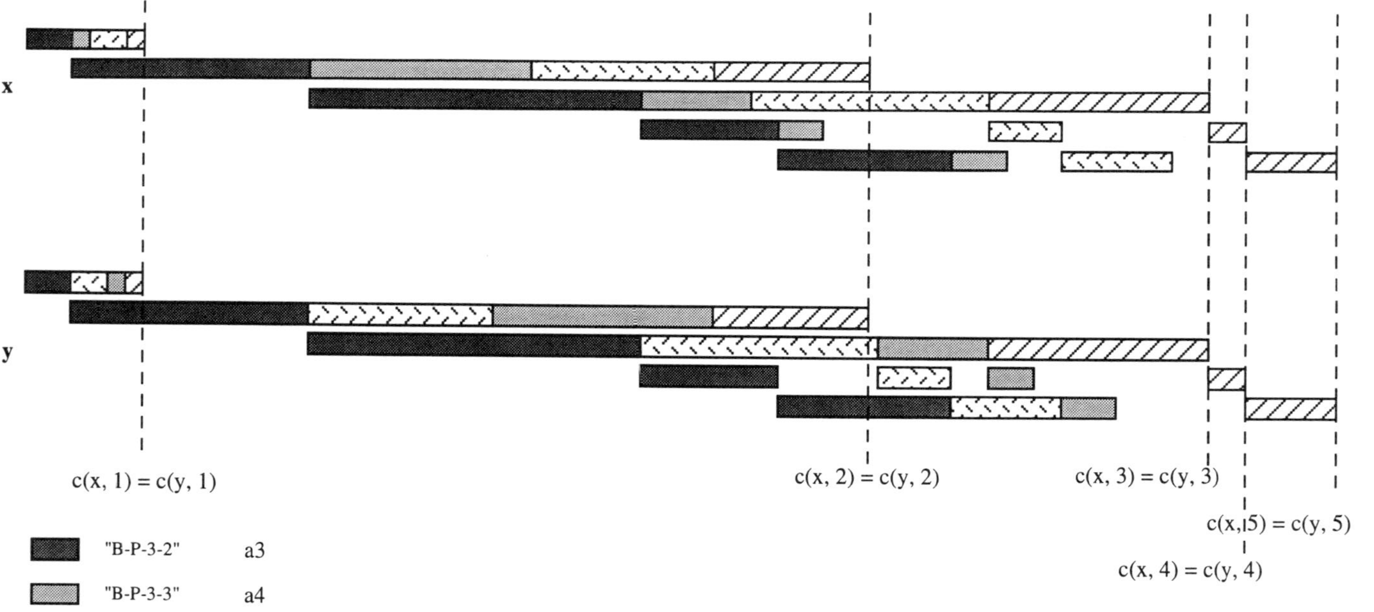

FIG. 6. Potentially equivalent schedules x and y.

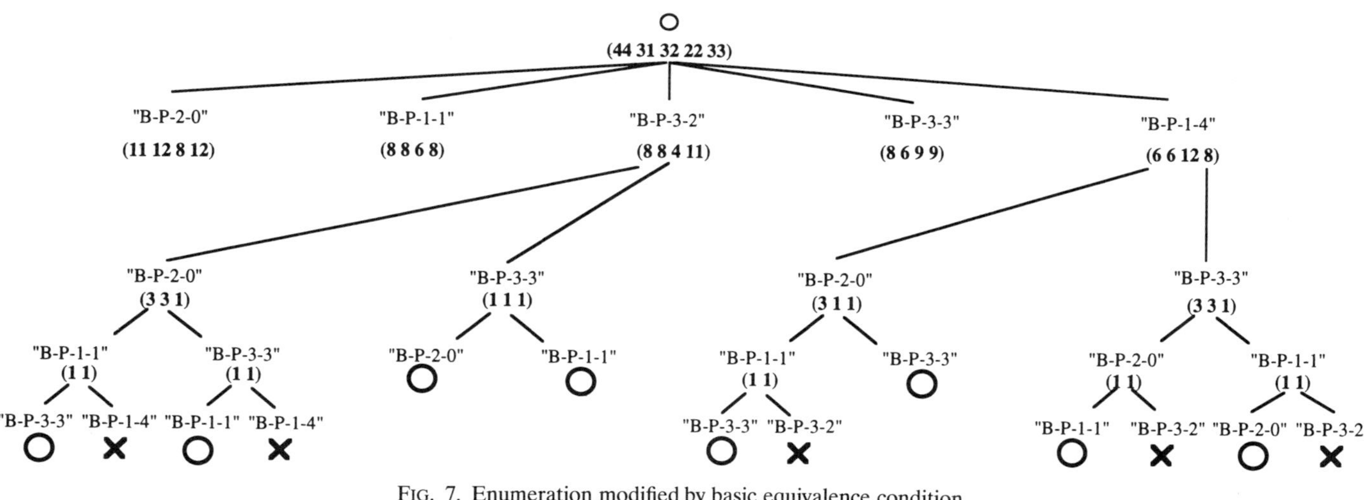

Fig. 7. Enumeration modified by basic equivalence condition.

577

of flowshop scheduling should enable us to see that indeed (x, y) are equivalent because the completion times on each machine are equal, i.e., $C(x, i) = C(y, i)$ for every machine, i, and hence for each arrangement of the remaining batches, the completion times will be equal, i.e., $C(xu, m) = C(yu, m)$ for all feasible completions, u.

The last step in the preceding argument, the use of our knowledge about flowshop scheduling, turns what had been a mainly *syntactic criterion* over the tree structure of the example, into a criterion based on state variables of (x, y). The state variable values, the completion times of the various flowshop machines, are accessible *before* the subtrees beneath x and y have been generated. Indeed, they determine the relationships between the respective elements of the subtrees (xu, yu). If we can formalize the process of showing that the pair (x, y) identified with our syntactic criterion, satisfies the conditions for equivalence or dominance, we will in the process have generated a "new" equivalence rule.

Using this equivalence rule in the example, we generate the branching structure of Fig. 7, where we can see the number of nodes generated has been reduced from 218 to 211. Our syntactic criterion, combined with ensuring that all the lower bounds are equal, will identify five pairs of potentially equivalent solutions in the example. All five of these pairs can be proved to be equivalent on the basis of a general theory of equivalence. Of these pairs, one would have enabled the elimination of a partial schedule of length three, which would have been expanded to give three more solutions, and the other four pairs each eliminate one solution for a total of seven.

The extension of this scheme to dominance conditions requires us to deal with the problem of *incomplete tree structure*. When certain objective and lower-bound values are unavailable, due to nodes being terminated before full expansion, we cannot be certain that the (x, y) pair satisfies the criterion for dominance. To solve this problem, we have to extend and relax the syntactic criteria. Such extensions are beyond the scope of this chapter, and for more details, the reader is referred to Realff, (1992).

C. LOGICAL ANALYSIS OF PROBLEM-SOLVING EXPERIENCE

The analysis of problem-solving experience that has taken place so far has been based on finding subproblems within an existing branching structure that, when solved, will produce subtrees that satisfy the definitions of dominance and equivalence. As we have noted, this is insufficient for generating new dominance and equivalence conditions because we want to avoid the subproblem solution, and not use it as the means of proof.

Thus, the next step in the problem-solving analysis is to use information about the *domain of the problem*, in this case flowshop scheduling, and information about dominance and equivalence conditions that is pertinent to the overall *problem formulation*, in this case as a state space, to convert the experience into a form that can be used in the future problem-solving activity.

We can now identify constraints on the "logical analysis" which must be satisfied for it to produce the desired dominance and equivalence conditions.

1. *Validity.* The reasoning involved in this phase must be logically justifiable; that is, the final conclusion should follow deductively from the facts of the example and from the theory of the domain. If we do not ensure this, we may be able to derive conditions on the two solutions that do not respect either the structure of the domain or the example. The use of these conditions could then invalidate the optimum-seeking behavior of the branch-and-bound algorithm.

2. *Sufficiency.* Given the example, the theory we employ to deduce the conditions must guarantee that the conditions on equivalence or dominance will be satisfied, since the example itself is simply an instance of the particular problem structure, and may not capture all the possible variations. The sufficient theory will thus end up being *specialized* by the example, but since we have asserted the need for validity of this process, it will still be guaranteed to satisfy the abstract necessary and sufficient conditions on dominance and equivalence.

D. SUFFICIENT THEORIES FOR STATE-SPACE FORMULATION

Before showing how the logical analysis will be carried out, it is useful to describe the sufficient theory we will be using for the specific flowshop example. This theory is not restricted to flowshop scheduling, but applies to many state-space problems.

We have characterized our equations and variables as inter- and intra-situational. In transitioning from one state to another, we can view the intersituational equations as constraining the values of the next state's variables. Viewed as constraints, we can define the "looseness" or "tightness" of the constraints that values of a state's variables place on the next state. Thus, for example, since the start-times of the units in the next state are constrained to be larger than the end-times of the same unit in the current state, the constraint will be "tighter" if the end-times are greater in one state than in another. The question then arises, what about the intrasituational constraints on the variables, such as the end-times of the previous unit in the same state?

By definition, the intrasituational variables must eventually be expressed in terms of the intersituational variables. It can be shown (Realff, 1992) that as long as we guarantee that the intersituational variables are at least as loosely constrained in x and y, the intrasituational variables will be, also. The final issue concerns intersituational variables that are constrained via an equality constraint.

In these cases there is no well defined notion of a "looser constraint," the choice is then either to force those variables to be equal in x and y, or to find some path from their value to a constraint on another inter- or intrasituational variable and thus be able to show that their values in x, y should obey some ordering based on these other constraints. This topic is the subject of current research, but is not limiting in the flowshop example, since no such constraints exist. Lastly, it is not enough to assert conditions on the state variables in x and y, since we have made no reference to the discrete space of alternatives that the two solutions admit. Our definition of equivalence and dominance constrains us to have the same set of possible completions. For equivalence relationships the previous statement requires that the partial solutions, x and y, contain the same set of alphabet symbols, and for dominance relations the symbols of x have to be equal to, or a subset of those of y. Thus our sufficient theory can be informally stated as follows:

> If every variable that is the output of an intersituational constraint is at least as loosely constrained in a state x and state y, and x has a subset of, or equal set of, the alphabet symbols of y, then x dominates y.

1. General Qualities of Sufficient Theories

In general, we would like our sufficient theory to have the following qualities:

1. *Simplicity.* The complexity of the sufficiency conditions will tend to translate into complex dominance and equivalence conditions. This complexity can take the form of either a large number of predicates to guarantee the conclusion or complex predicates.

2. *Efficacy.* We must balance making the sufficient theories simple versus making them effective. If, for example, the relative ordering of the intrasituational variables or some function of the state variables, such as the difference between the end-times of successive units, is the driving

force behind the problem structure, then the sufficient theory will be unable to explain why the observed relationship between x and y should hold in general.

3. *Generality.* The more abstract the general theory, the wider the class of problems to which it will apply, and hence the higher the likelihood of being able to transfer knowledge acquired in one class of problems to another, related class. However, we have to construct problem domain theories that connect the information from the specific problem-solving experience to the general theories. The more abstract the theory, the more effort required to do this. On the other hand, if the theory is very specific, then it will probably admit few generalizations, and apply in few problem domains.

4. *Modularity.* Since we would like to use the sufficient theory in a variety of contexts and problems, we need a theory that was easy to extend and modify depending on the context. In our state-space formulation the sufficient theory is couched in terms of constraints on variables. This theory gives us the opportunity to modularize its representation, partitioning the information necessary to prove the looseness of one type of constraint from that required to prove the looseness of a different constraint type. The ability to achieve modularity is a function not only of the theory but also of the representation, which should have sufficient granularity to support the natural partitioning of the components of the theory.

The next section will focus on the representation necessary to express this sufficient theory to the computer, so that it can automatically carry out the reasoning associated with analyzing the examples selected by the syntactic criteria presented in this section. Section V will describe the learning methodology, which, using the representation of Section IV, will generate the new dominance and equivalence conditions.

IV. Representation

This section details the different aspects of the representation we have adopted to describe the problem solutions and the new control knowledge generated by the learning mechanism. Throughout the section we will continue to use the flowshop scheduling problem as an illustration. The section starts by discussing the motives for selecting the horn clause form of first-order predicate calculus, and then proceeds to show how the representation supports both the synthesis of problem solutions and their analysis. The section concludes with a description of how the sufficient theory for state-space models can be expressed in a general way using the proposed representation. The "horn clause form" is described in the paragraphs that follow.

It is beyond the scope of this chapter to explain the syntax and semantics of first-order predicate logic, and the reader is referred to Lloyd's text (1987) for a general introduction. However, it is useful to provide some details on the horn clause form.

A logical clause is a disjunction of terms, and a horn clause is one where, at most, one term is positive. In propositional logic, where only literals are allowed horn clauses have the following form:

$$\neg p \vee \neg q \vee \neg r \vee s.$$

This statement is logically equivalent to

$$p \wedge q \wedge r \Rightarrow s.$$

In predicate calculus, where we are allowed function and predicate symbols that take one or more arguments, the form is

$$\neg p(x) \vee \neg q(y) \vee \neg r(x, y) \vee s(x, y),$$

where implicitly, the x, y variables are allowed to range over the entire domain; i.e., they are universally quantified.

We have chosen this representation for a variety of reasons:

1. Theoretically it has been shown (Thayse, 1988) that the DDP formalism is closely related to a simpler form of horn clause logic, i.e., the propositional calculus. This would suggest that we could use the horn clause form to express some of the types of knowledge we are required to manipulate in combinatorial optimization problems. The explicit inclusion of state information into the representation, necessitates the shift from the simpler propositional form, to the first-order form, since we wish to parsimoniously represent properties that can be true, or take different values, in different states. By limiting the form to horn clauses, we are striving to retain the maximum simplicity of representation, whilst admitting the necessary expressive power.

2. First-order predicate calculus admits proof techniques that can be shown to be *sound* and *complete* (Lloyd, 1987). The soundness of the proof technique is important because it ensures that our methodology will not deduce results that are invalid. We are less concerned with completeness, because in most cases, although the proof technique will be complete, the theory of dominance or equivalence we have available will be incomplete for most problems. Restricting the first-order logic to be of horn clause form, enables the employment of SLD resolution, a simpler

proof technique than full resolution (Robinson, 1965). [*Note*: SLD stands for linear resolution with selection function for definite clauses (Lloyd, 1987)].

3. First-order horn clause logic is the representation that has been adopted by workers in the field of *explanation-based learning* (Minton et al., 1990). Hence, using this representation allows us take advantage of the results and algorithms developed in that field to carry out the machine learning task.

A. Representation for Problem Solving

The representation introduced in the previous subsection must now be utilized to express the information derived from the problem-solving experience, and required to derive the new control knowledge. Our first step will be to define the types of predicates we require to manipulate the properties of the branching structure and the theory that is needed to turn those properties into useful dominance and equivalence conditions.

In Section II, we presented the computational model involved in branching from a node, σ, to a node σa_i. In this model, it was necessary to *interpret* the alphabet symbol a_i, and ascribe it to a set of properties. In the same way, we have to interpret σ as a state of the flowshop, and for convenience, we assigned a set of state variables to σ that facilitated the calculation of the lower-bound value and any existing dominance or equivalence conditions. Thus, we must be able to manipulate the variable values associated with state and alphabet symbols. To do this, we can use the distinguishing feature of first-order predicates, i.e., the ability to parameterize over their arguments. We can use two place predicates, or binary predicates, where the first place introduces a variable to hold the value of the property and the second holds the element of the language, or the string of which we require the value. Thus, if we want to extract the lower bound of a state σ, we can use the predicate (*Lower-bound* ?g $[\sigma]$) to bind ?g to the value of the lower bound of σ. This idea extends easily to properties, which are indexed by more than just the state itself, for example, unit-completion-times, v, which are functions of both the state and a unit

$$(\text{unit} - \text{completion} - \text{time } ?v \text{ k } [\sigma])$$

where k can range over the number of units in the flowshop.

We must now connect these various predicates to be able to derive state variables from other state variables. The process of transitioning from one state to another, can be modeled by the deductive form of the horn clauses. This is a common strategy in artificial intelligence, and is formalized by *situational calculus* (Green, 1969).

However, we do not need the full complexity of situational calculus, which allows for multiple operators to be applied to a given state, because we have only one way of going from one state to another, i.e., branching. Branching has the effect of updating the state properties and the partial string, σ. We will "borrow" the notation of Prolog (Clocksin and Mellish, 1984) to indicate a list of items, where $[?a.?\sigma]$ parses the list into the head of the list, $?a$, and the rest of the list, $?\sigma$. Thus, our horn clause *implications* (*rules*) will have the form

$$\left. \begin{array}{l} p(?v_1, \ldots, ?\sigma) \\ q(?v_2, \ldots, ?\sigma) \\ r(?v_i, \ldots, ?a) \\ (?i_1 = h_1(?v_1, \ldots, ?v_n)) \\ (?\omega = h_\omega(?v_1, \ldots, ?v_n, ?i_1, \ldots, ?i_m)) \end{array} \right\} \Rightarrow p(?\omega, \ldots, [?a.?\sigma])$$

Notice that the left-hand side of this rule contains two types of clauses. The first type is the variable values of the current state and those necessary to compute the new state, while the second, represented by "$=$" computes the value of the variable in the new state. This last clause enables the procedural information about how to compute the state variables to be attached to the reasoning. We must, however, be careful about how much of the computation we hide procedurally, and how much we make explicit in the rules. The level to which computation can be hidden will be a function of the theories we employ to try to obtain new dominance and equivalence conditions. If we do not hide the computation, we will be able to explicitly reason about it, and thus may find simplifications or redundancies in the computation that will lead to more computationally efficient procedures.

We can further subdivide the implications of the above form into two categories; *intersituational* and *intrasituational*:

1. The first type of implications involve clauses with $?\sigma$ on the left-hand side and $[?a.?\sigma]$ on the right-hand side, indicating a branching step.
2. The second type of implications involve clauses with the same state on the left- and right-hand sides of the implication, indicating the derivation of a state variable value from other variables in the state.

These implications types correspond closely to the definitions of state variable types we gave in Section II, E. Those variables that appear in the clause of the right-hand side of an implication that is intersituational are intersituational variables; those variables that appear on the right-hand side of an implication that is intrasituational are intrasituational variables.

1. Horn Clauses in the Synthesis of Branching Structures

Let us now illustrate how the horn clause for of predicate logic can be used for the unfolding of the branching structure and the synthesis of new solutions.

In the flowshop example, when we branch, we have to calculate the start-times of the new batch to be scheduled. Here are the rules that can accomplish this:

$$Axiom\text{-}1: (End\text{-}time \ 0 \ ?j \ [\ ])$$
$$Axiom\text{-}2: (End\text{-}time \ 0 \ 0 \ ?\sigma)$$

a. Intersituational Implications (Rules).

$Rule\text{-}1$: $(\text{End-time } ?e_1 \ 1 \ ?\sigma) \Rightarrow (\text{Start-time } ?e_1 \ 1 \ [?a.?\sigma])$

$$Rule\text{-}2: \ \left. \begin{array}{l} (?j = ?i - 1) \\ (End\text{-}time \ ?e_\sigma \ ?i \ ?\sigma) \\ (End\text{-}time \ ?e_j \ ?j \ [?a.?\sigma]) \\ (\geq ?e_j \ ?e_\sigma) \end{array} \right\} \Rightarrow (Start\text{-}time \ ?e_j \ ?i \ [?a.?\sigma])$$

$$Rule\text{-}3: \ \left. \begin{array}{l} (?j = ?i - 1) \\ (End\text{-}time \ ?e_\sigma \ ?i \ ?\sigma) \\ (End\text{-}time \ ?e_j \ ?j \ [?a.?\sigma]) \\ (\geq ?e_\sigma \ ?e_j) \end{array} \right\} \Rightarrow (Start\text{-}time \ ?e_\sigma \ ?i \ [?a.?\sigma])$$

There are several important points about these rules.

1. *Explicit conditioning.* We could have written the following rule:

$$\left. \begin{array}{l} (?j = ?i - 1) \\ (End\text{-}time \ ?e_\sigma \ ?i \ ?\sigma) \\ (End\text{-}time \ ?e_j \ ?j \ [?a.?\sigma]) \\ (?s = \max(?e_j \ ?e_\sigma)) \end{array} \right\} \Rightarrow (Start\text{-}time \ ?s \ ?i \ [?a.?\sigma])$$

However, by doing so we hide, in the computation of the maximum, a potentially important symbolic constraint between the values of the end-times of the previous unit and the current unit.

2. *Bidirectionality.* During the branching step, we could use these rules to deduce the new value of the start-time, by reasoning from the antecedents to the conclusion, i.e., the facts in the state associated with $?\sigma$ are known such as the unit end-times. We can use these facts to establish the new facts in $[?a.?\sigma]$ by opportunistically applying our rules. However, actually carrying out the updating in this way would be extremely inefficient, and a purely procedural approach is sufficient. During the logical analysis of partial solutions we would start with facts about $[?a\ ?\sigma]$ and be able to deduce constraints on the earlier partial solution variable values.

2. Horn Clauses in the Analysis of Branching Structure

The analysis of the branching structure turns the preceding deduction process around. We have all the facts available to us at the end of the solution synthesis, i.e., at the end of solving a particular problem. Our task is to select and connect subsets of those facts to prove new results that are useful for deriving new control information. In essence, we have to *turn facts about solutions and partial solutions at lower levels in the tree, into constraints on the properties of states and alphabet interpretations higher in the tree.*

For example, if our goal were to show that the end-time of the schedule, x, was equal to a particular value, but we wanted to verify this not for x directly, but for some ancestor of x, then we could use the following implication, in addition to our conditions for start-time deduction (see Section IV, A, 1, a) to carry out the reasoning.

a. Intrasituational Implication.

$$Rule\text{-}4 \quad \left.\begin{array}{l} (\textit{Start-time }?s\ ?j\ [?a.?\sigma\,]) \\ (\textit{Proc-time }?p\ ?j\ ?a) \\ (=?e(+?s\ ?p)) \end{array}\right\} \Rightarrow (\textit{End-time}?e\ ?j\ [?a.?\sigma\,]))$$

Figure 8 shows the trace of the application of *Axiom*-1, *Axiom*-2, *Rule*-1, *Rule*-2, *Rule*-3, and *Rule*-4 in our specific scheduling problem. The trace consists of a repeated pattern of rule applications *Rule*-4 followed by either *Rule*-2 or *Rule*-3. At each step the intrasituational rule converts an end-time to a start-time, and then the start-time is matched to the

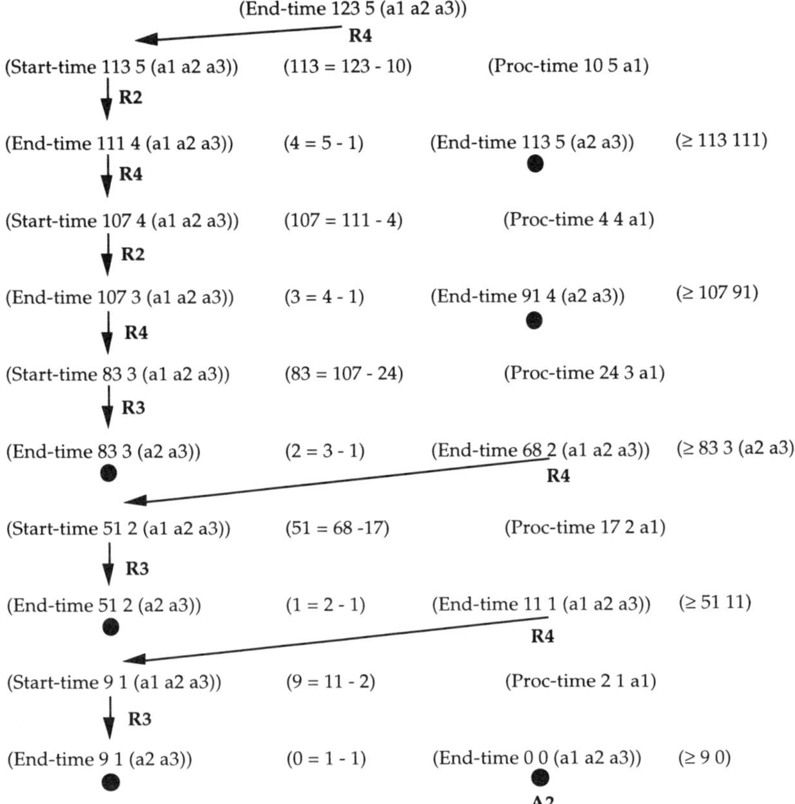

FIG. 8. Trace of deduction for proving the value of the end-time of batch a_1 on unit 5.

appropriate intersituational rule, to generate another end-time. This process is the reversal of the calculation of the state variables performed during the branching, where end-times were derived from start-times. The other rule antecedents fix the relationships between the relative sizes of end-times, and enable calculations to be performed. These conditions must be true of the example for the proof to be successful. We stopped the proof at the predicates

$$(\textit{End-time } 113\ 5[a_2 a_3]), (\textit{End-time } 83\ 2[a_2 a_3]),$$
$$(\textit{End-time } 51\ 2[a_2 a_3]), (\textit{End-time } 9\ 1[a_2 a_3])$$

Why did we not continue the deduction process beyond these particular predicates? Also, we continued to expand the end-time predicate,

$$(\text{end-time } 111\ 4[a_1 a_2 a_3])$$

which did not constrain the start-times. This lead to the derivation of a number of constraints between end-times that would appear to be unnecessary.

The first of these issues is known as *operationality* (Minton et al., 1990) and will be discussed in Section V. The second problem requires us to equip the deduction process with a means to know when deductions are irrelevant, or to change the way the conditions are expressed to avoid the problem altogether. In general, neither of these solutions can be accomplished in a domain independent way. However, the latter solution confines the domain dependency to the conditions themselves, rather than the deduction mechanism, and thus is preferred.

B. Representation for Problem Analysis

The focus of the representation so far, has been on giving the form of rules, which enable us to reason about the values of state variables. This, however, is only one part of the overall reasoning task. We must also represent the theoreies we are going to use to derive the new control knowledge.

The goal of the reasoning is to prove that two partial solutions are equivalent to one another, or that one dominates the other. To do this, we will start with conditions that contain

$$(equivalent \ ?x?y), (dominates \ ?x?y)$$

and terminate with predicates that are easy to evaluate within the branch-and-bound procedure.

The type of theories we will be using to prove dominance and/or equivalence of solutions will not be specific to the particular problem domain, but will rely on more general features of the problem formulation. Thus, for our flowshop example, we will not rely on the fact that we are dealing with processing times, end-times, or start-times, to formulate the general theory. The general theory will be in terms of sufficient statements about the underlying mathematical relationships, as described in Section III.

For example, consider trying to prove that one solution, x, will dominate another solution, y, if they both have scheduled the same batches, but the end-times of each machine are earlier in the partial schedule (partial solution), x, than in y. The general theory will not be couched in these terms, but more abstractly, in terms of the properties of binary operators, such as max, min, $+$, $-$. Specifically, the implications

$$A \leq B \wedge C \leq D \Rightarrow A + C \leq B + D$$

$$A \leq B \wedge C \leq D \Rightarrow \max(A, C) \leq \max(B, D)$$

would enable us to deduce the requisite sufficient conditions for dominance–equivalence relationships. The tree of algebraic manipulations and implications, shown in Fig. 9, is such a case in point.

We can thus think of our predicates as falling into two classes: (1) the *formulation-specific*, and (2) the *problem-specific*. The implications fall into three classes: (1) those that interconnect the general concepts of the formulation, (2) those that connect the general concepts of the formulation to the specific details of the problem, and finally (3) those that enable reasoning about the specific details of the problem. We have already described the predicates necessary for reasoning within the flowshop problem; thus the rest of this section will focus on the general predicates and their interconnection with the specific problem details.

1. Predicates for Problem Analysis

To support the analysis of the solutions we will introduce several new predicate types. We have already described the basic state model in Section II, E. Explicitly manipulating parts of this model is an important component of the reasoning required to derive new control conditions. We will introduce the predicate, "form?," which will enable us to analyze the form of the state update rules, or constraints between variables. Thus, for our flowshop example we have two basic types of constraints:

$$(form? \; ?c \; (?d \geq ?a))$$
$$(form? \; ?c \; (?d = \; + \; ?a?b))$$

These predicate structures can then be bound to the various instances of the constraints in the problem. For example, we would include among our facts declarations of the forms of the constraints, such as

$$(form? \; c_1 \; (s_{i+11} \geq e_{i1}))$$
$$(form? \; c_2 \; (s_{i+12} \geq e_{i+11}))$$

In addition to being able to identify the types of constraints, we must also be able to identify the types of variables. We introduce two predicates, "Intersituational?" and "Intrasituational?" to identify the variable types.

The definition of the abstract form of the constraints between states must now be linked to the actual values of the variables in the constraints

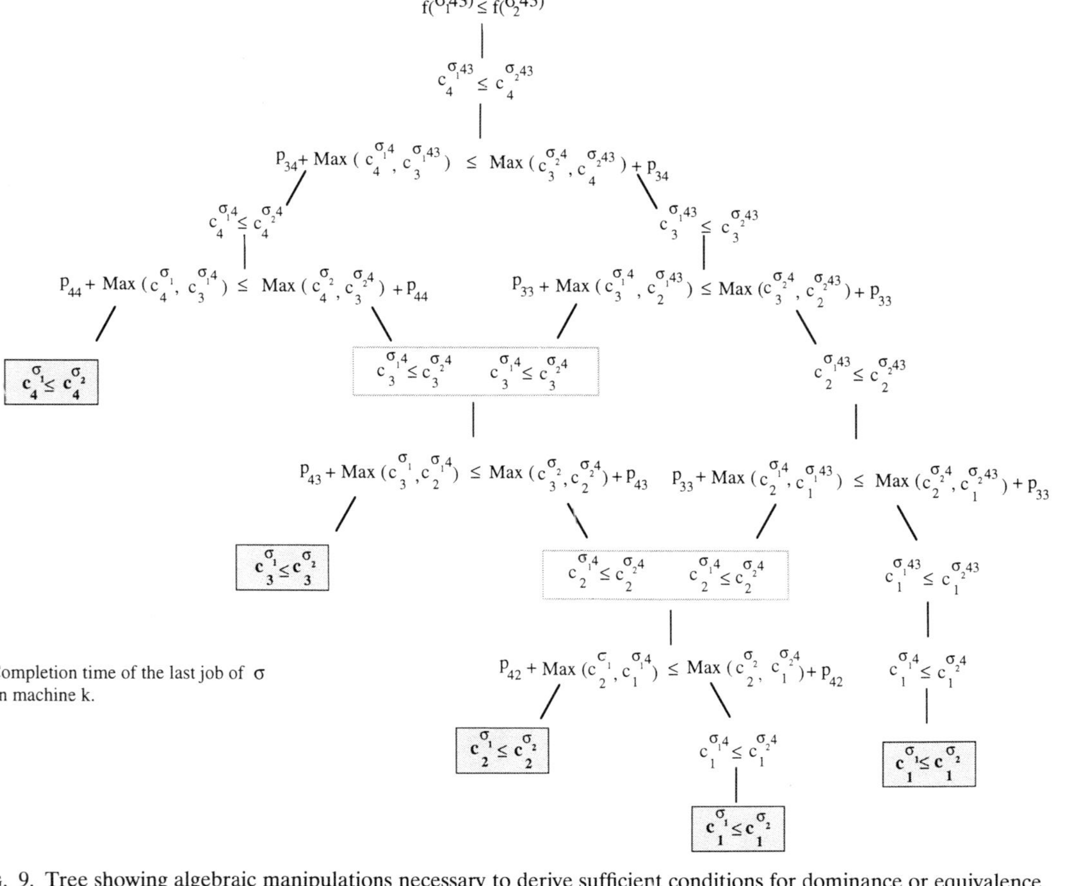

FIG. 9. Tree showing algebraic manipulations necessary to derive sufficient conditions for dominance or equivalence.

themselves, for different states. For this we use the predicates

$$(variable - in - con?\ ?c\ ?var\ ?partial - solution)$$
$$(State - variable - value\ ?val\ ?var).$$

These predicates could be created for the various states that are explored during the branch-and-bound algorithm, or an appropriate subset that has been identified during the example identification.

We then use a predicate for each variable type to be able to pattern match from the variable to its value:

$$(State - variable - value\ ?val\ (end\text{-}time\ ?val\ ?unit\ ?state))$$
$$(State - variable - value\ ?val\ (start\text{-}time\ ?val\ ?unit\ ?state)).$$

These predicates are clearly dependent on the specific problem. We will assume that they are available as basic facts, but we could continue to analyze them using the specific theory of the problem, to turn the start- and end-times of one state into constraints on processing times, and start- and end-times of other states.

We need to introduce two more types of predicates to support the sufficient theory used in the analysis. As we have stated in Section III, D, the sufficient theory rests on being able to prove that the constraints on one state are looser than those on another. The predicate that is used to express this is "looser $-$ constraint $-$ on $-$ variable?," which takes the form:

$$(looser - constraint - on - variable?\ ?var\ ?x\ ?y)$$

where $?x$ and $?y$ will be bound to particular partial solutions and the variable will be one that occurs on the left-hand side of the constraint. To prove that variables are more loosely constrained, we need to compare variable values in different states. Thus, we introduce a new predicate, *"less-equal"* which implements the predicate, *"looser-constraint-on-variable."* This predicate has the form:

$$(less\text{-}equal?\ ?a\ ?b\ ?x\ ?y)$$

where $?a\ ?b$ will be variables in the constraints and $?x\ ?y$ will be the partial solutions.

2. Expression of Sufficient Theory

With the basic predicate types in place we can now define the various implications that will allow us to express the sufficient theory. There are two key steps that have to be made. The first is to take an intersituational variable and figure out what the constraint on the variable is, which of the current state variables are involved in the constraint, and how these variables should be ordered for the constraint to be looser in x than in y. In our example there is a single constraint type, the greater than constraint, leading to the following rule:

$$\left.\begin{array}{l} (\textit{form? ?c } (?d \geq ?a)) \\ (\textit{Intersituational? ?a}) \\ (\textit{less-equal? ?a ?a ?x ?y}) \end{array}\right\}$$
$$\Rightarrow (\textit{looser-constraint-on-variable ?d ?x ?y})$$

For the second step we define predicates that analyze the relationships between variables, recognizing that we are interested only in comparing the actual numerical values of variables associated with solutions, $?x$ and $?y$, which are explicitly declared to be "operational." Thus, we create the following rule:

$$\left.\begin{array}{l} (\textit{Variable-in-con? ?a } ?v_1 \; ?x) \\ (\textit{Variable-in-con? ?b } ?v_1 \; ?y) \\ (\textit{Operational } ?v_1) \\ (\textit{Operational } ?v_2) \\ (\textit{State-variable-value } ?val_1 \; ?v_1) \\ (\textit{State-variable-value } ?val_2 \; ?v_2) \\ (> = \; ?val_1 \; ?val_2). \end{array}\right\} \Rightarrow (\textit{less-equal? ?a ?b ?x ?y})$$

We can now use the preceding implications (rules) to help build the general sufficient theory. As we stated in Section III, D, we have to ensure that all the intersituational variables of the next state are more loosely constrained in x than in y. Thus, our top-level implication (rule) is simply

$$\left.\begin{array}{l} (\textit{Output-intersituational-variables? ?v}) \\ (\textit{looser-constraints-on-variables? ?v ?x ?y}) \end{array}\right\} \Rightarrow (\textit{Dominates? ?x ?y})$$

Then we create a recursive definition of the second predicate to test each variable individually:

$$\left.\begin{array}{l} (\textit{looser-constraint-on-variable? ?var ?x ?y}) \\ (\textit{looser-constraints-on-variables? ?rest ?x ?y}) \end{array}\right\}$$
$$\Rightarrow (\textit{looser-constraints-on-variables? } (?var \; ?rest) \; ?x \; ?y)$$

The specific value of the first predicate would have been entered into the system from the results of the analysis of the structure of the con-

TABLE II

PREDICATES REQUIRED FOR PROOF OF DOMINANCE CONDITION

(*Dominates?* ?x ?y)
(*Output-intersituational-variables?* ?v)
(*looser-constraints-on-variables?* ?vars ?x ?y)
(*looser-constraint-on-variable* ?d ?x ?y)
(*Intersituational* ?a)
(*less-equal?* ?a ?b ?x ?y)
(*Variable-in-con?* ?b $?v_1$?y)
(*Operational* $?v_1$)
(*State-variable-value* $?val_1$ $?v_1$)
(*form?* ?c (?d?rel?a))

straints between two states in the branching structure. This is a predicate that connects the general theory to the specific details of the problem. It can be done manually or by the computer deducing the structure based on the definitions of variable types we gave in Section II, E.

Then if there are no variables, the following predicate is satisfied:

$$(looser\text{-}constraints\text{-}on\text{-}variables?\ (\ )\ ?x\ ?y)$$

This completes the representation of the sufficient theory required for the flowshop example. It consists of about 10 different predicates listed in Table II and configured in four different implications (rules). These predicates have an intuitive appeal, and are not complex to evaluate, thus the sufficient theory could be thought of as being "simple." The theory is capable of deriving the equivalence–dominance condition in flowshop problem. It is, however, expressed in terms that could be applied to any problem with that type of constraint. Thus it has *generality*, and since we can add new implications to deal with new constraint types, it has *modularity*.

Expressing this theory shifts the emphasis of creating specific knowledge for each problem formulation, to developing "pieces" of theory for more general problems. The method allows the computer to put these pieces together, based on the specific details of the problem, and the opportunities that the problem solving reveals.

V. Learning

In the previous sections, we presented three of the four components necessary to carry out the improvement of problem-solving performance for flowshop scheduling, using branch-and-bound algorithms. The first of these dealt with the declarative representation of the branch-and-bound algorithms, including the representation of both the solutions and control mechanisms. It also introduced a metric to quantify the efficiency of a branch-and-bound algorithm. The second component dealt with the methods that are needed in order to analyze the experience gained from the solution of specific problems and convert it into new knowledge of generic value. The third component covered the representation needed to analyze and generalize the selected portions of the acquired problem-solving experience.

In this section, we will briefly outline the method of *explanation-based learning* that we use to carry out the synthesis of new control conditions. This method has become well established within the AI community, and for more details, see Minton et al. (1990), Dejong (1990), and Mitchell et al. (1986).

The ultimate goal of the learning process is to transform a problem-solving experience into useful control knowledge that can be applied in future problem-solving episodes. We have already stated that the methodology should be sound, which translates to no dominance or equivalence rules being introduced that would violate the optimum-seeking behavior of the branch-and-bound algorithm. In addition, there are several other requirements that the learning methodology should try to include

1. *Generalizations derived from a few problem-solving instances*. Solving branch-and-bound problems is computationally expensive. Thus we would like to be able to achieve improvements in problem solving as rapidly as possible.
2. *Branching structure generalization*. It is unlikely that we will be solving problems with exactly the same length of alphabet all the time. Thus, the underlying branching structure on which the control conditions act is subject to change. It is important to be able to allow for these changes without having to re-learn the control rules.

A. Explanation-Based Learning

The following definition of explanation-based learning is taken from Minton et al. (1990).

Given:

1. Target concept definition. A concept definition describing the concept to be learned.
2. Training example. An example of the target concept.

3. Domain theory. A set of rules and facts to be used in explaining how the training example is an example of the target concept.
4. Operationality criterion. A predicate over concept definitions, specifying the form in which the learned concept must be expressed.

Determine:

A generalization of the training example that is a sufficient concept description for the target concept and that satisfies the operationally criterion.

The purpose of this section is to try and link these abstract components to the learning task which we have defined.

1. Target Concept Definition

The targets of our learning are new dominance and equivalence conditions. During problem solving each existing dominance or equivalence will pick out a subset of all the pairs of DDPs that appear in the branching structure. In the flowshop example, the pairs of DDPs are representative of partial schedules that share certain features, and whose properties obey certain relationships, such as having equal end-times and having the same set of batches scheduled. Thus, the target concept is defined by the set of all instances that belong to it; in other words, *our target concept is a dominance or equivalence condition, where the instances are all the pairs of DDPs for which the condition is true.*
Discrete decision processes (DDPs) are defined by the three parameters Σ, S, f. It is important to understand the *coverage* of the dominance and equivalence rules in terms of how much we can change the structure of the DDP before we invalidate the condition. It should be clear that any change to S or f will change the target concept definition, since we may no longer be able to guarantee that the necessary conditions on dominance and equivalence are satisfied. For example, if we were to redefine the objective of the flowshop problem to include some weighted value of the mean flowtime, or introduce a new property such as a lateness or earliness penalty, we could not be sure that $f(x) < f(y)$. Similarly, if we were to include extra conditions on the feasibility of a schedule, such as forbidding certain batches from following each other, we would again potentially disrupt the previous ordering on $f(x), f(y)$.
Having demonstrated that S, f must remain constant, we must now define how Σ is allowed to change. If we examine the necessary conditions on dominance and equivalence, Σ is not explicitly mentioned, but we are required to ensure that for all subsequent DDPs, certain properties hold. This suggests that, provided the *interpretation* of the alphabet remains the same, we can allow its length and thus the symbols it contains to vary. In

our scheduling example, this corresponds to allowing additional batches or batch types to be added to the alphabet, but we would not be allowed to change the interpretation of the alphabet symbol from being associated with a single batch to being a batch type.[2]

In summary, the branch-and-bound algorithm as defined in this chapter assumes that the semantics of "objective function," "feasibility," and "branching operation" are fixed with respect to the problem class. As long as their interpretations are not changed, the derived equivalence and dominance rules would remain valid.

Our last concern is for any change in the control information itself. Such change can be caused by a change in the lower-bound function, or a change in the dominance and/or equivalence conditions. In the former case, we must again abandon our existing dominances and equivalences, unless the altered lower-bound function, g', satisfies the condition

$$g'(x) < g'(y) \Rightarrow g(x) < g(y),$$

or, unless we abandon lower-bound consistency.

Changes in dominance conditions can, in very special circumstances, lead to possible expansion of more nodes, but *only* in situations where we have relaxed our conditions on dominance (for details, see Realff (1992)).

2. Training Example

A training example is an instance of the target concept, which in our case is a pair of DDPs, and their enumeration via the branch-and-bound algorithm. The information about the training example that the learning algorithm manipulates is dependent on the sufficient theory. For each DDP the problem-specific predicates will be constructed using variables, their values, and the constraints associated with the DDP. These problem-specific predicates given the values from DDPs constitute the *facts* about the example.

3. Domain Theory

The domain theory consists of the horn clauses that represent

1. The general implications of the sufficient theory (i.e., the state-space sufficient theory).
2. The implications that link the general theory to the specific problem structure
3. The facts associated with the training example.

[2]Note, however, in this case, this change would make a change in S also.

4. Operationality Criterion

In an earlier section, we had alluded to the need to stop the reasoning process at some point. The operationality criterion is the formal statement of that need. In most problems we have some understanding of what properties are easy to determine. For example, a property such as the processing time of a batch is normally given to us and hence is determined by a simple database lookup. The optimal solution to a nonlinear program, on the other hand, is not a simple property, and hence we might look for a simpler explanation of why two solutions have equal objective function values. In the case of our branch-and-bound problem, the operationality criterion imposes two requirements:

1. The predicates themselves must be easy to evaluate. Thus, we will restrict ourselves to simple comparison predicates, $(<, \leq, +, \geq, >)$, and predicates that extract properties from alphabet symbols and DDPs.
2. The predicates have to express fact about the DDPs for which we are trying to prove dominance or equivalence. This restriction is clearly necessary to avoid declaring the explanation complete before we have expressed the conditions at the appropriate level in the branching structure.

In choosing the definition of operationality, we are deciding the tradeoff between the number of predicates required to evaluate the condition versus the complexity of each predicate. This tradeoff further reduces to the cost of matching the facts to the predicates, versus the cost of evaluating a given match. If there are many different ways to match the facts to the predicates, but only one is successful, we will tend to have high match cost, unless we can find some ordering of the predicates, such that the unsuccessful ones are eliminated early. This topic has been the subject of much discussion; see Minton et al. (1990) for details.

To express operationality within our representation framework, we will use two methods:

1. Implicitly, we can express the fact that a predicate is operational by not including it on the right-hand side of an implication. In this case, it can never be further expanded or explained, and hence if it is not operational, we will fail to generate a satisfactory explanation. This definition makes it difficult to distinguish between the operationality of facts at one level of the branching structure versus another.
2. Explicitly, we can express the operationality by including a predicate (*Operational*), which does not appear as a consequent. We can then assert that certain facts are operational by including (*Operational fact* $- i$) in the database.

This explicit inclusion of operationality allows us to declare explicitly some facts about the pair (x, y), such as their state variable values, as operational. This will not permit the explanation to stop at other partial solutions, whose states have not been declared operational; thus we will use this approach.

B. EXPLANATION

So far we have described all the inputs to the explanation-based learning procedure. We must now describe a few of the basic feature of the algorithm itself.

The first step of the explanation-based learning paradigm is the construction of an explanation of why the training example is an example of the target concept. This explanation will be constructed by backward chaining from the instantiated target concept definition, through the horn clause implications, to reach a set of facts that are true of the example, and that satisfy the operationality criterion. In the backward chain process, we *unify* (Lloyd, 1987) the successively generated subgoals with the consequences of the horn clause implication.

This is high-level description of the explanation process can be illustrated in the flowshop example. To instantiate the target concept, we use the partial solution strings that represent x and y; call them σ_x, σ_y. Thus, our target concept, for example, becomes

$$(Dominates?\ \sigma_x\ \sigma_y).$$

We now search our horn clause implications for one whose consequent unifies with (Dominates? σ_x σ_y). The unification algorithm returns both whether a unification exists and the bindings generated by the algorithm. The bindings of a unification are the substitutions from the specific example for the variables in the consequent. Thus, to unify (Dominates? σ_x σ_y) with (Dominates? $?x$ $?y$), we can do so by binding $?x$ to σ_x and $?y$ to σ_y. Having found an implication whose consequent matches the goal, we backward-chain to the antecedents of the implication, and set these up as subgoals. In these antecedents, we carry the bindings of the consequent through to the corresponding variables of the antecedent. Thus, if the implication were

$$\left. \begin{array}{l} (\textit{Intersituational-variables?}\ ?v) \\ (\textit{looser-constraints-on-variables?}\ ?v\ ?x\ ?y) \end{array} \right\} \Rightarrow (Dominates\ ?x\ ?y)$$

then the subgoals would simply be

$$(\textit{Intersituational-variables? ?v})$$
$$(\textit{looser-constraints-on-variables? ?v } \sigma_x \; \sigma_y)$$

The antecedents are ordered in such a way that bindings, which are necessary to solve certain subgoals, are found before the subgoals are expanded. Thus, in the preceding case, binding for $?v$ will be sought before trying to solve the subgoal ($\textit{Looser-constraints-on-variables? ?v ?}\sigma_x$ $?\sigma_y$). Some of the antecedents may not appear as the consequences of other implications, and to satisfy them, we need to search for matching facts in the database. In the preceding example, we would have entered the intersituational variables of the problem formulation as a fact, and hence the subgoal ($\textit{Intersituational-variables? ?v}$) will be solved by binding $?v$ to a list of those variables. Figure 10 details the resulting explanation, including the unifications or bindings.

The explanation thus consists of a set of implications that connect the original goal, and subgoals, to the facts of the example. This set of implications is interconnected via the bindings that hold between the subgoal and the consequent of the implication used to solve it. The preceding qualitative description of the form of an explanation can be formalized in graph-theoretical terms (Mooney, 1990). For our purposes, the important feature of the explanation is its *structure*. The structure is the pattern of the inference and the bindings that enable us to connect the objects of the original goal to the facts at the leaves of the explanation.

1. Specific Explanation

The specific explanation structure for the flowshop problem is given in Fig. 10. In the example we have assumed that the sufficient condition is satisfied by having all the end-times of x less than or equal to those of y. Thus the proof begins by selecting the appropriate variable set, and proceeds to prove that each variable is more loosely constrained in x than in y. The intersituational variables in the flowshop problem are the start-times of the next state.

If we assume that the variables are ordered from the first unit to the last, we will begin by examining the start-times on the first unit. The start-times are not operational, but the end-times on the units for partial solution x, y are. The start-times of the first unit in the next state, which are equal to the end-times of the first unit in x and y are analyzed by using one of the less-equal implications. The end-times, which are operational, are then compared.

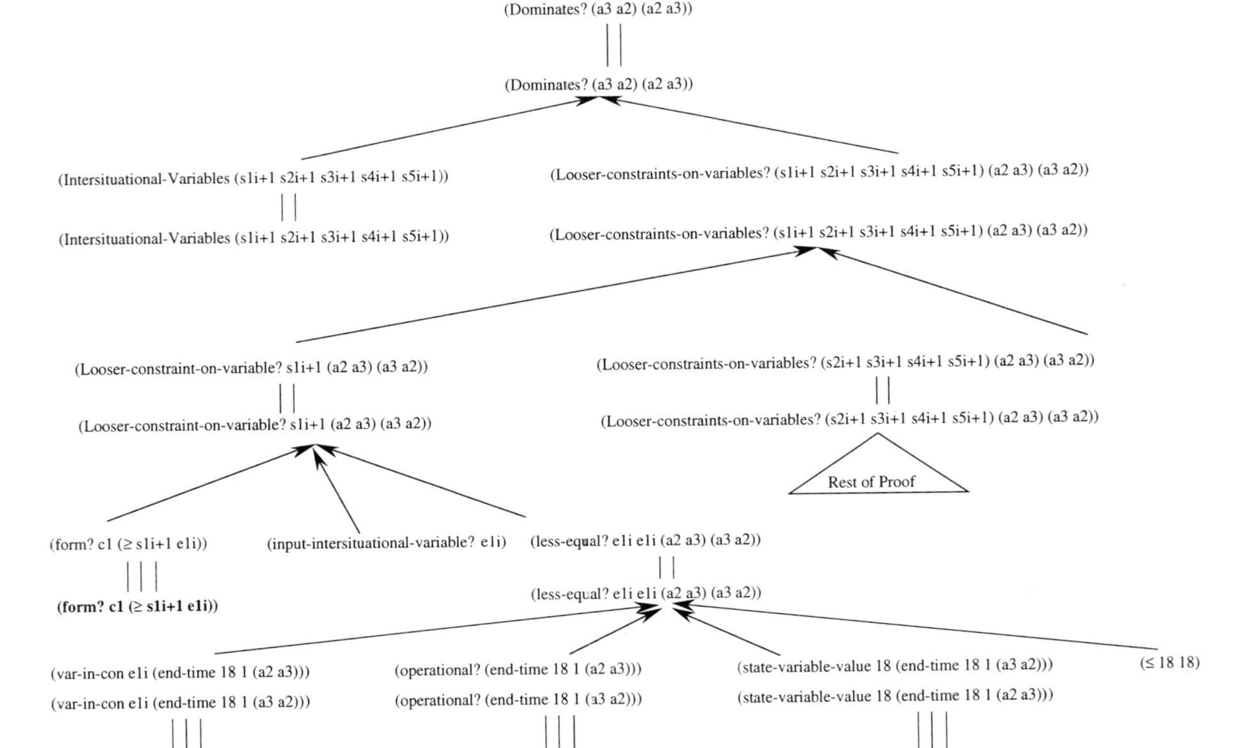

FIG. 10. Proof of simple dominance condition.

A similar analysis is carried out for the start-times of the second unit. Note that during the solution of the subgoal, it is possible that the constraint involving the end-time of the previous unit will be tried. This will fail since the end-time is not an input-intersituational variable. This process is repeated for each subsequent start-time, and in each case the subgoal is resolved the same way.

C. GENERALIZATION OF EXPLANATIONS

According to Mooney (1990), the process of generalization has an input an instance of an explanation; and as output, the explanation, divorced from the specific facts of the example, but retaining the structure implied by it.

We will divide the process of generalization into two parts. In the first, we will consider only the generalization of the explanation while leaving its structure unchanged. In the second, we will consider the generalization of the structure itself.

1. Variable Generalization

During the solution of a specific example we take the implications (rules) and instantiate them with the specific facts of the example. Thus part of the explanation might look like

$$
\left.
\begin{array}{l}
(\textit{Variable-in-con } e_{i1} \, (\textit{end-time } e_{i1} \, 18 \, (a_2 \, a_3)))\\[4pt]
(\textit{Variable-in-con } e_{i1} \, (\textit{end-time } e_{i1} \, 18 \, (a_3 \, a_2)))\\[4pt]
(\textit{Operational? } (\textit{end-time } e_{i1} \, 18 \, (a_2 \, a_3)))\\[4pt]
(\textit{Operational? } (\textit{end-time } e_{i1} \, 18 \, (a_3 \, a_2)))\\[4pt]
(\textit{State-variable-value } 18 \, (\textit{end-time } e_{i1} \, 18 \, (a_2 \, a_3)))\\[4pt]
(\textit{State-variable-value } 18 \, (\textit{end-time } e_{i1} \, 18 \, (a_3 \, a_2)))\\[4pt]
(< \, = 18 \; 18)
\end{array}
\right\}
$$

$$
\Rightarrow (\textit{less-equal? } e_{i1} \, e_{i1} \, (a_2 \, a_3)(a_3 \, a_2))
$$

and this would connect to the facts of the example. The explanation for why the variable $S_{i+1,1}$ is no more constrained in the solution $(a_2 \, a_3)$ than in $(a_3 \, a_2)$ is given in Fig. 11, which includes the specific facts and the relevant bindings of variables. From this specific instance of explanation, we generate the explanation structure, which removes the dependency on

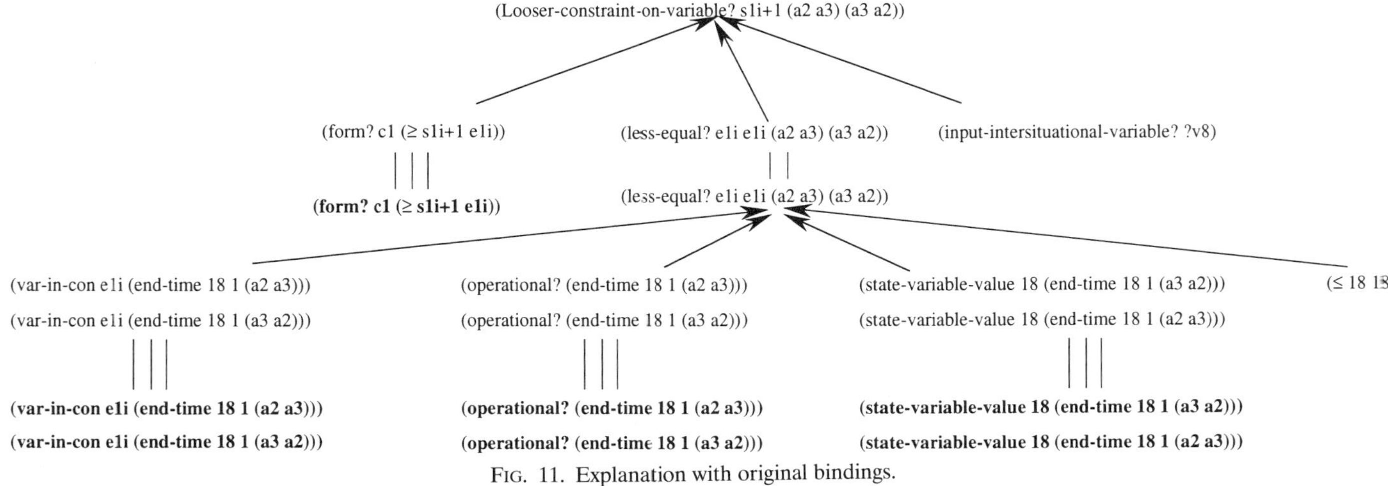

FIG. 11. Explanation with original bindings.

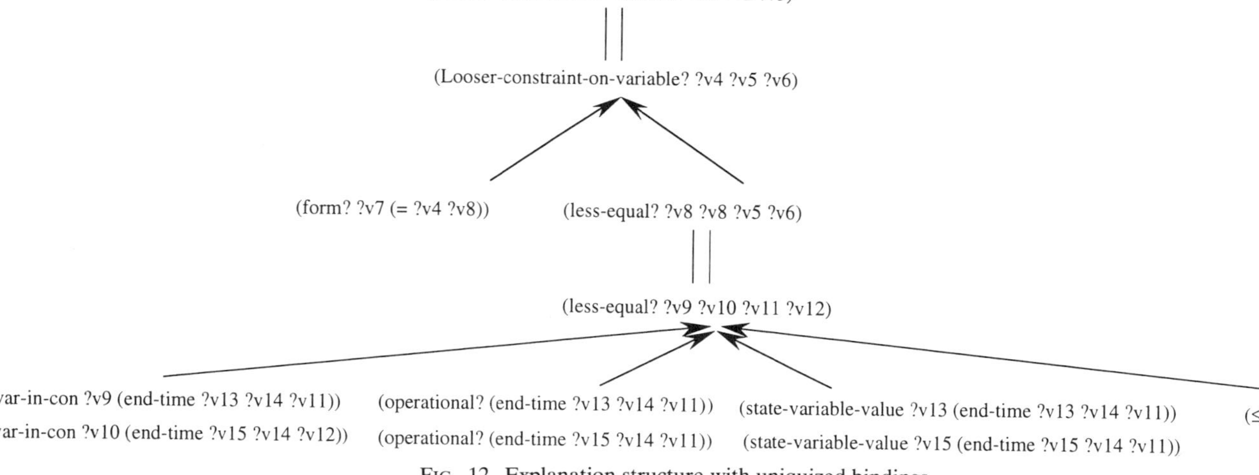

FIG. 12. Explanation structure with uniquized bindings.

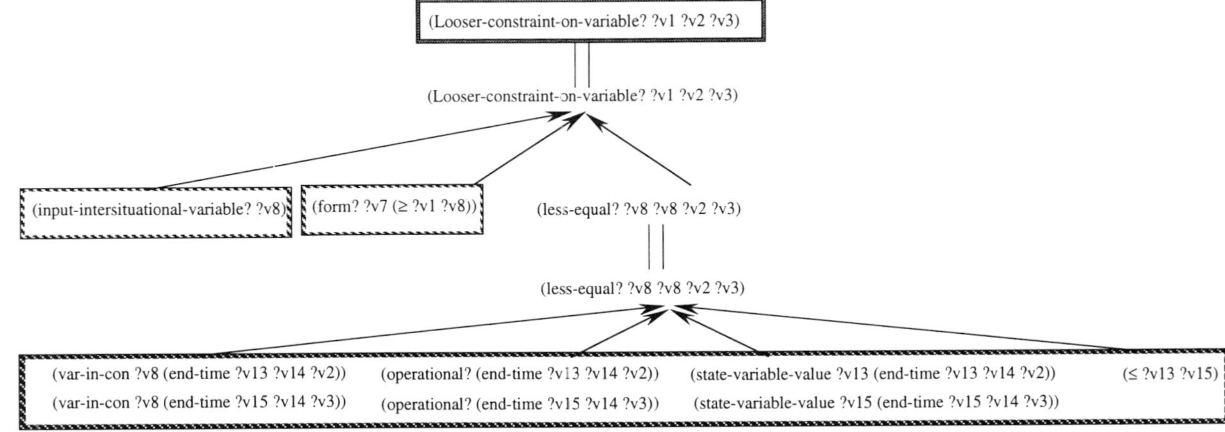

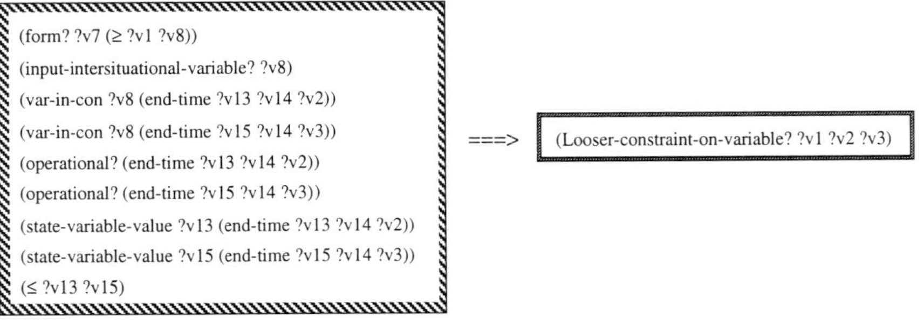

Final Condition - The leaves of the explanation and the top level goal.

Fig. 13. Explanation structure with unified bindings and the final rule.

TABLE III

GENERALIZATION ALGORITHM

Let γ be the null substitution { } for each equality between p_i and p_j in the explanation structure do

Let θ be the *most-general-unifier* of p_i and p_j
Let γ be $\gamma\theta$
for each pattern p_{kN} in the explanation structure do
replace p_k with $p_k\gamma$

---

the specific facts, and makes the arguments of the predicates unique (Fig. 12). Finally, we compute the most *general unifier*, so that at each antecedent consequent match the predicates are identical. The resulting generalized explanation is given in Fig. 13. Generalization of variables thus requires the computation of the most general unifier, which is the same as finding the unification, along with an algorithm for achieving the generalization, given in Table III, taken from Mooney (1990).

The purpose of creating and generalizing the explanation was to generate a way of identifying whether an instance—in this case, a pair of partial solutions—was a member of the target concept, i.e., dominance. The generalized explanations forms a "trace" from the difficult-to-evaluate predicate, at the root, *Dominates?* to a set of easier-to-evaluate predicates at the leaves. To form our condition, we gather up the leaves of the generalized explanation, and use them as antecedents to a new implication:

$$Leaf_1 \wedge Leaf_2 \wedge \ldots \Rightarrow goal\text{-}predicate.$$

In the preceding example:

$$
\begin{aligned}
&(form?\ ?con_1\ (?a \geq ?b)) \\
&(Input\text{-}intersituational\text{-}variable?\ ?b) \\
&(Variable\text{-}in\text{-}con?\ ?b\ (end\text{-}time\ ?val_1\ ?j\ ?x)) \\
&(Variable\text{-}in\text{-}con?\ ?b\ (end\text{-}time\ ?val_2\ ?j\ ?y)) \\
&(Operational\ (end\text{-}time\ ?val_1\ ?j\ ?x)) \\
&(Operational\ (end\text{-}time\ ?val_2\ ?j\ ?y)) \\
&(State\text{-}variable\text{-}value\ ?val_1\ (end\text{-}time\ ?val_1\ ?j\ ?x)) \\
&(State\text{-}variable\text{-}value\ ?val_2\ (end\text{-}time\ ?val_2\ ?j\ ?y)) \\
&(< =\ ?val_1\ ?val_2)
\end{aligned}
$$

$$\Rightarrow (Looser\text{-}constraint\text{-}on\text{-}variable\ ?a\ ?x\ ?y).$$

In this case, we have omitted one intermediate step of reasoning, but in general, there could be many steps between the original goal and the operational predicates. Any future application of the condition will not be required to perform those intermediate steps, or any of the associated search for implications with matching consequences.

2. Generalizations of the Proof Structure

The generalization procedure described above carefully preserves the structure of the proof; it does not attempt to take into account any repetitive structure, which might itself be capable of being generalized. In solving branch-and-bound problems, repetitive structure can occur very easily, since we are performing roughly similar functions as we branch from a parent node to a child, whatever level we are at in the tree.

Furthermore, in the context of flowshop scheduling we have repetitive calculations being performed not only at each level of the branching tree but also within a given node, since each unit essentially has its start-time and end-time calculated by identical procedures. The difficulty we face is understanding when generalization of the proof structure is justified, and when it is simply coincidental that certain elements of the proof have been repeated.

We will adopt the procedure due to Shavlik (1990), in which the justification for generalizing the proof structure is the existence of a recursive pattern of the application of the implications. A recursive pattern implies that at some point in proving a goal a subgoal of the same type, i.e., the same predicate, is generated which contains different arguments. This type of proof structure generalization will enable us to generalize certain facets of the problem structure. Since we have defined the parsing of the list of output intersituational variables recursively, we will be able to generalize the number of items in the list, or in this case the number of start-times. Generalizing the number of start-times implicitly enables us to have an arbitrary number of units in the flowshop. Figure 14 presents the overall structure of the proof.

The way we have stated the domain theory for the state-space representation has enabled us to avoid making explicit reference to the alphabet symbol properties. However, if in other formulations we need to refer to these properties, we would again use a recursive parsing of the list of symbols to enable generalization over the size of the alphabet.

The method that Shavlik developed to carry out the structure generalization is complex, and requires the introduction of an extension to the horn clause deduction scheme to allow the repetitive application of a

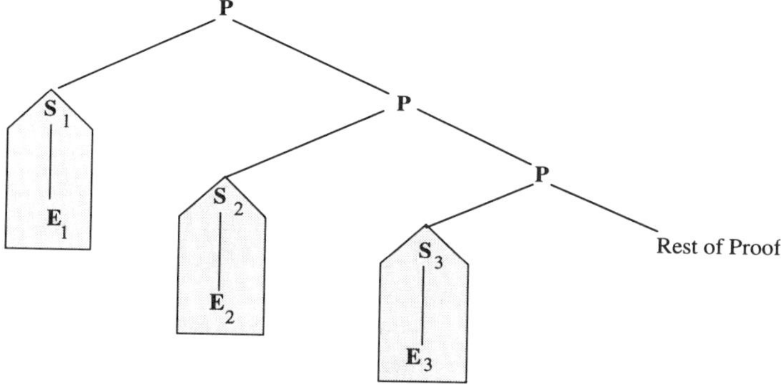

S   Start times of the next state - Intersituational Variables

E   End times of the current state, which are operational

Proof Structure Repeated but not in a sub-goal relationship

FIG. 14. Abstract structure of the proof with repeated subgoals.

group of implications without going through the general backward chaining procedure. The details can be found in Shavlik (1990).

VI. Conclusions

This chapter has presented a methodology for integrating machine learning into the branch-and-bound algorithm. The methodology has been shown to be capable of acquiring a simple dominance condition used in flowshop scheduling algorithms. In Realff (1992) the application of the methodology to a more general scheduling problem (Kondili et al. 1993) is presented. The general problem is cast as an MILP and a sufficient theory of dominance developed for MILP is used to find a dominance rule. Thus we have developed and represented sufficient theories for dominance in both a state-space problem-solving formulation and the traditional optimization formulation of an MILP, both of which cover a wide range of problems of interest to chemical engineers.

We believe this work can be extended in several directions. First, we have shown how to acquire the dominance condition expressed in first-order logic; the next step is to use automatic programming concepts (Barstow, 1986) to convert this expression back into the underlying programming language of the branch-and-bound algorithm, to speed up its application within problem solving. Second, we have concentrated on the theoretical aspects of improving the problem-solving performance. In addition to the theoretical improvements in efficiency, we must demonstrate empirical improvements in the execution time of the algorithm, or in the size of the problems tackled. The computer would *experiment* with the dominance and equivalence rules that it acquired, noting the changes in execution time solving the problem with or without the rules. It would try to detect features of the numerical data that make a given problem harder or easier to solve, using an appropriate empirical learning method such as that presented by Saravia (second chapter in this volume). This experience could then be converted into rules for selecting the appropriate configuration of the branch-and-bound algorithm.

This approach has opened up the possibility of a new type of optimization system, one that can learn from its own experience, test its conclusions, and configure itself to best solve the particular problem, based on its structure and on its data. By reducing some of the algorithm designer's synthetic burden of constructing new algorithms, or new extensions to existing general-purpose methods for each problem, allows the designer to concentrate on developing more widely applicable knowledge embodied in the sufficient theories of the system. It takes advantage of both the solution technology and problem-solving architectures of operations research used to *calculate* optimal answers, and the *reasoning* methods of artificial intelligence: a synthesis of a traditional numerical procedure and symbolic deductive techniques.

References

Abelson, H., Eisenberg, M., Halfant, M., Katzenelson, J., Sacks, E., Sussman, G.J., and Yip, K., Intelligence in scientific computing. *Commun. ACM* **32** (1989).

Baker, K.R., "Introduction to Sequencing and Scheduling." Wiley, New York 1974.

Baker, K.R., A comparative study of flow-shop algorithms. *Oper. Res.* **23**(1) (1975).

Barstow, D., A perspective on automatic programming, *In* "Readings in Artificial Intelligence and Software Engineering" (C. Rich and R.C. Waters, eds.), pp. 537–539. Morgan Kaufmann, San Mateo, CA, 1986.

Clocksin, W.F., and Mellish, C.S., "Programming in Prolog." Springer-Verlag, Berlin (1984).

Dantzig, G.B., "Linear Programming and Extensions." Princeton University Press, Princeton, NJ, 1963.

Dejong, G., and Mooney, R., Explanation-based learning: An alternate view. *Mach. Learn.* **1**, 145–176 (1986).

Garey, M.R., and Johnson, D.S., "Computers and Intractability: A Guide to the Theory of NP-Completeness." Freeman, New York, 1979.

Green, C.C., Application of Theorem proving to problem solving. *Proc. Int. Joint Conf. Art. Intell. 1st*, Washington, DC, 1969, pp. 219–239 (1969).

Ibaraki, T., The power of dominance relations in branch bound algorithms. *JACM* **24**(2), 264–279 (1977).

Ibaraki, T., Branch and bound procedure and state-space representation of combinatorial optimization problems. *Inf. Control* **36**, 1–27 (1978).

Jackson, J.R., Scheduling a Production Line to Minimize Maximum Lateness," Res. Rep. No. 43, Management Science Research Project. University of California, Los Angeles, 1955.

Karp, R.M., and Held, M. Finite-state processes and dynamic programming. *SIAM J. Appl. Math.* **15**, 698–718 (1967).

Kondili, E., Pantalides, C.C., and Sargent, R.H.W., A general algorithm for short-term scheduling of batch operations. I. MILP formulation. *Comput. Chem. Eng.* **17** (2) 211–228 (1993).

Kumar, V., and Kanal, L.N., A general branch and bound formulation for understanding and synthesizing and/or tree search procedures. *Art. Intell.* **21** 179–198 (1983).

Lagweg, B.J., Lenstra, J.K., and Rinnooy Kan, A.H.G., A general bounding scheme for the permutation flow-shop problem. *Oper. Res.* **26**(1) (1978).

Lloyd, J.W., "Foundations of Logic Programming," 2nd ed., Springer-Verlag, Berlin, 1987.

Minton, S., Carbonell, J., Knoblock, C., Kuokka, D., Etzioni, O., and Gil, Y., Explanation-based learning: A problem solving perspective in machine learning. *In* "Machine Learning: Paradigms and Methods" (J.G. Carbonell, ed). Massachusetts Institute of Technology/Elsevier, Cambridge and London, 1990.

Mitchell, T., Keller, R., and Cedar-Cabelli, S., Explanation-based generalization: A unifying view. *Mach. Learn.* **1**, 47–80 (1986).

Mooney, R.J., "A General Explanation-Based Learning Mechanism and its Application to Narrative Understanding." Morgan Kaufmann, San Mateo, CA (1990).

Nemhauser, G.L., and Wolsey, L.A., "Integer and Combinatorial Optimization." Wiley, New York (1988).

Nilsson, N.J., "Principles of Artificial Intelligence." Palo Alto, CA, 1980.

Rajagopalan, D., and Karimi, I.A., Completion times in serial mixed-storage multiproduct processes with transfer and set-up times. *Comput. Chem. Eng.* **13**(1/2), 175–186 (1989).

Realff, M.J., "Machine Learning for the Improvement of Combinatorial Optimization Algorithms: A Case Study in Batch Scheduling." Ph.D Thesis, MIT., Cambridge, MA, 1992.

Robinson, J.A., A machine-oriented logic based on the resolution principle, *JACM* **12**(1), 23–41 (1965).

Shah, N., Pantelides, C.C., and Sargent, R.W.H., A general algorithm for short-term scheduling of batch operations. II Computational issues. *Comput. Chem. Eng.* **17**(2), 229–244 (1993).

Shavlik, J.W., "Extending Explanation-Based Learning by Generalizing the Structure of Explanations." Morgan Kaufmann, San Mateo, CA, 1990.

Simon, H.A., Search and reasoning in problem solving. *Art. Intell.* **21**, 7–29 (1983).

Szwarc, W., Elimination methods in the $m \times m$ sequencing problem. *Nav. Res. Logist. Q.* **18**, 295–305 (1971).

Thayse, A., "From Standard Logic to Logic Programming. Wiley, New York, 1988.

Wiede, W., and Reklaitis, G.V., Determination of completion times for serial multiproduct processes. I-III. *Comput. Chem. Eng.*, **11**(4), 337–368 (1987).
Yip, K., "KAM: A System for Intelligently Guiding Numerical Experimentation by Computer." MIT Press, Cambridge, MA, 1992.

D

E

F

G

CONTENTS OF VOLUMES IN THIS SERIAL